T0190240

Communications in Computer and Information Science 1788

Rationale

The CCIS series is devoted to the publication of proceedings of computer science conferences. Its aim is to efficiently disseminate original research results in informatics in printed and electronic form. While the focus is on publication of peer-reviewed full papers presenting mature work, inclusion of reviewed short papers reporting on work in progress is welcome, too. Besides globally relevant meetings with internationally representative program committees guaranteeing a strict peer-reviewing and paper selection process, conferences run by societies or of high regional or national relevance are also considered for publication.

Topics

The topical scope of CCIS spans the entire spectrum of informatics ranging from foundational topics in the theory of computing to information and communications science and technology and a broad variety of interdisciplinary application fields.

Information for Volume Editors and Authors

Publication in CCIS is free of charge. No royalties are paid, however, we offer registered conference participants temporary free access to the online version of the conference proceedings on SpringerLink (http://link.springer.com) by means of an http referrer from the conference website and/or a number of complimentary printed copies, as specified in the official acceptance email of the event.

CCIS proceedings can be published in time for distribution at conferences or as post-proceedings, and delivered in the form of printed books and/or electronically as USBs and/or e-content licenses for accessing proceedings at SpringerLink. Furthermore, CCIS proceedings are included in the CCIS electronic book series hosted in the SpringerLink digital library at http://link.springer.com/bookseries/7899. Conferences publishing in CCIS are allowed to use Online Conference Service (OCS) for managing the whole proceedings lifecycle (from submission and reviewing to preparing for publication) free of charge.

Publication process

The language of publication is exclusively English. Authors publishing in CCIS have to sign the Springer CCIS copyright transfer form, however, they are free to use their material published in CCIS for substantially changed, more elaborate subsequent publications elsewhere. For the preparation of the camera-ready papers/files, authors have to strictly adhere to the Springer CCIS Authors' Instructions and are strongly encouraged to use the CCIS LaTeX style files or templates.

Abstracting/Indexing

CCIS is abstracted/indexed in DBLP, Google Scholar, EI-Compendex, Mathematical Reviews, SCImago, Scopus. CCIS volumes are also submitted for the inclusion in ISI Proceedings.

How to start

To start the evaluation of your proposal for inclusion in the CCIS series, please send an e-mail to ccis@springer.com.

Kanubhai K. Patel · K. C. Santosh · Atul Patel ·
Ashish Ghosh
Editors

Soft Computing and its Engineering Applications

4th International Conference, icSoftComp 2022
Changa, Anand, India, December 9–10, 2022
Proceedings

Editors
Kanubhai K. Patel
Charotar University of Science and Technology
Changa, Anand, Gujarat, India

Atul Patel
Charotar University of Science and Technology
Changa, Anand, India

K. C. Santosh
University of South Dakota
Vermillion, SD, USA

Ashish Ghosh
Indian Statistical Institute
Kolkata, India

ISSN 1865-0929 ISSN 1865-0937 (electronic)
Communications in Computer and Information Science
ISBN 978-3-031-27608-8 ISBN 978-3-031-27609-5 (eBook)
https://doi.org/10.1007/978-3-031-27609-5

This Springer imprint is published by the registered company Springer Nature Switzerland AG
The registered company address is: Gewerbestrasse 11, 6330 Cham, Switzerland

Preface

It is a matter of great privilege to have been tasked with the writing of this preface for the proceedings of The Fourth International Conference on Soft Computing and its Engineering Applications (icSoftComp2022). The conference aimed to provide an excellent international forum to the emerging and accomplished research scholars, academicians, students, and professionals in the areas of computer science and engineering to present their research, knowledge, new ideas, and innovations. The conference was held December 9–10, 2022, at Charotar University of Science & Technology (CHARUSAT), Changa, India, and organized by the Faculty of Computer Science and Applications, CHARUSAT.

There are three pillars of Soft Computing, viz. i) Fuzzy computing, ii) Neuro computing, and iii) Evolutionary computing. Research submissions in these three areas were received. The Program Committee of icSoftComp2022 is extremely grateful to the authors from 16 different countries including USA, UK, China, Germany, Portugal, Egypt, Tunisia, United Arab Emirates, Saudi Arabia, Bangladesh, Philippines, Finland, Malaysia, and South Africa; who showed an overwhelming response to the call for papers, submitting over 342 papers. The entire review team (Technical Program Committee members along with 3 additional reviewers) expended tremendous effort to ensure fairness and consistency during the selection process, resulting in the best-quality papers being selected for presentation and publication. It was ensured that every paper received at least three, and in most cases four, reviews. Checking of similarities was also done based on international norms and standards. After a rigorous peer review 36 papers were accepted, with an acceptance ratio of 10.53%. The papers are organised according to the following topics: Theory & Methods, Systems & Applications, and Hybrid Techniques. The proceedings of the conference are published as one volume in the Communications in Computer and Information Science (CCIS) series by Springer, and are also indexed by ISI Proceedings, DBLP, Ulrich's, EI-Compendex, SCOPUS, Zentralblatt Math, MetaPress, and Springerlink. We, in our capacity as volume editors, convey our sincere gratitude to Springer for providing the opportunity to publish the proceedings of icSoftComp2022 in their CCIS series.

icSoftComp2022 provided an excellent international virtual forum for the conference delegates to present their research, knowledge, new ideas, and innovations. The conference exhibited an exciting technical program. It also featured high-quality workshops, two keynotes, and six expert talks from prominent research and industry leaders. Keynote speeches were given by Dilip Kumar Pratihar (Indian Institute of Technology Kharagpur, India) and Witold Pedrycz (University of Alberta, Canada). Experts talks were given by Dimitrios A. Karras (National and Kapodistrian University of Athens, Greece), Massimiliano Cannata (University of Applied Sciences and Arts of Southern Switzerland (SUPSI), Switzerland), Maryam Kaveshgar (Ahmedabad University, India), Rashmi Saini (Govind Ballabh Pant Institute of Engineering and Technology, India), Saurin Parikh (Nirma University, India), and Krishan Kumar (National Institute of Technology Uttarakhand, India). We are grateful to them for sharing their insights on their latest research with us.

The Organizing Committee of icSoftComp2022 is indebted to R V Upadhyay, Provost of Charotar University of Science and Technology and Patron, for the confidence that he invested in us in organizing this international conference. We would also like to take this opportunity to extend our heartfelt thanks to the Honorary Chair of this conference, Kalyanmoy Deb (Michigan State University, USA), Janusz Kacprzyk (Polish Academy of Sciences, Poland), and Leszek Rutkowski (IEEE Fellow) (Czestochowa University of Technology, Poland) for their active involvement from the very beginning until the end of the conference. The quality of a refereed volume primarily depends on the expertise and dedication of the reviewers who volunteer with a smiling face. The editors are further indebted to the Technical Program Committee members and external reviewers who not only produced excellent reviews but also did so in a short time frame, in spite of their very busy schedules. Because of their quality work it was possible to maintain the high academic standard of the proceedings. Without their support, this conference could never have assumed such a successful shape. Special words of appreciation are due to note the enthusiasm of all the faculty, staff, and students of the Faculty of Computer Science and Applications of CHARUSAT, who organized the conference in a professional manner.

It is needless to mention the role of the contributors. The editors would like to take this opportunity to thank the authors of all submitted papers not only for their hard work but also for considering the conference a viable platform to showcase some of their latest findings, not to mention their adherence to the deadlines and patience with the tedious review process. Special thanks to the team of EquinOCS, whose paper submission platform was used to organize reviews and collate the files for these proceedings. We also wish to express our thanks to Amin Mobasheri (Editor, Computer Science Proceedings, Springer Heidelberg) for his help and cooperation. We gratefully acknowledge the financial (partial) support received from Department of Science & Technology, Government of India and Gujarat Council on Science & Technology (GUJCOST), Government of Gujarat, Gandhinagar, India for organizing the conference. Last but not least, the editors profusely thank all who directly or indirectly helped us in making icSoftComp2022 a grand success and allowed the conference to achieve its goal, academic or otherwise.

December 2022

Kanubhai K. Patel
K. C. Thomas
Atul Patel
Ashish Ghosh

Organization

Patron

R. V. Upadhyay	Charotar University of Science and Technology, India

Honorary Chairs

Kalyanmoy Deb	Michigan State University, USA
Janusz Kacprzyk	Polish Academy of Sciences, Poland
Leszek Rutkowski	Czestochowa University of Technology, Poland

General Chairs

Atul Patel	Charotar University of Science and Technology, India
George Ghinea	Brunel University London, UK
Pawan Lingras	Saint Mary's University, Canada

Technical Program Committee Chair

Kanubhai K. Patel	Charotar University of Science and Technology, India

Technical Program Committee Co-chairs

Ashish Ghosh	Indian Statistical Institute (ISI), Kolkata, India
K. C. Santosh	The University of South Dakota, USA
Deepak Garg	Bennett University, India
Gayatri Doctor	CEPT University, India

Advisory Committee

Arup Dasgupta	Geospatial Media and Communications, India
Valentina Emilia Balas	University of Arad, Romania
Bhuvan Unhelkar	University of South Florida Sarasota-Manatee, USA
D. K. Pratihar	Indian Institute of Technology Kharagpur, India
J. C. Bansal	Soft Computing Research Society, India
Narendra S. Chaudhari	Indian Institute of Technology Indore, India
Rajendra Akerkar	Vestlandsforsking, Sogndal, Norway
R. P. Soni	GLS University, India
Sudhir Kumar Barai	BITS Pilani, India
Suman Mitra	DAIICT, India
Devang Joshi	Charotar University of Science and Technology, India
S. P. Kosta	Charotar University of Science and Technology, India
Dharmendra T. Patel	Charotar University of Science and Technology, India

Technical Program Committee Members

Abhijit Datta Banik	IIT Bhubaneswar, India
Abdulla Omeer	Dr. Babasaheb Ambedkar Marathwada University, India
Abhineet Anand	Chitkara University, India
Aditya Patel	Kamdhenu University, India
Adrijan Božinovski	University American College Skopje, Macedonia
Aji S.	University of Kerala, India
Akhil Meerja	Vardhaman College of Engineering, India
Aman Sharma	Jaypee University of Information Technology, India
Ami Choksi	C.K. Pithawala College of Engg. and Technology, India
Amit Joshi	Malaviya National Institute of Technology, India
Amit Thakkar	Charotar University of Science and Technology, India
Amol Vibhute	MIT World Peace University, India
Anand Nayyar	Duy Tan University, Vietnam
Angshuman Jana	IIIT Guwahati, India
Ansuman Bhattacharya	IIT (ISM) Dhanbad, India

Dinesh Acharya	MIT, India
Divyansh Thakur	IIIT Una, India
Dushyantsinh Rathod	Alpha College of Engineering and Technology, India
E. Rajesh	Galgotias University, India
Gururaj Mukarambi	Central University of Karnataka, India
Gururaj H. L.	Vidyavardhaka College of Engineering, India
Hardik Joshi	Gujarat University, India
Harshal Arolkar	GLS University, India
Himanshu Jindal	Jaypee University of Information Technology, India
Hiren Joshi	Gujarat University, India
Hiren Mewada	Prince Mohammad Bin Fahd University, Saudi Arabia
Irene Govender	University of KwaZulu-Natal, South Africa
Jagadeesha Bhatt	IIIT Dharwad, India
Jaimin Undavia	Charotar University of Science and Technology, India
Jaishree Tailor	Uka Tarsadia University, India
Janmenjoy Nayak	AITAM, India
Jaspher Kathrine	Karunya Institute of Technology and Sciences, India
Jimitkumar Patel	Charotar University of Science and Technology, India
Joydip Dhar	ABV-IIITM, India
József Dombi	University of Szeged, Hungary
Kamlendu Pandey	VNSGU, India
Kamlesh Dutta	NIT Hamirpur, India
Kiran Trivedi	Vishwakarma Government Engineering College, India
Kiran Sree Pokkuluri	Shri Vishnu Engineering College for Women, India
Krishan Kumar	National Institute of Technology Uttarakhand, India
Kuldip Singh Patel	IIIT Naya Raipur, India
Kuntal Patel	Ahmedabad University, India
Latika Singh	Ansal University, Gurgaon, India
M. Srinivas	National Institute of Technology-Warangal, India
M. A. Jabbar	Vardhaman College of Engineering, India
Maciej Ławrynczuk	Warsaw University of Technology, Poland
Mahmoud Elish	Gulf University for Science and Technology, Kuwait
Mandeep Kaur	Sharda University, India

Manoj Majumder	IIIT Naya Raipur, India
Michał Chlebiej	Nicolaus Copernicus University, Poland
Mittal Desai	Charotar University of Science and Technology, India
Mohamad Ijab	National University of Malaysia, Malaysia
Mohini Agarwal	Amity University Noida, India
Monika Patel	NVP College of Pure and Applied Sciences, India
Mukti Jadhav	Marathwada Institute of Technology, India
Neepa Shah	Gujarat Vidyapith, India
Neetu Sardana	Jaypee University of Information Technology, India
Nidhi Arora	Solusoft Technologies Pvt. Ltd., India
Nilay Vaidya	Charotar University of Science and Technology, India
Nitin Kumar	National Institute of Technology Uttarakhand, India
Parag Rughani	GFSU, India
Parul Patel	VNSGU, India
Prashant Pittalia	Sardar Patel University, India
Priti Sajja	Sardar Patel University, India
Pritpal Singh	Jagiellonian University, Poland
Punya Paltani	IIIT Naya Raipur, India
Rajeev Kumar	NIT Hamirpur, India
Rajesh Thakker	Vishwakarma Govt Engg College, India
Ramesh Prajapati	LJ Institutes of Engineering and Technology, India
Ramzi Guetari	University of Tunis El Manar, Tunisia
Rana Mukherji	ICFAI University, Jaipur, India
Rashmi Saini	GB Pant Institute of Engineering and Technology, India
Rathinaraja Jeyaraj	National Institute of Technology Karnataka, India
Rekha A. G.	State Bank of India, India
Rohini Rao	Manipal Academy of Higher Education (MAHE), India
S. Shanmugam	Concordia University Chicago, USA
S. Srinivasulu Raju	VR Siddhartha Engineering College, India
Sailesh Iyer	Rai University, India
Saman Chaeikar	Iranians University, Iran
Sameerchand Pudaruth	University of Mauritius, Mauritius
Samir Patel	PDPU, India
Sandeep Gaikwad	Charotar University of Science and Technology, India
Sandhya Dubey	Manipal Academy of Higher Education (MAHE), India

Sanjay Moulik	IIIT Guwahati, India
Sannidhan M. S.	NMAM Institute of Technology, India
Sanskruti Patel	Charotar University of Science and Technology, India
Saurabh Das	University of Calcutta, India
S. B. Goyal	City University of Malaysia, Malaysia
Shachi Sharma	South Asian University, India
Shailesh Khant	Charotar University of Science and Technology, India
Shefali Naik	Ahmedabad University, India
Shilpa Gite	Symbiosis Institute of Technology, India
Shravan Kumar Garg	Swami Vivekanand Subharti University, India
Sohil Pandya	Charotar University of Science and Technology, India
Spiros Skiadopoulos	University of the Peloponnese, Greece
Srinibas Swain	IIIT Guwahati, India
Srinivasan Sriramulu	Galgotias University, India
Subhasish Dhal	IIIT Guwahati, India
Sudhanshu Maurya	Graphic Era Hill University, India
Sujit Das	National Institute of Technology-Warangal, India
Sumegh Tharewal	Dr. Babasaheb Ambedkar Marathwada University, India
Sunil Bajeja	Marwadi University, India
Swati Gupta	Jaypee University of Information Technology, India
Tanima Dutta	Indian Institute of Technology (BHU), India
Tanuja S. Dhope	Rajarshi Shahu College of Engineering, India
Thoudam Singh	NIT Silchar, India
Trushit Upadhyaya	Charotar University of Science and Technology, India
Tzung-Pei Hong	National University of Kaohsiung, Taiwan
Vana Kalogeraki	Athens University of Economics and Business, Greece
Vasudha M. P.	Jain University, India
Vatsal Shah	BVM Engineering, India
Veena Jokhakar	VNSGU, India
Vibhakar Pathak	Arya College of Engg. and IT, India
Vijaya Rajanala	SR Engineering College, India
Vinay Vachharajani	Ahmedabad University, India
Vinod Kumar	IIIT Lucknow, India
Vishnu Pendyala	San José State University, USA
Yogesh Rode	Jijamata Mahavidyalaya Buldhana, India
Zina Miled	Indiana University, USA

Additional Reviewers

Chintal Raval
Harshil Joshi
Meera Kansara

Contents

Theory and Methods

Theory and Methods

NAARPreC: A Novel Approach for Adaptive Resource Prediction in Cloud

Riddhi Thakkar(✉) and Madhuri Bhavsar

Institute of Technology, Nirma University, Ahmedabad, India
{18ftphde29,madhuri.bhavsar}@nirmauni.ac.in

Abstract. Cloud computing is one of the most widely used web-based technologies for providing computer resources or network infrastructure services. End-users submit a variety of workloads to cloud providers in the form of web applications, digital transaction services, mobile computing infrastructure, and graphical quality services, all with various QoS requirements expressed in SLA. Cloud is highly dynamic and scalable in nature. Dynamic resource planning is a critical task to ensure the Quality of Services (QoS) in a cloud environment. To optimally utilize cloud resources, it is required to accurately predict future resource demands in a real-time environment. In the cloud computing paradigm, resources like servers, networks, and cloud storage can be allocated to end-users dynamically based on their demand. Since the cloud workload is massive as well as heterogeneous in terms of various attributes and cloud resource demands fluctuate, cloud service providers (CSP) are required to efficiently furnish the available resources. In this paper, we proposed an adaptive approach named NAARPreC to predict future resource demands in cloud environments, which influence resource administration. NAARPreC employs long short-term memory (LSTM), a deep learning model, and auto-regressive integrated moving average (ARIMA), a statistical analysis model. The proposed model significantly improves accuracy and execution time.

Keywords: Cloud computing · Elasticity · Resource utilization · Deep learning

1 Introduction

Cloud computing is a web-based model that provides resources related to information technology to end-users and enterprises. Users can get on-demand access to resources and services through a cloud environment based on a pay-per-use paradigm. The resources can be servers, development platforms, or applications

K. K. Patel et al. (Eds.): icSoftComp 2022, CCIS 1788, pp. 3–16, 2023.
https://doi.org/10.1007/978-3-031-27609-5_1

as services. Cloud computing is incredibly cost-effective and saves a significant amount of money. The primary advantage of cloud computing is the remote storage and processing of data rather than in local systems, so in case of a system failure, the data and application will remain safe and secure.

One of the preliminary motives of the cloud service provider (CSP) is optimal cloud resource utilization and maximum revenue generation. Thus, CSP serves more requests from users and also satisfies QoS requirements [7]. Cloud resource management is vital for providing high-performance cloud services to end customers. In general, resource management is classified as reactive and proactive [6]. With the reactive approach, it is not possible to handle a sudden burst of workload in real-time, which may lead to a violation of the SLA and QoS agreement [16]. Proactive techniques, on the other hand, overcome this problem by allocating the necessary resources in advance of their need. They may leverage the prediction model for forecasting future workloads and recognizing potential resource usage patterns. As a result of the successful forecast, performance degradation can be avoided, and idle resources can be decreased, so CSP can increase the profit [3]. However, proactive resource planning is not an easy task, where the fluctuation in input workload may cause the following issues [12,16]:

1. Over-provisioning: Resources are assigned to the service in excess of demand, resulting in wastage, and higher costs for the customer
2. Under-provisioning: Service requests are not fulfilled due to resource unavailability
3. Oscillation: As a result of auto-scaling, a mix of over-provisioning and under-provisioning issues arise

A considerable amount of work is done in the field of workload prediction in cloud computing using artificial intelligence (AI), neural network (NN), and deep learning (DL) models [4,19]. However, these approaches take more time to train the model and are more complex, which reduces the adaptability for different cloud environments [12]. To overcome these issues and to achieve better accuracy in less time with less complexity, this work proposes a framework based on auto-regressive integrated moving average (ARIMA) and long short-term memory (LSTM) models. LSTM model is capable of learning from long-term dependencies in data [6]. ARIMA model is also used for time series forecasting. However, if a large number of random variations are present in data, and if data follows a non-linear relationship, then the ARIMA model is not accurate. To overcome these issues and get high accuracy concerning less complexity and time, an adaptive approach using LSTM and ARIMA models is proposed in this work.

1.1 Cloud Resource Prediction Challenges

Figure 1 shows major challenges for resource prediction in cloud computing.

1. Complexity: A prediction model should be less complex

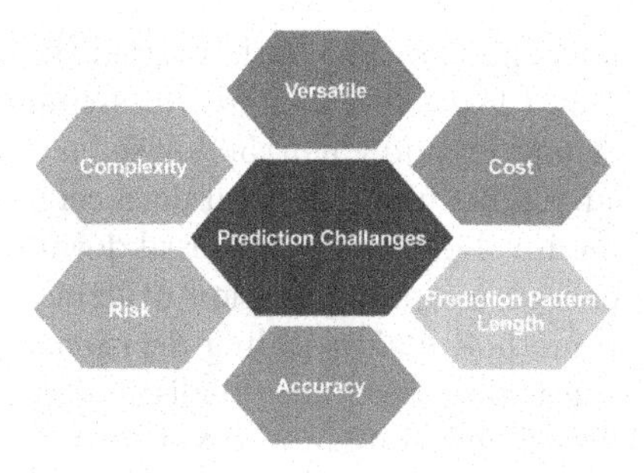

Fig. 1. Prediction challenges

2. Versatile: The model should be versatile enough to predict the future resources for different users, having resources with different configurations.
3. Risk parameter: To comprehend and deduce the risk parameters affecting over-provisioning and under-provisioning is a challenging task.
4. Cost: The overall cost of resource prediction should be less.
5. Accuracy: The proposed framework should be accurate enough so that CSP can make a profit with accurate resource management.
6. Prediction pattern length: Determining the pattern length is a difficult task. Unfit length leads the model to learn from only specific patterns only.

1.2 Contribution

We have proposed an adaptive novel approach called NAARPreC to predict the future cloud resource demand. In this approach, LSTM and ARIMA models are utilized. Bitbrains [17] dataset is used to train and predict future resource requirements. The proposed framework is less complex to implement and more versatile in nature due to its simplicity. This framework will help CSP to operate the resources optimally.

1.3 Organization

Section 4 elaborates the technical and conceptual background for the proposed framework. Section 5 and 6.1 depicted the framework architecture and corresponding flow diagram. Section 7 discussed the prediction results as well as time comparison. The last Sect. 8 concludes the research work and the future work which can be carried out in this area.

2 Related Work

Suresh et al. [20] developed an enhanced load balancing solution based on particle swarm optimization, which selects the best resource for the least amount of money. Using the kernel fuzzy c-means approach, their program classified cloud services into different clusters. The authors used the cloud simulation tool to assess their approach and learned that it reduces completion time, memory consumption, and cost. Xiong Fu and Chen [5] devised a two-step deep learning-based approach for resource scheduling in cloud data centers to cut power costs. Their suggested technique involves two steps: breakdown of the user load models into many jobs and energy cost minimization utilizing a deep learning-based strategy. Their proposed method makes scaling decisions dynamically based on learning from service request patterns and a realistic figure mechanism.

Sukhpal et al. [18] examined contemporary research in resource planning strategies such as scheduling algorithms, dynamic resources, and autonomously resource scheduling and provisioning. The authors classified resource monitoring systems using QoS criteria like responsiveness, resource consumption, pricing, and SLA breaches, among other things. They also presented upcoming research challenges in the area of autonomous methodologies in the cloud.

Gong et al. [8] developed an adaptive dynamic resource algorithm based on the control concept. For optimizing resource consumption and achieving quality standards, the authors proposed a hybrid approach integrating adaptive multi-input and multi-output (MIMO) control and radial basis function (RBF) neural network. The CPU and RAM are allocated to cloud services according to demand changes and quality standards in their suggested methodology.

For flexible cloud service delivery, Rafael Moreno-Vozmediano et al. [13] suggested a hybrid autoscaling system with machine learning-based and decision analysis. Their solution relied on the SVM classifier regression method to forecast the webserver workload based on past data. Furthermore, the proposed method made use of a queueing model to determine the number of cloud systems that should be deployed based on the projected load. The SVM-based regression method achieved better prediction accuracy than some other conventional projection models, according to the simulation data.

Bi et al. [2] proposed a new approach for estimating resource utilization and turnaround time for various workload patterns of web applications in data centers using implicit workload variables. Their approach used autonomous learning techniques to discover latent patterns in historic access logs to estimate resource requirements. The authors evaluated the approach using a variety of benchmarking applications, indicating that it beats existing methods in terms of predicting CPU, ram, bandwidth utilization, and reaction time. To meet future resource requirements, Amekraz et al. [1] presented prediction-based resource assessment and supply strategies utilizing a combination of neural networks and linear regression, whereas our work exclusively analyses time series and forecasting models. Mouine et al. [14] suggested a unique dynamic control strategy based on continuous supervised learning for flexible resource scheduling in the cloud's global market while contending with ambiguity.

In general, most recent research has relied on heuristic-based techniques to decide scaling choices and resource scheduling for cloud applications. The heuristic-based approaches are still not adequate for the supply of cloud resources to manage diverse cloud workloads because most workloads posted by users to cloud providers are varied with different quality demands. Although several meta-heuristic-based strategies for tackling large-scale cloud applications have already been employed. However, additional work is needed to supply cloud services on-demand effectively. To lower the service-level agreement (SLA) margin requirement, Rosa et al. [15] suggested a workload estimator paired with ARIMA and dynamic error compensation. Tseng et al. [21] introduced a genetic algorithm based prediction approach for speed and ram consumption of virtual and physical machines, which outperforms the grey model in terms of forecast accuracy under stable and unstable tendencies.

3 Motivation

Unbounded resource demand for computational activities is a key difficulty in cloud computing. Not unexpectedly, earlier work has produced several strategies for efficiently providing cloud resources. However, a forecast of future resource usage of impending computational processes is required to implement a comprehensive dynamic resource forecasting model. Resource management entails dynamic resource scaling up or down in response to present and future requirements. As we know, the demand for cloud computing is increasing very rapidly in every domain. CSPs are required to have a robust mechanism to deliver seamless services to the end-user or customer. If the CSP fails to attain resource demand, SLA will get violated. Such circumstances motivate us to perform research in this area, allowing the CSP to get the future requirements and avail the resource to users in no time. This way, users will also receive the cloud services swiftly.

4 Background

The proposed framework is designed using ARIMA and LSTM models, which are then configured as per the logic. The conceptual background for both models is described in the below section.

4.1 Statistical Model: ARIMA

ARIMA model is a form of regression analysis that gauges the strength of one changing variable relative to other dependent variables. It is a standard statistical model to predict future values based on past values. If a statistical model predicts possible trends based on previous values, it is called autoregressive. Each of these methods is being used to fit time series analysis, identify complex patterns in data and accurately predict future data points. When data show evidence of non-stationarity in the perspective of the mean, ARIMA models are used to execute an initial finite difference step one or more times to minimize the non-stationarity of the mean function. For smoothing the time series data ARIMA model uses lagged moving averages. It assumes that the future trend will resemble based on past trends. Following are the major components of the ARIMA model [9]:

1. Autoregression (AR): It refers to that the changing variable is dependent upon its prior and its own lagged values. AR(1) indicates the current value is based on the immediately previous value, while AR(2) indicates that the current value is based on the last two values.
2. Integrated (I): It represents differences between raw data and previous values.
3. Moving average (MA): It is an indicator to analyze and trend direction or to determine resistance levels and support (Fig. 2).

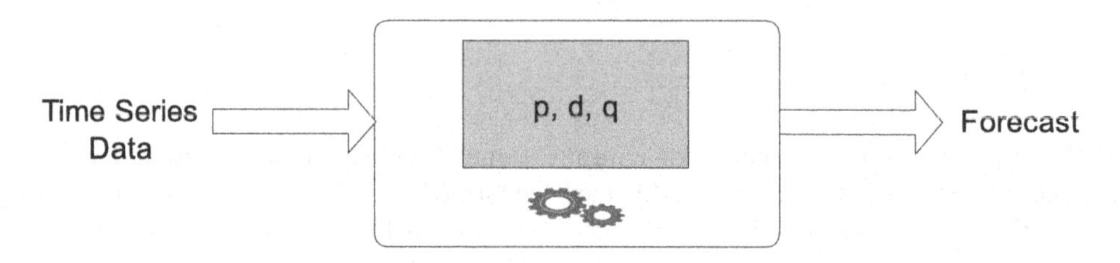

Fig. 2. Basic ARIMA model

Figure 2 depicts the basic ARIMA model. The parameters of the ARIMA model are defined as p,d, and q, with the model stated as ARIMA(p,d,q). Here, p denotes the number of lag observations in the model. d symbolizes the difference between raw observations and past data. q represents the size of the moving average window.

4.2 LSTM

Long short-term memory (LSTM) is an advanced architecture of recurrent neural networks (RNN), which can remember and predict long sequences. LSTM features backpropagation rather than normal feed-forward networks. It can handle not only individual data points but large data streams as well. The LSTM unit consists of a cell, an input, an output, and a forget gate. The cell's three gates transport data in and out, and store values for arbitrary lengths of time. As there may be an unpredictable length of time between significant occurrences in time series data, LSTM networks excel at classifying, evaluating, and forecasting time series data over the period. LSTM is a special kind of RNN model that can learn long-term dependencies in the dataset. The LSTM model increases the memory of RNN [11].

Figure 3 indicates inputs in orange circles, point-wise operators in green circles, neural network layers in yellow boxes, and cell states in blue circles. The LSTM module has three gates and cell states, which allow the model to forget, selectively learn, and retain information from each of the units. Cell states allow the LSTM model to flow information through units. Every unit has a forget gate and input and output gates. These gates can add or remove information from the cell. The forget gate decides the amount of data from past cell states to overlook

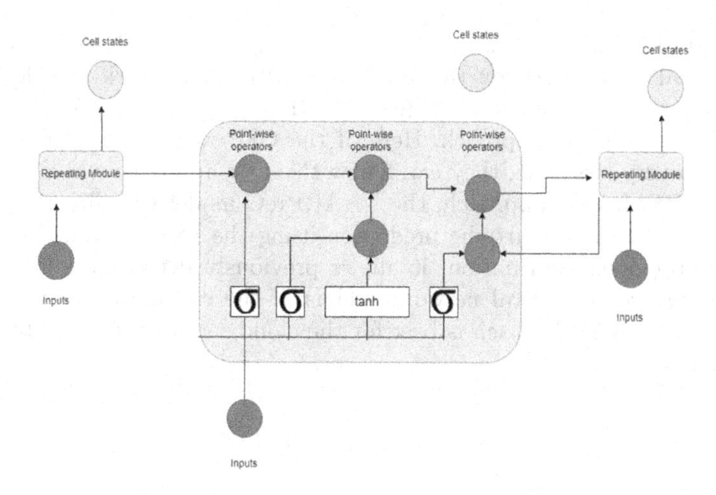

Fig. 3. LSTM model

using the sigmoid function. The input gate performs point-wise multiplication of sigmoid and tanh functions to control the information flow to the current cell. The output gate determines the input passed to the subsequent hidden state.

5 System Model

Figure 4 shows the general flow of NAARPreC framework. Initially, the framework will perform data pre-processing on raw data. For evaluation of the model, Bitbrain [17] dataset is used, which consists of CPU usage, memory usage, and disk throughput of 1750 VMs.

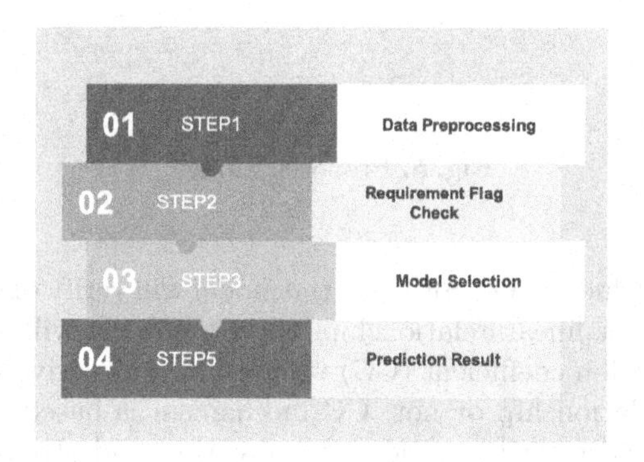

Fig. 4. Basic flow

Figure 5 represents, NAARPreC, architecture for resource prediction. NAARPreC will use one boolean type of variable named Requirement Flag (RF).

This RF variable can have values like SET, which is equal to 1, and NOT SET, which is equal to 0. Based on the value of RF, the framework will decide the whole flow for future prediction. When the RF value is SET, NAARPreC will follow a fast prediction approach. Here, if the CSP wants the future resource prediction in very less time, they can follow this first approach where they have to SET the RF. In this approach, the NAARPreC model will directly pick the pre-processed data and start the prediction using the LSTM model. As LSTM retains the previous information, it allows previously determined information to be used in present neural networks. Thus, if the end-user demand is more focused on time, this approach is best for the cloud service providers to predict in less time.

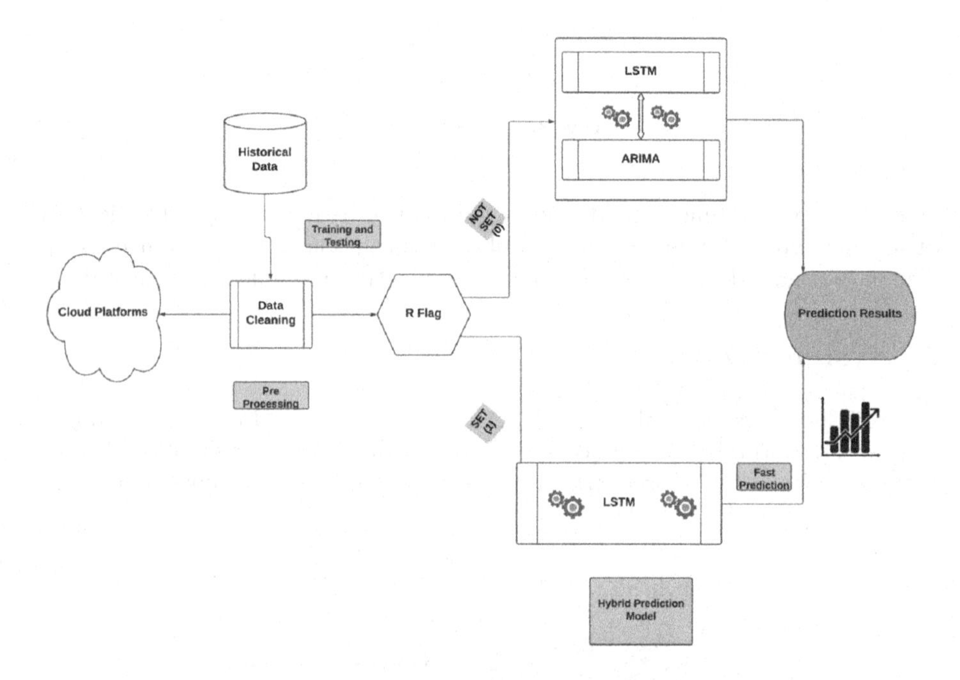

Fig. 5. Proposed model

When the RF value is NOT SET (0), the model will verify one more condition. If the data follows a linear relationship, the framework will pick the ARIMA model. The correlation coefficient (CC) value is used to derive whether the data follows a linear relationship or not. CC mechanism is based on the numerical field to find out the positive or negative direction and strength of the linear relationship. For finding the CC value, sample and mean value of variables X and Y are required, where Sx and Sy is the standard deviation of variable X and Y. CC values can be between –1 to +1. If the value is near zero, it is considered a weak linear relationship and vice versa.

$$CC = \frac{1}{n-1} \sum \frac{(x-X)}{S_x} \frac{(y-Y)}{S_y} \tag{1}$$

Here, X represents the value of an independent variable, Y represents the value of a dependent variable, and n represents the number of observations. S_x and S_y is the standard deviation of variable X and Y.

ARIMA model will give better results when data follow a linear relationship. Therefore, to achieve better accuracy and future prediction in less time, NAARPreC will follow the ARIMA model.

In the case of nonlinear data, we can not get a success rate for future prediction with only the ARIMA model. As differencing (d) is helpful for prediction using the ARIMA model, and for linear data, the difference value is considered as zero, which means ARIMA (p,0,q). Thus, it should be considered as ARIMA (p,q). Here, in this scenario, we will go with the other hybrid approach, ARIMA+LSTM. First, the filtered data will go to the ARIMA model. Here in this phase, it will get some residuals, then again pass this residual to the next phase, where the prediction will happen on the LSTM model. After completion of this phase, the result will be generated. The complete flow is depicted in Fig. 6.

6 Proposed Algorithm

This section discusses the NAARPreC algorithm and its flow. Pre-processing, the model section for prediction, and dynamic resource forecasting are the three primary elements in the NAARPreC framework.

Algorithm 1. Adaptive resource prediction

```
START
Input: Real time data
Output: Future resource predictions
Initialization: Dataset = Pre-Processed cloud data
  procedure NAARPreC(RF)
    if (RF) then
      Future resource prediction using LSTM model
      LSTM(Dataset)
    else
      if Dataset is linear then
        Future resource prediction using ARIMA model
        ARIMA(p,d,q)
      else
        Prediction using Hybrid model
        Hybrid(Dataset)
      end if
    end if
  end procedure
```

6.1 Schematic Representation of NAARPreC

This section discusses the schematic representation of NAARPreC. First, pre-processing is performed on the complete dataset. The pre-processed data is passed to the prediction model based on the requirement flag check. For quick prediction, the LSTM model approach will be executed, and we will get the prediction results in a very short time compared to the ARIMA + LSTM approach. If the focus is more on accuracy and not on time, then the framework will check if the data follows a linear relationship or not, and based on that, the next path will be selected. Thus, if the data is for the ARIMA model, it will get executed, and the future prediction will be generated. In either case, the hybrid approach consisting of ARIMA and LSTM models will be executed. Here, the ARIMA + LSTM approach takes more execution time, while achieving higher accuracy.

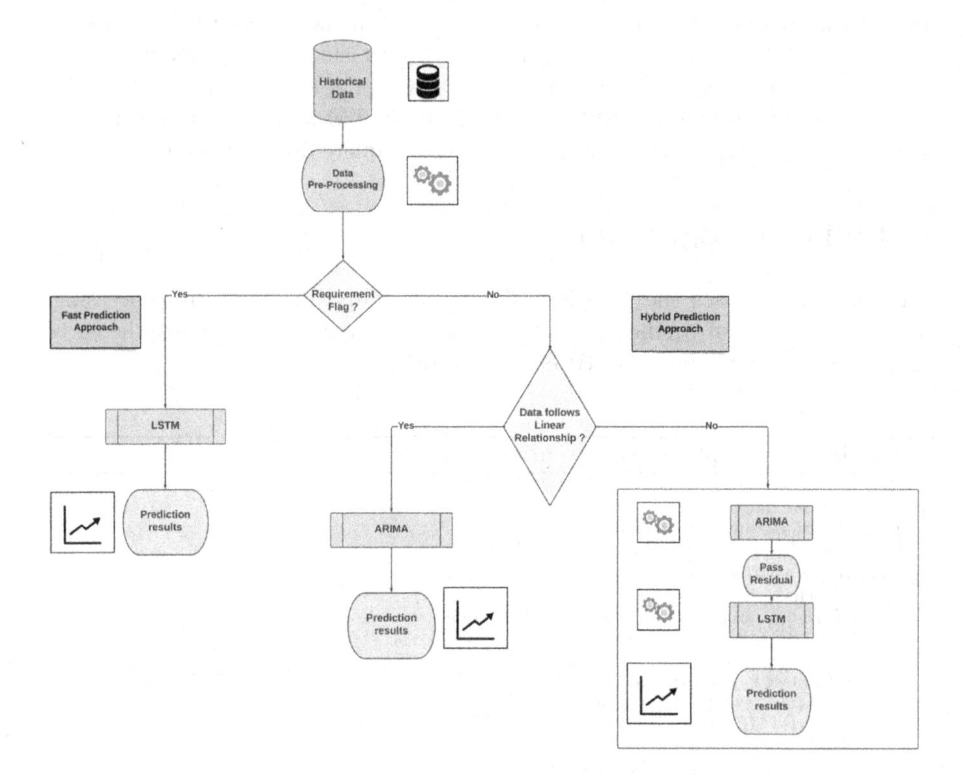

Fig. 6. Flowchart of proposed framework

7 Experimental Results and Analysis

To analyze the performance of our proposed model, we conducted experiments on ARIMA, LSTM, and hybrid approaches. The Bitbrains [17] cloud database is used to analyze all three paths of the proposed approach. To evaluate the prediction accuracy of the NAARPreC, Mean Absolute Error (MAE), and Root Means Square Error (RMSE), error metrics are operated on.

7.1 Dataset

To evaluate the performance of the NAARPreC, BitBrain [17] dataset is used. The dataset contains over 1 million data values. The dataset contains CPU usage, memory usage, network transmitted throughput, and disk read throughput attributes.

7.2 Evaluation Metrics

The performance of the proposed model is evaluated by examining RMSE and MAE. The lower the value of evaluation metrics, the better the model is. These evaluation metrics are calculated as follows: [10].

- Root Mean Squared Error (RMSE): RMSE calculates the squared root of the average of the squared difference between the actual and predicted values by the proposed model. The higher the value of RMSE, the larger the difference between the actual and predicted values.

$$RMSE = \sqrt{\frac{\sum (x - X)^2}{N}} \tag{2}$$

  Here, x represents actual value, X represent predicted value, and N represents the number of observations.
- Mean Absolute Error (MAE): MAE calculates the average of the difference between the actual and predicted values.

$$MAE = \frac{\sum |(y - Y)^2|}{N} \tag{3}$$

  Here, y represents actual value, Y represent predicted value, and N represents the number of observations.

7.3 Result Discussion

In the cloud, the workload data is not stable, and it can be of variable length. Hence, we compared the amount of time taken by the LSTM and the ARIMA. Thus, in the case of RF being set, the model will quickly generate the prediction result, as shown in Fig. 7a. For less than 500k inputs, the LSTM model is taking more than 50% less time than the ARIMA model. After increasing the inputs to 30% ARIMA model takes 90% more time.

We predicted the future resource requirements using the proposed approach for three attributes CPU utilization, memory usage, and disk utilization, which are important in the resource allocation as well as management field.

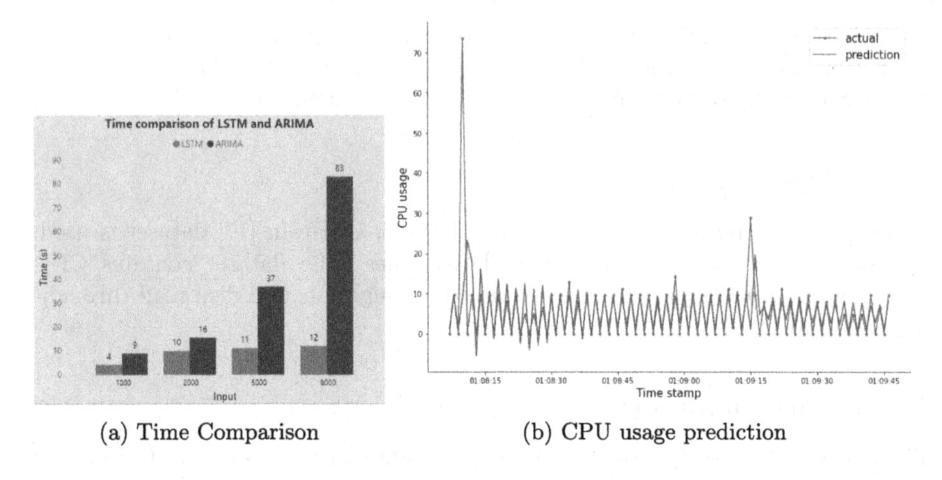

(a) Time Comparison (b) CPU usage prediction

Fig. 7. Results

Figure 7 shows that the proposed approach achieves high accuracy (more than 90%) with minimum execution time for predicted CPU usage. From Fig. 8a, we observe that the disk throughput is getting the best accuracy for a given instance. Moreover, future memory usage is not accurately predicted as depicted in Fig. 8b.

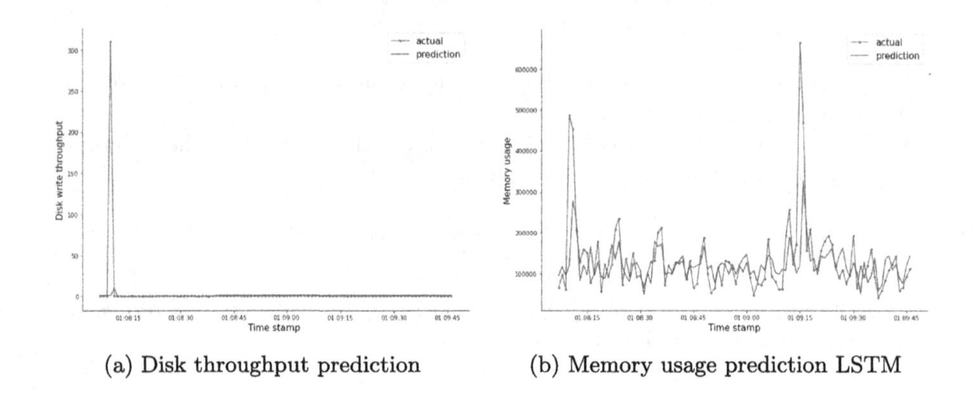

(a) Disk throughput prediction (b) Memory usage prediction LSTM

Fig. 8. Results

8 Conclusion and Future Scope

In this research work, different algorithms and models for predicting future cloud resource requirements are studied. We presented an adaptive solution based on LSTM and ARIMA models through which future resource requirements like CPU usage, memory usage, and throughput can be predicted. In this work, we carried out a series of tests and experiments to achieve a higher success rate for the fluctuating workload. Through the hybrid approach, higher accuracy for the prediction can be achieved. However, it is more time-consuming. The LSTM model gave the prediction results in less time, although it is not suitable for all types of data. It is analyzed that, if data follows a linear relationship, the ARIMA model approach gives the best prediction results. The proposed workload prediction model can forecast both seasonal and irregular task patterns, helping to reduce the wastage of resources. In the future, NAARPreC can be enhanced to predict the wide range of different attributes and distributed services. We will evaluate NAARPreC on diverse real-world cloud workloads.

Acknowledgements. Implementation of this work has partially been completed by Shiv Patel, Nirma University.

References

1. Amekraz, Z., Hadi, M.Y.: A cluster workload forecasting strategy using a higher order statistics based ARMA model for IAAS cloud services. Int. J. Netw. Virt. Organ. **26**(1–2), 3–22 (2022)
2. Bi, J., Li, S., Yuan, H., Zhou, M.C.: Integrated deep learning method for workload and resource prediction in cloud systems. Neurocomputing **424**, 35–48 (2021)
3. Chen, J., Wang, Y.: A hybrid method for short-term host utilization prediction in cloud computing. J. Electr. Comput. Eng. **2019** (2019)
4. Duc, T.L., García Leiva, R., Casari, P., Östberg, P.O.: Machine learning methods for reliable resource provisioning in edge-cloud computing: A survey. ACM Comput. Surv. **52**(5), 1–39 (2019)
5. Xiong, F., Zhou, C.: Predicted affinity based virtual machine placement in cloud computing environments. IEEE Trans. Cloud Comput. **8**(1), 246–255 (2017)
6. Gao, J., Wang, H., Shen, H.: Machine learning based workload prediction in cloud computing. In: 2020 29th International Conference on Computer Communications and Networks (ICCCN), pp. 1–9. IEEE (2020)
7. Ghobaei-Arani, M., Jabbehdari, S., Pourmina, M.A.: An autonomic resource provisioning approach for service-based cloud applications: a hybrid approach. Future Gener. Comput. Syst. **78**, 191–210 (2018)
8. Gong, S., Yin, B., Zheng, Z., Kai-yuan, C.: An adaptive control method for resource provisioning with resource utilization constraints in cloud computing. Int. J. Comput. Intell. Syst. **12**(2), 485 (2019)
9. Hillmer, S.C., Tiao, G.C.: An arima-model-based approach to seasonal adjustment. J. Am. Stat. Assoc. **77**(377), 63–70 (1982)
10. James, G., Witten, D., Hastie, T., Tibshirani, R.: An Introduction to Statistical Learning. Springer, Heidelberg (2013). https://doi.org/10.1007/978-1-0716-1418-1
11. Li, Y.F., Cao, H.: Prediction for tourism flow based on LSTM neural network. Procedia Comput. Sci. **129**, 277–283 (2018)
12. Masdari, M., Khoshnevis, A.: A survey and classification of the workload forecasting methods in cloud computing. Cluster Comput. **23**(4), 2399–2424 (2020)
13. Moreno-Vozmediano, R., Montero, R.S., Huedo, E., Llorente, I.M.: Efficient resource provisioning for elastic cloud services based on machine learning techniques. J. Cloud Comput. **8**(1), 1–18 (2019)

14. Mouine, E., Liu, Y., Sun, J., Nayrolles, M., Kalantari, M.: The analysis of time series forecasting on resource provision of cloud-based game servers. In: 2021 IEEE International Conference on Big Data (Big Data), pp. 2381–2389. IEEE (2021)
15. Rosa, M.J.F., Ralha, C.G., Holanda, M., Araujo, A.P.F.: Computational resource and cost prediction service for scientific workflows in federated clouds. Future Gener. Comput. Syst. **125**, 844–858 (2021)
16. Shahidinejad, A., Ghobaei-Arani, M., Masdari, M.: Resource provisioning using workload clustering in cloud computing environment: a hybrid approach. Cluster Comput. **24**(1), 319–342 (2021)
17. Shen, S., Van Beek, V., Iosup, A.: Statistical characterization of business-critical workloads hosted in cloud datacenters, pp. 465–474. IEEE (2015)
18. Singh, S., Chana, I., Singh, M.: The journey of QoS-aware autonomic cloud computing. IT Prof. **19**(2), 42–49 (2017)
19. Song, B., Yao, Yu., Zhou, Yu., Wang, Z., Sidan, D.: Host load prediction with long short-term memory in cloud computing. J. Supercomput. **74**(12), 6554–6568 (2018)
20. Suresh, A., Varatharajan, R.: Competent resource provisioning and distribution techniques for cloud computing environment. Cluster Comput. **22**(5), 11039–11046 (2019)
21. Tseng, F.-H., Wang, X., Chou, L.-D., Chao, H.-C., Leung, V.C.M.: Dynamic resource prediction and allocation for cloud data center using the multiobjective genetic algorithm. IEEE Syst. J. **12**(2), 1688–1699 (2017)

One True Pairing: Evaluating Effective Language Pairings for Fake News Detection Employing Zero-Shot Cross-Lingual Transfer

Samra Kasim(✉)

Johns Hopkins University, Baltimore, MD 21218, USA
skasim3@jhu.edu

Abstract. Fake news poses a great threat to democracy, human rights, health and more. Its viral proliferation, especially in low-resource languages, necessitates automated means for fake news detection be employed to combat the scourge. The improvements demonstrated by cross-lingual language models such as XLM-RoBERTa (XLM-R) and Multilingual BERT (mBERT) in cross-lingual understanding provide an opportunity to develop language-independent models, utilizing zero-shot cross-lingual transfer (i.e., training a model on a dataset in one language and applying it to a dataset in another language with no additional training), to address limitations posed by the lack of training data in low-resource languages. This paper demonstrates that, for particular language pairings, it is possible to employ zero-shot cross-lingual transfer for fake news detection in full-text articles. Utilizing Support Vector Machine and Neural Network classifiers for fake news detection and XLM-R and mBERT embeddings achieves average F1 scores of 0.90 for an Urdu monolingual model evaluated on Bengali fake news dataset; 0.68 for a Bengali monolingual model evaluated on an English fake news dataset; 0.67 for Spanish and Urdu monolingual models on English fake news datasets; 0.67 for Bengali monolingual model evaluating a Spanish fake news dataset. This paper also demonstrates that low-resource language pairings outperform pairings with high-resource language models for fake news detection in low-resource languages.

Keywords: NLP · Cross-lingual language models · Fake news · Low-resource language · Zero-shot cross-lingual transfer · XLM-RoBERTa · Multilingual BERT

1 Introduction

The growth in online content has enabled greater access to knowledge than ever before, but it has also led to the proliferation of misinformation. In 2018, misinformation posted to Facebook in Myanmar resulted in genocide committed against

K. K. Patel et al. (Eds.): icSoftComp 2022, CCIS 1788, pp. 17–28, 2023.
https://doi.org/10.1007/978-3-031-27609-5_2

the country's Rohingya minority [15]. In the days leading up to the 2020 American election, Spanish-language election misinformation thrived online with 43 Spanish-language posts alone generating 1.4 million social media interactions [4]. A Washington Post headline summarized the issue: "Misinformation online is bad in English. But it's far worse in Spanish" [19]. Although only a quarter of the web content is in English [7], most of the Natural Language Processing (NLP) efforts have focused on high-resource languages, like English, while most of the nearly 7,000 languages in the world are considered low-resource due to the lack of available training data for NLP tasks [14]. Additionally, when research is done on misinformation in low-resource languages, it is focused on monolingual language models.

This paper addresses the above challenges by utilizing training datasets in Bengali, Filipino, and Urdu, which are considered low-resource languages, in addition to using English and Spanish datasets for model training. Experiments also account for variation in script as Bengali text is in the Bengali alphabet, Urdu is in the Arabic script, and the remaining languages are in Latin alphabet. While most studies involving fake news detection focus on comparing a language to English or a language group to each other and English, this paper focuses on full-text fake news detection to gather and analyze the diverse languages listed above. The zero-shot cross-lingual transfer experiments were conducted using pre-trained XLM-RoBERTa (XLM-R) and Multilingual BERT (mBERT) models, and Support Vector Machine (SVM) and Neural Network (NN) models were used for fake news detection. The remainder of the paper is structured as follows. Section 2 discusses the related research. Section 3 outlines the experimental approach, including details on the datasets, classification algorithms, language transformers, and the processing and classification pipeline. Section 4 analyzes the experiments' results, and finally, Sect. 5 outlines the conclusions and opportunities for future work.

2 Related Works

In a survey on NLP for Fake News detection, Oshikawa, Qian, and Wang [16], highlight the unavailability of datasets for fake news detection as a considerable challenge. Regarding entire-article datasets, the authors state that there are few sources for manually labeled entire-article datasets as compared to short claims datasets. However, full-text articles across five languages, which were generously made publicly available by researchers conducting monolingual fake news detection research, are utilized in the experiments described in Sect. 3 and are discussed in detail in Sect. 3.2.

Noting the scarcity of news articles in low-resource languages, Du, Dou, Xia, Cui, Ma, and Yu [9], in their study of cross-lingual COVID-19 fake news detection, conducted several cross-lingual experiments. In one such experiment, they trained two cross-lingual fake news detection models leveraging Multilingual-BERT (mBERT) to encode English and Chinese data and found that multilingual models were ineffective in cross-lingual fake news detection. Section 4, however, demonstrates that XLM-R generated cross-lingual representations are an effective means of detecting fake news in full-text articles using models for particular language pairs.

Conneau et al. [5] conducted unsupervised training of cross-lingual embeddings at very large scale using the CommonCrawl Corpus. To develop XLM-R, which is a transformer-based multilingual masked language model, they applied a SentencePiece model directly to raw text data for 100 languages. They observed that XLM-R performed well on low-resource languages and concluded that XLM-100 and mBERT do not model low-resource languages as well as XLM-R. The experiments conducted for this paper demonstrate below that monolingual models trained with low-resource languages performed better than high-resource language trained monolingual models in most instances.

In 2019, Bhatia et al. [3] demonstrated the use of multilingual models in detecting hate and offensive speech across English, Hindi, and Marathi language tweets. The researchers trained the multilingual models on a combination of all languages, including the target language. The research described below takes a different approach. It trains each model only on full-text articles in one language and applies each monolingual model to full-text articles in a different target language.

3 Methods

3.1 Definitions

This paper formally defines fake news as articles containing misrepresentations, misinformation, and falsehoods and real news as articles containing legitimate and verifiable information.

3.2 Datasets

The following datasets were used for this study. Duplicate and empty entries were removed during data processing from the original datasets, and the number of instances used for this paper are listed in Table 1. For monolingual model training, 10% of the dataset was used for hyperparameter tuning and the remainder was split into 80% training and 20% testing with 10-fold cross-validation.

Table 1. Number of articles in each dataset with Bengali having the most instances and Urdu the least.

Language	Real	Fake
Bengali	7,202	1,299
English	128	123
Spanish	624	624
Filipino	1,496	1,509
Urdu	495	399

1. Bengali: 50,000 instance dataset of entire text news articles in the Bengali alphabet developed for a monolingual fake news detection study. True articles were collected from 22 mainstream Bangladeshi news portals, and fake news is classified as any article containing false/misleading claims, clickbait, satire/parody. The original dataset has 1,299 labeled fake articles, 48,678 labeled true articles. This study utilized a subset of the original dataset by selecting all of the articles labeled as fake and randomly selecting 7,202 articles labeled as real to mitigate for the imbalanced dataset [12].
2. English: Data developed by Horne and Adali contains manually checked real labeled articles from Buzzfeed news and randomly collected political news articles labeled, real, satire, or fake. The original dataset contained 128 labeled real news articles and 123 labeled fake news articles. Articles labeled satire were not used for this study [11].
3. Spanish: The original dataset contains 1,248 instances that were collected between November 2020 and March 2021 from newspaper and fact-checking websites. The topics cover science, sport, politics, society, COVID-19, environment, and international news [2,10,17].
4. Filipino: The dataset was originally developed for a monolingual fake news study and consisted of 1,603 labeled real news articles and 1,603 labeled fake news articles in the Latin alphabet. There were duplicates in the original dataset, which were removed in processing, resulting in 1,496 labeled true news articles and 1,509 labeled fake news articles. Fake news articles are from recognized fake news sites and were verified by a fact-checking organization, Verafiles, and by the National Union of Journalists in the Philippines. The real news articles are from mainstream news websites in the Philippines [6].
5. Urdu: Developed for a fake news detection study in Urdu covering technology, business, sports, entertainment, and health. The original dataset contains 400 labeled fake news articles and 500 labeled true news articles in Arabic script. Duplicate articles were removed from the dataset during processing resulting in 495 labeled true news articles and 399 labeled fake news articles. The labeled real articles are from mainstream news sites like BBC Urdu, CNN Urdu, and Express News while the labeled fake news articles are written by journalists and are fake versions of real articles [1].

3.3 Cross-Lingual Language Models

This section describes the cross-lingual transformers implemented for the experiments. As described by Conneau et al. [5], XLM-R outperformed mBERT in F1 scores on MLQA question answering where models were trained on an English dataset and then evaluated on seven languages resulting in an average F1 score of 70.7 for XLM-R and 57.7 for mBERT.

XLM-R. XLM-R is a transformer-based masked learning model trained on over a 100 languages and 2.5TB of filtered CommonCrawl data. This study utilized the pre-trained xlmr.large model that uses BERT-large architecture and contains

560 million parameters, a vocabulary size of 250,000 sub-words, and has 1,024 dimensions [5]. The following are the sizes of the monolingual training corpus utilized in XLMR-training: 300.8 GiB for English, 53.3 GiB for Spanish, 8.4 GiB for Bengali, 5.7 GiB for Urdu, and 3.1 GiB for Filipino [5]. Additionally, XLM-R utilizes a SentencePiece tokenizer.

mBERT. mBERT is also a transformer based model that was trained on 104 top languages on Wikipedia using masked language modeling. For this paper, the Multilingual BERT-base model was used, which has 110 million parameters, a shared vocabulary size of 110,000, and 768 dimensions. Unlike XLM-R, mBERT uses WordPiece embeddings. The maximum sizes of the monolingual corpus used for training mBERT are: 22.6 GB for English, 5.7 GB for Spanish, 0.4 GB for Bengali, 0.2 GB for Urdu, and 0.1 GB for Filipino [20]. In the paper introducing BERT, the authors demonstrated that concatenating the last four hidden layers resulted in a Dev F1 of 96.1 and outperformed the last hidden layer, which had a Dev F1 of 94.9 [8]. For this paper, the experiments used an averaged concatenation of the last four hidden layers as well as the last layer's pooled output.

3.4 Classification Algorithms

This section describes the implementation of a statistical supervised algorithm and a neural network.

Support Vector Machines. The study utilizes the SVM supervised machine learning algorithm for classification. SVM is well-suited to data classification and pattern recognition problems and is often applied in fake news detection studies. SVM constructs a maximum margin hyperplane to linearly separate and group data points into classes. If data is not linearly separable, then kernel functions can be leveraged to apply transformations that map data into a new space. Since neural networks have a tendency to overfit small datasets, SVM served well as a contrast to the neural network implementation. The implementation used for the experiments used a linear kernel function.

Neural Networks. Since small datasets were used in this study, a simple sequential neural network with two hidden layers was used for classification to prevent overtraining the model. The first hidden layer had the same number of nodes as the number of dimensions in each instance, i.e., 1,024 for XLM-R and 768 for mBERT. There was a 10% dropout rate after the first hidden layer. The second hidden layer had half the number of nodes as the first hidden layer followed by a 50% dropout rate. Both hidden layers used Rectified Linear Unit (ReLu) activation. The output layer had one node and used sigmoid activation. The compilation implemented binary cross-entropy for loss and the Adam optimizer.

3.5 Pipeline Architecture

This section details the architecture (see Fig. 1) used for transforming raw data into processable text that is then transformed via cross-lingual embeddings and used to generate monolingual models. The models are then used to classify articles as fake or real news.

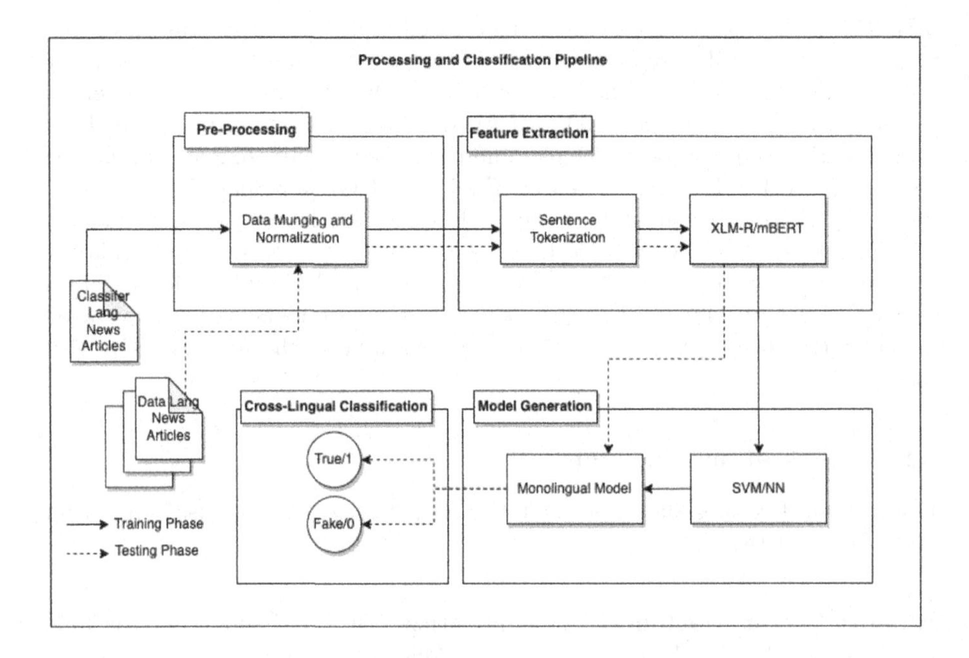

Fig. 1. Pipeline Architecture: From raw data to classified articles

Pre-processing. In pre-processing, raw text files for every language were deduplicated within a text and normalized across languages using Pandas. Since not all language datasets contained the same information, following the methodology implemented by Du, Dou, Xia, Cui, Ma, and Yu. [9], only the article text was investigated. The articles were classified as one (1) indicating an article labeled as real/true news, and zero (0) indicating an article labeled as fake.

Feature Extraction. Each article in the training set was tokenized using sentence tokenization libraries for the specific language (except Urdu, which does not have a sentence tokenization library at the time of this paper and so was split using common punctuation marks). Natural Language Toolkit's (NLTK) English sentence tokenizer was used for English and Filipino [13]. For Spanish, the NLTK Spanish tokenizer was used [13], and the Bengali Natural Language

Processing (BNLP) tokenizer was used for Bengali [18]. Some articles, particularly in Urdu, had long sentences. However, XLM-R only allows encoding of sentences with length less than 513 tokens, so any sentence exceeding the limit was divided into two and then recursively split until the resulting sentences met the XLM-R sentence length requirement and then processed separately.

After sentence tokenization, each sentence was processed through XLM-R or mBERT for encoding. For XLM-R, the last layer feature for each sentence is extracted. The result is a multi-row matrix with 1,024 dimensions. Using Pytorch, this matrix is averaged along the horizontal axis and results in a 1×1024 vector representing a sentence. Further, all the sentence vectors for a document are averaged along the horizontal axis and result in a 1×1024 axis representing a document. For mBERT, each tokenized sentence was parsed by mBERT into WordPiece representation. The pooled output of 768 dimensions was extracted and averaged to create a 1×768 vector to represent a sentence. Then every sentence in the document was averaged to create a 1×768 representation of the full-text article. Additionally, for mBERT, the last four hidden layers for each sentence were also captured. These were concatenated and averaged into a 1×768 vector. Then, same as above, each sentence vector was averaged so that there was 1×768 representation for each full-text article.

Monolingual Model Generation. 90% of each dataset was used for monolingual training and testing using 10-fold cross-validation (the remaining 10% was used for hyperparameter tuning) to capture test results in Table 2. Then, the monolingual models were trained on the entire dataset, resulting in five monolingual models for each of the three types of vector representation (i.e., XLM-R, mBERT pooled output, mBERT hidden layers output) for SVM and NN, respectively. This resulted in a total of 30 monolingual models.

Zero-Shot Cross-Lingual Classification: As a final step, each language's monolingual SVM or NN model was run against every other language dataset. Finally, Scikit-Learn's metrics library enabled calculation of F1, precision, recall, and accuracy metrics.

4 Results and Discussion

This section discusses the results of the monolingual and cross-lingual experiments.

4.1 Monolingual Results

As demonstrated in Table 2, the Bengali monolingual classifier outperformed every other language's monolingual classifier. This dataset was the most imbalanced dataset with 85% of the dataset comprising real news articles. After Bengali, the Filipino dataset, which was the second largest dataset used in this

Table 2. F1 results for monolingual models with low-resource languages achieving the highest scores

Language	XLM-R		mBERT pooled		mBERT hidden		Average
	F1 SVM	F1 NN	F1 SVM	F1 NN	F1 SVM	F1 NN	
Bengali	**0.98**	**0.98**	**0.97**	**0.97**	**0.98**	**0.98**	**0.98**
English	0.76	0.94	0.77	0.53	0.80	0.79	0.77
Spanish	0.81	0.86	0.76	0.72	0.79	0.69	0.77
Filipino	0.89	0.91	0.86	0.88	0.88	0.93	0.89
Urdu	0.80	0.85	0.69	0.69	0.73	0.72	0.75
Average	0.85	0.91	0.81	0.76	0.84	0.82	0.83

research, had the second highest F1 average overall while Urdu had the worst F1 average at 0.75. As expected, larger datasets performed better in monolingual classification. In addition, NN XLM-R model outperformed all other classification models likely because XLM-R SentencePiece encoding retains greater sentence context than mBERT's WordPiece encoding. However, for mBERT models, SVM models outperformed NN models.

4.2 Cross-Lingual Results

Unique to the Urdu dataset is that, although, it is the second smallest dataset with 881 total articles and achieved the worst F1 monolingual score among the monolingual classifiers, the Urdu monolingual classifier performed the best against other language datasets (see Table 3). Urdu monolingual classifier achieved an average score of 0.90 on the Bengali dataset, 0.67 on the English dataset and 0.60 each on the Spanish and Filipino dataset. It was the only dataset where although the real-labeled news articles were generated from mainstream websites, the fake news articles were written by real journalists to mimic fake news. Possibly by using this approach, the journalists writing the fake news articles distilled the essence of what makes an article fake and reduced the noise otherwise present in these fake news articles, thus making these artificially fake articles good predictors of fake news in other languages.

The second best performing classifier was Bengali, which achieved an average F1 score of 0.62 across the other languages. Bengali monolingual model performed particularly well against the English dataset and achieved an average F1 of 0.67 across the different classification models and 0.72 with XLM-R embeddings and SVM classifier. This is particularly interesting because Bengali and Filipino have different scripts and do not share a common language ancestry.

Another key finding is that if a monolingual classifier performs well on a particular language dataset, that does not mean the same will be true in reverse. For example, the average F1 score for a Urdu monolingual classifier against Filipino datasets was 0.60, but the F1 score for a Filipino monolingual classifier against the Urdu dataset was only 0.29.

Table 3. F1 scores for cross-lingual tests with Urdu monolingual model outperforming all other models

Classifier-data	XLM-R		mBERT pooled		mBERT hidden		Average
	F1 SVM	F1 NN	F1 SVM	F1 NN	F1 SVM	F1 NN	
Bengali-English	0.68	0.67	0.68	**0.69**	0.67	0.67	**0.68**
Bengali-Spanish	**0.72**	**0.68**	**0.69**	0.58	0.68	0.67	0.67
Bengali-Filipino	0.67	0.66	0.57	0.41	0.43	0.66	0.57
Bengali-Urdu	0.58	0.56	0.57	0.42	0.47	**0.70**	0.55
English-Bengali	0.03	0.00	0.47	0.36	0.02	**0.38**	0.21
English-Spanish	0.44	0.29	0.60	0.52	0.12	0.34	**0.39**
English-Filipino	0.16	0.03	0.55	**0.53**	**0.42**	0.36	0.34
English-Urdu	**0.46**	**0.38**	**0.68**	0.39	0.04	0.52	0.41
Spanish-Bengali	**0.95**	0.31	0.01	0.03	0.50	**0.92**	0.45
Spanish-English	0.70	**0.70**	**0.67**	**0.52**	**0.68**	0.76	**0.67**
Spanish-Filipino	0.67	0.66	0.27	0.23	0.57	0.56	0.49
Spanish-Urdu	0.40	0.72	0.00	0.02	0.70	0.24	0.35
Filipino-Bengali	0.06	0.30	0.13	0.17	**0.92**	**0.87**	0.41
Filipino-English	0.40	0.70	0.02	0.37	0.47	0.66	0.44
Filipino-Spanish	**0.67**	**0.67**	**0.51**	0.48	0.67	0.66	**0.61**
Filipino-Urdu	0.23	0.11	0.22	**0.53**	0.05	0.58	0.29
Urdu-Bengali	**0.92**	**0.89**	**0.92**	**0.76**	**0.94**	**0.94**	**0.90**
Urdu-English	0.67	0.67	0.67	0.67	0.67	0.67	0.67
Urdu-Spanish	0.67	0.67	0.67	0.67	0.21	0.69	0.60
Urdu-Filipino	0.66	0.66	0.48	0.50	0.63	0.66	0.60
Average	0.57	0.54	0.49	0.47	0.52	0.66	0.54

Shared Language Ancestry and History. The Filipino monolingual classifier had the worst average performance across the different languages with an F1 score of 0.43. However, the one bright spot was that the Filipino classifier performed well against the Spanish dataset with a score of 0.61. This is likely due to language transfer because of the shared colonial history between Spanish and Filipino given that the Spanish colonial period in the Philippines lasted for over 300 years. Similarly, Urdu and Bengali also share a colonial history with English as a result of British colonial rule between 1700 and 1947. Perhaps it is not surprising, then, that Bengali and Urdu classifiers had an average F1 score of 0.68 and 0.67 respectively against the English language dataset. These results demonstrate a shared history between languages results in language transfer, which positively impacts F1 scores.

However, the English monolingual classifier performed poorly against Bengali and Urdu achieving an average F1 score of 0.21 and 0.41 respectively. This is because English is the present-day *lingua franca* and it is common to use English vocabulary in foreign language articles, but not vice-versa.

Further, English was the worst performing monolingual classifier in the cross-lingual tests. However, its best performance (F1 of 0.39) was against the Spanish dataset and likewise, the Spanish monolingual classifier achieved its best average F1 score against the English dataset. These results indicate that shared language ancestry, which results in similar grammatical structures and shared root words also positively impacts F1 scores.

Impact of Cross-Lingual Embeddings and Classification Algorithms on Cross-Lingual Results. For monolingual experiments, the XLM-R cross-lingual embeddings with NN classifier had the highest average F1 performance. However, for cross-lingual tests, the highest F1 scores were obtained with the averaged output of mBERT's last four hidden layers. The NN classifier had the best average F1 performance and achieved an F1 score of 0.94 for Urdu-Bengali, 0.92 for Spanish-Bengali, 0.87 for Filipino-Bengali, and 0.70 for Bengali-Urdu monolingual classifier-dataset pairs. Since XLM-R uses SentencePiece encoding and mBERT uses WordPiece encoding, it was expected that the former would outperform the latter in classification results because XLM-R retains more of a sentence's context than mBERT. This bore true for the monolingual results, but it appears that retaining the context of a sentence is not as important when conducting cross-lingual experiments since the grammatical structures between languages can vary considerably.

5 Conclusion

Overall, the cross-lingual tests did not achieve F1 scores as high as the monolingual tests. This result is not surprising since the experiments did not account for different domains in the news datasets for each language. The datasets were created by different researchers and some are more heavily focused on political news (English) while others focus more on entertainment or sports (Urdu). However, full-text fake news detection is a notoriously difficult task [9] and the results of this paper's experiments are encouraging because they demonstrate that: 1) it is possible to detect fake news in low-resource languages using zero-shot cross-lingual transfer in particular language pairs; 2) models trained in low-resource languages are particularly adept at identifying fake news in other low-resource languages; 3) full-text articles contain a lot of information, some real and some fake, and to process the articles, the cross-lingual embeddings for each sentence were averaged to form a one-row vector, and then every sentence's embedding was averaged to form a one-row vector that comprised an article. Yet these averaged vectors retained the basic characteristics of the article that enabled them to be classified as real or fake; and 4) as the Urdu monolingual classifier demonstrates, what matters is the quality not the quantity of the dataset. Future research will focus on identifying other language pairs that perform well together for zero-shot cross-lingual transfer and also the role of universal domain adaptation in improving performance of zero-shot cross-lingual transfer.

References

1. Amjad, M., Sidorov, G., Zhila, A., Gómez-Adorno, H., Voronkov, I., Gelbukh, A.: "bend the truth": Benchmark dataset for fake news detection in Urdu language and its evaluation. J. Intell. Fuzzy Syst.: Appl. Eng. Technol. **39**, 2457–2469 (2020)
2. Aragón, M.E., et al.: Overview of MEX-A3T at IberLEF 2020: fake news and aggressiveness analysis in Mexican Spanish. In Notebook Papers of 2nd SEPLN Workshop on Iberian Languages Evaluation Forum (IberLEF) (2020). http://ceur-ws.org/Vol-2664/mex-a3t_overview.pdf
3. Bhatia, M., et al.: One to rule them all: towards joint indic Language Hate Speech Detection. pre-print arXiv:2109.13711 (2021). https://arxiv.org/abs/2109.13711
4. Bing, C., Culliford, E., Dave, P.: Spanish-language misinformation dogged Democrats in U.S. election (2020). https://reut.rs/3Pp0Gf6. Accessed 17 Nov 2020
5. Conneau, A., et al.: Unsupervised cross-lingual representation learning at scale. In: Proceedings of the 58th Annual Meeting of the Association for Computational Linguistics, pp. 8440–8451 (2020). https://doi.org/10.18653/v1/2020.acl-main.747
6. Cruz, J.C.B.B., Cheng, C.: Evaluating language model finetuning techniques for low-resource languages (2019)
7. Department, S.R.: Most common languages used on the internet as of January 2020, by share of internet users (2022). https://www.statista.com/statistics/262946/share-of-the-most-common-languages-on-the-internet/. Accessed 7 July 2022
8. Devlin, J., Chang, M.W., Lee, K., Toutanova, K.: BERT: pre-training of deep bidirectional transformers for language understanding. In: Proceedings of the 2019 Conference of the North American Chapter of the Association for Computational Linguistics: Human Language Technologies, vol. 1 (Long and Short Papers), pp. 4171–4186 (2019). https://doi.org/10.18653/v1/N19-1423
9. Du, J., Dou, Y., Xia, C., Cui, L., Ma, J., Yu, P.S.: Cross-lingual COVID-19 fake news detection. In: 2021 International Conference on Data Mining Workshops, pp. 859–862 (2021). https://doi.org/10.1109/ICDMW53433.2021.00110
10. Gómez-Adorno, H., Posadas-Durán, J., Enguix, G., Capetillo, C.: Resumen de fakedes en iberlef 2021: tarea compartida para la detección de noticias falsas en español. Procesamiento de Lenguaje Natural **67**, 223–231 (2021). https://doi.org/10.26342/2021-67-19
11. Horne, B.D., Adali, S.: This just. In: Fake News Packs a Lot in Title, Uses Simpler, Repetitive Content in Text Body, More Similar to Satire than Real News. pre-print arXiv:1703.09398 (2017). https://arxiv.org/abs/1703.09398
12. Hossain, M.Z., Rahman, M.A., Islam, M.S., Kar, S.: BanFakeNews: a dataset for detecting fake news in Bangla. In: Proceedings of the 12th Language Resources and Evaluation Conference, pp. 2862–2871 (2020). https://aclanthology.org/2020.lrec-1.349
13. Loper, E., Bird, S.: NLTK: the natural language toolkit. pre-print arXiv:cs/0205028 (2002). https://arxiv.org/abs/cs/0205028
14. Meng, W., Yolwas, N.: A review of speech recognition in low-resource languages. In: 2022 3rd International Conference on Pattern Recognition and Machine Learning (PRML), pp. 245–252 (2022). https://doi.org/10.1109/PRML56267.2022.9882228
15. Mozur, P.: A Genocide Incited on Facebook, With Posts from Myanmar's Military (2018). https://www.nytimes.com/2018/10/15/technology/myanmar-facebook-genocide.html. Accessed 15 Oct 2018

16. Oshikawa, R., Qian, J., Wang, W.Y.: A survey on natural language processing for fake news detection. In: Proceedings of the 12th Language Resources and Evaluation Conference, pp. 6086–6093 (2020). http://aclanthology.lst.uni-saarland.de/2020.lrec-1.747.pdf
17. Posadas-Durán, J.P.F., Gómez-Adorno, H., Sidorov, G., Escobar, J.J.M.: Detection of fake news in a new corpus for the Spanish language. J. Intell. Fuzzy Syst. **36**, 4869–4876 (2019). https://doi.org/10.3233/JIFS-179034
18. Sarker, S.: BNLP: natural language processing toolkit for bengali language. pre-print arXiv:2102.00405 (2021). https://doi.org/10.48550/ARXIV.2102.00405, https://arxiv.org/abs/2102.00405
19. Valencia, S.: Misinformation online is bad in English. But it's far worse in Spanish (2021). https://wapo.st/3Cdy1H0. Accessed 28 Oct 2020
20. Wu, S., Dredze, M.: Are all languages created equal in multilingual BERT? In: Proceedings of the 5th Workshop on Representation Learning for NLP, pp. 120 130 (2020). https://doi.org/10.18653/v1/2020.repl4nlp-1.16

FedCLUS: Federated Clustering from Distributed Homogeneous Data

Sargam Gupta and Shachi Sharma(✉)

Department of Computer Science, South Asian University, New Delhi, India
shachi@sau.int

Abstract. An emerging trend in the field of machine learning presently is Federated Learning (FL) that enables to build a global model without sharing private data distributed among multiple data owners. Extensive research conducted on FL is mostly restricted to deep learning methods utilizing labeled data. However, voluminous big data generated today is unlabeled. The paper presents a federated clustering method, called FedCLUS, that allows to learn clusters from distributed homogeneous data without sharing it. The distinguishing characteristic of the FedCLUS method is its ability to retain, merge and discard original clusters sent by data owners. Also, FedCLUS can identify any number of arbitrary shaped clusters that are robust to outliers. The feasibility and performance of the FedCLUS is tested on many diverse datasets.

Keywords: Federated clustering · Homogeneous data · Performance analysis · Unsupervised federated learning

1 Introduction

Technological advancements like connected devices, Internet of Things (IoT), cloud computing, social networks, virtual reality etc. has made acquisition and processing of big data possible. Machine Learning (ML) algorithms are popular means for analyzing collected data. Broadly, ML algorithms are classified as centralized and distributed. For executing centralized machine learning algorithms, the prime requirement is consolidation of data whereas distributed machine learning divides data across multiple nodes. Hence, distributed machine learning algorithms provide lucrative solution for processing big data. However, it still necessitates collection of data at one place, just like centralized machine learning, followed by its distribution. This prevents applications in healthcare, banking and many other sectors to share their private and secure data for learning an efficient collaborative model.

In year 2016, a novel concept of FL has been proposed by Google [1] allowing a collaborative model to be built from distributed data without sharing it thereby preserving privacy and security of the data. Assuming M data owners

K. K. Patel et al. (Eds.): icSoftComp 2022, CCIS 1788, pp. 29–41, 2023.
https://doi.org/10.1007/978-3-031-27609-5_3

$\{D_1, D_2, \ldots, D_M\}$, Yang *et al.* [2] define FL as a process of constructing collaborative model M_{Fed} with accuracy A_{Fed} such that

$$|A_{Fed} - A_{Cen}| < \delta \tag{1}$$

where A_{Cen} is the accuracy of centralized machine learning on data $D = D_1 \cup D_2 \cup \cdots \cup D_M$ and δ is a non-negative real number. The FL algorithm is said to tolerate δ-accuracy loss [2]. The terms node and client are used interchangeably to represent data owner. Mostly, two categories of FL are being investigated - one, where data at various clients have same features but different samples, called horizontal federated learning and second, called vertical federated learning, where clients have different feature space.

In past few years, extensive research is being conducted on FL. The first algorithm 'FedAvg' [1] has been proposed for training deep neural networks where weights are calculated by local nodes and sent to a central server. The aggregation of the weights received from multiple clients is then performed by the server using simple averaging function. After aggregation, the server sends updated weights to all local nodes and the process is repeated till the desired accuracy is achieved. Subsequently, many enhancements of FedAvg have been explored to reduce communication cost between local clients and central sever, volume of data exchange, indirect leakage of data and to increase its accuracy [2]. However, the focus still remains on deep neural networks and learning from labeled data with a few exceptions [3–5].

The big data generated from devices and other sources is often unlabeled necessitating unsupervised learning methods for its analysis. In relation to federated unsupervised learning, the available literature is limited. Some of the notable works comprise of k-Fed algorithm by Dennis *et al.* [6], combination of popular k-means and mini batch k-means clustering algorithms by Triebe and Rajagopal [7], Generative Adversarial Networks (GAN) based approach for clustering by Chung *et al.* [8]. However, there are following drawbacks of these approaches:

1. It is an established fact that k-means cannot find arbitrary shaped clusters in the data.
2. The prime task of the server is aggregation of client updates which is missing in [6,7].
3. The thorough performance analysis of proposed federated clustering methods in terms of cluster sizes, number of clusters etc. have not been performed.
4. GANs are known to suffer from unstable training making training process harder [9].

Hence, it becomes imperative to develop a federated clustering method that overcomes aforementioned pitfalls.

The paper contributes by proposing a new horizontal federated clustering method called FedCLUS that (i) can generate arbitrary shaped clusters from distributed data (ii) provides a sound aggregation procedure at the server with capability to merge and split clusters. A detailed performance analysis of FedCLUS algorithm is also performed. The paper is organized into six sections.

A detailed discussion on related work is presented in Sect. 2. The system model and problem considered in this paper are described in Sect. 3. The solution of the problem is provided by proposing FedCLUS method in Sect. 4. Section 5 contains experiments conducted to validate FedCLUS and to assess its performance. The last Sect. 6 contains concluding remarks.

2 Related Work

A federated learning approach to analyze unlabeled data, FedUL, by assigning surrogate labels is proposed by Lu *et al.* [10]. A surrogate model is trained using supervised FL algorithm and the final model is obtained from it using transformation function. Nour and Cherkaoui [11] suggest to divide large unlabeled dataset at a client into subgroups by maximizing dissimilarity in a subset and then use these subgroups for training.

One of the most popular unsupervised learning method is clustering [12, 13]. A true federated clustering algorithm proposed is k-Fed by Dennis *et al.* [6] which is the federated version of popular k-means algorithm. k-Fed is one shot algorithm in which clients first divide their respective data into groups and compute centroids using a distance metric. The centroids are shared with central server by each client. The server then executes k-means algorithm on collected centroids and sends back the newly learnt centroids to all clients. k-Fed treats each reported centroid as a representative data point and thus its accuracy can be ascertained only when a large number of centroids are reported by the clients. As also pointed out in [8], k-Fed is unable to capture complex cluster patterns such as the one present in real-world image datasets. Hence, Chung *et al.* [8] develop UIFCA algorithm by extending the iterative federated clustering algorithm (IFCA) proposed by Ghosh *et al.* [14] using GANs. With the strong assumption that each client owns data of a cluster (representing a group of users or devices), IFCA uses federated deep neural networks following an iterative process. Hence, IFCA is a federated supervised learning algorithm. UIFCA lifts the assumption of IFCA but follows a similar approach for unlabeled data using GANs. Xie *et al.* [15] also employ GANs on unlabeled brain image data to find clusters in federated environment. Combining distributed k-means at server and mini-batch k-means on clients, a solution is built for federated clustering in [7].

Even though the work in [6–8, 15] contribute towards learning from unlabeled data, it suffers from the drawbacks discussed in Sect. 1. Also, the experiments in related works have been conducted using classification datasets. Our work attempts to alleviate these drawbacks.

3 System Model and Problem Definition

It is a well known fact that FL suffers from high communication cost if a good model is to be trained [2]. To eliminate this problem in unsupervised learning domain, it has been proposed in [6] to adopt a simpler approach where only

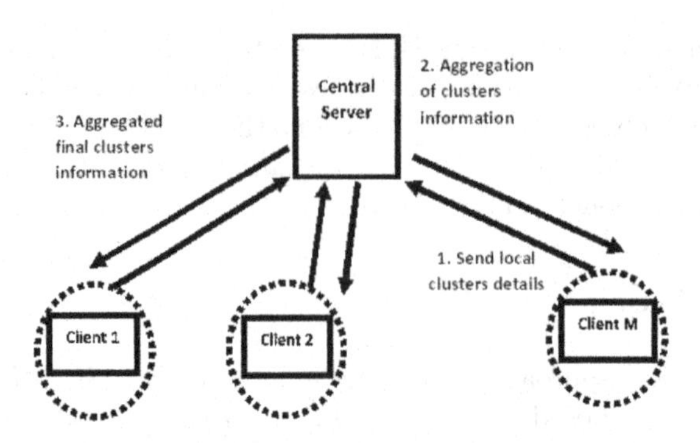

Fig. 1. System model for federated clustering.

one communication cycle is required between clients and server. Likewise, in our work, we consider the system shown in Fig. 1 where clients first execute a clustering algorithm on their respective local unlabeled data and send some properties of the clusters like number of clusters, size of clusters, radii, centroids, densities etc. to the server. The server aggregates this information using an aggregation function and returns a final set of clusters to the clients. The clients can then reassign data points to the updated clusters and also use them for grouping new data points.

Let the dataset D can be partitioned into K clusters. The cluster collection $C = \{C_1,\ C_2,\ \ldots,\ C_K\}$ is hidden from the server. The client i after executing clustering algorithm at its local data generates clusters $\{c_{ij}\}$ where j refers to cluster index. Here, we have not made any assumption on the number of clusters generated by the client. It is to be highlighted that server and clients reside at different locations and connected by a computer communication network. For example, in banking application, the server can be in the premise of regulatory bank while clients can be located at individual banks. The problem then can be defined as to develop a method to combine all $c_{ij},\ i = 1,\ 2,\ \ldots,\ M,\ \forall\ j$, so as to recover the original collection of clusters C. This task is performed by the server. The next section presents a solution to this problem.

4 FedCLUS: The Proposed Horizontal Federated Clustering Method

Every FL method works in two parts: one executes on clients and other on a central server. The client part of proposed FedCLUS method is explained in subsection below.

Algorithm 1. Client Side Algorithm

```
1: d_i ← Split dataset among clients
2: C ← DBSCAN(d_i, MinPts, Eps)
3: for each unique c_i in C do
4:     Function SizeCentroidClus (d_i, c_i)
5:     c_i.size ← 0
6:     v ← 0
7:     for each data x_i in c_i do
8:         if c_i.label == c_i then
9:             c_i.size = c_i.size + 1
10:            v = v + x_i.value
11:            c_i.centroid = v * (1/c_i.size)
12:        end if
13:    end for
14:    Function RadiusClus (c_i, c_i.centroid)
15:    for each data x_i in d_i do
16:        if x_i.label == c_i then
17:            x_i.radius = max(eucd dist(x_i, c_i.size))
18:        end if
19:    end for
20:    Function DensityClus (c_i.radius, c_i.size)
21:    n ← D.features + 1
22:    areaconst = (2 * pi^n /2)/gamma(n/2)
23:    c_i.area = areaconst * c_i.size
24:    c_i.density = c_i.area/c_i.size
25: end for
26: return updated C
```

4.1 Client Side Algorithm

The two major limitations of k-Fed [6] algorithm are:

- Each cluster centroid is treated as a new data point by the main server and k-means is run on them. As pointed out by Chung [8], such a simple approach fails to capture complex cluster structures.
- It is well known that k-means can effectively find circular clusters and fails to discover arbitrary shaped clusters. Hence, k-Fed too is limited to find circular circles. Besides, number of clusters should be known a prior in k-Fed.

We aim to overcome these limitations by modifying DBSCAN [16] for clients side execution. DBSCAN has many advantages over k-means algorithm such as self-determination of number of clusters, detection of arbitrary shape clusters and robustness to outliers [16]. The DBSCAN algorithm has been modified to include the centroid, size (i.e. number of data points), radius and density of the cluster. The centroids are calculated by taking the mean of all the data points belonging to the same cluster in all dimensions. For the computation of radius, the distance of all the data points from the centroid is calculated and the maximum of these distances is taken. For the calculation of density, the

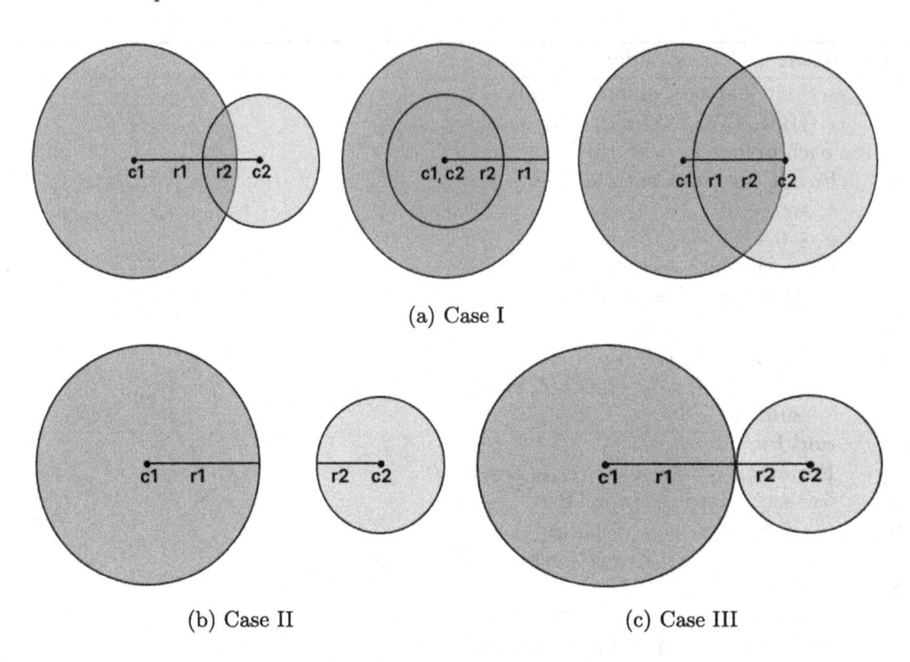

(a) Case I

(b) Case II

(c) Case III

Fig. 2. Different scenarios for aggregation at server.

area of a cluster needs to be determined which for arbitrary shaped cluster is difficult. Hence, we approximate the cluster area A by n-sphere area where n is the number of features as [17]

$$A = \frac{2\pi^{n/2}}{\Gamma\left(\frac{n}{2}\right)} R^{n-1} \tag{2}$$

where R is the radius of the sphere. Density is then calculated by dividing cluster area by its size. The modified DBSCAN algorithm is presented in Algorithm 1.

4.2 Server Side Algorithm

The server receives only cluster details from each client and not the actual data points.

Algorithm 2. Server Side Algorithm

```
1: DF ← set of all clients' clusters
2: D ← distance between clusters table
3: UDF ← set of all updated clusters
4: for each c_i in DF do
5:     Function Compute Distance (DF)
6:     for each c_i in DF do
7:         for each c_j in DF do
8:             d_ij = eucd dist(k_i.centroid, k_j.centroid)
9:             D ← d_ij
10:        end for
11:    end for
12: end for
13: sort D in ascending order of d_ij
14: Function Aggregation(DF, D)
15: for each d_ij in D do
16:    if c_i.radius + c_j.radius > d_ij then
17:        //Overlapping clusters
18:        if c_i.density > c_j.density then
19:            if c_i.radius > c_j.radius then
20:                if c_i.size > c_j.size then
21:                    if c_j.size < MinPts then
22:                        // Retain only Cluster i
23:                        UDF ← c_i
24:                    else
25:                        // Retain both Clusters
26:                        UDF ← c_i, c_j
27:                    end if
28:                else
29:                    UDF ← MergeClus(c_i, c_j)
30:                end if
31:            else
32:                UDF ← MergeClus(c_i, c_j)
33:            end if
34:        else
35:            if c_i.densityh < c_j.density then
36:                if c_i.radius < c_j.radius then
37:                    if c_i.size < c_j.size then
38:                        if c_i.size < MinPts then
39:                            // Retain only Cluster j
40:                            UDF ← c_j
41:                        else
42:                            // Retain both Clusters
43:                            UDF ← c_i, c_j
44:                        end if
45:                    else
```

```
46:                 UDF ← MergeClus(c_i, c_j)
47:             end if
48:         else
49:             UDF ← MergeClus(c_i, c_j)
50:         end if
51:     else
52:         if (c_i.radius + c_j.radius) < d_ij then
53:             //Far Clusters
54:             // Retain both Clusters
55:             UDF ← c_i, c_j
56:         else
57:             // Touching clusters
58:             nsize = c_i.size + c_j.size
59:             nrad = d_ij/2
60:             ncentroid = mid(c_i.centroid + c_j.centroid)
61:             narea = nradius * areaconst
62:             ndensity = narea/nsize
63:         end if
64:     end if
65:     end if
66:   end if
67: end for
68: S ← Set of near cluster centroids
69: for each c_i in UDF do
70:   for each c_j in UDF do
71:     if eucd dist (c_i, c_j) < Eps then
72:       S ← c_j
73:     end if
74:   end for
75:   UDF ← (c_j) where c_j.size is max in S
76: end for
77: return UDF
```

The aggregation function on the server utilizes this information. There are many possibilities such as clients may report same clusters or completely distinct clusters. The aggregation function on the server should be designed to remove inconsistencies and conflicting information to generate the most optimal set of clusters. To this end, we narrowed down to three cases when two clusters are considered: (i) sum of radii exceeds distance between centroids, (ii) sum of radii is less than the distance between centroids, and (iii) sum of radii is equal to distance between centroids.

The first case implies overlapping clusters. Hence, we can compare three quantities viz. their densities, radii and sizes. If one cluster is dominant in all the three quantities then the dominant cluster is retained and the other one is discarded. In case, both clusters have comparable values of three quantities then they are merged to create a single cluster. This mechanism also addresses the scenario when one cluster is completely contained in the other. The Fig. 2(a) depicts this case. The second case represents that clusters are far from each other. Hence, both the clusters are retained as shown in Fig. 2(b). In last case, the clusters touch each other as in Fig. 2(c). Hence, they can be merged by taking the mid point of their radii as the new centroid and half of the sum of their radii as the new radius. The distance measure used is Euclidean distance. However, other distance measures can also be considered in future work.

The aggregation function addresses aforementioned three cases so as to decide whether to merge and create a new cluster, discard clusters or retain clusters. This forms the heart of the server aggregation algorithm as presented in Algorithm 2 and 3. The merging algorithm taking two clusters as input forms a new merged cluster. The size of new merged cluster is the sum of number of points in both input clusters and radius of the merged cluster is sum of input clusters radii. The new centroid is mid point of the line joining centroids of input clusters. For the calculation of the new density, we use (2).

Algorithm 3. Merging Algorithm

//Merges Clusters
Function MergeClus(c_i, c_j)
$n \leftarrow D.features + 1$
$areaconst = (2 * pi^n/2)/gamma(n/2)$
$nsize = c_i.size + c_j.size$
$nrad = c_i.radius + c_j.radius$
$ncentroid = mid(c_i.centroid + c_j.centroid)$
$narea = nradius * areaconst$
$ndensity = narea/nsize$
return ncentroid, nsize, nrad, ndensity

Table 1. Datasets Description

Dataset name	#Records	#Features	#Clusters
S1 [18]	5000	2	15
Unbalanced [19]	6500	2	8
Asymmetric [20]	1000	2	5
D31 [21]	3100	2	31

The validation of FedCLUS method and its performance analysis is carried out in the next section.

5 Experiments and Results

5.1 Datasets and Experiments

In contrast to previous works on federated clustering and other unsupervised learning methods, we have chosen many popular benchmarking datasets specific to clustering domain for validating the FedCLUS method and assessing its

Table 2. Experiment results.

Dataset	DBSCAN		FedCLUS		
	#Clusters	Average cluster size	#Clients	#Clusters	Average cluster size
S1	16	26.19	2	16	25.00
			4	15	25.33
			6	16	24.53
			8	14	26.29
			10	13	23.77
Unbalanced	8	803.50	2	8	803.50
			4	8	803.50
			6	8	709.25
			8	8	655.12
			10	8	541.00
Asymmetric	5	192.40	2	5	195.40
			4	5	116.40
			6	5	72.60
			8	5	40.83
			10	5	30.00
D31	31	80.42	2	30	85.36
			4	30	84.44
			6	32	88.40
			8	30	79.56
			10	30	75.33

effectiveness. The importance of benchmarking datasets in ascertaining robustness of a method has been discussed by Pasi *et al.* [22]. Motivated by this, we have included four benchmarking datasets in our study. The description of selected datasets is given in Table 1. The S dataset contains Gaussian overlapping spherical and non-spherical clusters whereas the A dataset majorly contains the spherical clusters in which the cluster size and deviation remains constant among all clusters. Unbalanced dataset is a mixture of dense and sparse clusters. The Asymmetric dataset consists of clusters varying in shape, separability and sparsity. The D31 dataset is a large dataset compared to others with randomly placed clusters. Thus, these four datasets provide enough diversity to test FedCLUS.

We specifically do not partition datasets among clients to create identical and independently distributed (iid) or non-iid environments. Rather, a dataset is divided equally and randomly among clients. As we start increasing the number of clients, the data becomes more distributed and non-iid conditions automatically satisfy. The code for both client and server side algorithms have been developed in Python. Centralized DBSCAN is used as baseline algorithm to compare performance of FedCLUS. The experiments are carried out on AMD Ryzen 5 5500U system with Radeon Graphics 2.10 GHz and 8 GB RAM.

5.2 Results of Performance Analysis

The prior works on federated clustering adopt approaches similar or close to supervised learning and hence consider supervised learning datasets and performance metrics. We differentiate by considering standard clustering datasets and focus on metrics specific to clustering domain like number of clusters, cluster sizes and computation of centroids. The results of experiments are summarized in Table 2. It can easily be noted that FedCLUS method is able to generate clus-

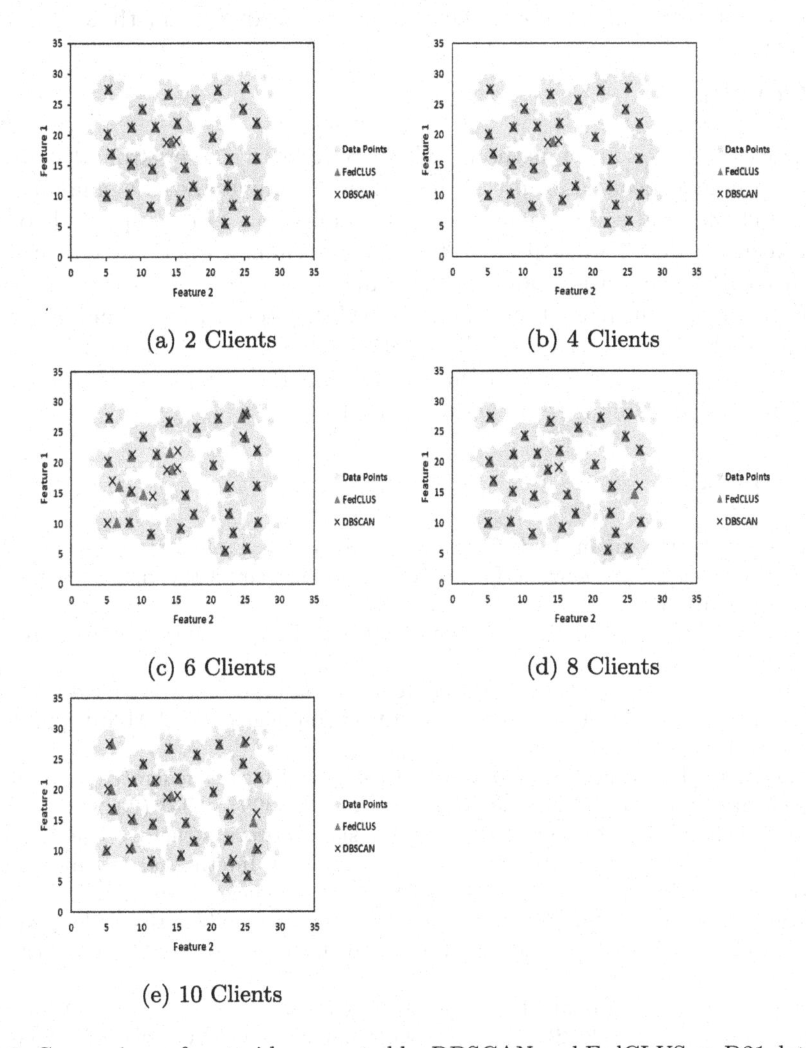

(a) 2 Clients

(b) 4 Clients

(c) 6 Clients

(d) 8 Clients

(e) 10 Clients

Fig. 3. Comparison of centroids computed by DBSCAN and FedCLUS on D31 dataset.

ter count close to centralized DBSCAN in majority of the cases. In two cases, i.e. for Unbalanced and Asymmetric datasets, the number of clusters generated by FedCLUS remains constant as the clients are increased. However, smaller average cluster size is observed as the number of clients increases. This is primarily because high distribution of data leads to emergence of non-iid characteristic and more data points are discarded by the clients as noise. We have also encountered scenarios in our experiments where some clients do not report any cluster. In spite of this, FedCLUS is able to produce desired number of cluster. This proves the robustness of FedCLUS method.

Since we have used benchmarking datasets with two features, we plot data points of D31 dataset along with centroids generated by DBSCAN and FedCLUS in Fig. 3. FedCLUS is able to create same centroids for all the clusters irrespective of number of clients. This proves the soundness of FedCLUS method.

6 Conclusion

The paper presents a novel method, FedCLUS, to learn clusters from distributed unlabeled homogeneous data preserving privacy in one round of communication between clients and server. This saves communication cost as opposed to previous works. FedCLUS has been tested under scenarios when clients can have data from multiple clusters and its performance is compared with centralized DBSCAN algorithm. It is observed that FedCLUS performs well and generates same number of clusters as centralized DBSCAN. The robustness of FedCLUS is also ensured by checking it on various benchmarking datasets. In future, we aim to test FedCLUS on real experimental setup.

References

1. Konecny, J., McMahan, H.B., Ramage, D., Richtárik, P.: Federated optimization: Distributed machine learning for on-device intelligence (2016). https://arxiv.org/abs/1610.02527
2. Yang, Q., Fan, L., Yu, H.: Federated Learning: Privacy and Incentive. Springer (2020)
3. Li, Q., Wen, Z., He, B.: Practical federated gradient boosting decision trees. In: The Thirty-Fourth AAAI Conference on Artificial Intelligence (AAAI-20), pp. 4642–4649 (2020)
4. Yamamoto, F., Ozawa, S., Wang, L.: eFL-Boost: Efficient federated learning for gradient boosting decision trees. IEEE Access **10**, 43954–43963 (2022)
5. Ng, I., Zhang, K.: Towards federated bayesian network structure learning with continuous optimization. In: Proceedings of the 25th International Conference on Artificial Intelligence and Statistics (AISTATS), Valencia, Spain (2022)
6. Dennis, D.K., Li, T., Smith, V.: Heterogeneity for the win: One-shot federated clustering. In: Proceedings of the 38th International Conference on Machine Learning (2021)
7. Triebe, O.J., Rajagopal, R.: Federated K-means: Clustering algorithm and proof of concept. (2022). https://github.com/ourownstory/federated_kmeans/blob/master/federated_kmeans_arxiv.pdf

8. Chung, J., Lee, K., Ramchandran, K.: Federated unsupervised clustering with generative models. In: AAAI (2022)
9. Saxena, D., Cao, J.: Generative adversarial networks (GANs): challenges, solutions, and future directions. ACM Comput. Surv. **54**(3), 1–42 (2022)
10. Lu, N., Wang, Z., Li, X., Niu, G., Dou, Q., Sglyama, M.: Federated learning from only unlabeled data with class-conditional-sharing clients. In: ICLR (2022)
11. Nour, B., Cherkaoui, S.: Unsupervised data splitting scheme for federated edge learning in IoT networks (2022). https://arxiv.org/abs/2203.04376
12. Sharma, S., Bassi, I.: Efficacy of Tsallis Entropy in clustering categorical data. In: IEEE Bombay Section Signature Conference (IBSSC), Mumbai, India, July (2019)
13. Sharma, S., Pemo, S.: Performance analysis of various entropy measures in categorical data clustering. In: International Conference on Computational Performance Evaluation (ComPE), Shillong, India, (2020)
14. Ghosh, A., Chung, J., Yin, D., Ramchandran, K.: An efficient framework for clustered federated learning. In: 34th Conference on Neural Information Processing Systems (NeurIPS), Vancouver, Canada (2020)
15. Xie, G., et al.: FedMed-GAN: Federated domain translation on unsupervised cross-modality brain image synthesis. (2022). https://arxiv.org/abs/2201.08953
16. Ester, M., Kriegel, H., Sander, J., Xu, X.: A density-based algorithm for discovering clusters in large spatial databases with noise. In: KDD'96: Proceedings of the Second International Conference on Knowledge Discovery and Data Mining, pp. 226–231 (1996)
17. Henderson, D.G.: Experiencing geometry: on plane and sphere. Prentice Hall (1995)
18. Fränti, P., Virmajoki, O.: Iterative shrinking method for clustering problems. Pattern Recogn. **39**(5), 761–765 (2006)
19. Rezaei, M., Fränti, P.: Set-matching measures for external cluster validity. IEEE Trans. Knowl. Data Eng. **28**(8), 2173–2186 (2016)
20. Rezaei, M., Fränti, P.: Can the number of clusters be determined by external indices? IEEE Access **8**(1), 89239–89257 (2020)
21. Veenman, C.J., Reinders, M.J.T., Backer, E.: A maximum variance cluster algorithm. IEEE Trans. Pattern Anal. Mach. Intell **24**(9), 1273–1280 (2002)
22. Fränti, P., Sieranoja, S.: K-means properties on six clustering benchmark datasets. Appl. Intell. **48**(12), 4743–4759 (2018). https://doi.org/10.1007/s10489-018-1238-7

On Language Clustering: Non-parametric Statistical Approach

Anagh Chattopadhyay[1], Soumya Sankar Ghosh[2](✉), and Samir Karmakar[3]

[1] Indian Statistical Institute, Kolkata, India
anagh72@gmail.com
[2] VIT Bhopal University, Bhopal, India
soumya.sankar@vitbhopal.ac.in
[3] Jadavpur University, Kolkata, India
samir.karmakar@jadavpuruniversity.in

Abstract. Any approach aimed at pasteurizing and quantifying a particular phenomenon must include the use of robust statistical methodologies for data analysis. With this in mind, the purpose of this study is to present statistical approaches that may be employed in non-parametric non-homogeneous data frameworks, as well as to examine their application in the field of natural language processing and language clustering. Furthermore, this paper discusses the many uses of non-parametric approaches in linguistic data mining and processing. The data depth idea allows for the centre-outward ordering of points in any dimension, resulting in a new non-parametric multivariate statistical analysis that does not require any distributional assumptions. The concept of hierarchy is used in historical language categorisation and structuring, and it aims to organize and cluster languages into subfamilies using the same premise. In this regard, the current study presents a novel approach to language family structuring based on non-parametric approaches produced from a typological structure of words in various languages, which is then converted into a Cartesian framework using MDS. This statistical-depth-based architecture allows us to use data-depth-based methodologies for robust outlier detection, which is extremely useful in understanding the categorization of diverse borderline languages and allows for the re-evaluation of existing classification systems. Other depth-based approaches are also applied to processes such as unsupervised and supervised clustering. This paper, therefore, provides an overview of procedures that can be applied to non-homogeneous language classification systems in a non-parametric framework.

Keywords: Non-parametric statistics · Language structuring · Lexical statistics · MDS

K. K. Patel et al. (Eds.): icSoftComp 2022, CCIS 1788, pp. 42–55, 2023.
https://doi.org/10.1007/978-3-031-27609-5_4

1 Introduction

The goal of data clustering is to divide a set of n items into groups, which can be represented as points in a d dimensional space or as a $n*n$ similarity matrix. Due to the lack of a standard definition of a cluster and the fact that it is task- or data-dependent, numerous clustering algorithms have been developed, each with a different set of presumptions about cluster formation. Parametric and non-parametric techniques can be used to categorise the suggested methodologies. As the name implies, non-parametric methods make no distributional assumptions about the data, in contrast to parametric procedures. [13, 14] introduced this non-parametric model in many forms for applications, while [17] provides a complete study of non-parametric approaches in the statistical paradigm. This model in particular doesn't call for any assumptions about the distribution of the data or the data points, making it possible to provide much more flexible and reliable inference and classification approaches that may be used with various intricate or simple data structures. Non-parametric statistics are of particular importance when the provided data is insufficient or unsuitable for applying distributional assumptions in order to infer distributional characteristics.

To better understand how languages are dispersed globally while maintaining linguistic convergence, the current research project focuses on employing this non-parametric model in natural language clustering. The study will be broken into four sections to do this. Section 2 will concentrate mostly on the idea of data depth and how it relates to linguistics. The data and methodological part of the discussion will be added in Sect. 3. The paper will then investigate the framework that is essential for offering a holistic explanation of language clustering in Sect. 4.

2 Data Depth and Linguistic Clustering

Data depth is a term that is part of the non-parametric approach to multivariate data analysis. It provides one way to arrange multivariate data. Central-outward ordering is the name given to this kind of arrangement. A depth function is essentially any function that yields a "reasonable" central-outward ordering of points in a multidimensional space. The fundamental concept underlying data depth is how "close" a certain point is with a measure of the distribution centre. Many other depth-based approaches have been described in the literature, including multivariate paradigms [4, 9, 16] and uni-variate paradigms ([2, 3], etc.). [7] makes a remark on non-parametric classification based on data depth and suggests a fast classification approach based on non-parametric statistics. According to [19], statistical depth is a function that has the following:

- affine transformation invariance
- maximality at the centre of symmetry of the distribution for the class of symmetric distributions
- monotonicity relative to the point with the highest depth
- vanishing at infinity

This will result in a function that recognizes "typical" and "outlier" observations, as well as a quantile generalization for multivariate data. Data depth, because of its non-parametric nature, allows a great deal of freedom for data analysis and categorization and is therefore quite general. It is a notion that provides an alternate idea or methodology for assessing the centre/centrality in a data frame, as well as a measure of centrality.

It is this data depth property of being able to distinguish between "typical" and "outlier" observations that help us achieve our goal of outlier detection in language families, as well as the general non-parametric properties and tools that can be used to study the structure of the various languages in a "language space," which is especially useful as we are not aware of any distributional properties of the languages in that space. Non-parametric approaches, as their name implies, do not require data to be parameterized, and are thus particularly effective in situations like this, where no distributional structure is present. We have considered Romance languages for the sake of this study, which is a subgroup of the Italic branch of the Indo-European language family. The main languages of this group are French, Italian, Spanish, Portuguese, and Romanian. For a better understanding, consider Fig. 1

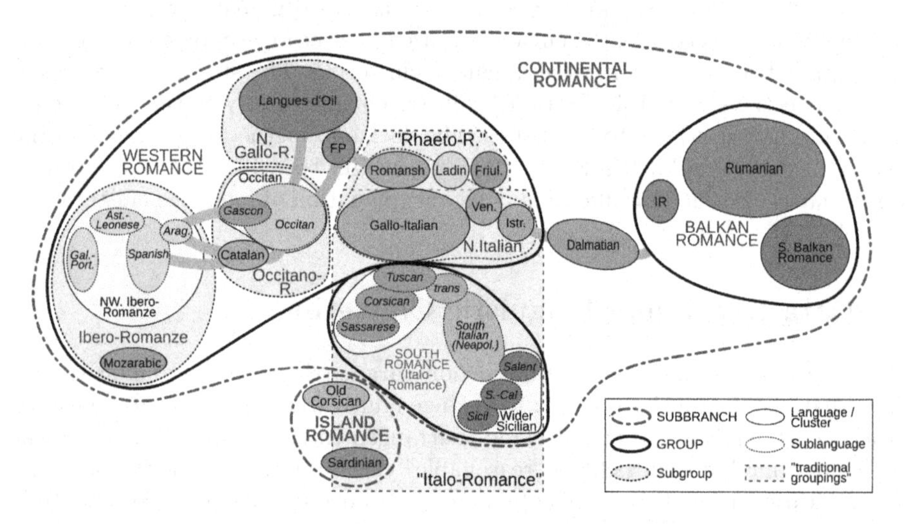

Fig. 1. Chart of Romance languages based on structural and comparative criteria, not on socio-functional ones. FP: Franco-Provençal, IR: Istro-Romanian

3 Data Collection and Methodology

The overall methodology in this paper can be understood in the following step-by-step manner:-

1. We begin by creating sets of the most frequently used terms in the languages. We are interested in languages (in our case, Romance languages) that correlate to a set of fixed and common meanings.

2. The distance matrix relating to the distance between any pair of languages is then calculated using the Levenshtein distance metric. This stage lays the groundwork for creating a language space with the right amount of dimensions. The use of various non-parametric approaches is based on this language space.
3. Dimension scaling, which is done using MDS (multidimensional scaling), is the next stage in creating the language space. The current work then embeds the points belonging to various languages in an abstract Cartesian system (with appropriate scaling measures).
4. Finally, we use appropriate non-parametric measures to analyze the numerous outlier-based features of this particular structure.

The following subsections deal with each of these steps in the aforementioned order.

3.1 Data Collection

The primary step is the collection of data,which has been done primarily from the *Serva* corpus [1] and from the database of [15,18].The *Serva* corpus contains the most common word corresponding to a given language and common meaning, particularly for Romance languages. This exhaustive list contains 60 languages and around 110 words. For the sake of our current paper, we have taken a small part of the data set which has been shown in Table 1. Although the use of a single word meaning and knowledge of the typological variations between the corresponding words in various languages aids us in understanding localised evolutionary properties corresponding to that word meaning, it leads to many erroneous conclusions when used to understand the general differences between various languages as whole entities. Nevertheless, this unpredictability may be consistently retrieved to a significant extent, which leads to a strong understanding of the language space and can be evened out by using a lot of words.

Table 1. A peek of the data set

Word	Classical.Latin	Meg.Romanian	Ist.Romanian	Aromanian	Romanian
all	omnis	tot	tot	tut	tot
ashes	cinis	tsanusa	ceruse	cinuse	cenusa
bark	cortex	coaja	cora	coaje	scoarta
belly	venter	foali	tarbuh	pintic	burta
big	grandis	mari	mare	mare	mare
bird	avis	pul	pul	puliu	pasare
bite	mordere	mutscu	mucca	miscu	musca

3.2 Distance Matrix

A string metric for quantifying the difference between two sequences is the Levenshtein distance, which is a sort of edit distance. Following this, the Levenshtein distance between two words is defined as the minimum number of single edits that possess the ability to change one word into the other([8]). It has been widely used in the linguistic literature for phrasal and typological differentiation in a variety of contexts, such as in the works of [10–12]. In these works, it has been adopted to measure language phonetic variance and to quantify dialectal differences.

The Levenshtein distance between two strings p and q (of length $|p|$ and $|q|$ respectively) which is written as $\text{lev}(p, q)$ and is defined in 1

$$\text{levenshtein.dist(p,q)} = \begin{cases} |p| & \text{if } |q| = 0 \\ |q| & \text{if } |p| = 0 \\ \text{levenshtein.dist(tail(p),tail(q))} & \text{if head(p)=head(q)} \\ 1 + \min \begin{cases} \text{levenshtein.dist(tail(p),q)} \\ \text{levenshtein.dist(p,tail(q))} \\ \text{levenshtein.dist(p,q)} \end{cases} & \text{otherwise} \end{cases} \tag{1}$$

The first letter of the word, in 1, denotes as *head* whereas remaining letters are termed as *tail* once the head has been removed. The key element in the distance matrix related to the languages is this distance.

For a given word meaning, we calculate the Levenshtein distance between all language pairings and offer a matrix that corresponds to the word meaning. As one might anticipate, the matrix is symmetric and all of its diagonal elements are equal to zero. For each word's meaning, a distinct Levenshtein matrix is obtained. Two levels can be used for language-based differentiation: a localized level that examines each Levenshtein distance matrix corresponding to a particular word meaning and a globalized level that is an algebraic function of all the Levenshtein distances obtained across all word meanings in all samples. We shall concentrate on the aforementioned worldwide structure due to the overall justification below.

The use of a single word meaning to construct the distance matrix has already been noted as having the potential to yield inaccurate results, particularly because the similarity structure of words corresponding to a single meaning can reveal disproportionate similarity and dissimilarity among languages that are both close to and far apart, especially due to the predominance of chance causes of similarity (or dissimilarity).

We therefore average the distance matrices derived from the multiple-word meanings applied, resulting in a logical and reliable distance matrix that reflects the similarity structure between the languages.

The Table 2 below displays the final distance matrix:

3.3 On Law of Large Numbers and Sampling Procedures

When only a small number of common words are taken into account, this property corresponds to the lack of interpretability of distance matrices. However, when a large number of common words are taken into account, the interpretability and validity of the same can be inferred or derived using an analogous form of the law of large numbers, such as in an appropriate language space that takes into account the various languages. Consider the following for better understanding:

If θ_i is the location where i^{th} language falls in our Cartesian plane, with

$$\delta_{ij} = \theta_i - \theta_j$$

the reference to the population parameter relating to the average Levenstein distance between the most prevalent term corresponding to the two languages with a common meaning, averaged over all feasible shared meanings, and allowing $\hat{\delta}_{ij,n}$,$\hat{\theta}_{i,n}$,$\hat{\theta}_{j,n}$ to be the corresponding sample counterparts obtained from our collected data with say n elements, then the current work will achieve the following by adhering to the law of big numbers' premise:

$$\hat{\delta}_{ij,n} \rightarrow \delta_{ij} \quad \text{as} \quad n \rightarrow \infty$$

.

The law of large numbers motivates us to choose a large number of common words between the languages of interest in order to derive robust and interpretable results with acceptable levels of error because any two well-established, widely-spoken, and recognized languages have a large number of common words in most senses.

3.4 Interpreting the Distance Matrix

A comprehensive hierarchical clustering is used to check the authenticity and meaning of the produced distance matrix. The results of which are shown in Fig. 2 (corresponding to average linkage clustering) and Fig. 3 (corresponding to complete linkage clustering).

In complete-link (or complete linkage) hierarchical clustering, we merge the two clusters with the least merger diameter (or: the two clusters with the smallest maximum pairwise distance).

In single-link (or single-linkage) hierarchical clustering, we merge the two clusters with the least distance between their two closest members in each step (or: the two clusters with the smallest minimum pairwise distance).

The current work at this point observes that the obtained hierarchical clustering-based dendrogram captures the actual structure of the Romance family of languages to a significant extent (and we also note that various dialects of the same language are close), and that aids us in confirming the validity of the distance matrix, as it becomes crucial and important in the steps mentioned in the following sections. With this understanding, if we compare Fig. 1 with Fig. 2 and Fig. 3, results will show that the outcomes are pretty comparable, in

Table 2. A principal submatrix corresponding to the obtained distance matrix

	Late.Classical.Latin	Megleno.Romanian	Istro.Romanian	Aromanian	Romanian	Dalmatian	Friulian
Late.Classical.Latin	0.00	3.69	3.73	3.67	3.64	3.75	3.67
Megleno.Romanian	3.69	0.00	2.44	1.76	2.11	3.30	3.25
Istro.Romanian	3.73	2.44	0.00	2.39	2.03	3.65	3.02
Aromanian	3.67	1.76	2.39	0.00	1.76	3.40	3.27
Romanian	3.64	2.11	2.03	1.76	0.00	3.29	3.04
Dalmatian	3.75	3.30	3.65	3.40	3.29	0.00	3.02
Friulian	3.67	3.25	3.02	3.27	3.04	3.02	0.00

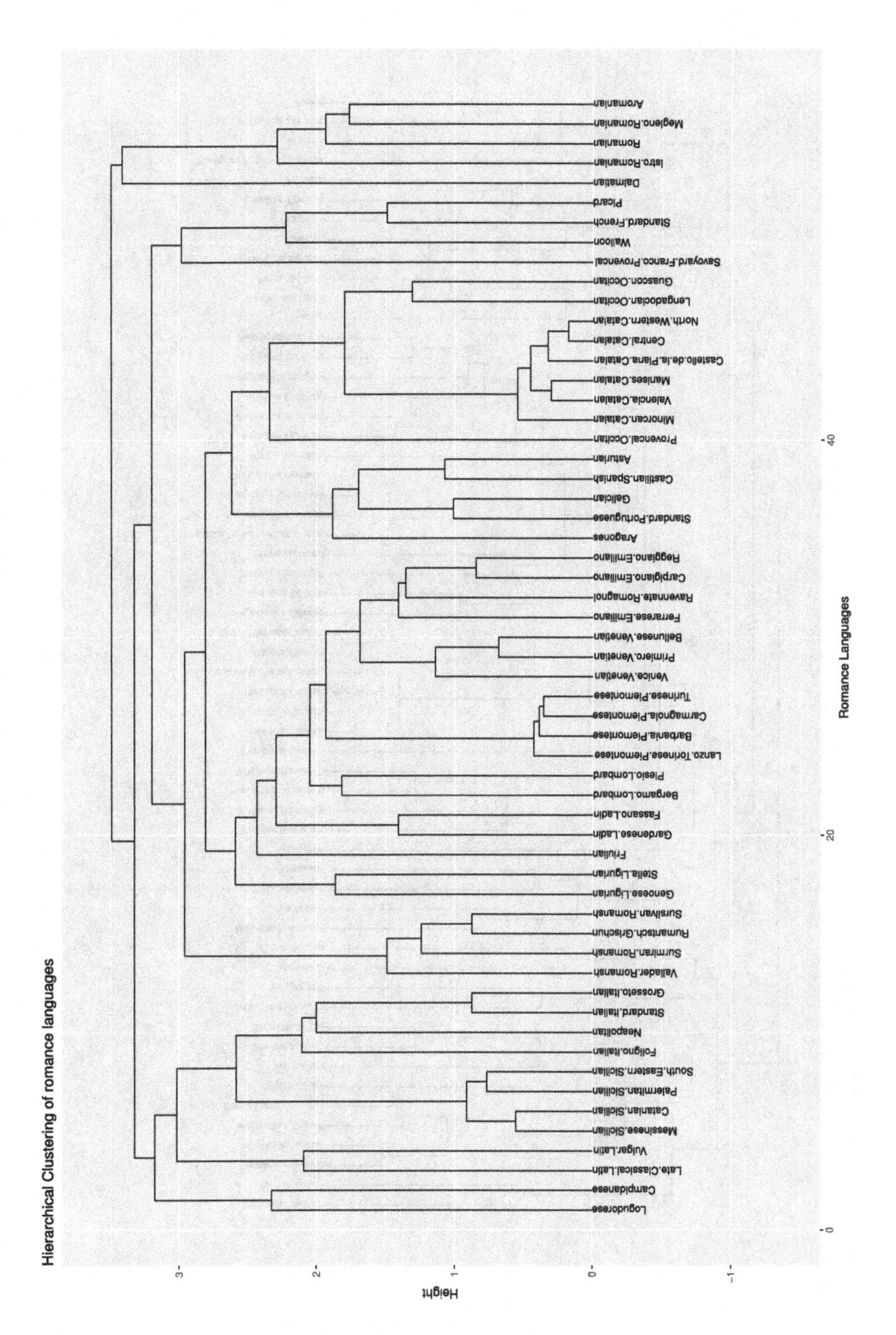

Fig. 2. Hierarchical clustering(we note the lack of any measure of centricity (which can be useful in an Historical Linguistic point of view) (Average method)

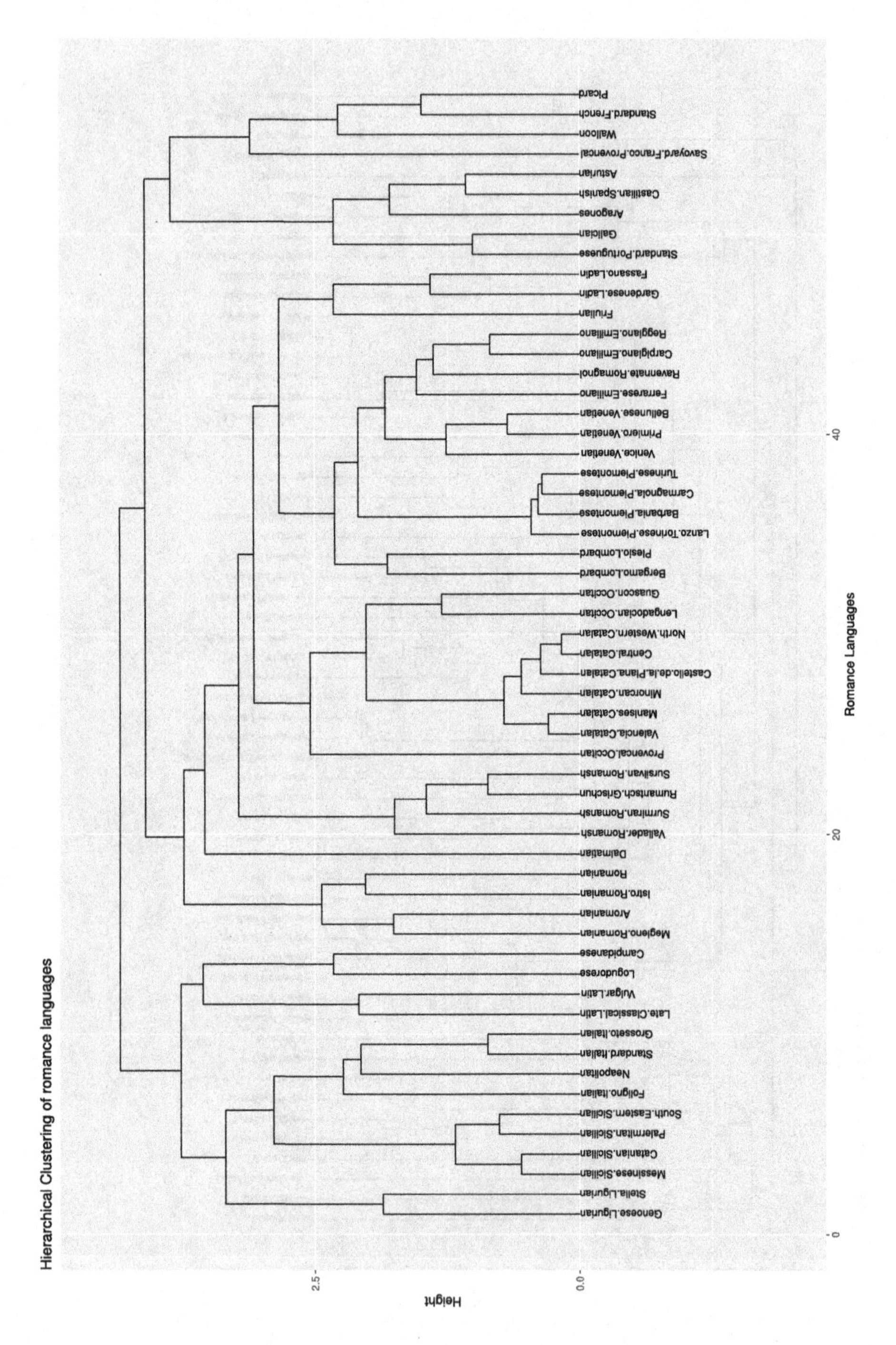

Fig. 3. Hierarchical clustering(we note the lack of any measure of centricity(which can be useful in a Historical Linguistic point of view)(Complete method)

spite of the fact that Fig. 1 was obtained using historical procedures. The law of large numbers is helpful in the context of linguistic clustering, which brings us to the study's conclusion in this regard. It is worth mentioning that the suggested methodology has the ability to retrieve a sizable amount of data on both their typological and historical commonalities. This means that any future use of non-parametric analytical methodologies is justified because of the resemblance between observed and generated hierarchies, which tells us that this structure captures a considerable quantity of information.

We must now try to embed the data set-specifically, the language arrangement a Cartesian plane in addition to visualising it using the language distance matrix. Consequently, we apply multidimensional scaling with the appropriate dimensions (obtained by scaling). Both conventional and non-parametric MDS can be employed by implementing R's fundamental architecture.

The Fig. 4 portrays two dimensional MDS for the cluster of languages:

4 Application and Discussion

In the previous section, we constructed and explained a structure made up of embedded points in a two-dimensional Cartesian plane that are related to Romance languages and can be applied to various non-parametric statistical models. The embedding maintains the distance between the various Romance languages, which is why the applications of non-parametric techniques are illustrated in the following subsections.

4.1 Outlier Detection

The application of data-depth-based approaches to language group distinction is one of the most essential aspects of their use. This essentially aids us in developing a statistically sound methodology for determining whether a language belongs to a specific family based on the degree of typological distinctions. We use the `Spatial.Depth` functionality at R for outlier detection, especially for languages which cannot be easily put into a given language family. We used the above functionality at R, and we added the languages *Hindi* and *Sanskrit* to the list of romance languages. **We note that functionality detected the two languages as outliers at 0.05 level.**

The depth structure of the space of languages is shown in Fig. 5 and Fig.6

4.2 Unsupervised L1-Depth Based Clustering

Depth-based clustering has been studied and is usually implemented for its robustness and interpretability [5]

We employ the L-1 depth in this case for the clustering process and use the functionality `TDDclust` in R which is the trimmed version of the clustering algorithm based on the L1 depth proposed by [6]. The paper segments all the observations in clusters, and assigns to each point z in the data space, the L1

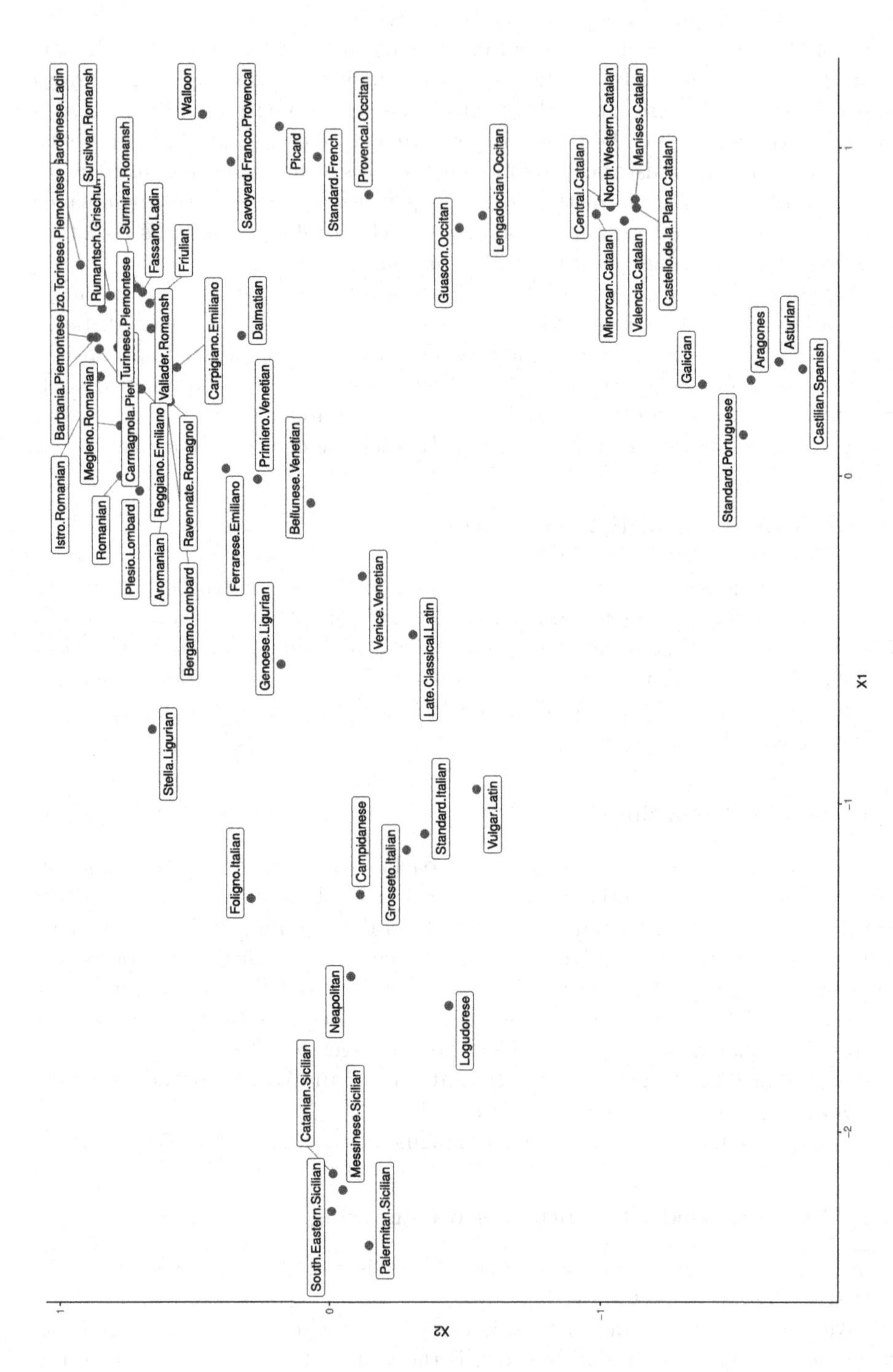

Fig. 4. Embedding in a Cartesian plane by using MDS

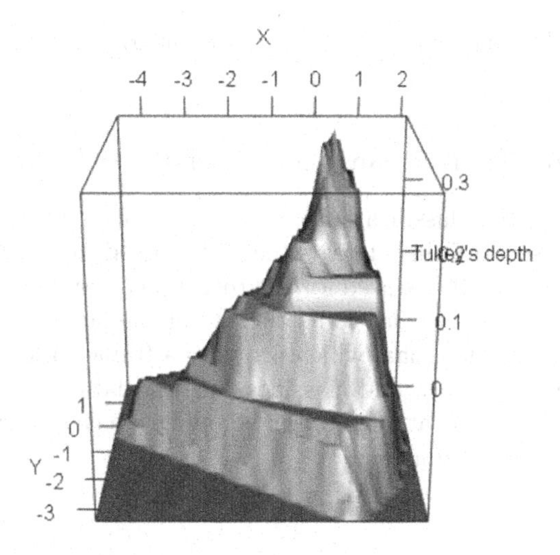

Fig. 5. A 3-D perspective of the depth structure of the Languages

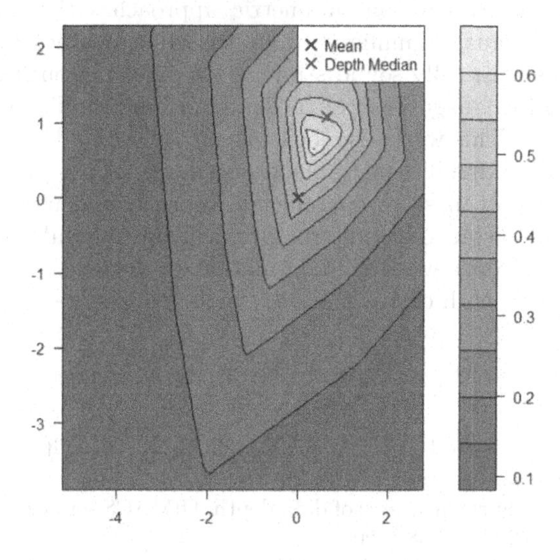

Fig. 6. A 2-D heat-map of the depth structure of the Languages

depth value regarding its cluster. A trimmed procedure is incorporated to remove the more extreme individuals of each cluster (those with the lowest depth values). This methodology allows us to decide the number of clusters we deem suitable for the given language structure, and subsequently allows us the flexibility to cluster,w.r.t the various romance language sub-group structures in the literature. The result for the case of three clusters centred at *Campidanese*, *Vallader Romansh*, and *Valencia Catalan*. Application of k-NN non-parametric Regression and PAM was also done, which yielded similar results. Each of these methodologies yields insights towards the inherent groups within the Romance family itself.

4.3 Supervised Classifications and Other Possible Generalizations

Depth-based Supervised classifications can also be done by training with appropriate language-based training data (which comprised about 80% of the total data and the remaining 20% is used for the testing) corresponding to each of the clusters in the proposed classification structure of the languages.

We use the package `ddalpha`, which provides a framework for the utilisation of the inputted training data and classifies the test-data w.r.t various types of depth. 82.3% classification rate was obtained, when we classified among Eastern and Western Romance languages.

5 Conclusion

This research demonstrates non-parametric approaches that can aid in the understanding of language families (in our case the romance languages). The paper develops a statistically sound strategy for detecting outliers in the Indo-European family of languages, as well as distinguishing Hindi and Sanskrit from Romance languages. This work also looks at supervised and unsupervised clustering of linguistic sub-families using a non-parametric approach, with promising findings. The success of these non-parametric approaches demonstrates the utility of a Levenshtein distance-based structure, which is not only computationally efficient but also delivers a wealth of information on the behaviour and evolution of different languages with only a modestly high sample size.

References

1. Romance language word lists. http://people.disim.univaq.it/~serva/languages/55+2.romance.htm. Accessed 23 Dec 2021
2. Aloupis, G.: Geometric measures of data depth. DIMACS series Disc. Math. Theor. Comput.er Sci. **72**, 147–158 (2006)
3. Dyckerhoff, R., Mosler, K., Koshevoy, G.: Zonoid data depth: Theory and computation. In: Prat, A. (eds) COMPSTAT, pp. 235–240, Heidelberg, Physica-Verlag HD (1996)
4. He, X., Wang, G.: Convergence of depth contours for multivariate datasets. Ann. Stat. **25**(2), 495–504 (1997)
5. Jeong, M.H., Cai, Y., Sullivan, C.J., Wang, S.: Data depth based clustering analysis. In: Ali, M., Newsam, S., Ravada, S., Renz, M., Trajcevski, G., (eds). In: Proceedings of the 24th ACM SIGSPATIAL International Conference on Advances in Geographic Information Systems, pp. 1–10, New York, USA, (2016). Association for Computing Machinery
6. Jörnsten, R.: Clustering and classification based on the l1 data depth. J. Multivar. Anal. **90**(1), 67–89 (2004)

7. Lange, T., Mosler, K., Mozharovskyi, P.: Fast nonparametric classification based on data depth. Stat. Papers **55**(1), 49–69 (2014)
8. Levenshtein, V.I.: Binary codes capable of correcting deletions, insertions, and reversals. Soviet Phys. Doklady **10**(8), 707–710 (1966)
9. Liu, R.Y., Parelius, J.M., Singh, K.: Multivariate analysis by data depth: descriptive statistics, graphics and inference. Ann. Stat. **27**(3), 783–858 (1999)
10. Nerbonne, J., Heeringa, W.: Measuring dialect distance phonetically. In: Proceedings of the Third Meeting of the ACL Special Interest Group in Computational Phonology (SIGPHON-97), pp. 11–18. ACL Anthology (1997)
11. Nerbonne, J., Heeringa, W., Kleiweg, P.: Edit distance and dialect proximity. In: Sankoff, D., Kruskal, J., (eds) Time Warps, String Edits and Macromolecules: The theory and practice of sequence comparison, pp. v–xv. CSLI Press, Stanford, CA, (1999)
12. Nerbonne, J., Heeringa, W., Van den Hout, E., Van der Kooi, P., Otten, S., Van de Vis, W.: Phonetic distance between dutch dialects. In: G. Durieux, W., Gillis, D.S., (eds), CLIN VI: Proceedings of the Sixth CLIN Meeting, pp. 185–202, Antwerp, Centre for Dutch Language and Speech (UIA) (1996)
13. Richard Savage, I.: Nonparametric statistics: a personal review. Sankhyā: Indian J. Stat., Ser. A (1961–2002), **31**(2), 107–144 (1969)
14. Siegel, S.: Nonparametric statistics. Am. Stat. **11**(3), 13–19 (1957)
15. Swadesh, M.: Towards greater accuracy in lexicostatistic dating. Int. J. Am. Ling. **21**(2), 121–137 (1955)
16. Vardi, Y., Zhang, C.-H.: The multivariate l1-median and associated data depth. Proc. National Acad. Sci. **97**(4), 1423–1426 (2000)
17. Wasserman, L.: All of Nonparametric Statistics. Springer, New York, USA (2006)
18. Wichmann, S., Rama, T., Holman, E.W.: Phonological diversity, word length, and population sizes across languages: The asjp evidence. Linguistic Typol. **15**(2), 177–197 (2011)
19. Zuo, Y., Serfling, R.: General notions of statistical depth function. Ann. Stat. **1** 461–482 (2000)

Method Agnostic Model Class Reliance (MAMCR) Explanation of Multiple Machine Learning Models

Abirami Gunasekaran[1], Minsi Chen[1](✉), Richard Hill[1], and Keith McCabe[2]

[1] School of Computing and Engineering, University of Huddersfield, Huddersfield, UK
M.Chen@hud.ac.uk

[2] Planning and Business Intelligence, University of Huddersfield, Huddersfield, UK

Abstract. Various Explainable Artificial Intelligence (XAI) methods provide insight into the machine learning models by quantitatively analysing the contribution of each variable to the model's predictions globally or locally. The contribution of variables identified as (un)important by one method's explanation may not be identified as the same by another method's explanation for the same machine learning (ML) model. Similarly, the important feature of many well performing ML models that fit equally well on the same data (which are termed as Rashomon set models) may not be the same as each other. While this is the case, providing the explanation based on a single model in the lens of a specific explanation method would be biased over the model/method. Hence, a framework is proposed to describe the consensus variable importance across multiple explanation methods for many almost-equally-accurate models as a method agnostic explanation for the model class reliance. Empirical experiments are carried out on the COMPAS dataset with six XAI (the Sage, Lofo, Shap, Skater, Dalex and iAdditive) methods for verifying whether an inadmissible feature becoming an (un)important feature is consistent across multiple explanation methods and getting the consensus explanation. The results demonstrate the efficiency of the method agnostic model class reliance explanation and its coverage to the model reliance range of all the almost-equally-accurate models of the model class.

Keywords: XAI · Ensembled explanation · Feature importance · Rashomon set

1 Introduction

The recent strategies under the XAI umbrella are mostly model agnostic. It means that irrespective of the ML model type and the internal structure, the explanation methods provide the explanation for the model's decisions. One such technique is the feature importance method [1]. These methods [2–8] can be plugged into any ML model to know the learning behaviour of the model in terms of feature importance. Here, the learning behaviour represents the order of important features on which the model takes its prediction. These model-agnostic methods require only the input and the predicted output of the model for providing the feature importance explanation.

K. K. Patel et al. (Eds.): icSoftComp 2022, CCIS 1788, pp. 56–71, 2023.
https://doi.org/10.1007/978-3-031-27609-5_5

The feature importance can be defined as a quantitative indicator that quantifies how much a model's output changes with respect to the permutation of one or a set of input variables [9]. The computation of these variable importance values is operationalized in different ways. The importance of the variables can be quantified by introducing them one by one, called feature inclusion [8] or by removing them one by one from the whole set of features, called feature removal [2]. The model can be retrained several times [11] for each of the input feature inclusions/removals or multiple retraining can be avoided [12] by handling the absence of removed features or the inclusion of new features. For that, any supplementary baseline input [13], conditional expectations [14], the product of marginal expectations [15], approximation with marginal expectations [3] or replacement with default values [2] can be used.

Though all these methods explain the feature importance behind the decisions of the model, the explanation obtained from a method may not be similar to the explanation of another method for the same model [17, 34]. This would confuse the analyst as which explanation should be trusted when different explanations are obtained. Unfortunately, there is no clear, standard principle to choosing the appropriate explanation method.

There may be many but different ML models that can fit equally well and produce almost similar accurate predictions on the same data. But the feature which is most important to one such model may not be an important feature for another well performing model [19].

In such a scenario, providing the explanation based on a single ML model using a specific explanation method would be biased (unfair) over the model/method. To this end, a novel explanation method is proposed to provide a method agnostic explanation across various method explanations of multiple almost-equally-accurate models. These near-optimal models [29] are termed as Rashomon set [19]. Instead of selecting a single predictive model from a set of well performing models and providing the explanation for it, the proposed method offers an explanation across multiple methods to cover the feature importance of all the well performing models in the model class.

The rest of the work is structured as follows: Sect. 2 reviews the related works, Sect. 3 deals with the proposed method, Sect. 4 speaks about the experiments, results and discussion, and Sect. 5 presents the conclusion.

2 Related Works

A plethora of strategies under XAI is developed for providing explanations for the black-box models. Among them, the major attention is being received by the feature importance methods. These methods [3–8, 11] aim to explain a single model's variable importance values by permutating the variables. The methods can give explanations as local feature importance [2] for a single instance or as global importance [4] for the entire data set.

Rashomon Effects: Initially, the problem of model multiplicity where multiple models fit on the data are equally good but different models was raised by [10]. There is no clear reason to choose the 'best' model among all those almost-equally-accurate models [22]. Moreover, the learning behaviour of the models varies among themselves. It means that the feature that is important for one model may not be important for another model. Hence, to avoid a biased explanation of a single model, the comprehensive explanation for all the well performing models (Rashomon set models) is given as a range of explanation by [19].

In line with [19], the authors of [22] expanded the Rashomon set concept by defining the cloud of the variable importance (VIC) values for the almost-equally-accurate models and visualizing it with the variable importance diagram (VID). The VID informs that the importance of a variable gets changed when another variable becomes important.

Aggregating over a set of competing equally good models would reduce the non-uniqueness [10]. Based on this concept, the authors of [29] generated a set of 350 near-optimal logistic regression models on the COMPAS dataset, aggregated the models' feature importance values and presented the explanation a less biased importance explanation for the model class than a single model's biased explanation. Similarly, by ensembling the Rashomon set models using prior domain knowledge, the authors [30] correct the biased learning of a model. If the Rashomon set is large, the models contained in the set could exhibit various desirable properties [31]. Also, the authors observe that the model performance does not necessarily vary across different algorithms even though the ratio of Rashomon set models on the dataset is small.

All these works aim on solving the bias that arises from multiple models (Rashomon set) rather than considering the bias that comes for a model from multiple methods.

Explanation Evaluation and Ensembling: The common evaluating measures found in the literature for ensembling explainable approaches are stability [32, 34, 37], (in)fidelity [18, 37], consistency [32, 35], informativeness [33] and comparison metric [36]. Though the explanation methods provide varying explanations for the same model, no principled way could be found in the literature to get a consensus explanation across various methods. A framework [32] proposes the ensembled explanation of several model agnostic algorithms based on the consistency and stability scores with the aim to provide an ensembled explanation independent of the XAI algorithms. Similarly, a unifying framework for understanding the feature removal-based explanation methods is introduced in [7]. The authors showed the relationship of how the methods are related to one another in providing the explanation. It does not combine the explanations of the various methods into one explanation but offers comparable explanations of those methods. At the same time, by comparing the various method explanations for a model, the most representative knowledge of the data set is obtained through the common explanations from the various methods [34]. All the ensembling explanation works focus on the multiple explanations for a single model rather than model multiplicity.

A unified explanation across multiple methods has not been extensively studied and the research works related to the Rashomon set focus on the explanations that vary across the multiple models rather than across multiple methods. Hence, a framework is proposed to address the explanation bias happening across multiple methods for the multiple almost-equally-accurate models. The work is motivated to find the answer to the following research questions:

RQ1. while various explanation methods are applied on multiple well performing models to get the feature importance explanations, will the feature which is projected as (un)important by one explanation method be agreed by other methods?

RQ2. Is getting a consensus explanation that is consistent across the various applied methods for multiple almost-equally-accurate models possible?

3 Proposed Method

This section presents the proposed method for obtaining the method agnostic ensembled explanation of various almost-equally-accurate ML models. The processes involved in obtaining the model agnostic model class reliance range using the MAMCR framework are depicted in Fig. 1.

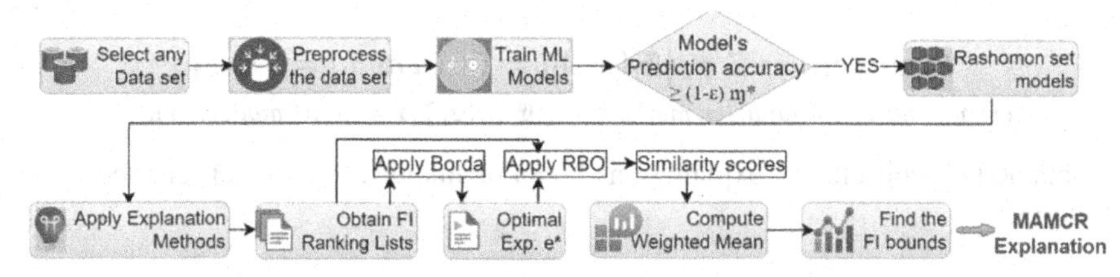

Fig. 1. The Process pipeline of Model Agnostic Model Class Reliance (MAMCR) framework

3.1 Models Building

Let $(X, Y) \in \mathbb{R}^{p+1}$, where $p > 0$, $X \in \mathbb{R}^p$ is the random vector of p input variables and $Y \in \mathbb{R}^1$ is the output variable. The process pipeline starts with the modelling of a class of multiple ML models on the pre-processed data of tabular type. As per the No Free Lunch theorem [21], there is no single ML model that is considered as best for solving the problems. Consequently, multiple ML models can be fitted on the same data set to verify the model's performance. This set of prespecified predictive models is referred to as model class [19].

$$Model\ class\ M = (\mathfrak{m}_t, \{t = 1, 2, \ldots, m\}) \tag{1}$$

where, M is a model class that consists of m models. Each model can take the input X and convert it to response Y. Each model's performance is assessed in terms of its prediction accuracy. The model class can be built with a set of regression algorithms. In that case, the model performance is assessed in terms of R^2 value.

3.2 Finding the Rashomon Set Models

From the multiple fitted models of the model class M, the almost-equally-accurate models form the Rashomon set (Ɍ). A Rashomon set is constructed based on a benchmark model $\mathfrak{m}^*$and a nonnegative factor ε as follows:

$$Ɍ(\varepsilon, \mathfrak{m}^*, M) = \{ \mathfrak{m} \in M \mid \mathfrak{m}(X) \geq (1-\varepsilon)\, \mathfrak{m}^*(X) \} \tag{2}$$

Selection of $\mathfrak{m}^*$ with possible maximum accuracy and ε with a small positive value helps to search for the models whose prediction accuracy are not less than the $(1-\varepsilon)$ factor of $\mathfrak{m}^*$ accuracy and to construct the Ɍ models i.e., Ɍ(M).

3.3 Obtaining Model Reliance Values and Ranking Lists

The model reliance [19] (or feature importance) indicates how much a model relies on a variable for making its predictions. The model reliance on the variable k (mr^k) is measured by the quantity of change in the model's performance with and without the variable k, where k = 1, 2, …, p. The more the change in the model performance, the higher the importance of that variable in the model's prediction contribution.

Different state-of-the-art explanation methods are selected to apply to each Rashomon set model to obtain their model reliance on p variables. Any global explanation method that returns the explanation in the form of feature importance can be chosen.

$$Explanations\ E = \{\mathfrak{E}_i(\mathfrak{m}_j) | i = 1\ to\ n\ and\ j = 1\ to\ r\} \tag{3}$$
$$where\ \boldsymbol{n} = no.of\ explanation\ methods\ \{\mathfrak{E}_1, \mathfrak{E}_2, .., \mathfrak{E}_n\},\ \boldsymbol{r} = no.of\ \boldsymbol{models}\ \text{in}\ \mathcal{R}(\mathcal{H})$$

The obtained model reliance explanations E can be mapped to a model reliance vector as follows:

$$MRV_n(\mathfrak{m}) = (mr_n^1(\mathfrak{m}), (mr_n^2(\mathfrak{m}), \ldots, (mr_n^p(\mathfrak{m})) \tag{4}$$

where $mr_n^p(\mathfrak{m})$ represents the model reliance of the model $\mathfrak{m}$ on variable p that is obtained from the explanation method n. The model reliance vector values are also mapped to model reliance ranking lists as follows:

$$\begin{aligned} E_{MRR} &= \{[MRR_1, MRR_2, \ldots, MRR_r]\} \\ &= \{[e_1(\mathfrak{m}_1), e_2(\mathfrak{m}_1), \ldots, e_n(\mathfrak{m}_1)], [e_1(\mathfrak{m}_2), e_2(\mathfrak{m}_2), \ldots, e_n(\mathfrak{m}_2)], \ldots, \\ &\quad [e_1(\mathfrak{m}_r), e_2(\mathfrak{m}_r), \ldots, e_n(\mathfrak{m}_r)]\} \end{aligned} \tag{5}$$

The explanation $E_{MRR}[1] = MRR_1 = [e_1(\mathfrak{m}_1), e_2(\mathfrak{m}_1), \ldots, e_n(\mathfrak{m}_1)]$is the set of model reliance ranking lists obtained for the 1st model ($\mathfrak{m}_1$) from $\boldsymbol{n}$ explanation methods. The $e_n(\mathfrak{m}_1)$ shows the feature ranking list for the model $\mathfrak{m}_1$ obtained from the nth explanation method. For example, the order can be represented as follows,

$$e_n(\mathfrak{m}_1) = [f_1, f_3, f_4, f_p, \ldots, f_2]$$

where f_1 is the name of the input feature that has the highest importance value than all other variables $f_2, f_3, f_4, \ldots, f_p$. The model reliance ranking list follows the order $f_1 > f_3 > f_4 > f_p >, \ldots, > f_2$, where variable f_2 has the least importance among the p variables.

3.4 Finding the Reference Explanation $e^*$and Consistent Explanations

Various methods that operationalize the feature importance computation may not produce the same explanation for a model [34]. The explanations not only differ in the ranking order but also in the computed model reliance values. Despite the variances, no clear reason could be found in the literature for selecting a specific explanation method. As pointed out by [16], if the results of different techniques point to the same conclusion, they very likely reflect the real aspects of the underlying data. Therefore, a reference explanation reflecting the commonly found feature order among the different methods' explanations of a model should be discovered. This reference explanation captures the optimal feature order by aggregating all the explanations' feature ranking preferences using the modified Borda Count method [23].

$$e_j^* = Borda(E_{MRR}[j]), \, where \; j = 1 \; to \; r \; models \tag{6}$$

The Borda function returns the result as an aggregated model reliance ranking order i.e., e_1^*captures the optimal ranking order of the features from the n explanations of the 1st model. Likewise, for each model, a reference explanation is aggregated from the corresponding model's explanations from n methods. This leads to a totally $\boldsymbol{r}$ number of reference explanations for the Rashomon set models $R(M)$.

To quantify the consistency of several methods in producing similar explanations to the model, the methods' explanations for the model are compared against the reference explanation. To find the consistency score, a ranking similarity method needs to be applied. The existing statistical method such as Kendall's τ [24] is considered inadequate to this problem because the ranking lists may not be conjoint. On the other hand, the Rank-Biased Overlap (RBO) [28] could handle the ranking lists even though the lists are incomplete. The RBO similarity between two feature ranking order lists R_1 and R_2 is calculated using the following equation as per [28].

$$RBO(R_1, R_2, p) = (1 - p)\sum\nolimits_{d=1}^{\infty} p^{d-1}.A_d$$
$$A_d = \frac{|R_1 \cap R_2|}{d} \tag{7}$$

The RBO similarity value ranges from 0 to 1, where 0 indicates no similarity between the feature ranking order lists and 1 indicates complete similarity. The p parameter ($0 < p < 1$) defines the weight for the top features to be considered. The parameter A_d defines the agreement of overlapping at depth d. The intersection size of the two feature ranking lists at the specified depth d is the overlap of those 2 lists (Refer to Eqs. 1–7 in [28]).

A similarity score is computed between the model's various explanations and the corresponding reference explanation and is referred to as *optimal similarity*. It is calculated as follows,

$$OPTIMAL\_SIM_{i,j} = RBO\left(e_{i,j}, e_j^*\right)$$
$$where \; i = 1 \; to \; n \; methods; \; j = 1 \; to \; r \; models \tag{8}$$

The $OPTIMAL\_SIM_{i,j}$ defines how much the explanation ($e_{i,j}$) from method i for the model j (m_j) is similar in complying with the reference explanation e_j^*, in terms of feature order. The *OPTIMAL_SIM* value is computed for all the method explanations of each model. Therefore, $\boldsymbol{n} \times \boldsymbol{r}$ similarity scores are obtained totally, that is, each explanation method gets a consistency score for each model.

3.5 Computing the Weighted Grand Mean (θ)

Among the various explanations of the Rashomon set models, the optimal similarity scores of the methods are calculated based on the method explanations' compliance with the corresponding model's reference explanation. This score shows the degree of similarity that the method has, in explaining the model's optimal learning behaviour.

Since the different explanation methods produce different feature importance coefficients for each feature, the model has varying levels of reliance on a feature. Therefore, a grand mean (θ) across several methods should be estimated. For that, a weighted mean [38] is to be implemented. To weigh the feature importance values that are computed by each method for a model, the optimal similarity score is used. For each feature, the weighted mean of the feature importance values based on the methods' optimal similarity score as weight is calculated by,

$$\theta_{j,k} = \frac{\sum_{i=1}^{n} OPTIMAL\_SIM_{i,j} * mr_i^k(m_j)}{\sum_{i=1}^{n} OPTIMAL\_SIM_{i,j}} \quad (9)$$

$$where\ k = 1\ to\ p\ features\ ;\ j = 1\ to\ r\ models$$

The grand mean of the feature k of the model j ($\theta_{j,k}$) is calculated by adding the product of the optimal similarity score of the 1 to n methods with its computed feature importance value for the k feature (mr_1^k *to* mr_n^k) and dividing the result with the sum of n methods' weights (i.e., optimal similarity scores of n methods). The grand mean is computed for all the p features for each Rashomon set model. Therefore, $p \times r$ weighted mean feature importance values are obtained.

3.6 Method Agnostic Model Class Reliance (MAMCR) Explanation

The method agnostic model class reliance explanation of the Rashomon explanation set is given as a comprehensive reliance range for each variable based on the reliance of all the well performing models under *n* explanation methods.

The model class reliance of all the p variables can be given as a range of lower and upper bounds of weighted feature importance values. The lower and upper bounds of the model class reliance for each variable can be defined as,

$$MCR^k = [MCR^{k-}, MCR^{k+}],\ \text{k} \in \text{p variables} \quad (10)$$

$$MCR^{k-} = \min_{\theta}\ \theta_{j,k}\ ,\quad MCR^{k+} = \max_{\theta}\ \theta_{j,k} \quad (11)$$

$$\text{where } r = |R(\mu)|$$

The range [MCR^{k-}, MCR^{k+}] of variable k represents that if the MCR^{k-} value is low, the variable k is not important for any almost-equally-accurate models in the Rashomon set models [$R(\mu)$]whereas if the MCR^{k+} is high, then the variable k is most important for every well performing model in $R(\mu)$. Thus, the MCR provides a method agnostic variable importance explanation for all the well performing models of the Rashomon set.

4 Experiments and Results

In this section, the concept of the proposed method is illustrated with the experiments on the 2-year criminal recidivism prediction dataset[1] which was released by ProPublica to study the COMPAS (Correction Offender Management Profiling for Alternative Sanctions) model that was used throughout the US Court system. The dataset consists of 7214 defendants (from Broward County of Florida) with 52 features and one outcome variable, which is 2-year recidivism. Among the 52 features, 12 are date type to denote jail-in and out, offence, and arrest dates, 21 are personal data identifiers such as first and last name, age, sex, case numbers and descriptions and other features are mostly numeric values such as no. of days in screening, in jail, from compas, etc. The framework is not limited to this data but is flexible enough to support any dataset.

In the analysis of the Race variable's contribution to predicting the 2-year recidivism, the authors [22] say that there are some well performing models which do not rely on inadmissible features like Race and gender. Additionally, for the same data set, the authors [29] report that the explanation based on a single model is biased over the inadmissible feature 'Race', whereas the grand mean of multiple models' feature importance values does not highlight the feature as an important feature for the majority of the models. To ensure whether these claims will be consistent across multiple methods' explanations and to answer the research questions as well, the same dataset used by [22, 29] with similar a setup (with 6 features - age, race, prior, gender, juvenile crime, and current charge - of all the 7214 defendants) is taken for the analysis.

To make the outcome prediction, the logistic regression model class is used in the analysis with 90% (6393) training data and 10% (721) test data as in [29]. The Stratified 5-fold cross-validation is used to train and validate the multiple models. The total trained models and the selected models to the Rashomon set are shown in Fig. 2. The reasonable sample of Rashomon set models (350) are obtained from the total trained (2665) models by filtering the models whose prediction accuracy are above the accuracy threshold $(1-\varepsilon)\mathfrak{m}^* = 0.6569$, where $\mathfrak{m}^*$ accuracy $= 0.6914$ and $\varepsilon = 5\%$. Those models form the Rashomon set.

To obtain the explanations for models' decisions, the iAdditive[2] and other 5 state-of-the-art XAI methods [3, 4, 7, 25, 26] based on the feature importance approach are applied to the Rashomon set models $[\mathfrak{R}(\mathfrak{H})]$. Normalization is applied to each method's computed importance values for each model. The model reliance rankings for each model are also obtained (E_{MRR}). Figure 3 shows the various methods' model reliance ranking range of the Rashomon set models grouped by each feature of the COMPAS dataset.

The distribution of feature importance ranks that are obtained from different methods illustrates the variation found in the various method explanations. Let's consider the 'Race' feature's rank explanations. The Shap [3], Skater [4] and iAdditive methods' ranks span from 1–6 for the models, whereas for the other 3 methods, the range is from 2–6. It means, as per the former methods' explanations, there are some models which consider the 'Race' feature as their most important (1st rank) feature. But in the view

---

[1] https://www.propublica.org/datastore/dataset/compas-recidivism-risk-score-data-and-analysis.

[2] iAdditive is an in-house XAI software tool.

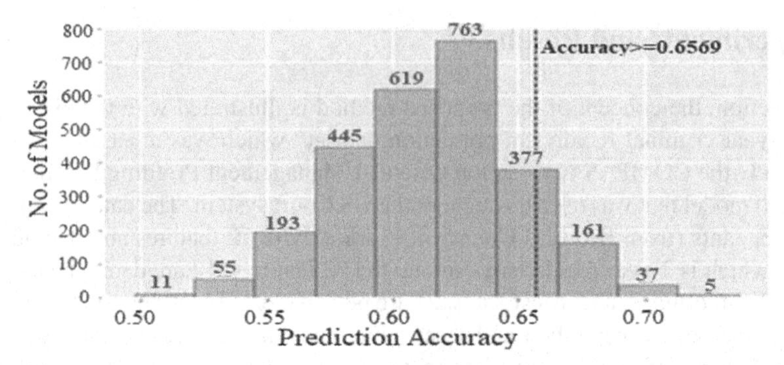

Fig. 2. The prediction accuracy frequency of all the trained models. The accuracy threshold $(1-\varepsilon)\mathfrak{m}^* = 0.6569$, where $\mathfrak{m}^* = 0.6914$ and $\varepsilon = 5\%$ is used to search for the Rashomon set models ($\mathcal{R}$). Models with an accuracy level above the threshold value are only included in the $\mathcal{R}$.

of the latter methods, for none of the models, 'Race' is the 1st priority feature. Let's take the 'Juvenile crime' feature. As per the Sage [7] method explanations, the'crime' feature is the most important feature for most of the models, whereas, for the Shap and iAdditive methods, the median ranks lie in 4th and 5th positions, respectively. The Skater and Lofo [26] methods have similar 3rd rank position to the feature and the Dalex [25] method stood in between the Sage and Skater rank positions by giving 2nd rank.

From this, it could be observed that for the same models, these methods provide different feature importance explanations (in the form of computed values and ranks as well). If any one of the methods is selected to provide the explanation for a well performing model, it could end up in a method-dependent explanation of that model. It means that the explanation would be biased over the specific method. Therefore, to get a consensus explanation for the almost-accurate models over all the applied explanation methods, the model agnostic model class reliance (MAMCR) explanation method is to be implemented.

Firstly, a reference explanation e* is aggregated from the corresponding explanations of 6 methods for each model to reflect the common feature ranking order. These reference explanations reflect the optimal learning for all the models in the Rashomon set (see Fig. 4). To quantify the consistency of various explanations obtained from multiple methods, the corresponding reference explanation (e*) is compared against each model's method-wise explanation.

Next, for each model of the Rashomon set, the weighted average is computed for all the features based on the method's consistency score. The method explanation which complies well with the optimal explanation will contribute more to the average model reliance value. For each of the six variables of the 350 models, the grand means ($\theta_{j,k}$) are computed using Eq. 9 based on the concern method's consistency/optimal similarity scores.

The method agnostic model class reliance explanation (MAMCR) for the multiple almost-accurate models based on multiple methods' explanations is presented as a range. The lower and upper bounds [MCR^-, MCR^+] of each variable's grand mean are selected as the model class reliance for all the models in the Rashomon set. The method agnostic

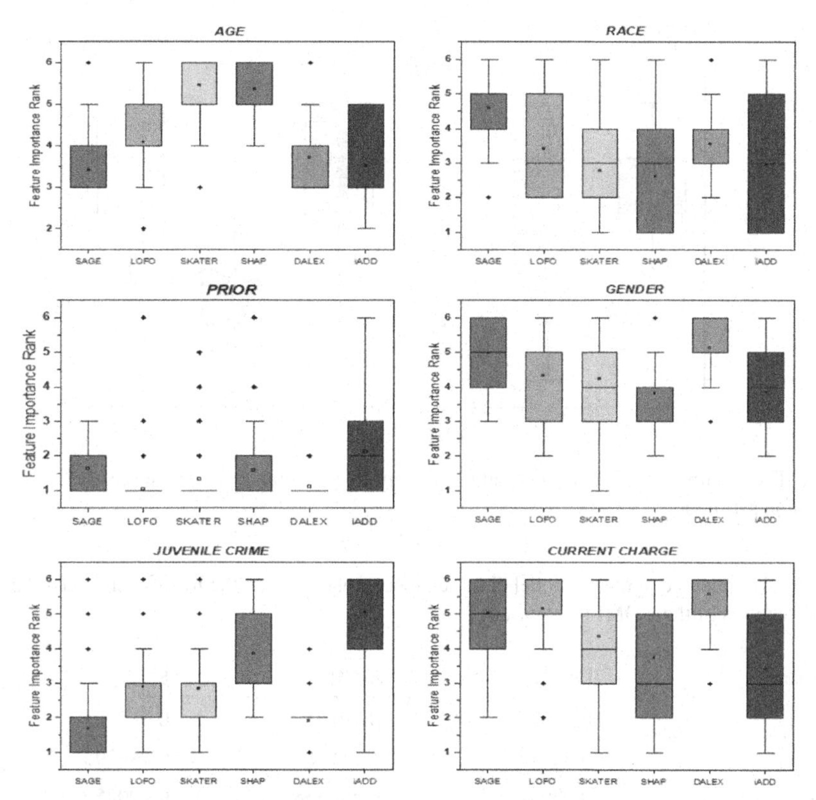

Fig. 3. Model reliance/feature importance rankings obtained from the 6 explanation methods for the COMPAS dataset. A box plot showing the range of ranks allocated by each method for the 350 Rashomon set models for a feature is shown in each panel. The difference in the feature rankings illustrates the variations found in the various method explanations.

MCR is shown in Table 1. In that, the high MCR^- value (e.g., 0.08) indicates that the *Prior* feature is used by all the models and the low MCR^+ value (e.g., 0.10) indicates that the *Age* feature is least used by all the models.

4.1 Discussion

Various methods' explanations are compared against the 'Race' feature's importance. The distribution of many models' model reliance is shown in Fig. 5. The number of models that falls on the feature importance range is displayed on each bar in the histogram. As per the Sage [7] explanation, the 'Race' feature is not at all an important feature used by most of the models. It could be observed from Fig. 5a that 324 models out of 350 are given the feature importance value as less than 0.1. This informs that the Race feature is not an important feature for the 324 models. It complies well with the claim of [29]. On the other hand, it is not true based on other methods' explanations. From Figs. 5b–5e, it could be observed that there are many models that rely on the 'Race' feature from the moderate to high range, whereas Fig. 5f is consistent with Fig. 5a. It alerts us that

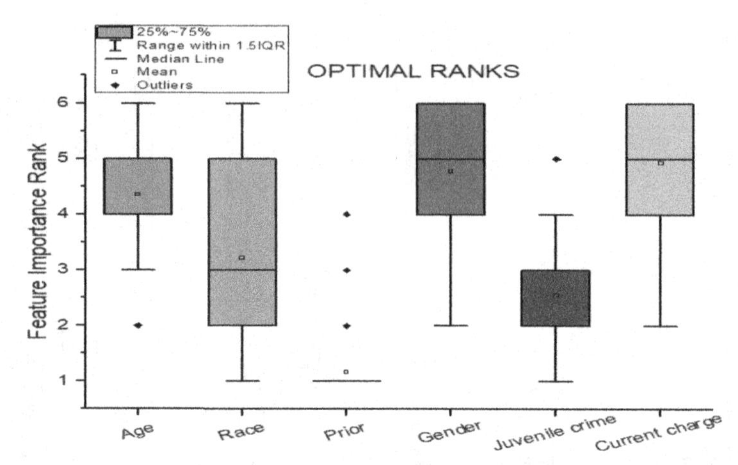

Fig. 4. The feature-wise rank distribution of optimal reference explanations (e*) for 350 Rashomon set models.

Table 1. The method agnostic model class reliance explanation of the Rashomon set models for the six features of the COMPAS dataset.

Features	[MCR$^-$	MCR$^+$]	STD
Age	0.023774	0.103612	0.015621
Race	0.021176	0.33566	0.089296
Prior	0.08947	0.698398	0.090259
Gender	0.017584	0.188301	0.039289
Juvenile crime	0.074106	0.426745	0.054915
Current charge	0.017144	0.236485	0.041713

the explanation obtained from a method is not necessarily the same as the one obtained from another method for the model.

This addresses the first research question (RQ1) that while multiple explanation methods are applied on multiple well-performing models for getting the feature importance explanations, the feature which is projected as (un)important by one explanation method is not necessarily agreed by another method. Therefore, the identified importance of the feature depends completely on the method that is applied for obtaining the explanation.

While comparing the method explanations of each feature (see, Fig. 3), no two methods could be identified in producing a similar explanation pattern in all the feature explanations. For example, the Skater and Shap method explanations for the Age feature resemble the same pattern except for the outlier. Similarly, the Sage and Dalex are in a similar pattern on the same variable. The same methods could not be found with similar patterns in other feature explanations. For example, the Skater and Shap methods have contrasting explanation patterns in Juvenile crime feature, whereas the Skater and Lofo

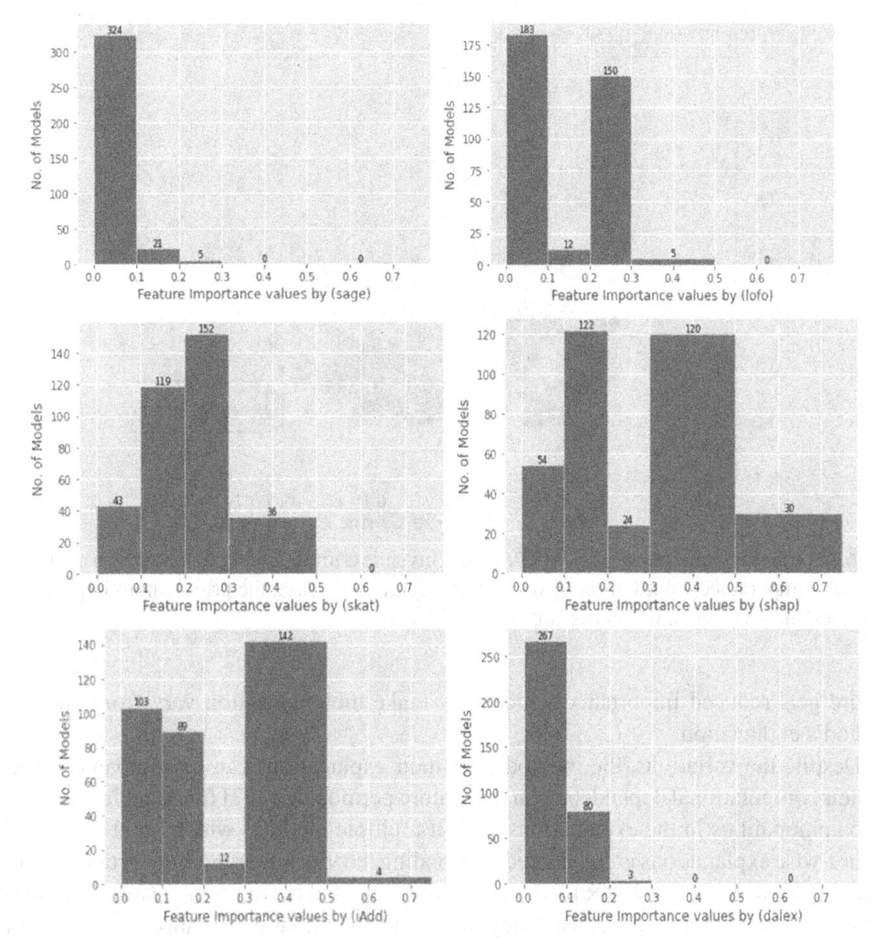

Fig. 5. The feature importance values of the 'Race' feature for 350 Rashomon set models, grouped by each method (5a. Sage, 5b. Lofo, 5c. Skater, 5d. Shap, 5e. iAdditive and 5f. Dalex). The data label of each bar shows how many models lie within the feature importance bin range.

methods exhibit a similar pattern. One of the possible reasons observed for the variation could be that a feature becomes the most important when another variable becomes the least important [22]. It is illustrated in Fig. 6.

Figure 6 shows the feature importance values computed for Juvenile crime and Prior features by the 6 methods for the 350 almost-accurate models. Each point in the plot represents a model's reliance on those variables. When the Prior feature importance (y-axis) of the models reaches its maximum value such as above 0.6, the crime feature importance (x-axis) of them is below ≈0.35 (shown within a box). When the crime feature's importance of a model reaches above 0.8 or around 1, its Prior importance is very low such as less than ≈0.15. It indicates that the feature Prior is the most important feature of a model when the Juvenile crime is less important than the Prior feature. So, if a method allocates a feature with high importance in its explanation obviously another

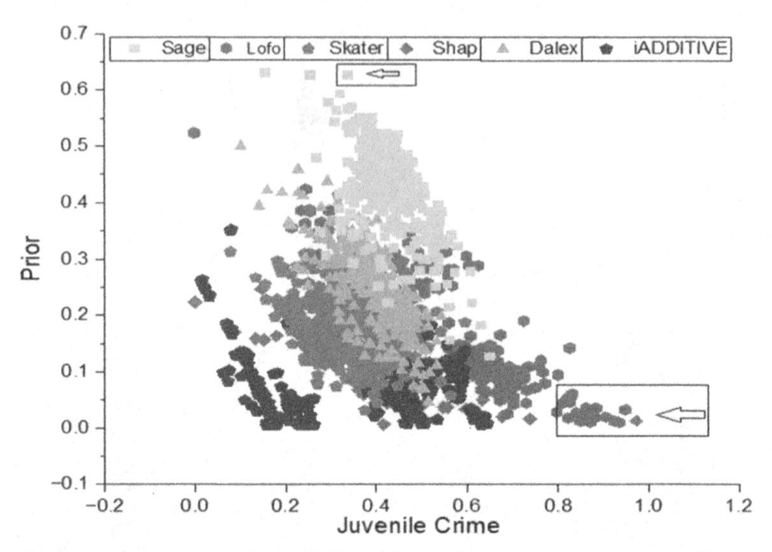

Fig. 6. The feature importance values of Prior and Juvenile crime features computed by 6 methods. While the importance values of the Juvenile crime feature increase, the prior feature importance decreases and vice versa, which is emphasised with a box.

feature gets reduced importance which may make the explanation vary from another method's explanation.

Despite the variations, the methods and their explanations can be compared based on their computational dependency on the feature permutation [27] function. Identifying the commonalities in the explanations [20] of multiple methods which point to similar feature-wise explanations is considered as revealing the true importance of the underlying data [16]. Hence, the MAMCR method finds the weighted mean for the feature explanations based on the method's consistency in producing similar explanations and through which it provides a comprehensive range for the multiple almost-equally-accurate models. It represents the feature-wise model reliance bounds for all the well-performing models of the pre-specified model class that are computed by the pre-specified methods.

To validate the MAMCR explanation bound suitability to all well performing models, a new, almost-equally-accurate test model is created using the same model class (i.e., Logistic regression) algorithm with random sampling data. This model's accuracy is verified against the Rashomon set threshold (0.6569). The explanations from the six methods are obtained for the model and the grand mean of each variable is found. The test model's feature importance which is plotted along with the MAMCR bounds is displayed in Fig. 7. It elucidates that the test model's feature importance of all the variables lies within the MAMCR boundary values. Thus, the second research question (RQ2), finding the consistent explanation across multiple explanation methods for the almost-equally-accurate models, is addressed through the MAMCR framework by obtaining the method agnostic MCR bounds.

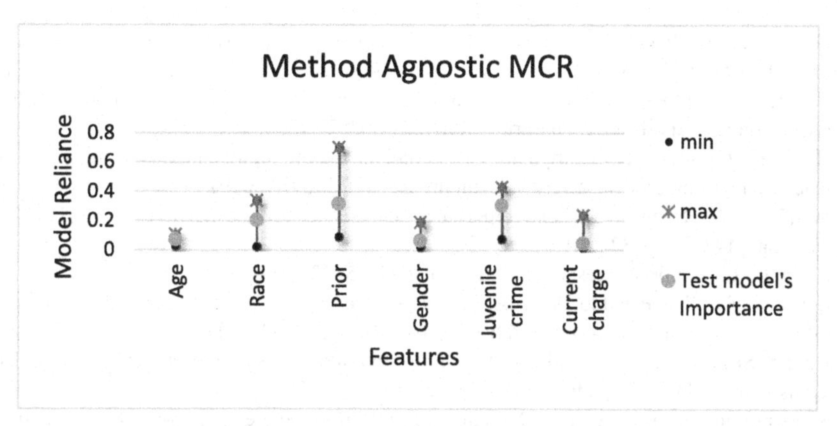

Fig. 7. The feature importance values of a Test model's features along with the MAMCR bounds. The test model's importance values lie within the MAMCR explanation range.

5 Conclusion

The experiments conducted on the COMPAS data set alert us that the method's explanation which highlights a feature as most important may not be projected as such by another method. These inconsistencies in the generated explanations by different explanation methods for the Rashomon set models motivated the proposal of a novel framework for discovering consistent explanations across multiple explanation methods. It provided a method agnostic explanation as a model class reliance for the multiple almost-equally-accurate models. The efficiency of the method agnostic MCR explanation is illustrated by describing the comprehensive variable importance value range for all the well performing models of the pre-specified model class across multiple explanation methods.

In this work, the explanation methods that return the feature importance values as a global explanation are only considered for the explanation ensembling. The future work can be extended for the instance-wise explanations and for other explanation output formats as well.

References

1. Adadi, A., Berrada, M.: Peeking inside the black-box: a survey on explainable artificial intelligence (XAI). IEEE access **6**, 52138–52160 (2018)
2. Ribeiro, M.T., Singh, S., Guestrin, C.: Why should i trust you?" Explaining the predictions of any classifier. In: Proceedings of the 22nd ACM SIGKDD International Conference on Knowledge Discovery and Data Mining. pp. 1135–1144 (2016)
3. Lundberg, S.M., Lee, S.-I.: A unified approach to interpreting model predictions. In: Advances in Neural Information Processing Systems, pp. 4765–4774 (2017)
4. Choudhary, P., Kramer, A.: datascience.com team: datascienceinc/Skater: Enable Interpretability via Rule Extraction (BRL) (v1.1.0-b1). Zenodo (2018). https://doi.org/10.5281/zenodo.1198885
5. Mateusz, S., Przemyslaw, B.: Explanations of model predictions with live and breakdown packages. R J. **10** (2018). https://doi.org/10.32614/RJ-2018-072

6. Gosiewska, A., Biecek, P.: iBreakDown: Uncertainty of Model Explanations for Nonadditive Predictive Models. arXiv preprint arXiv:1903.11420 (2019)
7. Covert, I., Lundberg, S., Lee, S.I.: Feature Removal Is a Unifying Principle for Model Explanation Methods. arXiv preprint arXiv:2011.03623 (2020)
8. Horel, E., Giesecke, K.: Computationally efficient feature significance and importance for machine learning models. arXiv preprint arXiv:1905.09849 (2019)
9. Wei, P., Lu, Z., Song, J.: Variable importance analysis: a comprehensive review. Reliab. Eng. Syst. Saf. **142**, 399–432 (2015)
10. Breiman, L.: Random forests. Mach. Learn. **45**(1), 5–32 (2001)
11. Lei, J., G'Sell, M., Rinaldo, A., Tibshirani, R.J., Wasserman, L.: Distribution-free predictive inference for regression. J. Am. Statist. Assoc. **113**(523), 1094–1111 (2018)
12. Robnik-Šikonja, M., Kononenko, I.: Explaining classifications for individual instances. IEEE Trans. Knowl. Data Eng. **20**(5), 589–600 (2008)
13. Sundararajan, M., Taly, A., Yan, Q.: Axiomatic attribution for deep networks. In: International Conference on Machine Learning, pp. 3319–3328, PMLR (2017)
14. Štrumbelj, E., Kononenko, I.: Explaining prediction models and individual predictions with feature contributions. Knowl. Inform. Syst. **41.3**, 647–665 (2014)
15. Datta, A., Sen, S., Zick, Y.: Algorithmic transparency via quantitative input influence: Theory and experiments with learning systems. In: 2016 IEEE Symposium on Security And Privacy (SP), pp. 598–617 (2016)
16. Gifi, A.: nonlinear multivariate analysis (1990)
17. Kobylińska, K., Orłowski, T., Adamek, M., Biecek, P.: Explainable machine learning for lung cancer screening models. Appl. Sci. **12**(4), 1926 (2022)
18. Yeh, C.-K., Hsieh, C.-Y., Suggala, A., Inouye, D.I., Ravikumar, P.K.: On the (in) fidelity and sensitivity of explanations. In: Proceedings of the NeurIPS, pp. 10 965–10 976 (2019)
19. Fisher, A., Rudin, C., Dominici, F.: All models are wrong, but many are useful: learning a variable's importance by studying an entire class of prediction models simultaneously. J. Mach. Learn. Res. **20**(177), 1–81 (2019)
20. Jamil, M., Phatak, A., Mehta, S., Beato, M., Memmert, D., Connor, M.: Using multiple machine learning algorithms to classify elite and sub-elite goalkeepers in professional men's football. Sci. Rep. **11**(1), 1–7 (2021)
21. Wolpert, D.H.: The supervised learning no-free-lunch theorems. In: Roy, R., Köppen, M., Ovaska, S., Furuhashi, T., Hoffmann, F. (eds.) Soft Computing and Industry, p. 2542. Springer, London, U.K (2002). https://doi.org/10.1007/978-1-4471-0123-9_3
22. Dong, J., Rudin, C.: Exploring the cloud of variable importance for the set of all good models. Nature Mach. Intell. **2**(12), 810–824 (2020)
23. Lin, S.: Rank aggregation methods. Wiley Interdiscipl. Rev. Comput. Statist. **2**(5), 555–570 (2010)
24. Kendall, M.G.: Rank correlation methods (1948)
25. Baniecki, H., Kretowicz, W., Piatyszek, P., Wisniewski, J., Biecek, P.: dalex: responsible machine learning with interactive explainability and fairness in python. J. Mach. Learn. Res. **22**(1), 9759–9765 (2021)
26. Erdem, A.: https://github.com/aerdem4/lofo-importance. Accessed 22 July 2022
27. Covert, I., Lundberg, S.M., Lee, S.I.: Explaining by removing: a unified framework for model explanation. J. Mach. Learn. Res. **22**, 209–211 (2021)
28. Webber, W., Moffat, A., Zobel, J.: A similarity measure for indefinite rankings. ACM Trans. Inf. Syst. **28**, 4 (2010)
29. Ning, Y., et al.: Shapley variable importance cloud for interpretable machine learning. Patterns 100452 (2022)

30. Hamamoto, M., Egi, M.: Model-agnostic ensemble-based explanation correction leveraging rashomon effect. In: 2021 IEEE Symposium Series on Computational Intelligence (SSCI), pp. 01–08. IEEE (2021)
31. Semenova, L., Rudin, C., Parr, R.: A study in Rashomon curves and volumes: A new perspective on generalization and model simplicity in machine learning. arXiv preprint arXiv: 1908.01755 (2019)
32. Bobek, S., Bałaga, P., Nalepa, G.J.: Towards model-agnostic ensemble explanations. In: Paszynski, M., Kranzlmüller, D., Krzhizhanovskaya, V.V., Dongarra, J.J., Sloot, P.M.A. (eds.) ICCS 2021. LNCS, vol. 12745, pp. 39–51. Springer, Cham (2021). https://doi.org/10.1007/978-3-030-77970-2_4
33. Nguyen, T.T., Le Nguyen, T., Ifrim, G.: A model-agnostic approach to quantifying the informativeness of explanation methods for time series classification. In: Lemaire, V., Malinowski, S., Bagnall, A., Guyet, T., Tavenard, R., Ifrim, G. (eds.) AALTD 2020. LNCS (LNAI), vol. 12588, pp. 77–94. Springer, Cham (2020). https://doi.org/10.1007/978-3-030-65742-0_6
34. Fan, M., Wei, W., Xie, X., Liu, Y., Guan, X., Liu, T.: Can we trust your explanations? Sanity checks for interpreters in Android malware analysis. IEEE Trans. Inf. Forensics Secur. **16**, 838–853 (2020)
35. Ratul, Q.E.A., Serra, E., Cuzzocrea, A.: Evaluating attribution methods in machine learning interpretability. In: 2021 IEEE International Conference on Big Data (Big Data) pp. 5239–5245 (2021)
36. Rajani, N.F., Mooney, R.J.: Ensembling visual explanations. In: Escalante, H.J.,et al. (eds.) Explainable and Interpretable Models in Computer Vision and Machine Learning. TSSCML, pp. 155–172. Springer, Cham (2018). https://doi.org/10.1007/978-3-319-98131-4_7
37. Velmurugan, M., Ouyang, C., Moreira, C., Sindhgatta, R.: Evaluating Explainable Methods for Predictive Process Analytics: A Functionally-Grounded Approach. arXiv preprint arXiv:2012.04218 (2020)
38. Bland, J.M., Kerry, S.M.: Weighted comparison of means. BMJ **316**(7125), 129 (1998)

Association Rules Based Feature Extraction for Deep Learning Classification

Ruba Kharsa and Zaher Al Aghbari(✉)

Department of Computer Science, University of Sharjah, Sharjah, UAE
{U21103616,zaher}@sharjah.ac.ae

Abstract. The number of extracted features from medical data, such as computer-aided diagnosis, has been known to be too large and affects the performance of the used classifiers. Moreover, the large number of input features affect the accuracy of the classifiers, such as the traditional machine learning classifier. Therefore, in this paper, we proposed the use of association rules to select features from medical data, which result in dimensionality reduction of the input feature space. The selected features become the input to a deep neural network, particularly ResNet, which is known for its high accuracy of classification results. The conducted experiments prove that the use of association rules to select the most representative features and the use of deep neural networks as a classifier outperformed other traditional machine learning models in terms of accuracy of classification.

Keywords: Association rules mining · Feature selection · Deep learning classification · Resnet · Neural networks · Classification

1 Introduction

A fundamental step in data mining is to select representative features from raw data. The purpose of feature extraction has two main aims. The first aim is to represent each data object with a small, but effective, number of features, which would improve the efficiency of the classifier. The second aim is to remove the features that are not representative of the data object, and thus improve the accuracy of classification. The feature selection methods are categorized into three approaches: filter, embedded, and wrapper.

The *filter* approach selects features based on some general metric, such as correlation coefficient or information gain. This approach does not depend on a predictive model and thus it is relatively fast. Consequently, the filter approach is favored when the number of raw features is large. The *embedded* approach selects the representative features during the training of a predictive model. An example

K. K. Patel et al. (Eds.): icSoftComp 2022, CCIS 1788, pp. 72–83, 2023.
https://doi.org/10.1007/978-3-031-27609-5_6

of the embedded approach is the decision tree model. That is the feature selection is embedded into the building of a model, such as a classifier. On the hand, in the *wrapper* approach, the feature selection process is wrapped around the classifier. That the feature selection process uses the same classification model to identify the representative features. An example of the wrapper approach is the sequential feature selection algorithm, or the genetic algorithm.

Existing feature selection methods reduce the feature space into a smaller one, however the classification accuracy based on the reduced set of features may deteriorate. Some existing works show only a slight improvement of the classification accuracy. In this paper, due to the large number of raw features, we propose the use of a filter approach to select the representative features. In particular, association rules mining is used to select the representative features of data objects, such as medical images, from the set of raw features. Association rules have been used to select features, as in [23], however these features were used with traditional machine learning classifiers, such as a decision tree. Although the reduce feature set improved the performance of the decision tree, but the improvement of the classification accuracy was not significant.

Classification is the task of predicting the labels of new unknown observations based on previous knowledge of similar data. Researches have been studying data classification for decades, utilizing efforts, time, and money to construct models and enhance their prediction and classification accuracy. All the effort spent on the studies of classification resulted in considerable evolution of data mining classification approaches. Particularly, the improvement in classification over the years made a revolutionary change in the human perspective of computers' capabilities in various areas. For example, computer Vision (CV) is one of the most benefited fields from this change as all the tasks associated with CV are considered as types of classification (e.g., face recognition, voice recognition, item recognition, and text recognition) [1–5]. Other fields such as data streams management [7] and sensor data processing [8,9] have also benefited.

Moreover, medical diagnosis has become more accurate and precise, assisting doctors, helping with early detection of diseases, saving lives, and adding new hidden insights on popular diseases. There are many models and approaches to classification, such as Random Forest (RF), K Nearest Neighbor (KNN), Support Vector Machines (SVM), Neural Networks (NN), and Deep Learning (DL). However, using any of these models directly on the data without any data pre-processing is usually inefficient and produces imprecise results. Datasets should be studied thoroughly, analyzed, and cleaned by handling missing values and outliers as well as eliminating unrelated data and attribute. Afterwards, the representative feature are selected, which are then used as input for the classifier. In this paper, we investigate the use association rules to select the reduced set of representative features to be used as an input to a deep learning classifier. This proposed model enhances the accuracy of classification of medical data, such as image data. Particularly, this paper uses ResNet, which is a deep learning model to the classify data objects, such as medical images, based on the reduced set of representative features. Our experiments shows the effectiveness of the proposed model, which outperforms other competing predictive models such as RF, KNN, SVM, and NN.

The rest of the paper is organized as follows. In Sect. 2, we review the related research. Background information regarding the association rules and deep learning models are presented in Sect. 3. Section 4 discuss the details of the proposed model. The experiments are presented and discussed in Sect. 5. Finally, the paper is concluded in Sect. 6.

2 Literature Review

This section discusses previous studies that utilize mining algorithms to increase the efficiency of classification models. These studies proposed various methods in diverse fields like medicine, intruder detection, and malware detection. They also demonstrated the effectiveness of using association rules in enhancing the classification accuracy. According to previous studies, the main two approaches of using association rules algorithms with deep learning classification are feature selection and analysis of the classification models.

2.1 Association Rules for Feature Selection

The curse of dimensionality is a challenging issue that researchers have been trying to avoid. Many studies propose various techniques to reduce the dimensionality of the datasets. A traditional technique to reduce dimensionality is Principle Component Analysis. However, there have been many feature selection techniques to reduce dimensionality like InformationGain, Pearson correlation coefficient, and association rule mining. Vougas et al. [14] constructed a DL model to predict the therapeutic response of a particular drug on specific cancer cell lines. They collected the data from massive cancer datasets of drugs, genes, and cancer cell lines. After that, they organized and combined the data to build their dataset. The created dataset consisted of 1001 rows describing the cell lines and over 60000 column attributes.

To select a small set of features, the Apriori algorithm [10] is applied to extract association rules, which consist of the rules for gene-to-drug, tissue-to-drug, drug-to-drug. Subsequently, they used the association rules as input to the deep learning model and evaluated the performance by comparing their results with the state-of-the-art random forest classifiers. Vougas et al. demonstrated the classification power of their model by achieving high accuracy and specificity. Boutorh & Guessoum [15] also used association rules as a feature selection method for a neural network model, which classifies Single Nucleotide Polymorphisms (SNPs) for breast cancer diagnosis. They extracted rules using the Genetic Algorithm and used these rules as an input for the DLM. Boutorh & Guessoum achieved the highest accuracy by using 42 features (i.e., SNPs). Karabatak & Ince [6] used a similar approach that used deep learning neural network as a model for classifying breast cancer and used the Wisconsin dataset for training and testing the model. Association rules algorithm is used for dimensionality reduction. This method was able to reduce the number of features from nine to four using the Apriori algorithm [10]. Karabatak & Ince performed 3-fold cross-validation to assess the model and achieved high classification accuracy. Inan et al. [16] performed almost the same approach as [6] with the same dataset and the same problem; however, they used a hybrid method of feature selection, utilizing both Apriori and PCA. They achieved higher classification accuracy.

2.2 Association Rules for Result Analysis

A deep learning model is considered as a black box. Even though deep learning models perform well, it not easy to explain how these results were achieved. The association rules mining can assist in analyzing the results and extracting interpretable and easy-to-understand association rules to explain the process of deep learning classification. Tihilina et al. [17] built an Intruder Detection System, IDS, using a deep learning classifier and association rule algorithm. They used Recurrent Neural Networks (RNN) to perform classification and FP-growth to perform analysis. A streaming data API was used to train the RNN in real-time. After that, data pre-processing techniques was applied for removing outliers, normalizing the data, and reducing dimensionality by ignoring features that have a low correlation with other attributes. The performance was evaluated by measures like recall, precision, F1-score, and Receiver Operating Characteristic (ROC). Lastly, association rule mining was employed for the reasoning of the proposed classifier.

Yuan et al. [18] proposed a model for Android malware detection and called it DroidDetector. DroidDetector can detect whether an android application is malware or benign by performing static and dynamic analysis to extract 192 of the app's features and input them into the DL model for classification. Yuan et al. used the Deep Belief Networks (DBN) as the DL model, trained it on 20,000 benign apps from Google Play Store and 1760 malware from Contagio Community and Genome Project. Then, the model was evaluated using precision, recall, and overall accuracy. The DBN achieved maximum accuracy of 96.76% and average accuracy of more than 95%. Also, the DBN outperformed other machine learning techniques like SVM, Naïve Bayes, and logistic regression. Yuan et al. utilized association rules to perform in-depth analysis on the features exploited by the DBN.

2.3 Other Methods

Eom et al. [19] used association rules with a deep learning model to cluster protein interactions and to discover novel knowledge from the functional categories of the proteins, their set of features, and interaction parameters. They derived association rules from the trained model using genetic algorithms and used them to perform clustering. Montaez et al. [20] built a model that classifies obesity by extending the Genome-wide association studies, which reveal associations between specific genes and particular diseases. They assessed the model using specificity, sensitivity, ROC, and the Area under the ROC curve (AUC). Zhang et al. [21] employed DBN to detect traffic accidents using the data from social media. Three main steps, Feature Selection, Classification, and Evaluation, were performed.

3 Background

3.1 Association Rule Mining

Association Rule Mining (ARM) is part of the descriptive Data mining and Knowledge Discovery process (KDD). ARM assists researchers in extracting hidden knowledge and patterns from tremendous amounts of data in the form of Association Rules (ARs).

Agrawal et al. [10] first introduced the problem of ARM as a Market Basket Analysis to learn about patterns and associations in the purchases. Agrawal et al. described the formal model of ARM as follows. Let $I = \{i_1, i_2, i_3, \dots i_n\}$ where I is an itemset that contains n items. Let D be a dataset of transactions where $D = \{t_1, t_2, t_3, \dots t_m\}$ and any transaction $t_i \subseteq I$. The implication $X \Rightarrow Y$ describes an AR, where $X, Y \subseteq I$ and $X \cap Y = \emptyset$. The two measurements that describe the AR are support and confidence. $p(X \cup Y)$ is the support s that measures the frequency of the rule by finding the percentage of transactions in the database that contains both X and Y to the whole number of transactions. Confidence c, on the other hand, determines the strength of the rule by calculating the proportion of the transactions containing both X and Y to the number of transactions containing X. The formula for the confidence is $\frac{s(X \cup Y)}{s(X)}$. The ARM algorithms filter the rules and choose only the useful ones based on the minimum support and confidence.

Although ARM started as an algorithm for extracting ARs in transaction databases, it became popular in various fields and areas like medicine, entertainment, user experience, and fraud detection because it proved its significance and reliability. ARs are easy to interpret, explain, and understand. These characteristics of ARs make them a convenient choice in analyzing the data and using the derived knowledge in enhancing and improving different areas [11].

The main issue in ARM was the intensive computational power required to generate the frequent itemset by finding each combination of items. This problem has been the focus of researchers for decades. Researchers have developed many algorithms that reduce the time and effectively find the frequent itemset (e.g., Apriori, Frequent Pattern Tree, Sampling, Partitioning).

3.2 Neural Networks

Artificial Neural Networks (ANN) are highly interconnected layers of neurons inspired by the human brain. Multi-layered Perceptron Neural Network (MLPNN) is the simplest form of ANN architecture. The MLPNN has three layers: input layer, hidden layer, and output layer. Figure 1 illustrates the MLP where x_i is an input signal to the neuron i of the input layer, and w_{ij} is the weight of the connection between the input signal x_i and the hidden neuron j. Neuron j sums the multiplication of its input signals and their corresponding weights then calculates the output y_j as a function of the sum.

$$y_j = f\left(\sum x_i w_{ij}\right) \tag{1}$$

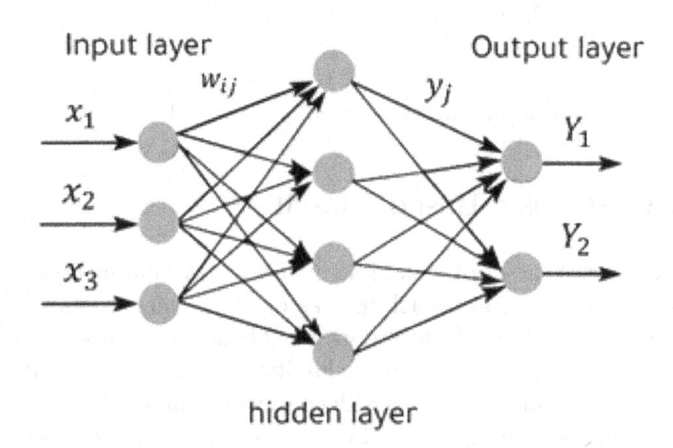

Fig. 1. Multilayer perceptron neural network

$f(.)$ is the activation function which can be a binary step function, a linear function, or a sigmoid function, etc. Similarly, the neurons in the output layer calculate their outputs Y using the same equation. The training stage of the NN adjusts the weights of the connection by an algorithm like backpropagation until it reaches the desirable outcome Y. To design an effective deep learning NN classifier, a suitable structure of the model, e.g. number of layers, appropriate activation function, and number of neurons in each layer, should selected [12].

3.3 Deep Learning Residual Networks, ResNet

DLRN is an advanced form of deep learning models, which adds extra blocks and organizes them differently based on the selected model. The proposed model uses ResNet to implement the deep learning NN. ResNet has been one of the most popular deep learning NN especially in image classification. He et al.'s [13] motivation for introducing ResNet is that the more you concatenate layers to NN models with just activation and batch normalization, the more they become difficult to train and the higher tendency for accuracies to drop. The solution to this problem is with the Residual Block that is illustrated in Fig. 2, which uses "Skip Connection". This technique of skip connection adds the input to the resulting output from a sequence of layers. The residual block allows for deeper training of the deep learning NN without affecting its performance.

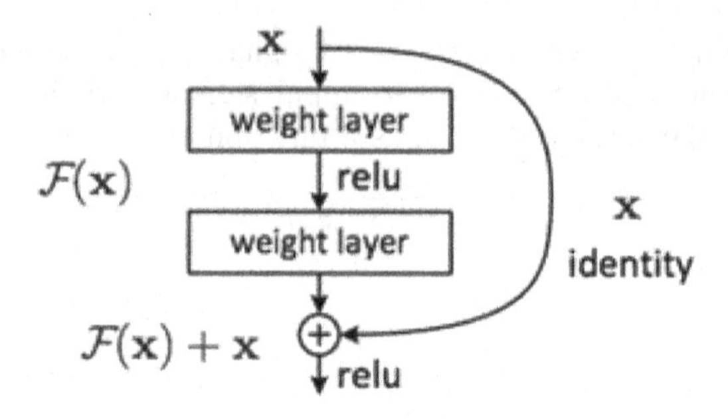

Fig. 2. Skip connection of ResNet [13]

4 AR with ResNet Based Classifier

This section discusses the approach and the used datasets for implementing the experiments. The proposed approach uses association rules to select the reduced set of representative features from a dataset. Then, the reduced set of features is used as an input to a deep learning classifier (see Fig. 3). In this paper, we used ResNet as a classifier, which is a deep learning model that classifies data objects, such as medical images, based on the reduced set of features.

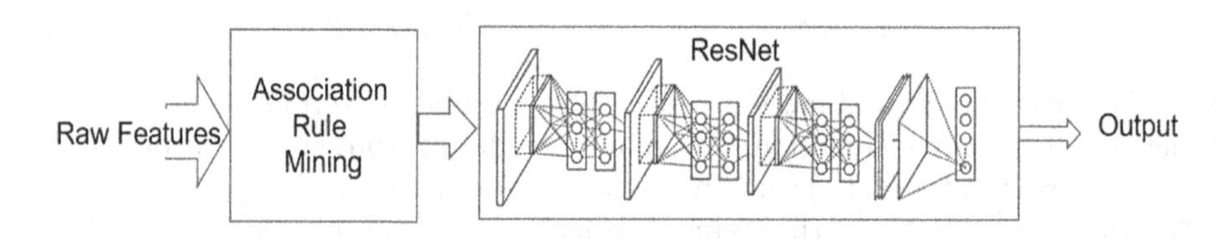

Fig. 3. Block diagram of proposed approach

4.1 Association Rules

Two types of association rules, AR1 and AR2, are produced. The first type AR1 consists of the association rules between independent variables without considering the class. A high correlation between attributes is called multicollinearity and it results in worse classification. Therefore, in this type, the Apriori algorithm finds association rules that assist in eliminating any attribute that is highly correlated with any other attributes. That means the consequent of the rules that satisfy the min-support and confidence is not necessary for the classification because it is already represented in the antecedent. We will set the min-support as 0.9 and the min-confidence as 1. For example, if a and c are independent variables and we find a rule $a \rightarrow c$ that has a support of 0.9 and confidence of 1, then we suppose that c is not significant to the prediction and can be eliminated.

The second type AR2 consists of the large itemsets for each target class. The attributes in the large itemset are considered as the relevant features that define the class. The relevant attributes for each class are found and used as input to the classification models. A min-support of 0.9 is considered to find the large itemsets in each class. When using AR1, the unrelated columns are simply dropped, but when using AR2, the related columns are selected. When AR1 and AR2 are used, the irrelevant attributes are dropped and then the relevant ones are chosen.

4.2 Classification Models

Traditional feedforward NN with few hidden layers can represent the features of an input object. However, as the number of features increase and the associations between the features become more complex, the few layers of the feedforward NN might lead to overfitting. Consequently, this might decrease the accuracy of classification. Therefore, deep learning NN have found a solution to this problems by going deeper. That is by increasing the number of layers. For example, in AlexNet the number of layers is increased to 5, in VGG architecture the number of layers is increased to 19, and in InceptionV1 they are increased to 22.

However, increasing the number of layers makes the training of the network harder and time consuming. Moreover, as the number of layers become large, the gradient becomes smaller and smaller due to its repetitive multiplication and thus it saturates or even degrades. Therefore, ResNet solves this problem of vanishing gradient by adding a controlled smalled loss value (see Fig. 2) to compensate the loss of gradient value. ResNets can go very deep, that is the number of layers can be increased to a large number, without being affected by the gradient loss and thus be able to accurately represent the complex associations of the input objects. That results in high classification accuracy of the proposed model.

This paper constructs a classification model using ResNet, which takes its input from the selected representative features by association rules. Other models like NN, RF, KNN, and SVM are used to compare results.

4.3 Datasets

All the datasets used in this research paper are obtained from the UCI Machine Learning Repository [22]. The models were trained and tested on two datasets. The first dataset is the Dermatology dataset, which is used to predict the type of the erythemato-squamous disease that consists of 6 classes. The dataset has 366 instances. The total number of features that describe each instance is 34. These features are divided into clinical features and pathological features. Most of these features have three values.

The second dataset is the Breast Cancer Wisconsin (Diagnostic) dataset [22]. This dataset is another version of the original Breast cancer dataset, which contains only 9 categorical attributes. The Diagnostic version of the dataset contains 31 numerical attributes derived from 10 main attributes measured using images from the breast mass. Breast Cancer dataset was collected at the Hospitals of the Wisconsin–Madison university. The main purpose of the dataset is to diagnose whether the tumor is malignant (class 4) or benign (class 2) from the measured attributes, which can provide early detection of breast cancer.

5 Experiments and Results

All the experiments were conducted using Python via the Google Collaboratory environment.

To evaluate the proposed classification model on the breast cancer and the dermatology datasets, three-Fold Cross-Validation was used. The dataset is split into 3 partitions. In each iteration, one of the partitions is used as a testing dataset and the remaining two parts are used for training the model. This allows all observations to contribute to the evaluation of the classifiers. The three-Fold Cross-Validation provides better assessment than a simple train-test-split.

The results of the experiments confirm the feasibility of the proposed classification model that includes selecting a reduced set of representative features through using association rules and then the reduced set of features are fed into a deep learning classifier, such as ResNet.

5.1 AR1

After finding the frequent itemset using the Apriori algorithm on the independent data, it was found that some attributes, such as area_se (i.e., 5), in the breast cancer dataset is highly correlated with other attributes and thus they can be removed. Similarly, some attributes in the dermatology dataset were removed either due to being highly correlated with other attributes, or not related. For example, perifollicular_parakeratosis (i.e., 27) is an unrelated attribute found by AR1 and thus removed.

Table 1. Attributes from AR2 for breast cancer dataset.

Class	AR2 Attributes
Malignant	6, 12, 13, 14, 15, 16, 17, 18, 20, 21
Benign	2, 4, 5, 7, 8, 9, 12, 13, 14, 15, 16, 17, 18, 19, 20, 21, 22, 24, 25, 27, 28, 30, 31
Whole attributes (24)	2, 4, 5, 6, 7, 8, 9, 12, 13, 14, 15, 16, 17, 18, 19, 20, 21, 22, 24, 25, 27, 28, 30, 31

5.2 AR2

The frequent attributes from the AR2 experiments are depicted in Tables 1 and 2.

Table 2. Attributes from AR2 for dermatology dataset.

Class	AR2 Attributes
1	6, 7, 8, 12, 13, 15, 25, 27, 28, 29, 30, 31, 33
2	5, 6, 7, 8, 9, 10, 11, 12, 15, 20, 22, 24, 25, 26, 27, 29, 30, 31, 33
3	7, 9, 10, 11, 14, 15, 20, 21, 22, 23, 24, 30, 31
4	6, 7, 8, 9, 10, 11, 12, 13, 15, 20, 21, 22, 23, 24, 25, 27, 29, 30, 31, 33
5	5, 6, 8, 9, 10, 11, 12, 13, 14, 20, 22, 23, 24, 25, 26, 27, 29, 30, 31, 33
6	5, 6, 8, 12, 13, 15, 20, 21, 22, 23, 24, 25, 26, 27, 29, 33, 34
Whole attributes (25)	5, 6, 7, 8, 9, 10, 11, 12, 13, 14, 15, 20, 21, 22, 23, 24, 25, 26, 27, 28, 29, 30, 31, 33, 34

5.3 Classifier Implementation and Results

ResNet was created using pytorch wide-deep library, RFC, KNN, and SVM was built using Scikit-learn python library, and NN was constructed using TensorFlow Keras. The NN has a basic architecture that consists of three layers (input, hidden, and output), uses a Relu activation function, softmax output layer, and Adam optimizer. The NN was trained for 200 epochs.

5.4 Results of Breast Cancer and Dermatology Datasets

The classification accuracies of the Breast Cancer and Dermatology datasets shown in Tables 3 and 4 are the average of 20 runs of the 3-fold cross-validation. After each run, the mean of the 3-Fold accuracies was recorded. Then, at the end of the 20 runs, the average and standard deviation of all means were recorded for each classifier. Table 3 and 4 summarize the results for the dermatology and the breast cancer datasets, respectively. In the dermatology dataset, the use of ARs was effective and enhanced the performance of all the classifiers except for the SVM. On the contrary, some of the classifiers did not favor using feature selection with the breast cancer dataset.

Table 3. Results of dermatology dataset.

Classifier	With AR1 (33 features)	With AR2 (25 features)	With AR1 and AR2 (24 features)	Without ARs (34 features)
NN	97.786 (+/− 0.333)	97.101 (+/− 0.3713)	97.087 (+/− 0.323)	97.109 (+/− 0.365)
RFC	97.540 (+/− 0.386)	97.109 (+/− 0.413)	97.072 (+/− 0.616)	97.172 (+/− 0.344)
KNN	82.240 (+/− 2.842)	83.606 (+/− 1.421)	83.606 (+/− 1.421)	81.967 (+/− 1.421)
SVM	68.852 (+/− 1.421)	65.846 (+/− 0.0)	65.846 (+/− 0.0)	69.125 (+/− 2.842)
ResNet	96.448 (+/− 0.634)	97.349 (+/− 0.335)	97.445 (+/− 0.618)	96.366 (+/− 0.54)

Table 4. Results of breast cancer diagnosis dataset.

Classifier	With AR1 (30 features)	With AR2 (24 features)	With AR1 and AR2 (23 features)	Without ARs (31 features)
NN	91.917 (+/− 3.286)	91.784 (+/− 1.477)	89.499 (+/− 3.914)	92.056 (+/− 1.261)
RFC	96.249 (+/− 0.295)	94.221 (+/− 0.367)	94.290 (+/− 0.258)	94.214 (+/− 0.257)
KNN	93.145 (+/− 1.421)	92.794 (+/− 2.842)	93.321 (+/− 0.0)	92.969 (+/− 1.421)
SVM	91.917 (+/− 0.0)	92.093 (+/− 1.421)	91.917 (+/− 0.0)	91.093 (+/− 1.421)
ResNet	95.334 (+/− 0.779)	94.622 (+/− 0.49)	95.824 (+/− 1.00)	95.624 (+/− 0.676)

6 Conclusion

In this paper, we proposed a classification model that leverages association rules to select a reduced set of representative features, which are fed into ResNet, a deep learning classifier. We compared the proposed classifier model with four other classifiers. We applied two types of association rule mining approaches, AR1 and AR2, on two datasets to reduce the number of features. These classifiers are trained using different numbers of attributes. (i.e., reduced with AR1, reduced with AR2, reduced with AR1 and AR2, and without reduction). The results were reported after finding the mean and standard deviation of the accuracies from 20 runs of the experiments. The results varied from one dataset to another and from one classifier to another. Overall, the dermatology dataset was in favor of using feature selection by ARs, and the results improved after the reduction. These experiments show that the proposed model achieves high accuracy as compared to other classifiers.

References

1. Elzobi, M., Al-Hamadi, A., Al Aghbari, Z., Dings, L., Saeed, A.: Gabor wavelet recognition approach for off-line handwritten Arabic using explicit segmentation. In: Choras, R.S., (eds.) Image Processing and Communications Challenges 5. Advances in Intelligent Systems and Computing, vol. 233, pp. 245–254. Springer, Heidelberg (2014). https://doi.org/10.1007/978-3-319-01622-1_29
2. Aghbari, Z., Makinouchi, A.: Semantic approach to image database classification and retrieval. NII J. **7**(9), 1–8 (2003)
3. Aghari, Z., Kaneko, K., Makinouchi, A.: Modeling and querying videos by content trajectories. In: 2000 IEEE International Conference on Multimedia and Expo. ICME2000. Proceedings. Latest Advances in the Fast Changing World of Multimedia, vol. 1, pp. 463–466. (Cat. No. 00TH8532). IEEE (2000)
4. Aghari, Z., Kaneko, K., Makinouchi, A.: Modeling and querying videos by content trajectories. In: 2000 IEEE International Conference on Multimedia and Expo. ICME2000. Proceedings. Latest Advances in the Fast Changing World of Multimedia, vol. 1, pp. 463–466. (Cat. No. 00TH8532). IEEE (2000)
5. Dinges, L., Al-Hamadi, A., Elzobi, M., Al Aghbari, Z., Mustafa, H.: Offline automatic segmentation based recognition of handwritten Arabic words. Int. J. Sig. Process. Image Process. Pattern Recogn. **4**(4), 131–143 (2011)
6. Karabatak, M., Ince, M.: An expert system for detection of breast cancer based on association rules and neural network. Elsevier (2009)

7. Al Aghbari, Z., Kamel, I., Awad, T.: On clustering large number of data streams. Intell. Data Anal. **16**(1), 69–91 (2012)
8. Abu Safia, A., Al Aghbari, Z., Kamel, I.: Phenomena detection in mobile wireless sensor networks. J. Netw. Syst. Manage. **24**(1), 92–115 (2016)
9. Al Aghbari, Z., Kamel, I., Elbaroni, W.: Energy-efficient distributed wireless sensor network scheme for cluster detection. Int. J. Parallel Emergent Distrib. Syst. **28**(1), 1–28 (2013)
10. Agrawal, R., Imieliński, T., Swami, A.: Mining association rules between sets of items in large databases, pp. 207–216 (1993). https://doi.org/10.1145/170035.170072
11. Palanisamy, S.: Association rule based classification (2006). Accessed 20 Oct 2021. https://digital.wpi.edu/downloads/2v23vt44r
12. Abraham, A.: Artificial neural networks. Handbook of measuring system design (2005)
13. He, K., Zhang, X., Ren, S., Sun, J.: Deep residual learning for image recognition. In: Proceedings of the IEEE Conference on Computer Vision and Pattern Recognition, pp. 770–778 (2016)
14. Vougas, K., et al.: Deep learning and association rule mining for predicting drug response in cancer. A personalised medicine approach, p. 070490. BioRxiv (2017)
15. Boutorh, A., Guessoum, A.: Classication of SNPs for breast cancer diagnosis using neural-network-based association rules. In: 2015 12th International Symposium on Programming and Systems (ISPS), pp. 1–9. IEEE (2015)
16. Inan, O., Uzer, M.S., Yılmaz, N.: A new hybrid feature selection method based on association rules and PCA for detection of breast cancer. Int. J. Innov. Comput. Inf. Control **9**(2), 727–729 (2013)
17. Thilina, A., et al.: Intruder detection using deep learning and association rule mining. In: 2016 IEEE International Conference on Computer and Information Technology (CIT), pp. 615–620. IEEE (2016)
18. Yuan, Z., Lu, Y., Xue, Y.: DroidDetector: android malware characterization and detection using deep learning. Tsinghua Sci. Technol. **21**(1), 114–123 (2016)
19. Eom, J.-H., Zhang, B.-T.: Prediction of protein interaction with neural network-based feature association rule mining. In: King, I., Wang, J., Chan, L.-W., Wang, D.L. (eds.) ICONIP 2006. LNCS, vol. 4234, pp. 30–39. Springer, Heidelberg (2006). https://doi.org/10.1007/11893295_4
20. Montaez, C.A.C., Fergus, P., Montaez, A.C., Hussain, A., Al-Jumeily, D., Chalmers, C.: Deep learning classification of polygenic obesity using genome wide association study SNPs. In: 2018 International Joint Conference on Neural Networks, pp. 1–8 (IJCNN). IEEE (2018)
21. Zhang, Z., He, Q., Gao, J., Ni, M.: A deep learning approach for detecting traffic accidents from social media data. Transp. Res. Part C: Emer. Technol. **86**, 580–596 (2018)
22. Dua, D., Graff, C.: UCI machine learning repository. Irvine, CA: University of California, School of Information and Computer Science (2019). http://archive.ics.uci.edu/ml (Accessed 16 Dec 2021)
23. Qu, Y., Fang, Y., Yan, F.: Feature selection algorithm based on association rules. J. Phys. Conf. Ser. **1168**(5), 052012 (2019). IOP Publishing (2019)

Global Thresholding Technique for Basal Ganglia Segmentation from Positron Emission Tomography Images

Zainab Maalej(✉), Fahmi Ben Rejab, and Kaouther Nouira

Institut Supérieur de Gestion de Tunis, Université de Tunis, Tunis, Tunisia
zainab.maalej@isg.rnu.tn

Abstract. The basal ganglia is a small brain structure in the brainstem that plays a crucial role in the pathogenesis of Parkinson's Disease (PD). It processes the different signals coming from the cortex, enabling the right execution of voluntary movements. The decrease of dopaminergic neurons of the substantia nigra provokes multiple changes affecting the whole basal ganglia network. This leads systematically to patients disability and affects their quality of life progressively. Positron Emission Tomography (PET) is an ideal tool to detect the amount of changes in the brain as it produces detailed quantitative information of the PD progression. In this regard, Artificial Intelligence (AI) proved notable and fundamental changes in the way we detect and treat abnormalities in medical images. Image segmentation techniques, part of AI, are significant in detecting changes in medical images. Thus, accurate PET image segmentation is necessary for follow-ups and treatment planning to enhance the health status of different patients. For this reason, in this paper, we aim to improve the detection of PD progression by using intelligent technique such as global thresholding segmentation. This technique has been tested onto 110 different PET images and evaluated with the corresponding ground truth which were segmented manually. We tested three multiple threshold values and evaluated the segmentation performance in each case using Dice Similarity Coefficient (DSC) and mean Intersection over Union (mIoU) metrics. The results obtained indicate that global thresholding technique have reached higher performance using 150 as threshold with 0.7701 of DSC and 0.6394 of mIoU.

Keywords: Parkinson's Disease · Segmentation · Global thresholding · PET

1 Introduction

Parkinson's Disease (PD) is a chronic condition affecting the central nervous system. It is the second most common neurodegenerative disease after Alzheimer's

Supported by Université de Tunis, Institut Supérieur de Gestion de Tunis (ISGT), BESTMOD Laboratory.

K. K. Patel et al. (Eds.): icSoftComp 2022, CCIS 1788, pp. 84–95, 2023.
https://doi.org/10.1007/978-3-031-27609-5_7

Disease and affects 8.5 million individuals globally as of 2017 [7]. The onset of this disease is caused by a progressive loss of dopaminergic neurons in the substantia nigra pars compacta located in the basal ganglia region of the brain [10]. The primary role of these neurons is to provide dopamine which is a biochemical molecule enabling coordination of movement. When the amount of dopamine produced in this region decreases, communication between neurons in order to coordinate body movement is blocked. This affects systematically the acceleration and velocity of movements and leads to shaking, difficulties in walking, speaking, writing, balance and many other simple tasks. These debilitating symptoms get worse over time and affect adversely the quality of life. PD is generally diagnosed with the manifestation of these motor symptoms. They could not appear until the patient has lost about from 50% to 70% of their neurons [1]. Although there are multiple treatments as deep brain stimulation and dopamine-related medication, the patient still suffers a progressive increase in symptom severity over time. Hence, early detection as well as evolution detection might play a crucial role for disease management and treatment. The diagnosis of this disease is usually supported by analyzing the structural changes in brain through neuroimaging techniques of different modalities. In particular, Positron Emission Tomography (PET) has proved its success in detecting the dopaminergic dysfunction in PD. Furthermore, it can monitor PD progression as reflected by modifications in brain levedopa and glucose metabolism as well as dopamine transporter binding [15].

PET image is a nuclear imaging technique that allows in vivo estimations of several physiological parameters especially neuroreceptor binding that enables a deep understanding of PD pathophysiology [8]. Generally, using Flourodopa F18 (F-Dopa) for PET which is a radioactive tracer can help visualizing the nerve endings of all dopaminergic neurons. When injected, F-Dopa will be absorbed by these neurons located in basal ganglia. However, when these neurons become damaged or die, a decline in the striatal F-Dopa uptake will occur [15]. It has recently been proved that F-Dopa PET can provide a reliable biological marker and accurate means for monitoring the PD progression. As the PD progresses over time, more dopaminergic neurons become damaged, thus less F-Dopa uptake which causes a volume decrease of basal ganglia visualized in PET images. In other word, when we observe that the size of basal ganglia region has decreased over time, we approve that the PD has been evoluted. However, human interpretation of this region in PET images can demonstrate inconsistency and interobserver variability due to the complexity of anatomy, the variation of the region of interest representation and the ability difference of the physicians.

In this respect, Artificial Intelligence (AI) technologies are progressively prevalent in society and are becoming to be applied to healthcare [2]. They have the ability to enhance the decision making performance of several medical tasks such as patient care, administrative procedures, improving communication between physicians and patients and transcribing medical documents. Multiple research studies have already proved that AI can sometimes outperform humans at key healthcare decision making tasks [2]. One of the medical practice domains in which AI is intended to have fundamental impact is the diagnostic analysis of several medical images [13]. Its application in this domain provides a second opinion that assists physicians in the detection of abnormalities and measuring disease progress more efficiently. Automatic segmentation techniques which are considered as part of AI have a tremendous effect in the interpretation of medical images. Hence, segmenting basal ganglia region in PET images can enhance the performance of detecting the PD progression by improving the physicians' capabilities and decreasing the time required for accurate analysis. Among these techniques, global thresholding (binary thresholding) which is an intuitive and popular segmentation technique can be applied in this case. Its primary objective is to convert a grayscale image into a binary one through defining all pixels greater than a specific value to be foreground and all remaining pixels as background [18].

Along these lines, the main goal of the present study is to integrate the relevance of AI technologies in digital image processing. In other words, we focus on applying global thresholding technique for basal ganglia segmentation using PET images and evaluate the performance of this technique based on some evaluation metrics. For comparison purposes, as there is no previous work that applies other segmentation techniques and focuses on the same dataset used in this paper, we tested another global thresholding technique called Otsu's thresholding [12] for the same dataset and compare the accuracy between these two techniques. In particular, we aim at: (i) collecting data from the database; (ii) constructing ground truth using manual segmentation for further evaluating the performance of segmentation for proper comparison ; (iii) applying thresholding technique for PET images segmentation; (iv) evaluating the performance of thresholding in basal ganglia segmentation.

The paper is structured as follows. Section 2 describes the related work of using thresholding for image segmentation. Section 3 illustrates an overview of our proposed approach. Section 4 presents and analyzes the experimental results. Concluding remarks are given in Sect. 5.

2 Related Works

Image segmentation is a central point of image processing. It is considered as a technique of separating an image into sub-component parts in order to extract some useful information. Medical images take a crucial part in health care because of its high impact and influence in the diagnosis as well as treatment. Segmenting the important segments of those images can undoubtedly enhance their analysis. In this context, several segmentation techniques have been proposed in the literature. The choice of a particular technique over others depends especially on the type and nature of the image. PET images provide detailed information about many diseases. It is able to detect emitted photons through radiotracer in abnormal cells in order to evaluate cancer, inflammation and infection. In this context, image segmentation in PET images is crucial to distinguish abnormal tissue from other surrounding areas. The literature has shown a large number of articles applying several segmentation techniques in PET images for multiple diseases. In particular, [24] proposed a novel automated pulmonary parenchyma segmentation method which is based on region growing algorithm using PET-CT (Computed Tomography) imaging. First, they performed image binarization and regions of interest were extracted. Then, the segmentation of lung was applied using region growing. After that, they removed noise for the surrounding of the region and used expansion operations for smoothing the lung boundary. Although its high performance, region growing method is by nature sequential and pretty expensive in either computational time and memory [14]. On the other hand, edge detection approaches have been also used by many researchers. Several methods for edge-based segmentation can be applied such as Canny, Sobel, Prewitt and Roberts. For instance, [5] performed Canny methods to segment liver PET images. Recently, [21] used sobel operator to extract the contour of the regions from PET images in order to classify pulmonary nodules. Despite the relevance of obtained results, edge detection methods are les immune to noise than other techniques [14].

Nevertheless, global thresholding may overcome the disadvantages of other techniques mentioned above as it does not require prior information of the input image and can work well with low computation complexity [14]. It has been widely used by many researchers in health care using different types of image. In [11], an automatic segmentation of pectoral muscles inmammogram images was proposed using global thresholding. The proposal achieved 92.86% segmentation accuracy. In [22], global thresholding was performed based on C-Y image as input while color thresholding was performed based on RGB image in order to segment Ziehl-Neelsen Tuberculosis bacilli slide images in sputum samples. Authors conducted a comparison study and demonstrated that global thresholding provides better results compared to the second technique. This is explained by the fact that global thresholding is capable of reducing more noise and sputum background from images. More specifically, notable studies applied global thresholding technique in PET images for segmenting lesions through various body regions in multiple diseases. We can mention, [23] that focused on non-small cell lung cancer, [4] on Sphere phantoms and [19] on larynx and oral cavity. Nevertheless, to the best of our knowledge, the application of this technique in PD using PET images didn't get much attention especially in segmenting basal ganglia region.

Therefore, as global thresholding segmentation technique showed high performance using many types of image for the diagnosis of numerous diseases, we aim mainly to apply this technique to test its accuracy in segmenting basal ganglia region from PET images in order to diagnose PD.

3 Proposed Approach

As demonstrated in Fig. 1, our proposed approach is composed of four fundamental components: preparing dataset, ground truth construction, thresholding segmentation and performance evaluation. Bellow, we describe the goal of each component in details.

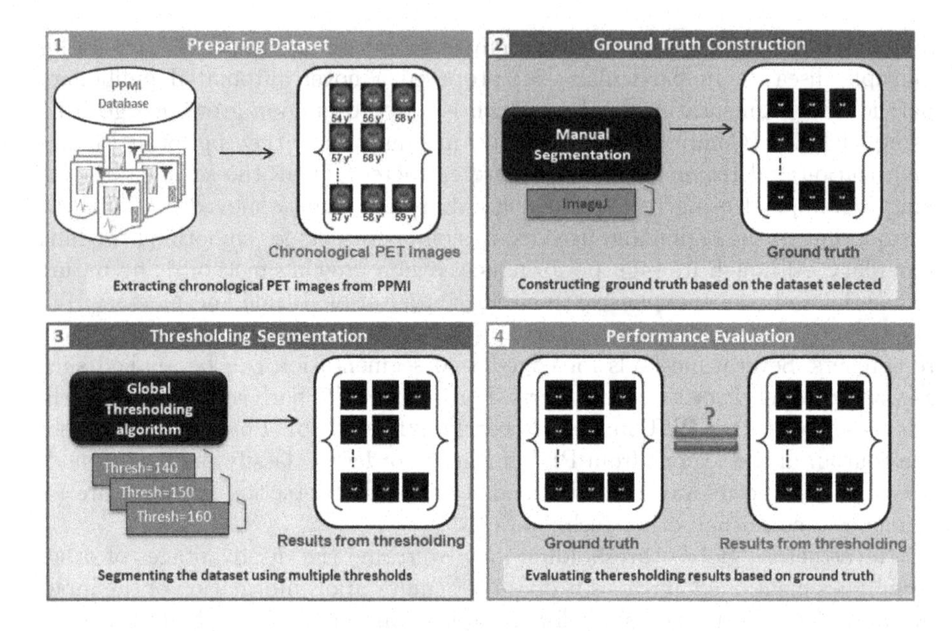

Fig. 1. Proposed approach for basal ganglia segmentation from PET images using global thresholding

3.1 Preparing Dataset

PET images used in the preparation of this paper were obtained from the Parkinson's Progression Markers Initiative (PPMI) [9] database. PPMI is an international, observational and large-scale clinical study. It is a collaborative effort of PD researchers which are expertise in PD study design and implementation, biomarker development and also in data management in order to confirm PD progression biomarkers. This database is composed of study data which are values describing many symptoms such as tremors, medical history, motor and non-motor symptoms, and also imaging data as MRI (Magnetic resonance imaging), PET and CT. PET were performed at the screening visit. Since there is a difference in all PET imaging systems in different centers, these images acquired a pre-processing step before publically sharing in the database site.

As we aim to study the progression of PD patients, we were based on only patients who have more than one PET image. Hence, a total of N = 110 chronological PET images were selected from PPMI (89 men, 21 women, age range 33–76). The size of all input images is 2048 $\times$ 2048 while the intensity varies from an image to another. The average value of the intensity is $\simeq$ 20.

3.2 Ground Truth Construction

The second step of our proposal consists in manually segmenting PET images for comparing them with thresholding algorithm-generated segmentations in terms of boundary or overlap differences. It is the state-of-the-art segmentation method to evaluate medical image segmentations issues. A wide range of software is available. For instance, ImageJ is an open software platform that had a large impact on the life sciences [20]. It continues to attract not only biologists but also computer scientists who desire to implement specific image processing algorithms. Therefore, focusing on the dataset prepared in the first step of our proposal, we aim to use ImageJ to manually segmenting PET images for further evaluation. We apply background with black color and the region of interest segmented in white. All segmented images are saved for later use in comparison with the global thresholding results.

3.3 Thresholding Segmentation

Global thresholding is an intuitive, simple and common segmentation technique. It converts a grayscale image into a binary image through assigning all pixels greater than a specific threshold value τ to be foreground and all other pixels as background [18]. The segmentation of an image $I(i,j)$ is defined as:

$$J(i,j) = \begin{cases} 1 \text{ if } I(i,j) > \tau \\ 0 \text{ if } I(i,j) \leq \tau \end{cases} \tag{1}$$

where $J(i,j)$ is the image resulted from the segmentation, pixels labeled with 1 refer to the region segmented and pixels labeled with 0 refer to background.

Generally, there is no consensus on the selection of a threshold value. Hence, an optimal threshold value determination remains a challenging task in thresholding technique. Several different methods exist to choose threshold such as the use of histogram in order to visualize pixel intensities and select the valley point as the exact value. Nevertheless, this method may be computationally expensive and may not deliver clearly defined valley points which allows a difficulty in the selection of an accurate threshold [17]. Accordingly, we aim to identify the threshold value by selecting multiple seed points of the region to be segmented (top, bottom and center of the region). For each seed point, we determine its intensity value. We found that all the intensity values vary between 140 and 160. Therefore, in order to get high accuracy in the segmentation process, the threshold value must belong to this range. The selection of threshold lower than 140 or greater than 160 will undoubtedly decrease the segmentation accuracy. Thus, we have chosen to test three different threshold values between this range which are 140, 150 and 160. In the next step, we applied global thresholding technique for all the images saved in the dataset previously using the three threshold values already identified. All the steps are summarized in Algorithm 1.

On the other side, Otsu's segmentation technique has also been selected as an effective thresholding technique. Unlike global thresholding (binary) that requires a selection of threshold value, Otsu's technique can automatically identify the optimal threshold value. It holds the variance between clusters as a criterion for the selection of optimal threshold. In this respect, to prove the performance of our proposal, global thresholding results will be compared to Otsu's results.

We aim to test these two techniques using the same dataset and compare their performance.

Algorithm 1. Global thresholding algorithm steps

Input: Original images from PPMI database
Output: Binary segmented images

1: Select multiple seed point (top, bottom and center of the region)
2: Find the intensity value of the selected pixels
3: $S =$ list of the selected threshold values
4: **while** $S \neq \emptyset$ **do**
5: Read Input
6: **for** each image $i \in Input$ **do**
7: Read i as grayscale
8: Thresholding i with $S_j \in S$
9: Save the resulted image in a previously created output file
10: **end for**
11: **end while**

3.4 Performance Evaluation

We evaluated the segmentation results using two different metrics: Dice similarity coefficient (DSC) [3] and mean Intersection over Union (mIoU), also called Jaccard Index [6] as they are among the most popular utilized metrics in medical image segmentation [16]. The main difference between them is that the mIoU penalizes both under and over segmentation more than DSC metric. Despite the relevance of these metrics, DSC is more used and frequent in the majority of scientific publications for medical image segmentation evaluation [16].

Dice Similarity Coefficient: It is one of the most broadly used quantitative metrics for segmentation evaluation accuracy. It measures the similarity between the ground truth (manual segmentation) and the resulted segmentation image (automatic segmentation). Given that the segmented volume is denoted by Sv_1 and the ground truth is indicated as Sv_2, the DSC is calculated as follow:

$$DSC(Sv_1, Sv_2) = 2\frac{|Sv_1 \cap Sv_2|}{|Sv_1| + |Sv_2|} \tag{2}$$

where the overlap of two volumes Sv_1 and Sv_2 indicates the sensitivity (True Positive Volume Fraction). The amount of false positive volume segmentation is calculated in the False Positive Volume Fraction (which indicates the specificity). The value of DSC may vary between 0.0 that means no overlap between the two shapes and 1.0 which indicates a perfect overlap. Larger values match with better spatial agreement between manually and automatically shapes.

Mean Intersection Over Union: It is a highly intuitive evaluation metric for measuring the accuracy of segmentation method. It gives the similarity between the predicted zone and the ground truth zone of an image. Given that the segmented zone is denoted by A and the ground truth is indicated as B, the mIoU is calculated as follow:

$$mIoU(A,B) = \frac{|A \cap B|}{|A \cup B|} \tag{3}$$

where the size of the intersection between A and B is divided by the union of two different regions. Similarly to DSC, the value of mIoU may vary between 0.0 and 1.0. Larger values confirm a better spatial agreement between manually and automatically segmented regions.

4 Experiments

We used the Python programming language (version 3.7.5) for the application of global thresholding algorithm.

4.1 Results and Discussion

We applied global thresholding with 110 different PET images. It converts these images which are grayscale imagery into binary imagery. Information contained has only two different values: black and white. We tested three different threshold values: 140, 150 and 160 separately. Then, we evaluated their performance using DSC and mIoU metrics in order to identify the most accurate threshold value. Figure 2 shows an example of an original PET image from PPMI, its ground truth and the corresponding segmented image. Images in the top row correspond to the original PET image from PPMI (A) and its ground truth using ImageJ (B). The bottom row (C), (D) and (E) match with segmentation using 140, 150 and 160 respectively. We notice that segmented images through thresholding are very comparable to the ground truth image. Table 1 shows the average DSC and mIoU over 110 PET images for each selected threshold value. It can be seen that global thresholding segmentation technique using PET images was highly accurate for all threshold values tested as all the performance values are greater than 0.6 for DSC as well as mIoU. Furthermore, the high performance is detected when using 150 as threshold value with DSC equal to 0.7701 and mIoU of 0.6394.

Furthermore, to the best of our knowledge, there is no previous work in literature that applies other segmentation techniques with the same dataset used in order to validate the accuracy of our proposal. In fact, Otsu's technique proved notable performance for segmentation tasks and may outperforms global thresholding as it can determine automatically a threshold value from the input images. To confirm or deny this assumption, we tested the segmentation performance with the same dataset using Otsu's thresholding technique. Table 2 shows the comparison performance with the high accuracy selected in global thresholding (threshold = 150) and Otsu's technique. It provided inaccurate performance since the average of DSC and mIoU were 0.00648 and 0.0337 respectively. To this end, comparing the performance of both techniques applied, we validate that the best accuracy is depicted in global thresholding. This can be explained by the fact that the automatic selection of threshold value by Otsu's technique does not provide the intensity value of basal ganglia region which is not the case for our proposal that searches first of all of the intensity value of this region and then applies global thresholding with an exact threshold value. This can conduct to a segmentation of only the region of interest whereas Otsu's segmentation leads to multiple regions segmentation other than basal ganglia zone in PET images.

Hence, our proposal using global thresholding proved promising results for basal ganglia segmentation. The output images are ready now to be used for calculating the size of basal ganglia segmented and verify if its size has decreased over time for each patient to finally deduce the progression of PD. This work proved that global thresholding is a relevant segmentation technique for PET images and is sufficient to obtain a satisfying diagnostic accuracy for PD progression.

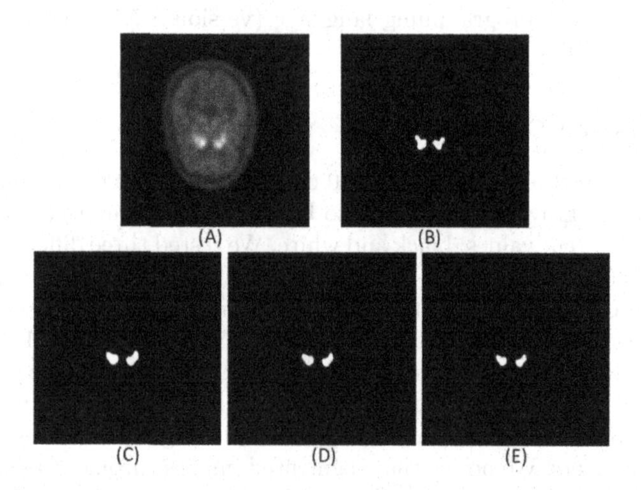

Fig. 2. Example of PET image segmentation using global thresholding

Table 1. Performance evaluation of global thresholding

Metrics	Threshold = 140	Threshold = 150	Threshold = 160
DSC	0.7437	0.7701	0.7666
mIoU	0.6135	0.6394	0.6308

Table 2. Performance evaluation comparison of global thresholding with Otsu's thresholding

Metrics	Methods	
	DSC	mIoU
Global thresholding (150)	0.7701	0.6394
Otsu's thresholding	0.0648	0.0337

4.2 Limits and Future Work

Despite the fact that our proposal which focuses on global thresholding for PET images achieved notable performance, this study has several limits. First, selecting the appropriate threshold value using multiple seed points may not be the effective method. Other threshold value could provide better performance than what we found. Second, global thresholding does not take into account the spatial details. Hence, it cannot guarantee that the segmented areas are contiguous. Third, the construction of ground truth should be conducted by experts in order to properly segment the image and minimize the error.

Therefore, in the near future, we aim to overcome these drawbacks using deep learning model in order to segment the basal ganglia region in PET images. This can obviously provide higher performance as deep learning implies less manual feature engineering, so we no longer required to identify a threshold value. Moreover, it can be trained effectively with large and complex datasets without time consuming.

5 Conclusion

In this paper, we focused on the advances of AI technologies in medical image processing in order to improve the decision making in detecting image abnormalities. Specifically, we aimed to develop a new proposal applying digital image processing including global thresholding to segment basal ganglia regions from PET images from PD patients through PPMI database. As we focus essentially on the detection of PD progression in multiple patients, we were based on only patients who have more than one PET image. Accordingly, we identified 110 chronological PET images from PPMI. These images were then segmented manually in order to construct the ground truth dataset for further evaluation. After that, we applied global thresholding focusing on the original chronological PET images to segment the basal ganglia region. We tested the performance of this technique using three different threshold values: 140, 150 and 160. This proposal provides significant performance for all threshold values. The best average DSC and mIoU over 110 subjects were 0.7701 and 0.6394 respectively. In addition, our proposal using global thresholding proved notable accuracy comparing to other global thresholding technique tested which is Otsu's segmentation. As a result, we can admit that global thresholding yields high accuracy in segmenting basal ganglia in PET images. The output images can be used now to calculate the size of the region segmented for detecting the PD evolution. Although this achieveness, we aim to reach higher accuracy for PET segmentation using deep learning model for the same dataset.

References

1. Cheng, H.C., Ulane, C.M., Burke, R.E.: Clinical progression in Parkinson disease and the neurobiology of axons. Ann. Neurol. **67**(6), 715–725 (2010)
2. Davenport, T., Kalakota, R.: The potential for artificial intelligence in healthcare. Future Healthcare J. **6**(2), 94 (2019)
3. Dice, L.R.: Measures of the amount of ecologic association between species. Ecology **26**(3), 297–302 (1945)
4. Hatt, M., Le Rest, C.C., Albarghach, N., Pradier, O., Visvikis, D.: Pet functional volume delineation: a robustness and repeatability study. Eur. J. Nucl. Med. Mol. Imag. **38**(4), 663–672 (2011)
5. Hsu, C.Y., Liu, C.Y., Chen, C.M.: Automatic segmentation of liver pet images. Comput. Med. Imaging Graph. **32**(7), 601–610 (2008)
6. Jaccard, P.: The distribution of the flora in the alpine zone. 1. New Phytol. **11**(2), 37–50 (1912)
7. James, S.L., et al.: Global, regional, and national incidence, prevalence, and years lived with disability for 354 diseases and injuries for 195 countries and territories, 1990–2017: a systematic analysis for the global burden of disease study 2017. Lancet **392**(10159), 1789–1858 (2018)
8. Loane, C., Politis, M.: Positron emission tomography neuroimaging in Parkinson's disease. Am. J. Transl. Res. **3**(4), 323 (2011)
9. Marek, K., et al.: The Parkinson progression marker initiative (PPMI). Prog. Neurobiol. **95**(4), 629–635 (2011)
10. Mostafa, T.A., Cheng, I.: Parkinson's disease detection using ensemble architecture from MR images. In: 2020 IEEE 20th International Conference on Bioinformatics and Bioengineering (BIBE), pp. 987–992. IEEE (2020)
11. Naz, S.I., Shah, M., Bhuiyan, M.I.H.: Automatic segmentation of pectoral muscle in mammogram images using global thresholding and weak boundary approximation. In: 2017 IEEE International WIE Conference on Electrical and Computer Engineering (WIECON-ECE), pp. 199–202. IEEE (2017)
12. Otsu, N.: A threshold selection method from gray-level histograms. IEEE Trans. Syst. Man Cybern. **9**(1), 62–66 (1979)
13. Park, S.H., Han, K.: Methodologic guide for evaluating clinical performance and effect of artificial intelligence technology for medical diagnosis and prediction. Radiology **286**(3), 800–809 (2018)
14. Patil, D.D., Deore, S.G.: Medical image segmentation: a review. Int. J. Comput. Sci. Mob. Comput. **2**(1), 22–27 (2013)
15. Pavese, N., Brooks, D.J.: Imaging neurodegeneration in Parkinson's disease. Biochim. Biophys. Acta. (BBA)-Mol. Basis Disease **1792**(7), 722–729 (2009)
16. Popovic, A., De la Fuente, M., Engelhardt, M., Radermacher, K.: Statistical validation metric for accuracy assessment in medical image segmentation. Int. J. Comput. Assist. Radiol. Surg. **2**(3), 169–181 (2007)
17. Raju, P.D.R., Neelima, G.: Image segmentation by using histogram thresholding. Int. J. Comput. Sci. Eng. Technol. **2**(1), 776–779 (2012)

18. Saha, P.K., Udupa, J.K.: Optimum image thresholding via class uncertainty and region homogeneity. IEEE Trans. Pattern Anal. Mach. Intell. **23**(7), 689–706 (2001)
19. Schinagl, D.A., Vogel, W.V., Hoffmann, A.L., Van Dalen, J.A., Oyen, W.J., Kaanders, J.H.: Comparison of five segmentation tools for 18f-fluoro-deoxy-glucose-positron emission tomography-based target volume definition in head and neck cancer. Int. J. Radiat. Oncol. Biolo. Physi. **69**(4), 1282–1289 (2007)
20. Schindelin, J., Rueden, C.T., Hiner, M.C., Eliceiri, K.W.: The imageJ ecosystem: an open platform for biomedical image analysis. Mol. Reprod. Dev. **82**(7–8), 518–529 (2015)
21. Teramoto, A., et al.: Automated classification of pulmonary nodules through a retrospective analysis of conventional CT and two-phase pet images in patients undergoing biopsy. Asia Ocean. J. Nucl. Med. Biol. **7**(1), 29 (2019)
22. Wahidah, M.N., Mustafa, N., Mashor, M., Noor, S.: Comparison of color thresholding and global thresholding for ziehl-Neelsen TB bacilli slide images in sputum samples. In: 2015 2nd International Conference on Biomedical Engineering (ICoBE), pp. 1–6. IEEE (2015)
23. Wanet, M., et al.: Gradient-based delineation of the primary GTV on FDG-pet in non-small cell lung cancer: a comparison with threshold-based approaches, CT and surgical specimens. Radiother. Oncol. **98**(1), 117–125 (2011)
24. Zhao, J., Ji, G., Han, X., Qiang, Y., Liao, X.: An automated pulmonary parenchyma segmentation method based on an improved region growing algorithmin pet-CT imaging. Front. Comp. Sci. **10**(1), 189–200 (2016)

Oversampling Methods to Handle the Class Imbalance Problem: A Review

Harsh Sharma[1] and Anushika Gosain[2(✉)]

[1] Amazon.com Services LLC, Austin, USA
[2] University of Texas at Dallas, Austin, Texas, USA
Anushika.Gosain@UTDallas.edu

Abstract. Many real-world applications, such as medical diagnostics, fraud detection etc., have a class imbalance problem where one class has less instances than the other. The non-uniform distribution of the dataset has a significant impact on the classification models' performance because they fail to detect a minority class instance. In this paper, we empirically reviewed five oversampling methods to address the class imbalance problem (CIP), including SMOTE (Synthetic Minority Oversampling Technique), Safe level SMOTE, SMOTE Tomek Links, Borderline SMOTE1 and Adaptive SMOTE (ADASYN), using four classification models: Decision Tree (DT), Support Vector Machine (SVM), Random Forest (RF) and K Nearest Neighbor (KNN). Different performance metrics such as accuracy, precision, recall, f1 score, and area under the curve (AUC) were also used. The experimental results showed that SMOTE Tomek Links technique outperformed the other methods for most of the datasets.

Keywords: Oversampling · SMOTE · SMOTE Tomek Links · Safe level SMOTE · ADASYN

1 Introduction

Class imbalance problem occurs in various disciplines when one class has less number of instances as compared to other class. Generally, a classifier ignores minority class and become biased in nature. The issue with the imbalanced dataset is that it effects the performance of the learning systems. However, the classifiers obtain high predictive accuracy over the negative class but poor predictive accuracy over the positive class [1].

Imbalanced datasets exist in many real-world domains such as medical diagnostics, text classification, information retrieval etc. In these domains, we are more interested in minority class. However, the classifiers behave undesirably [2].

Various approaches have been proposed to address the class imbalance problem. These approaches can be grouped into three categories [3]: External approaches, Internal approaches and hybrid approaches. In external approaches or data level approaches, first we balance the dataset and then model training is performed. This is done by re-sampling the dataset either by over-sampling the minority class or under-sampling the majority class [3]. Data level approaches are independent of the classifier's logic.

K. K. Patel et al. (Eds.): icSoftComp 2022, CCIS 1788, pp. 96–110, 2023.
https://doi.org/10.1007/978-3-031-27609-5_8

In Internal approaches or algorithm level approaches [3], we develop new classification algorithms or improve the existing one to handle the class imbalance problem. However, these approaches require the knowledge of the algorithms. Cost-sensitive approaches comes under this category. In Hybrid approaches, we combine the data level and algorithm level approaches to take the advantages of both the categories.

In this paper, we have reviewed five oversampling techniques such as SMOTE, Safe Level SMOTE, SMOTE Tomek Links, Borderline SMOTE1 and ADASYN to handle the class imbalance problem. Four classification models were used DT, SVM, RF and KNN on ten datasets to evaluate different performance measures.

The rest of the paper is organized as follows: Sect. 2 provides an overview of oversampling methods used. Different performance metrics are discussed in Sect. 3. Section 3 describes the datasets properties. Section 4 describes the experiment and results. Section 5 concludes the paper.

2 Oversampling Methods Used

This section presents an overview of five oversampling methods: SMOTE, Safe Level SMOTE, SMOTE Tomek Links, Borderline SMOTE1 and ADASYN.

Synthetic Minority Oversampling Technique (SMOTE). In 2002, N. V. Chawla et al. [4, 5] proposed SMOTE, an oversampling strategy to address the class imbalance problem. SMOTE considers minority class instances and creates synthetic instances by interpolating between the minority class instance and its k nearest neighbors. But SMOTE ignores the majority class while generating synthetic instances which can lead to class overlapping problems.

Safe Level SMOTE (SL). R C Bunkhumpornpat et al. [1] proposed Safe level SMOTE in 2009, which assigns a safe level to each positive class instance before generating synthetic examples. All synthetic instances are generated solely in safe regions because each synthetic instance is positioned closer to the largest safe level. Unlike SMOTE, Safe level SMOTE generates synthetic instances in safe positions which can help classifiers anticipate the minority class better.

SMOTE Tomek Links (TL). SMOTE Tomek Links was proposed by Batista et al. [6] in 2004 for imbalanced datasets. They proposed to apply Tomek link as the cleaning method to the oversample data. By removing instances from both the classes, a balanced dataset with well-defined class clusters can be produced.

Borderline SMOTE (BS). H. Han et al. [2, 7] in 2005, proposed two over-sampling techniques namely borderline SMOTE1 and borderline SMOTE2. We have used borderline SMOTE1 in this paper. In borderline SMOTE1 only borderline examples from minority class are oversampled.

Adaptive Synthetic Sampling Approach (ADASYN). H. He et al. [8, 9] in 2008, proposed ADASYN, a novel adaptive sampling approach for handling the class imbalance problem. ADASYN generates the minority class instances by considering their distributions. More synthetic samples are generated for minority class samples that are harder to learn as compared to those minority samples that are easier to learn. By focusing on examples that are more difficult to learn, it improves the imbalanced class learning.

3 Dataset

In this paper, we have used ten imbalanced datasets imported from inbuilt imbalanced_databases library in jupyter notebook. These datasets have no missing values. The instances with class label = 0 represent majority class and with class label = 1 represents minority class. Imbalance ratio value range is greater than 0 and less than 1. Table 1 shows the properties of the dataset.

Table 1. Dataset properties

Datasets	#Dimensions	Dataset size	#Minority class	#Majority class	Imbalance ratio
abalone-21_vs_8	9	581	14	567	0.02
car-good	7	1728	69	1659	0.04
page-blocks0	11	5472	559	4913	0.11
hepatitis	20	155	32	123	0.26
segment0	20	2308	329	1979	0.16
wisconsin	10	683	239	444	0.53
vowel0	14	988	90	898	0.10
hypothyroid	25	3163	151	3012	0.05
pima	9	768	268	500	0.53
yeast1	9	1484	429	1055	0.40

4 Experimental Results

This section presents presents the results of performance metrics of five oversampling methods such as SMOTE, Safe Level SMOTE, SMOTE Tomek Links, Borderline SMOTE1 and ADASYN on ten datasets using Jupyter notebook. Default settings are sued for the models. Tables 2, 3, 4 and 5 demonstrate the performance of DT, SVM, RF and KNN across each oversampling method. We have also compared the performance of DT, SVM, RF and KNN on the original imbalanced datasets. Performance results for each method across different evaluation metrics is shown along with the winning times. The best results are highlighted.

Table 2. Performance of decision tree classifier

Decision tree						
Datasets	Methods	Accuracy	Precision	Recall	F1	AUC
abalone-21_vs_8	NOSMOTE	0.97	0.55	0.60	0.55	0.79
	SMOTE	0.96	0.94	**0.98**	0.96	0.97
	SL	0.97	0.95	0.97	0.96	0.98
	TL	0.95	0.94	0.97	0.95	0.97
	BS	**0.99**	**0.99**	**0.98**	**0.99**	**0.99**
	ADASYN	0.95	0.94	0.97	0.95	0.96
car_good	NOSMOTE	0.85	0.04	0.22	0.07	0.87
	SMOTE	**0.93**	**0.90**	**1.00**	**0.94**	**0.93**
	SL	0.85	0.04	0.22	0.07	0.87
	TL	**0.93**	**0.90**	**1.00**	**0.94**	0.92
	BS	**0.93**	**0.90**	**1.00**	**0.94**	**0.93**
	ADASYN	**0.93**	**0.90**	**1.00**	**0.94**	**0.93**
page-blocks0	NOSMOTE	**0.96**	0.85	0.81	0.81	0.91
	SMOTE	**0.96**	0.95	**0.97**	**0.96**	**0.98**
	SL	0.94	0.88	0.89	0.88	0.93
	TL	**0.96**	**0.96**	0.96	**0.96**	**0.98**
	BS	0.95	0.94	0.96	0.95	0.97
	ADASYN	0.93	0.91	**0.97**	0.94	0.97
hepatitis	NOSMOTE	0.81	0.54	0.45	0.45	0.73
	SMOTE	0.82	0.82	0.84	0.81	0.84
	SL	0.64	0.52	0.44	0.47	0.66
	TL	**0.87**	0.88	**0.87**	**0.86**	0.87
	BS	0.85	0.85	0.86	0.83	0.86
	ADASYN	**0.87**	**0.89**	0.86	**0.86**	**0.89**
segment0	NOSMOTE	**0.99**	0.98	0.96	0.97	0.98
	SMOTE	**0.99**	**0.99**	**0.99**	**0.99**	**0.99**
	SL	0.96	**0.99**	0.86	0.91	0.98
	TL	**0.99**	**0.99**	**0.99**	**0.99**	**0.99**
	BS	**0.99**	0.98	**0.99**	**0.99**	**0.99**
	ADASYN	**0.99**	0.98	**0.99**	**0.99**	**0.99**
wisconsin	NOSMOTE	0.95	0.93	0.93	0.93	0.96

(continued)

Table 2. (*continued*)

Decision tree						
Datasets	Methods	Accuracy	Precision	Recall	F1	AUC
	SMOTE	**0.96**	**0.96**	**0.96**	**0.96**	**0.97**
	SL	0.94	0.94	0.94	0.94	0.95
	TL	**0.96**	**0.96**	**0.96**	**0.96**	0.96
	BS	0.95	**0.96**	0.94	0.95	0.95
	ADASYN	0.95	**0.96**	0.94	0.95	0.96
vowel0	NOSMOTE	**0.95**	0.87	0.67	0.71	0.83
	SMOTE	**0.95**	0.92	**0.99**	**0.95**	0.92
	SL	0.90	0.84	0.85	0.81	0.87
	TL	0.94	0.93	**0.99**	**0.95**	0.92
	BS	0.93	**0.94**	0.95	0.93	0.93
	ADASYN	0.94	**0.94**	0.95	0.94	**0.94**
hypothyroid	NOSMOTE	**0.97**	0.77	0.74	0.75	0.91
	SMOTE	0.96	**0.95**	0.97	**0.96**	**0.98**
	SL	0.93	0.90	0.95	0.92	0.97
	TL	0.96	**0.95**	0.98	**0.96**	**0.98**
	BS	0.94	0.90	**0.99**	0.94	0.96
	ADASYN	0.93	0.90	0.97	0.94	0.95
PIMA	NOSMOTE	0.73	0.66	0.52	0.57	0.77
	SMOTE	0.76	0.73	0.80	0.77	0.82
	SL	0.71	0.68	0.71	0.70	0.75
	TL	**0.79**	**0.76**	**0.85**	**0.80**	**0.85**
	BS	0.75	0.74	0.76	0.75	0.80
	ADASYN	0.75	0.72	0.81	0.76	0.80
yeast1	NOSMOTE	0.74	0.60	0.46	0.51	0.73
	SMOTE	0.75	**0.76**	0.74	0.75	**0.81**
	SL	0.66	0.64	0.61	0.62	0.70
	TL	**0.76**	0.75	0.78	0.77	0.80
	BS	0.73	0.69	**0.86**	**0.76**	0.79
	ADASYN	0.73	0.70	0.81	0.75	0.78
Winning Times	NOSMOTE	4	0	0	0	0

(*continued*)

Table 2. (*continued*)

Decision tree						
Datasets	Methods	Accuracy	Precision	Recall	F1	AUC
	SMOTE	5	5	**6**	6	**6**
	SL	0	1	0	0	0
	TL	**7**	**6**	**6**	**8**	4
	BS	3	4	5	4	3
	ADASYN	3	4	3	3	4

Table 3. Performance of Support vector machine classifier

Support vector machine						
Datasets	Methods	Accuracy	Precision	Recall	F1	AUC
abalone_21_vs_8	NOSMOTE	0.98	0.70	0.80	0.70	0.90
	SMOTE	**0.99**	0.98	**1.00**	**0.99**	**0.99**
	SL	0.97	0.95	0.98	0.96	0.98
	TL	**0.99**	0.98	**1.00**	**0.99**	**0.99**
	BS	**0.99**	**0.99**	0.99	**0.99**	**0.99**
	ADASYN	**0.99**	0.98	**1.00**	**0.99**	**0.99**
car_good	NOSMOTE	0.96	0.79	0.97	0.82	**0.99**
	SMOTE	**0.98**	**0.97**	**1.00**	**0.98**	**0.99**
	SL	0.96	0.79	0.97	0.82	**0.99**
	TL	**0.98**	**0.97**	**1.00**	**0.98**	**0.99**
	BS	**0.98**	**0.97**	**1.00**	**0.98**	**0.99**
	ADASYN	**0.98**	**0.97**	**1.00**	**0.98**	**0.99**
page-blocks0	NOSMOTE	**0.93**	0.81	0.46	0.57	0.93
	SMOTE	0.92	0.88	0.96	0.92	**0.97**
	SL	0.92	0.86	0.83	0.84	0.95
	TL	0.91	0.88	0.96	0.92	**0.97**
	BS	**0.93**	**0.89**	**0.99**	**0.94**	**0.97**
	ADASYN	0.92	0.87	0.98	0.92	0.96
hepatitis	NOSMOTE	0.78	0.55	0.53	0.49	0.70
	SMOTE	**0.92**	**0.89**	0.97	**0.93**	0.95
	SL	0.69	0.63	0.62	0.62	0.73

(*continued*)

Table 3. (*continued*)

Support vector machine						
Datasets	Methods	Accuracy	Precision	Recall	F1	AUC
	TL	0.91	0.87	**0.98**	0.92	0.95
	BS	0.90	0.88	0.94	0.90	0.93
	ADASYN	0.91	0.86	**0.98**	0.91	**0.97**
segment0	NOSMOTE	0.95	0.96	0.72	0.82	**0.99**
	SMOTE	**0.98**	**0.98**	0.98	**0.98**	**0.99**
	SL	0.94	0.97	0.78	0.86	0.98
	TL	**0.98**	**0.98**	**0.99**	**0.98**	**0.99**
	BS	0.93	0.92	0.94	0.93	0.98
	ADASYN	0.94	0.90	**0.99**	0.94	**0.99**
wisconsin	NOSMOTE	0.95	0.93	0.94	0.93	**0.99**
	SMOTE	**0.97**	**0.96**	0.98	**0.97**	**0.99**
	SL	0.94	0.94	0.93	0.93	**0.99**
	TL	**0.97**	**0.96**	0.98	**0.97**	**0.99**
	BS	**0.97**	**0.96**	0.97	**0.97**	**0.99**
	ADASYN	**0.97**	**0.96**	**0.99**	**0.97**	**0.99**
vowel0	NOSMOTE	0.96	0.89	0.75	0.75	**0.99**
	SMOTE	**0.99**	**0.98**	**1.00**	**0.99**	**0.99**
	SL	0.92	0.74	0.72	0.71	0.96
	TL	**0.99**	**0.98**	**1.00**	**0.99**	**0.99**
	BS	0.95	**0.98**	0.91	0.92	0.97
	ADASYN	0.96	**0.98**	0.94	0.95	**0.99**
hypothyroid	NOSMOTE	**0.97**	0.80	0.62	0.69	0.97
	SMOTE	**0.97**	**0.96**	0.98	**0.97**	**0.99**
	SL	0.94	0.92	0.95	0.93	0.98
	TL	**0.97**	**0.96**	0.98	**0.97**	**0.99**
	BS	**0.97**	**0.96**	**0.99**	**0.97**	**0.99**
	ADASYN	**0.97**	0.94	**0.99**	**0.97**	0.98
	NOSMOTE	0.73	0.66	0.50	0.57	0.79
PIMA	SMOTE	0.78	0.77	0.81	0.79	0.85
	SL	0.71	0.70	0.69	0.69	0.77
	TL	**0.81**	**0.80**	**0.83**	**0.81**	**0.87**
	BS	0.78	0.76	**0.83**	0.79	0.83
	ADASYN	0.78	0.76	**0.83**	0.79	0.83

(*continued*)

Table 3. (*continued*)

Support vector machine						
Datasets	Methods	Accuracy	Precision	Recall	F1	AUC
yeast1	NOSMOTE	**0.76**	0.66	0.42	0.51	0.76
	SMOTE	0.73	**0.72**	0.78	0.75	0.81
	SL	0.64	0.60	0.68	0.64	0.72
	TL	0.75	**0.72**	0.82	**0.76**	**0.82**
	BS	0.73	0.69	**0.84**	**0.76**	0.79
	ADASYN	0.72	0.68	0.83	0.75	0.78
Winning Times	NOSMOTE	3	0	0	0	4
	SMOTE	**7**	**7**	3	7	7
	SL	0	0	0	0	2
	TL	**7**	**7**	6	**8**	**9**
	BS	5	6	5	6	5
	ADASYN	4	3	**7**	4	6

Table 4. Performance of Random Forest classifier

Random forest						
Datasets	Methods	Accuracy	Precision	Recall	F1	AUC
abalone_21_vs_8	NOSMOTE	**0.98**	0.55	0.55	0.53	0.88
	SMOTE	0.96	0.96	0.97	0.96	**0.99**
	SL	0.97	0.96	0.97	0.97	**0.99**
	TL	0.96	0.96	0.96	0.96	**0.99**
	BS	**0.98**	**0.99**	**0.98**	**0.98**	**0.99**
	ADASYN	0.97	0.97	0.96	0.97	**0.99**
car_good	NOSMOTE	0.95	0.60	0.55	0.47	0.92
	SMOTE	**0.97**	**0.96**	**0.99**	**0.98**	**0.99**
	SL	0.95	0.59	0.58	0.39	0.92
	TL	**0.97**	**0.96**	**0.99**	0.97	**0.99**
	BS	**0.97**	0.95	**0.99**	**0.98**	**0.99**
	ADASYN	**0.97**	0.95	**0.99**	**0.98**	**0.99**
page-blocks0	NOSMOTE	0.96	0.84	0.79	0.81	0.96
	SMOTE	**0.97**	**0.96**	0.98	**0.97**	**0.99**
	SL	0.95	0.89	0.90	0.89	0.97

(*continued*)

Table 4. (*continued*)

Random forest						
Datasets	Methods	Accuracy	Precision	Recall	F1	AUC
	TL	**0.97**	**0.96**	0.98	**0.97**	**0.99**
	BS	**0.97**	**0.96**	0.97	**0.97**	**0.99**
	ADASYN	**0.97**	0.95	**0.99**	**0.97**	**0.99**
hepatitis	NOSMOTE	0.80	0.53	0.39	0.51	0.85
	SMOTE	**0.87**	0.89	0.91	0.88	0.95
	SL	0.67	0.62	0.59	0.62	0.71
	TL	0.85	0.89	**0.93**	**0.89**	0.95
	BS	**0.87**	**0.91**	0.88	**0.89**	**0.96**
	ADASYN	**0.87**	0.90	0.90	**0.89**	**0.96**
segment0	NOSMOTE	**0.99**	**0.99**	0.97	0.98	**0.99**
	SMOTE	**0.99**	**0.99**	**0.99**	**0.99**	**0.99**
	SL	0.97	0.96	0.95	0.95	**0.99**
	TL	**0.99**	**0.99**	**0.99**	**0.99**	**0.99**
	BS	**0.99**	**0.99**	**0.99**	**0.99**	**0.99**
	ADASYN	**0.99**	**0.99**	**0.99**	**0.99**	**0.99**
wisconsin	NOSMOTE	0.95	0.94	0.92	0.94	**0.98**
	SMOTE	**0.97**	**0.97**	0.96	**0.97**	**0.98**
	SL	0.96	0.96	0.96	0.96	**0.98**
	TL	**0.97**	**0.97**	**0.97**	**0.97**	**0.98**
	BS	**0.97**	0.96	**0.97**	**0.97**	**0.98**
	ADASYN	0.96	0.96	**0.97**	**0.97**	**0.98**
vowel0	NOSMOTE	0.95	0.79	0.65	0.66	0.97
	SMOTE	**0.99**	0.98	**0.99**	0.98	**0.99**
	SL	0.94	0.92	0.77	0.76	0.98
	TL	**0.99**	0.98	**0.99**	**0.99**	**0.99**
	BS	0.95	**0.99**	0.91	0.50	0.96
	ADASYN	0.96	**0.99**	0.91	0.92	0.98
hypothyroid	NOSMOTE	**0.98**	0.85	0.63	0.78	0.92
	SMOTE	**0.98**	**0.98**	0.98	**0.98**	**0.99**
	SL	0.96	0.95	0.97	0.96	0.98
	TL	**0.98**	**0.98**	**0.99**	**0.98**	**0.99**
	BS	**0.98**	**0.98**	0.98	**0.98**	**0.99**
	ADASYN	**0.98**	**0.98**	**0.99**	**0.98**	**0.99**

(*continued*)

Table 4. (*continued*)

Random forest						
Datasets	Methods	Accuracy	Precision	Recall	F1	AUC
PIMA	NOSMOTE	0.73	0.61	0.54	0.57	0.74
	SMOTE	0.77	0.77	0.79	0.77	0.86
	SL	0.64	0.62	0.64	0.64	0.70
	TL	**0.81**	**0.79**	**0.83**	**0.80**	**0.87**
	BS	0.76	0.76	0.80	0.77	0.84
	ADASYN	0.76	0.75	0.80	0.77	0.83
yeast1	NOSMOTE	0.72	0.54	0.46	0.48	0.72
	SMOTE	0.80	0.77	0.82	**0.80**	0.86
	SL	0.62	0.61	0.60	0.58	0.64
	TL	**0.81**	**0.79**	0.83	0.79	**0.88**
	BS	0.79	0.77	**0.84**	**0.80**	0.86
	ADASYN	0.78	0.76	0.81	**0.80**	0.86
Winning Times	NOSMOTE	3	1	0	0	2
	SMOTE	7	5	3	6	7
	SL	0	0	0	0	3
	TL	**8**	**7**	**7**	7	**9**
	BS	7	6	5	**8**	7
	ADASYN	5	3	5	7	7

Table 5. Performance of K nearest neighbor classifier.

K nearest neighbor						
Datasets	Methods	Accuracy	Precision	Recall	F1	AUC
abalone_21_vs_8	NOSMOTE	0.98	0.30	0.30	0.30	0.74
	SMOTE	0.98	0.96	0.99	0.98	**0.99**
	SL	0.97	0.95	0.99	0.97	0.98
	TL	0.98	0.96	**1.00**	0.98	**0.99**
	BS	**0.99**	**0.99**	0.98	**0.99**	**0.99**
	ADASYN	0.98	0.96	0.99	0.98	**0.99**
car_good	NOSMOTE	0.95	0.40	0.17	0.20	0.86

(*continued*)

Table 5. (*continued*)

K nearest neighbor						
Datasets	Methods	Accuracy	Precision	Recall	F1	AUC
	SMOTE	**0.96**	**0.93**	**1.00**	**0.96**	**0.99**
	SL	0.95	0.40	0.17	0.20	0.86
	TL	**0.96**	**0.93**	**1.00**	**0.96**	**0.99**
	BS	**0.96**	**0.93**	0.99	**0.96**	**0.99**
	ADASYN	**0.96**	**0.93**	0.99	**0.96**	**0.99**
page-blocks0	NOSMOTE	0.95	0.85	0.63	0.70	0.91
	SMOTE	**0.97**	0.95	**0.98**	**0.97**	**0.98**
	SL	0.94	0.89	0.88	0.88	0.95
	TL	**0.97**	**0.96**	**0.98**	**0.97**	**0.98**
	BS	**0.97**	**0.96**	0.97	**0.97**	**0.98**
	ADASYN	0.96	0.95	**0.98**	0.96	**0.98**
hepatitis	NOSMOTE	0.80	0.36	0.15	0.19	0.76
	SMOTE	**0.88**	0.84	**0.94**	**0.89**	**0.93**
	SL	0.72	0.68	0.59	0.63	0.75
	TL	0.87	**0.85**	0.93	0.88	**0.93**
	BS	0.87	**0.85**	0.92	0.88	0.92
	ADASYN	0.87	0.83	**0.94**	0.88	**0.93**
segment0	NOSMOTE	**0.99**	0.96	0.99	0.97	**0.99**
	SMOTE	**0.99**	0.98	**1.00**	**0.99**	**0.99**
	SL	0.97	0.94	0.95	0.94	**0.99**
	TL	**0.99**	0.98	0.99	**0.99**	**0.99**
	BS	**0.99**	**0.99**	0.99	**0.99**	**0.99**
	ADASYN	**0.99**	0.98	0.99	**0.99**	**0.99**
wisconsin	NOSMOTE	0.96	0.96	0.94	0.95	0.98
	SMOTE	**0.98**	**0.97**	**0.99**	**0.98**	0.98
	SL	0.97	0.96	0.97	0.97	0.98
	TL	**0.98**	**0.97**	0.98	**0.98**	**0.99**
	BS	**0.98**	0.96	**0.99**	**0.98**	0.98
	ADASYN	0.97	0.96	**0.99**	0.97	0.98
vowel0	NOSMOTE	**0.95**	0.86	0.67	0.71	0.91

(*continued*)

Table 5. (*continued*)

K nearest neighbor

Datasets	Methods	Accuracy	Precision	Recall	F1	AUC
	SMOTE	**0.95**	0.92	**1.00**	**0.96**	**0.96**
	SL	0.87	0.51	0.61	0.55	0.77
	TL	**0.95**	0.92	**1.00**	**0.96**	**0.96**
	BS	0.93	**0.95**	0.91	0.90	0.94
	ADASYN	0.93	**0.95**	0.93	0.92	0.94
hypothyroid	NOSMOTE	0.96	0.79	0.45	0.56	0.87
	SMOTE	**0.98**	0.96	**0.99**	**0.98**	**0.99**
	SL	0.96	0.94	0.98	0.96	0.97
	TL	**0.98**	0.96	**0.99**	**0.98**	**0.99**
	BS	**0.98**	**0.97**	**0.99**	**0.98**	**0.99**
	ADASYN	0.97	0.95	**0.99**	0.97	0.98
PIMA	NOSMOTE	0.71	0.67	0.40	0.49	0.73
	SMOTE	0.76	0.75	0.76	0.76	0.83
	SL	0.65	0.66	0.55	0.60	0.72
	TL	**0.79**	**0.79**	**0.79**	**0.79**	**0.86**
	BS	0.76	0.74	**0.79**	0.76	0.82
	ADASYN	0.76	0.74	0.78	0.76	0.83
yeast1	NOSMOTE	0.74	0.62	0.32	0.41	0.73
	SMOTE	0.79	0.77	0.81	0.79	0.86
	SL	0.64	0.63	0.54	0.58	0.70
	TL	0.80	**0.79**	0.82	0.80	**0.87**
	BS	**0.82**	0.78	**0.88**	**0.82**	0.86
	ADASYN	0.80	0.77	0.86	0.81	0.85
Winning Times	NOSMOTE	2	0	0	0	1
	SMOTE	7	2	7	7	7
	SL	0	0	0	0	1
	TL	7	6	6	7	10
	BS	7	7	4	7	5
	ADASYN	2	2	4	2	5

Figure 1 and Fig. 2 represent the F1 score, & Fig. 3 and Fig. 4 represent the AUC values of five oversampling methods on two datasets with DT, SVM, RF and KNN respectively. NOSMOTE represents the original dataset without oversampling.

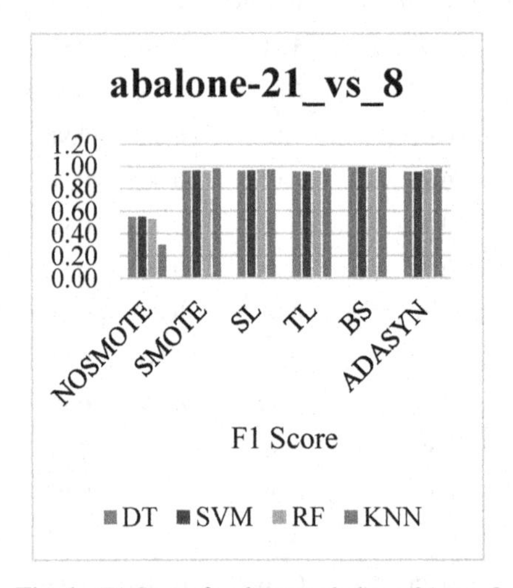

Fig. 1. F1 Score for dataset abalone-21_vs_8

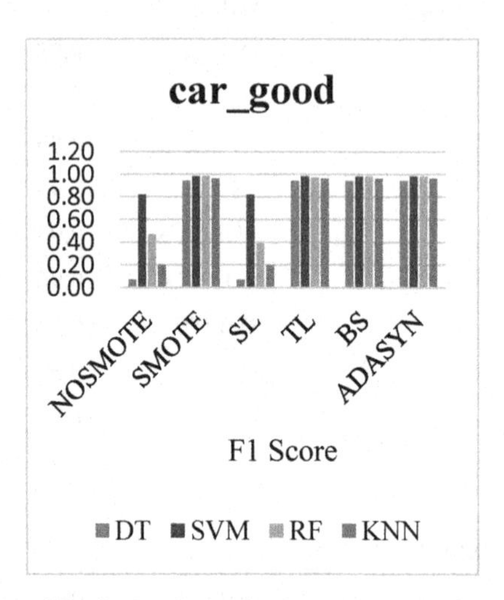

Fig. 2. F1 Score for dataset car_good

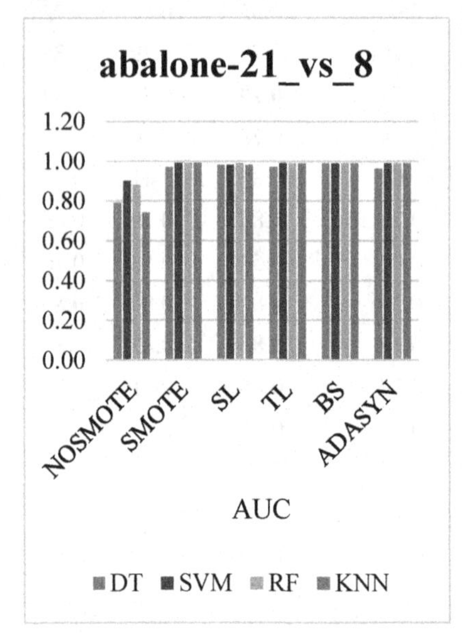

Fig. 3. AUC for dataset abalone-21_vs_8

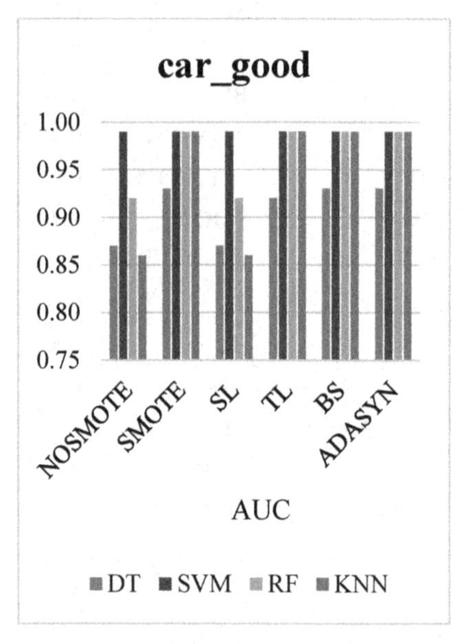

Fig. 4. AUC for dataset car_good

It was found that SMOTE Tomek Links method outperformed the other methods according to the winning times as SMOTE-Tomek Links method discovers the Tomek links and cleans them.

5 Conclusion

In this paper, we have reviewed five oversampling techniques SMOTE, Safe Level SMOTE, SMOTE Tomek Links, Borderline SMOTE1 and ADASYN to handle the class imbalance problem. The performances of the oversampling methods used to address class imbalance problem are then empirically compared by four classification models namely DT, SVM, RF and KNN. Various performance metrics are used such as accuracy, precision, recall, f1 score and AUC value in this paper. The results showed that SMOTE Tomek Links outperformed the other oversampling methods in terms of performance metrics for most of the datasets as SMOTE Tomek Links first performed the noise cleaning method and then generated the synthetic instances. For future work, the results can be validated on high dimensional datasets.

References

1. Bunkhumpornpat, C., Sinapiromsaran, K., Lursinsap, C.: Safe-level-smote: Safe-level-synthetic minority over-sampling technique for handling the class imbalanced problem. In: Theeramunkong, T., Kijsirikul, B., Cercone, N., Ho, TB. (eds.) Pacific-Asia Conference on Knowledge Discovery and Data Mining. LNAI, vol. 5476, pp. 475–482. Springer, Berlin (2009). https://doi.org/10.1007/978-3-642-01307-2_43
2. Han, H., Wang, W.Y., Mao, B.H.: Borderline-SMOTE: a new over-sampling method in imbalanced data sets learning. In: Huang, DS., Zhang, XP., Huang, GB. (eds.) International Conference on Intelligent Computing. LNCS, vol. 3644, pp. 878–887, Springer, Berlin (2005). https://doi.org/10.1007/11538059_91
3. Gosain, A., Sardana, S.: Handling class imbalance problem using oversampling techniques: a review. In: International Conference on Advances in Computing, Communications and Informatics, pp. 79–85 (2017)
4. Elyan, E., Moreno-Garcia, C.F., Jayne, C.: CDSMOTE: class decomposition and synthetic minority class oversampling technique for imbalanced-data classification. Neural. Comput. Appl. **33**(7), 2839–2851 (2020). https://doi.org/10.1007/s00521-020-05130-z
5. Chawla, N.V., Bowyer, K.W., Hall, L.O., Kegelmeyer, W.P.: SMOTE: synthetic minority over-sampling technique. J. Artific. Intell. Res. **16**, 321–357 (2002)
6. Batista, G.E., Prati, R.C., Monard, M.C.: A study of the behavior of several methods for balancing machine learning training data. ACM SIGKDD Explor. Newsl. **6**(1), 20–29 (2004)
7. Sun, Y., et al.: Borderline smote algorithm and feature selection-based network anomalies detection strategy. Energies **15**(13), 4751 (2022)
8. He, H., Bai, Y., Garcia, E.A., Li, S.: ADASYN: adaptive synthetic sampling approach for imbalanced learning. In IEEE International Joint Conference on Neural Network, pp. 1322–1328 (2008)
9. Datta, D., et al.: A hybrid classification of imbalanced hyperspectral images using ADASYN and enhanced deep subsampled multi-grained cascaded forest. Remote Sens. **14**(19), 4853 (2022)

10. Kaur, P., Gosain, A.: Comparing the behavior of oversampling and undersampling approach of class imbalance learning by combining class imbalance problem with noise. In: Saini, A., Nayak, A., Vyas, R. (eds.) ICT Based Innovations, AISC, vol. 653, pp. 23–30. Springer, Singapore (2018). https://doi.org/10.1007/978-981-10-6602-3_3
11. Kovács, G.: An empirical comparison and evaluation of minority oversampling techniques on a large number of imbalanced datasets. Appl. Soft Comput. **83**, 105662 (2019)
12. Upadhyay, K., Kaur, P., Verma, D.K.: Evaluating the performance of data level methods using KEEL tool to address class imbalance problem. Arab. J. Sci. Eng. 1–14 (2021). https://doi.org/10.1007/s13369-021-06377-x
13. Santoso, B., Wijayanto, H., Notodiputro, K.A., Sartono, B.: Synthetic over sampling methods for handling class imbalanced problems: a review. In: IOP Conference Series: Earth and Environmental Science, vol. 58. IOP Publishing (2017)
14. Sanni, R.R., Guruprasad, H.S.: Analysis of performance metrics of heart failured patients using python and machine learning algorithms. Global Trans. Proc. **2**(2), 233–237 (2021)

Analog Implementation of Neural Network

Vraj Desai(✉) and Pallavi G. Darji

Dharmsinh Desai University, College Road, Nadiad 387 001, Gujarat, India
vrajdesai8888@gmail.com

Abstract. As Moore's law comes to the end, to increase the speed and density of computation, analog based approach has to be revisited. The multiplication and addition operations can be easily done by digital circuits but the power consumption and area occupied by processing units will drastically increase. The implementation of the same operations performed by analog circuits not only provides continuous interrupt-free operations but also reduces area and power consumption considerably. This makes analog circuit implementation ideal for Neural networks (NN) processing. In this work, the single neuron has been realized by a Common Drain amplifier, Trans-Impedance Amplifier(TIA) and CMOS rectifier circuits. The 3×1 NN(3 input $\times$ 1 output), 3×3 NN, and two layers of 3×3 NN have been implemented using this single neuron and only forward propagation has been performed. The simulation has been done on LTSpice for 16 nm CMOS Technology and results have been compared with theoretical values. The implemented two-layer 3×3 NN is capable to work up to 61 MHz and found a 90% reduction in transistor count compared to the 8-bit Vedic multiplier NN.

Keywords: Neural network · Neuron · Multiplication · Addition · Artificial intelligence · Vedic multiplier

1 Introduction

An Artificial Neural Network (ANN) has become a promising solution to inject Artificial Intelligence in various applications like google search engine, mobile phones, Tesla car, robotics, pattern recognition, data mining and machine vision etc. It is an information processing archetype that learn by example and configured for such a specific applications.

It has a set of neurons organized in layers and each neuron in each layer does a mathematical operation that takes it's input, multiplies it by it's weights and then passes the sum through the activation function to the other neurons. Hence it is basically three step process: Fetch inputs, learn from it(training) and predict output. Learning processes are of two types: Forward Propagation and Back Propagation. In forward propagation, inputs are multiply by the random numbers known as weights and pass the results through a activation function to calculate the neuron's output. In Back Propagation, the difference between the

K. K. Patel et al. (Eds.): icSoftComp 2022, CCIS 1788, pp. 111–122, 2023.
https://doi.org/10.1007/978-3-031-27609-5_9

actual output and the expected output will be calculated which is known as error and this error is used to adjust the weights using gradient decent method. All of this process means higher and complex computations and goal of this work is to use analog circuits to reduce these overhead by effectively manipulating weight data. Researchers are constantly trying to minimise the time for learning process in NN as well as the energy consumption in its working.

This work is organized as follows. Section 2 covers background work and Sect. 3 illustrates architecture for single neuron. Section 4 shows circuit realization and theoretical calculation for 3×1, 3×3 and two layer of 3×3 NN. The comparison of simulation results with theoretical values as well as with other related work has been done in Sect. 5 and Conclusion is derived in Sect. 6.

2 Literature Survey

A 4×4 vedic multiplier had been proposed in [1] using specific algorithm in 90 nm CMOS technology and claimed for the lowest power consumption and least delay. Author in [2] had designed the low power analogue neuron using multiplier and programmable activation function circuit which had been successfully used in learning algorithms in back propagation model. The circuit had been implemented for 1 V power supply at 180 nm CMOS technology. An artificial neural network (ANN) model using the gradient descent method had been implemented in [3] on a 1.2 μm CMOS technology. The author in this had demonstrated optimized method to obtain the synaptic weights and bias values require for each neuron. The basic topology for optimal implementation of Neural Network in Analog domain can be created with a resistive cross bar pattern similar to a keyboard matrix [4,5]. It uses memristive cross-bar for matrix multiplication for forward and backward propagation but the same principle can be realized with resistor which can further be replaced by PCM memory for in memory computation.

Author [6] had realized neural network using a Gilbert Cell Multiplier for multiplication function and differential pair for activation function. In [7], author had used logarithmic amplifier circuit for multiplication which gives higher accuracy at a cost of area. The spiking neural network [8] uses a Neuromorphic circuit which made up of capacitors and integrator circuits to provide charge to the next layer to mimic a biological neuron.

3 Neuron Network Architecture

The function of the neuron is to add the multiplied input values to produce the activated output for the next neuron as shown in Fig. 1.

It is divided among three main stages: multiplication, addition and activation block. The Analog matrix multiplication has been performed by amplifier circuit and resistor(memristor). Since neural network is made up of multiple layers, this layers in analog circuit create loading effect. In order to prevent it, MOS based Common Drain amplifier is designed which works as a voltage buffer. The multiplier output(current) is fed to Trans-Impedance Amplifier(TIA) through

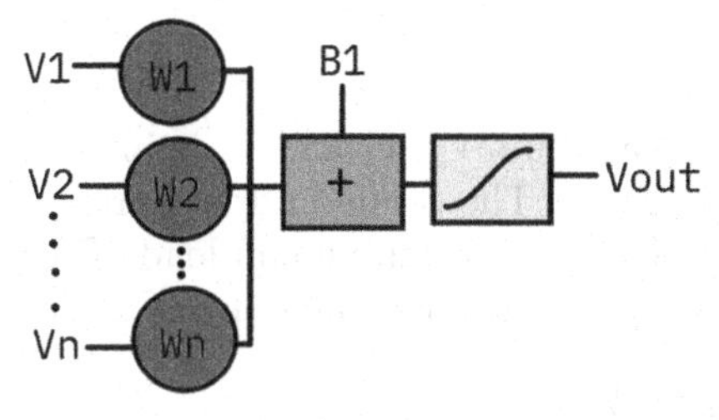

Fig. 1. Block diagram for single neuron

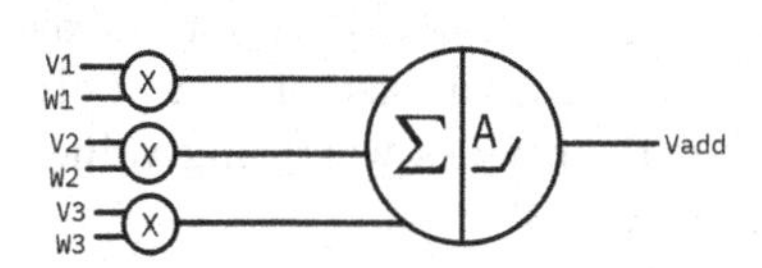

Fig. 2. Single neuron

the summing node. The Activation function is required to create non-linear approximation because the neural network works in the weighting factor range of –1 to +1. For this, Rectified Linear Unit(ReLU) function has been implemented by CMOS rectifier. This non-linear approximation provides output signal from 0 to $Vmax$, which also prevents the neuron to die. The Sigmoid Activation function can be used to replace the ReLU function which gives output from –1 to +1.

The Neural Network circuit has been implemented by 16 nm MOSFET, based on PTM model on LTSPICE simulator. The linear operating range for this Nano transistor has been decided by plotting transfer and drain characteristics.

3.1 Multiplier and Adder

Each neuron does addition and activation function as shown in Fig. 2. It carries inputs V_i and weighting factor W_i to generate output as,

$$V_{add} = V_1W_1 + V_2W_2 + V_3W_3 \tag{1}$$

The activation function may be rectifier, segmoid and tanh. The rectifier activation function output,

$$V_{add} = \begin{cases} V_{add} & V_{add} > 0 \\ 0 & V_{add} < 0 \end{cases} \tag{2}$$

3.2 Approach

Figure 3 shows a basic 3 × 3 neuron network architecture as a matrix multiplication, which has three inputs and three outputs. The Resistor R provides weighting factor W_{ij} and it is multiplied with the input voltage V_i where i = 1 to N. The input voltages are applied through voltage buffer for isolation from heavy load. The equivalent current corresponding to $V_i \times w_{ij}$ is converted to voltage by trans-impedance amplifier. This voltage is passed to the next stage or feedback through activation function which decides whether it is important for further processing in NN or not.

4 Implementation

The voltage buffer circuit with resistor for three inputs V_a, V_b and V_c is implemented by common drain circuit as shown in Fig. 4. The common drain circuit requires transistor size of 1 μm only to get desired g_m for a minimum load. The resistor R provides weighting values w or it may be a trained value in NN. The output I_{add} is,

$$I_{add} = \frac{V_a}{R_1} + \frac{V_b}{R_2} + \frac{V_c}{R_3} \tag{3}$$

This I_{add} is converted to V_{add} through trans-impedance amplifier to get Eq. 1. The trans-impedance amplifier is realized by two stage op-amp as shown in Fig. 5. The first stage is made up of cross coupled differential pair to obtain high gain and second stage provides single ended output [9–11]. The circuit is designed for gain of 3 with transistors size listed in Table 1. The CMOS Rectifier [12] is used as a ReLU(Rectifier Linear Unit) for activation function shown in Fig. 6. The working is as follows: when input is positive, M_2 and M_3 are on and output is shorted to ground through M_2. When input is negative M_1 and M_4 are on and through M_4 the output goes to ground.

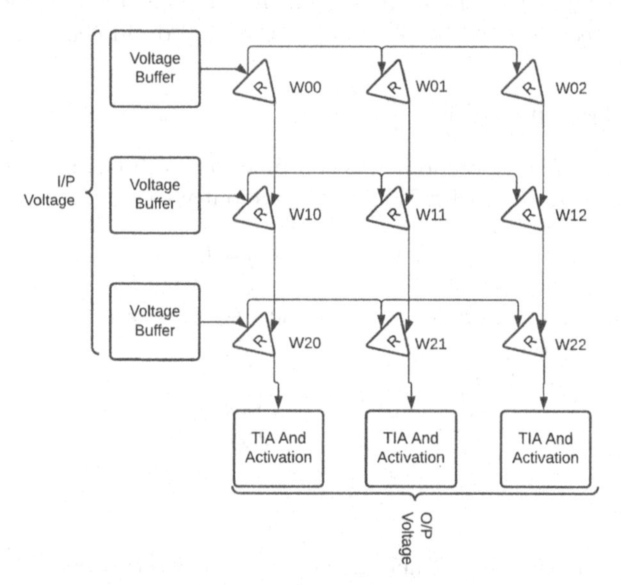

Fig. 3. A basic 3 × 3 neuron network architecture [4].

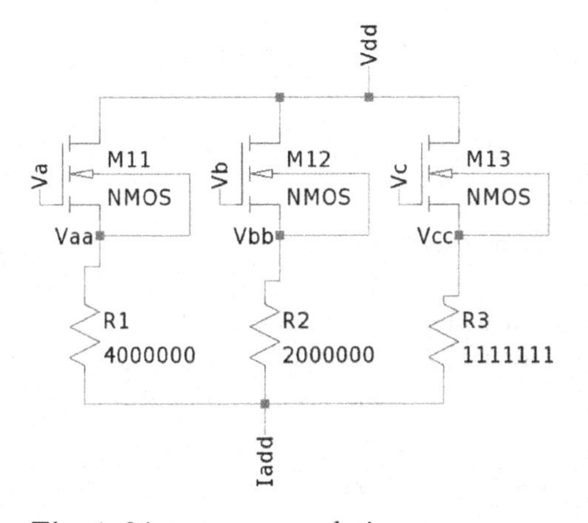

Fig. 4. 3 input common drain configuration

Table 1. Size of MOSFETs in TIA.

Name	Length (nm)	Width (nm)
M3	16	128
M4	16	128
M10	16	128
M11	16	128
M7	16	256
M8	16	256
M6	16	256
M5	16	256

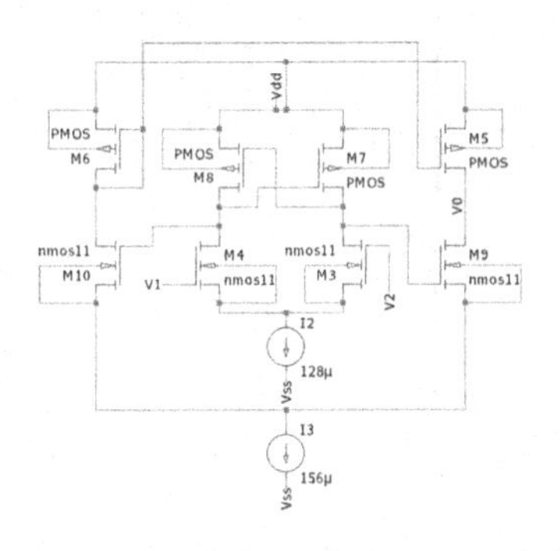

Fig. 5. Trans-impedance amplifier (TIA)

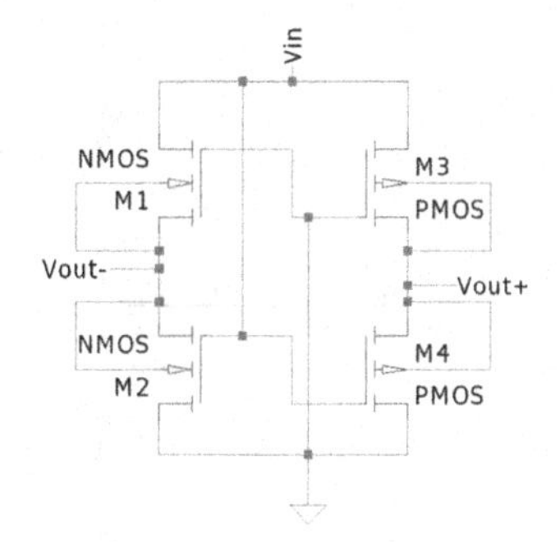

Fig. 6. CMOS rectifier for ReLU.

4.1 3×1 Positive Weight Matrix Multiplication

The Fig. 4, 5 and 6 are coupled together to form a neuron as shown in Fig. 1. Figure 7 shows the 3 × 1 matrix multiplication circuit with TIA and ReLU. It is a single neuron that receives three inputs V_a,V_b and V_c and generates single output V_{add}.

The 3 × 1 matrix multiplication circuit is implemented for 16nm CMOS technology in LTSpice for listed input voltages in Table 2. The obtained I_{add} for 40, 50 and 30 mV input and 4(0.25), 2(0.5) and 1(1) MΩ resistors(conductors) respectively through matrix multiplier, is 65 nA which is nearly match to the calculated values shown Eq. 4 which is 62 nA. The shown matrix calculation is in terms of voltage, but actually the common drain circuit converts the voltage into a weighted current listed in the Table 2, hence the sum of current is the output for the matrix multiplication but to feed it into the next layer a 1 to 1 conversion is perform from current to voltage through TIA.

$$\left[40 * 10^{-3}\ 50 * 10^{-3}\ 30 * 10^{-3}\right] * \begin{bmatrix} 0.25 \\ 0.5 \\ 0.9 \end{bmatrix} = \left[0.062\right] \quad (4)$$

Table 2. 3×1 matrix multiplication.

Input voltage (mV)	Weight	Voltage at buffer (mV)	Current through resistor (nA)
40	0.25	38.15	10.67
50	0.5	47.12	25.83
30	0.9	26.98	28.36

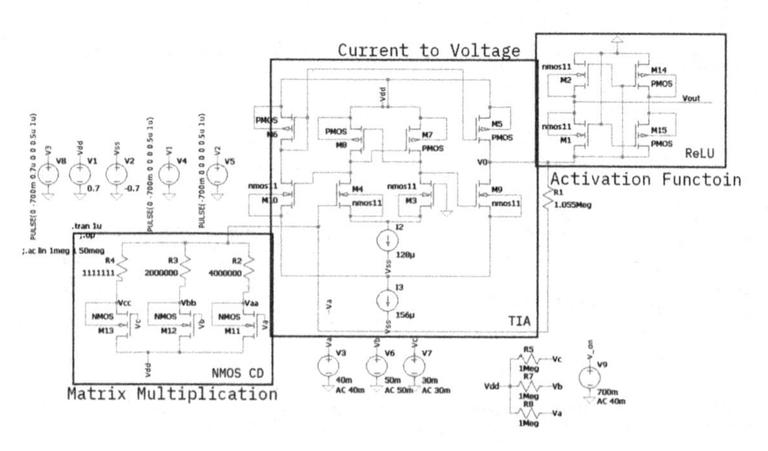

Fig. 7. 3×1 weight matrix multiplication.

4.2 3x1 Negative Weight Matrix Multiplication

The multiplication circuit, which is the common drain circuits for each input, in each odd layers in NN are realized by NMOS transistor while every even layers are realized by PMOS transistor due to negative output from TIA.

For negative weight a PMOS can be used to subtract the current at the node as shown Fig. 8 from NMOS and PMOS CD block. Its matrix operation is depicted in Eq. 5 and the voltage and current values are given in Table 3.

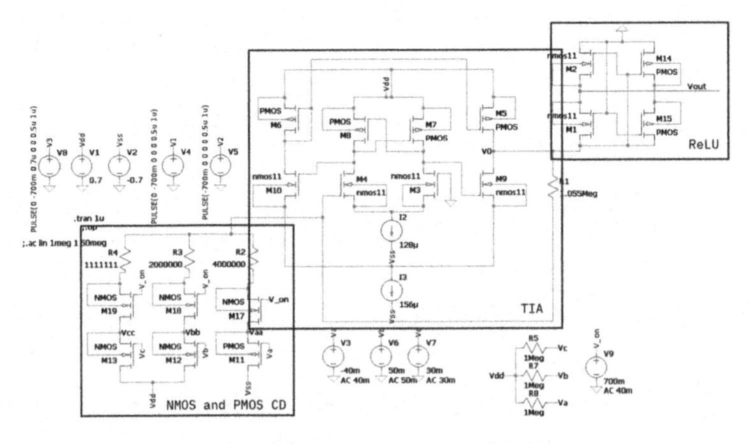

Fig. 8. 3 × 1 negative matrix multiplication.

$$\begin{bmatrix}40 * 10^{-3} \ 50 * 10^{-3} \ 30 * 10^{-3}\end{bmatrix} * \begin{bmatrix}-0.25 \\ 0.5 \\ 0.9\end{bmatrix} = \begin{bmatrix}0.042\end{bmatrix} \quad (5)$$

Table 3. 3 × 1 negative matrix multiplication table.

Input voltage (mV)	Weight	Voltage at buffer (mV)	Current through resistor (nA)
40	-0.25	-38.15	-8.72
50	0.5	47.16	25.13
30	0.9	27.05	27.13

4.3 3 × 3 Weight Matrix Multiplication

This work is further extended for two layers of 3 × 3 matrix multiplication.

The circuit in Fig. 9 takes 3 input and 3 × 3 weight matrix in the NMOS Common Drain(CD) block to produces 3 output from TIA and ReLU block that can be applied to the next layer. The operation has been performed as per given in Eq. 6 and the voltages and currents are tabulated in Table 4.

3 × 3 Matrix Multiplication :

$$\begin{bmatrix}40 * 10^{-3} \ 50 * 10^{-3} \ 30 * 10^{-3}\end{bmatrix} * \begin{bmatrix}0.2 \ 0.5 \ 0.8 \\ 0.2 \ 0.5 \ 0.8 \\ 0.2 \ 0.5 \ 0.8\end{bmatrix} = \begin{bmatrix}26 * 10^{-3} \ 60 * 10^{-3} \ 96 * 10^{-3}\end{bmatrix} \quad (6)$$

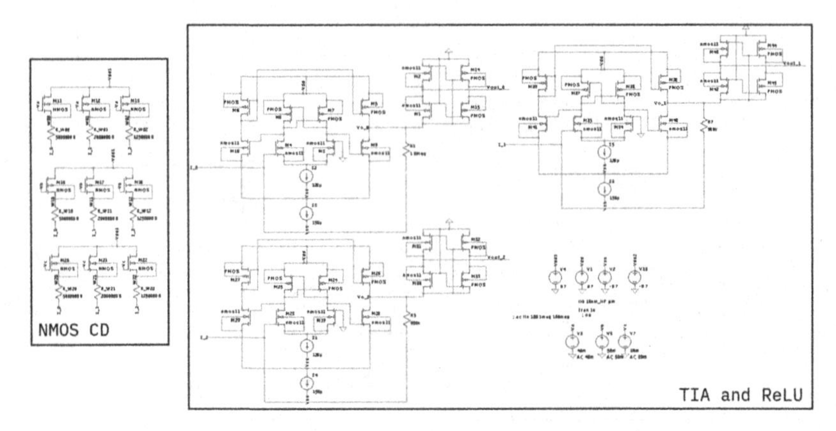

Fig. 9. 3×3 matrix multiplication.

Table 4. 3×3 matrix multiplication.

Input voltage (mV)	Weight	Voltage at buffer (mV)	Current through resistor (nA)
40	0.2	38.3	8.151
	0.5	37.4	20.9
	0.8	36.6	34.5
50	0.2	48.2	10.1
	0.5	47.1	25.7
	0.8	46.3	42.2
30	0.2	28.5	6.1
	0.5	27.7	16.06
	0.8	27.06	26.8

4.4 2 Layer of 3×3 Weight Matrix Multiplication

The Fig. 10 shows two layer of 3×3 neural network. For first(odd) layer, the NMOS CD and TIA and ReLU block is used while for second(even) layer in which negative output is generated, is applied to the PMOS CD block to perform the operation shown in Eqs. 7 and 8. The Voltages and Currents are listed in Table 5.

NMOS :-

$$\begin{aligned} \left[40 * 10^{-3}\ 50 * 10^{-3}\ 30 * 10^{-3}\right] * \begin{bmatrix} 0.2\ 0.5\ 0.8 \\ 0.2\ 0.5\ 0.8 \\ 0.2\ 0.5\ 0.8 \end{bmatrix} \\ = \left[24 * 10^{-3}\ 60 * 10^{-3}\ 96 * 10^{-3}\right] \end{aligned} \tag{7}$$

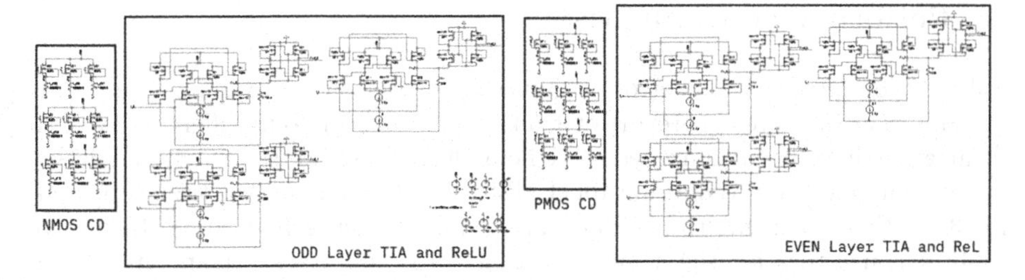

Fig. 10. 2 layer 3 × 3 matrix multiplication.

PMOS :-

$$\begin{aligned} &\left[24 * 10^{-3}\ 60 * 10^{-3}\ 96 * 10^{-3}\right] * \begin{bmatrix} 0.2\ 0.5\ 0.8 \\ 0.2\ 0.5\ 0.8 \\ 0.2\ 0.5\ 0.8 \end{bmatrix} \\ &= \left[36.4 * 10^{-3}\ 91 * 10^{-3}\ 145.6 * 10^{-3}\right] \end{aligned} \tag{8}$$

Table 5. 2 layer 3 × 3 matrix multiplication.

Input voltage (mV)	Weight	Voltage at buffer (mV)	Current through resistor (nA)
40	0.2	38.3	8.151
	0.5	37.4	20.9
	0.8	36.6	34.51
50	0.2	48.2	10.1
	0.5	47.1	25.7
	0.8	46.3	42.2
30	0.2	28.5	6.1
	0.5	27.7	16.06
	0.8	27.06	26.8
–26.5	0.2	–25.2	–5.5
	0.5	–24.3	–15.4
	0.8	–23.6	–27.4
–59.65	0.2	–57.6	–12.015
	0.5	–56.4	–31.57
	0.8	–55.5	–53.032
–95.79	0.2	–93.3	–19.1
	0.5	–91.8	–49.14
	0.8	–90.6	–81.157

The realized circuit functionality occupies minimum number of transistor count compared to the fastest digital multiplier which is a known as a vedic multiplier. Reduction in transistor count minimizes area as well as power consumption.

5 Performance Comparison

Table 6 shows less than 1% error in simulated and theoretical values. The power consumption and area in terms of transistor count for four different layer of Analog NN have been compared to vedic multiplier [13] in Table 7. It should be noted that one 8 bit vedic multiplier [13] uses 1638 number of transistor and for 3×1, three such multiplier units are required. Hence it requires 1638 × 3 = 4914 transistors for the same operation. The values are extrapolated from [13] for all other matrix multiplications. In analog, the addition is done on the node without any transistors, therefore no digital adder is required. Also In [1] power consumption at different voltage levels had been shown and 0.9 V is used for comparison. For 8-bit multiplier, two such 4-bit multipliers are required and therefore one 4-bit in [1] uses 31.48 μW, hence 8 bit will use 125.92 μW and such the extrapolated value is shown in Table 7.

It is found that two layer of 3 × 3 Analog multiplication has been performed with 96% and 7% reduction in transistor count and power dissipation respectively compared to vedic multiplier [1,13].

As seen the analog circuit uses much lower die area compare to digital counterpart, such circuits may be suited for edge devices where area is a tight constraint. Since it comes at the cost of error, its effect can be eliminated in neural network through training of it in the presence of noise. The main advantage of such analog circuit design compare to a software based approach are speed of operation and power consumption.

Table 6. Theoretical vs simulated values.

Name	Theoretical value (mV)	Simulated value (mV)
3 × 1 matrix	62	62.058
3 × 1 negative matrix	42	41.987
3 × 3 Matrix	24	26.5
	60	59.6
	96	95.7
2 3 × 3 matrix	24	26.5
	60	59.6
	96	95.7
	36.4	35.94
	91	90.99
	145.6	144.95

Table 7. Comparison between analog and digital matrix multiplier.

Name	Analog power (uW)	Digital power (uW) [1]	Analog transistor count	Vedic multiplier transistor count [13]
3 × 1 matrix	398	377.76	258	4914
3 × 1 negative matrix	401	377.76	258	4914
3 × 3 Matrix	1,192	1,133.28	642	14,742
2 3 × 3 matrix	2,436	2,266.56	1192	29,484

6 Conclusion

The basic architecture of a neuron is presented using Common Drain, Resistor, TIA and CMOS Rectifier analog circuits. The circuit for 3×1, 3×3, and two layers of 3×3 have been implemented, which generate output with an error voltage of 1 mV and simulation results has been compared to theoretical values for a maximum voltage of 700 mV. The circuit is capable to work up-to 61 MHz. The reduction in area and power dissipation is 96% and 7% respectively compared to vedic multiplier [1,13]. And this is obtained at a cost of output error of ± 1 mV.

References

1. Jie, L.S., Ruslan, S.H.: A 4×4 bit vedic multiplier with different voltage supply in 90 nm CMOS technology. Int. J. Integr. Eng. **9**, 114–117 (2017)
2. Ghomi, A., Dolatshahi, M.: Design of a new CMOS low-power analogue neuron. IETE J. Res. **64**, 1–9 (2017). https://doi.org/10.1080/03772063.2017.1351315
3. Santiago, A.M., Hernández-Gracidas, C., Rosales, L.M., Algredo-Badillo, I., García, M., Orozco-Torres, M.C.: CMOS implementation of ANNs based on analog optimization of n-dimensional objective functions implementation of anns based on analog optimization of n-dimensional objective functions. Sensors **21**, 7071 (2021). https://doi.org/10.3390/s21217071
4. Hasan, R., Taha, T.M., Yakopcic, C.: On-chip training of memristor based deep neural networks. In: 2017 International Joint Conference on Neural Networks (IJCNN), pp. 3527–3534 (2017). https://doi.org/10.1109/IJCNN.2017.7966300
5. Krestinskaya, O., Salama, K.N., James, A.P.: Learning in memristive neural network architectures using analog backpropagation circuits. IEEE Trans. Circ. Syst. I: Regular Papers **66**(2), 719–732 (2019). https://doi.org/10.1109/TCSI.2018.2866510
6. Yammenavar, D.B., Gurunaik, V., Bevinagidad, R., Gandage, V.: Design and analog VLSI implementation of artificial neural network. Int. J. Artif. Intell. Appl. **2**, 96–109 (2011). https://doi.org/10.5121/ijaia.2011.2309
7. Kawaguchi, M., Ishii, N., Umeno, M.: Analog neural network model based on improved logarithmic multipliers. In: 2022 12th International Congress on Advanced Applied Informatics (IIAI-AAI), pp. 378–383 (2022). https://doi.org/10.1109/IIAIAAI55812.2022.00082

8. Srivastava, S., Rathod, S.S.: Silicon neuron-analog CMOS VLSI implementation and analysis at 180 nm. In: 2016 3rd International Conference on Devices, Circuits and Systems (ICDCS), pp. 28–32 (2016). https://doi.org/10.1109/ICDCSyst.2016.7570617
9. Razavi, B.: The cross-coupled pair - part i [a circuit for all seasons]. IEEE Solid-State Circ. Maga. **6**, 7–10 (2014). https://doi.org/10.1109/MSSC.2014.2329234
10. Razavi, B.: The cross-coupled pair - part ii. IEEE Solid-State Circ. Maga. **6**, 9–12 (2014). https://doi.org/10.1109/MSSC.2014.2352532
11. Razavi, B.: The cross-coupled pair? part iii [a circuit for all seasons]. IEEE Solid-State Circ. Maga. **7**(1), 10–13 (2015). https://doi.org/10.1109/MSSC.2014.2369332
12. Raghuram, S., Priyanka, P., Nisarga, G.K.: CMOS implementations of rectified linear activation function. IEEE Solid-State Circ. Maga. **892**, 121–129 (2018)
13. Patro, A.K., Dekate, K.N.: A transistor level analysis for a 8-bit vedic multiplier. Int. J. Electron. Signals Syst. **2**, 2231–5969 (2012)

Designing Fog Device Network for Digitization of University Campus

Satveer Singh[✉] and Deo Prakash Vidyarthi

School of Computer and Systems Sciences, Jawaharlal Nehru University, New Delhi 110067, India

satveersingh.1339@gmail.com, dpv@mail.jnu.ac.in

Abstract. With the evolving digitization, services of Cloud and Fog make things easier which is offered in form of storage, computing, networking etc. The importance of digitalization has been realized severely with the home isolation due to COVID-19 pandemic. Researchers have suggested on planning and designing the network of Fog devices to offer services nearby the edge devices. In this work, Fog device network design is proposed for a university campus by formulating a mathematical model. This formulation is used to find the optimal location for the Fog device placement and interconnection between Fog devices and the Cloud (Centralized Information Storage). The proposed model minimizes the deployment cost and the network traffic towards Cloud. The IBM CPLEX optimization tool is used to evaluate the proposed multi-objective optimization problem. Classical multi-objective optimization method, i.e., Weighted Sum approach is used for the purpose. The experimental results exhibit optimal placement of Fog devices with minimum deployment cost.

Keywords: Fog computing · Cloud computing · Fog device placement · Multi-objective optimization · University digitization

1 Introduction

Cloud computing is a well-known term for everyone in the era of heavy computer usage and digitalization. It is a method of accessing information and applications over the internet instead of keeping and maintaining them on the personal system. One can access the Cloud services like stored data, applications, development tools from anywhere and anytime using internet. The Cloud service provider stores data and applications at a centralized location known as Cloud datacenters. As these centralized datacenters are oblivious and located very far from end-users, these are referred as Cloud. Cloud computing is a business model and therefore limited demands from the end-users are not profitable for the cloud service providers in terms of time and cost.

Recently, Fog computing has emerged as an advanced computing technology for processing the limited and time-critical demand by the consumers. It is an extension of the Cloud services at the edge of the network in terms of the Fog devices such as smart routers, switches, and limited capacity machines. Fog computing is decentralized

K. K. Patel et al. (Eds.): icSoftComp 2022, CCIS 1788, pp. 123–134, 2023.
https://doi.org/10.1007/978-3-031-27609-5_10

and heterogeneous in terms of its functionality and capacity. The integration of Cloud and Fog computing offers a more suitable and viable architecture to handle every type of demand. Cloud datacenter can be used as backup machinery and for analyzing the pattern of information to make the system faster in response.

1.1 Network Architecture of Fog-integrated Cloud

Many researchers have presented the three-layer architecture of the Fog-integrated Cloud in [10, 14–16, 22]. Figure 1 shows a three-layer network architecture of Fog-integrated Cloud computing.

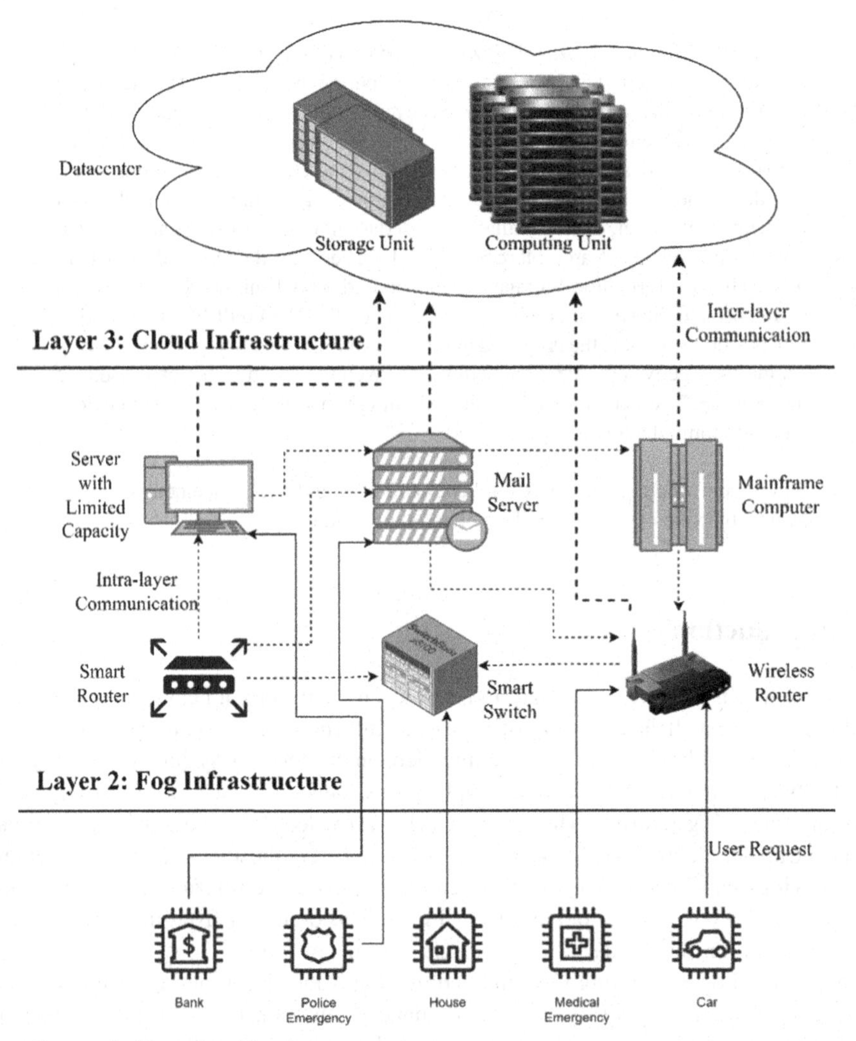

Fig. 1. Three-layer fog-cloud architecture

The first layer includes the user applications, like bank finance applications, medical equipment such as ECG machine at hospital, a house with smart TV and refrigerator, a car with GPS connectivity etc., equipped with internet connectivity. These applications can connect to Fog devices or Cloud datacenter via local or wide area network, respectively. Layer two consists of multiple heterogeneous Fog devices with limited storage and processing capacity. These Fog devices can communicate with each other on local area network and need a wide area network to communicate with Cloud datacenter. The Cloud infrastructure layer consists of a centralized datacenter, which has multiple physical machines with an extensive capacity for storage and processing.

Many researchers have developed several models to digitalize and make smart the heath related services in [7, 11, 23], and [1]. They have placed smart gateways in between end devices and Cloud infrastructure. Several smart parking models are proposed in [2, 4, 21], and [3]. Using these smart parking models, they have reduced CO_2 emission, the time need to find a space for parking and also maximized the profit of parking authority. In [5, 12, 20], and [19], smart waste management systems are proposed in integrated environment of Fog and Cloud. They have used some shortest path algorithms for waste collection. Some models are fully automatic and implemented in real environment.

Several researchers have proposed network design of Fog devices considering various objectives in [6, 8, 17, 18, 24], and [13]. In some designs, they formulate their problem into mathematical form. They have also found out optimal placement positions for Fog devices, but, still, none focused on university digitalization. This work proposes a university digitalization model by formulating a mathematical model.

An appropriate plan, to construct the network of Fog devices, is essential for the providers because it reduces their deployment cost and consumers will have a better quality of experience. Designing of optimal Fog network is a very challenging task because of heterogeneous and decentralized nature of Fog infrastructure. To address this, a linear mathematical formulation is modelled.

Fog device network design has received relatively less attention so far. Furthermore, to our knowledge, no previous study has addressed the Fog Device Network Design Problem (FDNDP) for a university campus. During COVID-19 pandemic, the need for the university digitalization has been vigorously felt for the easy access to campus resources over the internet. The contribution of this work is towards the University digitization which can be summarized as follows.

- A mathematical model is formulated to solve the multi-objective optimization problem.
- The model selects an optimal location for the Fog node placement and workload offloading at the best available Fog device.
- Two objectives; minimization of deployment cost and the traffic towards the Cloud are considered.
- CPLEX optimization tool of IBM is used to implement the proposed model.
- A Weighted Sum Multi-Objective Optimization Method (WS-MOOM) is used.

The outline of this paper is as follows. The proposed work, with mathematical formulation, is presented in Sect. 2. Experimental results are presented and analyzed in Sect. 3. The concluding remarks are made in Sect. 4.

2 The Proposed Work

The preliminaries, required for understanding of the proposed work and the mathematical formulation, are described in this section.

2.1 Set of Network Entities

User Request Category (URC). In this work, it is suggested to group the request, received from the users, which is termed as User Request Category. For example, in the proposed model of university digitalization, six fundamental categories of services are considered to process the batch of user requests; library, hostels, administration, hospital, and canteen. URC denotes here a collection of requests of a user application.

- U, set of URCs which generates requests for services offered by Fog devices. Each URC packet $u \in U$ has an aggregated number of Processing Elements (PE), Main Memory size (RAM), and Storage Memory (SM) demands, and packet size.
  - m_u, the amount of RAM required by an URC $u \in U$.
  - e_u, the number of PE required by an URC $u \in U$.
  - s_u, the amount of SM required by an URC $u \in U$.
  - R_u, the number of requests generated by an URC $u \in U$.
  - $RSize_u$, the average request size generated by an URC $u \in U$.
  - n_u, the link speed of an URC $u \in U$.
  - T_u, the total packet size sending from an URC $u \in U$.

Fog Device Types (FDT). The Fog devices are heterogeneous in processing and hardware configuration. This work categorizes Fog devices based on their configuration like PE, RAM, and SM.

- F, set of FDTs that are used to deploy in the network. Each FDT $f \in F$ have different specifications.
  - M_f, the RAM size available in a fog device of type $f \in F$.
  - E_f, the number of PEs available in a fog device of type $f \in F$.
  - S_f, the SM size available in a fog device of type $f \in F$.
  - I_f, the number of fog devices available of type $f \in F$.
  - $Cost_f$, the cost (Rs.) of fog device of type $f \in F$.

Connection Link Type (CLT). Different types of connection links are considered in this network design to connect Fog devices and CIS. Each CLT has their specific bandwidth capacity and cost. These connection links are used to communicate between Fog devices and the CIS to handle data synchronization and backup purposes.

- L, set of CLTs that are used to make connections between the Fog devices and the CIS. Each CLT $l \in L$ have different bandwidth capacities, and cost.

- BW_l, the bandwidth capacity of CLT $l \in L$.
- $Cost_l$, the cost (Rs./meter) of CLT $l \in L$.

Possible Placement Locations (PPL). Five placement locations are considered.

- P, set of PPLs for the installation of fog devices. Each PPL $p \in P$ have different renting cost.
  - $Cost_p$, the cost (Rs.) of PPL $p \in P$.

Resource Demand Types (RDT). Six fundamental services, offered by a university, are considered. The demand for resource size may vary depending on the type of user request. Each packet, generated by a URC, may demand PE, RAM, and SM. An URC packet is a collection of requests generated from a particular category. Each URC has an average request size, which helps in packet size calculation. The selection of a Fog device type depends on the URC packet size.

2.2 Constants

In this work, two constant values are considered to compute the traffic and propagation delay.

- r, the average percentage of traffic going from Fog devices to CIS.
- t, the speed of light (in meter per second).

2.3 Functions

- $Dist_{ab} = Distance(a, b)$, the Euclidean distance between two-points a and b of x, y coordinates in the plane.
- $T_u = R_u * RSize_u$, T_u indicates the total packet size of an URC. The total packet size can be calculated using the number of requests generated (R_u) multiply by the average request size ($RSize_u$) of an URC.

2.4 Decision Variables

- $A_{fp} = \begin{Bmatrix} 1 & if\,f\ Fog\ device\ of\ type\ f \in F\ is\ placed\ at\ PPL\ p \in P \\ 0 & Otherwise \end{Bmatrix}$
- $B_{up} = \begin{Bmatrix} 1 & if\,f\ the\ URC\ is\ served\ by\ a\ Fog\ device\ placed\ at\ PPL\ p \in P \\ 0 & Otherwise \end{Bmatrix}$
- $C_u = \begin{Bmatrix} 1 & if\,f\ the\ URC\ u \in U\ is\ served\ by\ the\ CIS \\ 0 & Otherwise \end{Bmatrix}$
- $D_{pl} = \begin{Bmatrix} 1 & if\,f\ a\ PPL\ is\ connected\ to\ the\ CIS\ using\ link\ of\ type\ l \in L \\ 0 & Otherwise \end{Bmatrix}$

2.5 The Mathematical Model

Objectives. In this FDNDP formulation, Eq. (1) and (2) are used to minimize the deployment cost and traffic towards the CIS, respectively.

Minimize Deployment Cost

$$min\left[\sum_{p\in P}\sum_{f\in F}A_{fp} * Cost_p \sum_{p\in P}\sum_{f\in F}A_{fp} * Cost_f \sum_{p\in P}\sum_{l\in L}D_{pl} * Cost_l * Dist_{p,CIS}\right] \quad (1)$$

Minimize Traffic

$$min\left[\sum_{u\in U}C_u * T_u + \sum_{p\in P}\sum_{l\in L}\sum_{u\in U}B_{up} * T_u * r\right] \quad (2)$$

where r indicates the percentage of traffic (1% in the experiment) forwarded from Fog devices to the CIS for data synchronization and backup purposes.

Constraints. The constraints, for the model, are as follows:

Unique Fog Device Placement. The constraint in Eq. (3) ensures that at most one Fog device of type $f \in F$ can be placed at a PPL $p \in P$. If left side equals zero, no Fog device is placed at that particular location.

$$\sum_{f\in F}A_{fp} \leq 1; \quad (\forall p \in P) \quad (3)$$

Unique Service Provider. The constraint in Eq. (4) indicates that a URC packet is served by only one service provider, i.e., Fog device or CIS.

$$\sum_{p\in P}B_{up} + C_u = 1, \quad (\forall u \in U) \quad (4)$$

Unique Link Placement. The constraint in Eq. (5) ensures that at most one connection link of type $l \in L$ can be used for communication between PPL selected for the Fog devices and the CIS. If Eq. (5) equals to zero, means corresponding PPL is not selected for the Fog device placement.

$$\sum_{l\in L}D_{pl} \leq 1; \quad (\forall p \in P) \quad (5)$$

Link Assignment. The constraint in Eq. (6), ensures that the total number of Fog devices placed and connection links used are equal.

$$\sum_{f\in F}A_{fp} = \sum_{l\in L}D_{pl}; \quad (\forall p \in P) \quad (6)$$

PE Capacity. The constraint in Eq. (7) ensures that the number of PEs required by an URC packet does not exceed the number of PEs of the corresponding Fog device.

$$\sum_{u \in U} B_{up} * e_u \leq \sum_{f \in F} A_{fp} * E_f; \quad (p \in P) \tag{7}$$

RAM Capacity. The constraint in Eq. (8) ensures that the RAM size required by an URC packet does not exceed the RAM capacity of the corresponding Fog device.

$$\sum_{u \in U} B_{up} * m_u \leq \sum_{f \in F} A_{fp} * M_f; \quad (p \in P) \tag{8}$$

SM Capacity. The constraint in Eq. (9) ensures that the SM size required by an URC packet does not exceed the SM capacity of the corresponding Fog device.

$$\sum_{u \in U} B_{up} * s_u \leq \sum_{f \in F} A_{fp} * S_f; \quad (p \in P) \tag{9}$$

Inventory Capacity. The constraint in Eq. (10) ensures that one cannot use number of Fog devices more than the available Fog devices.

$$\sum_{p \in P} A_{fp} \leq I_f; \quad (\forall f \in F) \tag{10}$$

Link Capacity. The constraint in Eq. (11) ensures that the portion of data sent from the Fog devices to CIS cannot exceed the bandwidth capacity of communication link.

$$\sum_{u \in U} B_{up} * T_u * r \leq \sum_{l \in L} D_{pl} * BW_l; \quad (\forall p \in P) \tag{11}$$

Finally, Eq. (12) to Eq. (15) indicates that the all four decision variables are Boolean variables.

$$A_{fp} \in \{0, 1\}; \quad (\forall f \in F, \forall p \in P) \tag{12}$$

$$B_{up} \in \{0, 1\}; \quad (\forall u \in U, \forall p \in P) \tag{13}$$

$$C_u \in \{0, 1\}; \quad (\forall u \in U) \tag{14}$$

$$D_{pl} \in \{0, 1\}; \quad (\forall l \in L, \forall p \in P) \tag{15}$$

2.6 The Weighted Sum Multi-objective Optimization Method

This classical method converts a multi-objective problem into a single objective function, using weight coefficients for each objective [9]. Thus, the modified objective function for FDNDP can be rewritten as in Eq. (16).

$$minimize\left(w1 * Cost_{norm} + w2 * Traffic_{norm}\right) \; In \; this, w1 + w2 = 1 \tag{16}$$

Here, $w1$, and $w2$ are weight coefficients; and $Cost_{norm}$ & $Traffic_{norm}$ are the normalized values of the objectives because of the scale differences. The normalized values of the objectives can be calculated using Eq. (17) and Eq. (18).

$$Cost_{norm} = (Cost - minCost)/(maxCost - minCost) \tag{17}$$

$$Traffic_{norm} = (Traffic - minTraffic)/(maxTraffic - minTraffic) \tag{18}$$

Here, *minCost* and *minTraffic* can be calculated by minimizing each objective separately. Similarly, *maxCost* and *maxTraffic* are calculated by maximizing. Figure 2 depicts the steps applied in the proposed FDNDP using Weighted Sum MOOM (WS-MOOM). In the first step, we randomly generate the input values of user requests and start the optimization process. Initially, the FDNDP is solved using values of $w1$ and $w2$, 1 and 0, respectively. The obtained optimal value is put into the Pareto optimal set and solved the model using updated values of $w1$ and $w2$ until the terminating condition is reached. After getting all the possible values of the Pareto set, we will choose the best minimum value as an optimal solution for our problem.

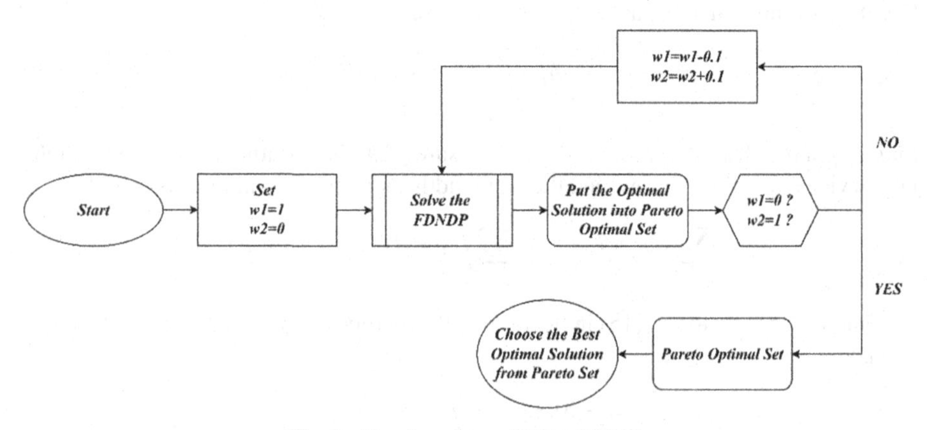

Fig. 2. The flowchart of WS-MOOM

3 The Experimental Evaluation

This section includes the experimental setup. IBM ILOG CPLEX Optimization Studio (Version: 20.1.0.0) is used to solve the proposed FDNDP. The simulation is performed on a system of 11th Gen Intel(R) Core (TM) i5-1135G7 at a speed of 2.40GHz with 16GB RAM.

3.1 Experimental Input

The input values of URC parameters are generated randomly based on a random function. Table 1 contains all relevant information of the input parameters of URC. We have considered only six URCs based on the number of applications in this FDNDP. A total of five data sets are randomly generated in this proposed problem, i.e., R10, R20, R40, R80, R160.

Table 1. The user application cluster input parameters

Simulation Area	5000×5000 m^2
Coordinates of URC's	(x, y) within (5000×5000) meter^2
Number of URC's (u)	6
Average Request Size of a URC ($RSize$)	[1000, 1200, 1500, 1800, 1100, 800] Bytes
RAM size required by a URC (m)	[50 to 1000] in Mega Bytes
Number of PEs required by a URC (e)	[2 to 32] in cores
SM size required by a URC (s)	[2 to 50] in Giga Bytes

The Fog devices are heterogeneous and have different configurations in terms of capacity. In this proposed work, we have considered four types of Fog devices. The characteristics of these Fog devices are mentioned in Table 2.

Table 2. The characteristics of the fog device types

Fog device type	No. of PEs (Cores)	RAM size (Gigabytes)	SM size (Gigabytes)	Cost ($Cost_f$)	Inventory (Quantity)
1	10	10	500	50000	3
2	20	20	1000	75000	2
3	30	30	2000	125000	2
4	50	50	5000	150000	1

Table 3. The characteristics of communication link types

Connection link type	Bandwidth capacity (Mbps)	Cost (per meter)
1	100	500
2	1000	1000
3	10000	2000

Table 3 includes the information on the communication link types which are used to connect each Fog device with the CIS. In this work, three types of communication links are used for the connection. Five possible placement locations are considered in

Table 4. The parameters of possible placement locations

No. of PPLs (p)	Coordinate x	Coordinate y	Cost (Thousands)($Cost_p$)
1	4090	4232	10
2	4921	1617	20
3	4582	4632	30
4	3940	3053	40
5	2796	2154	50

the campus. The x and y coordinates of PPL are fixed. A fixed cost of each location needs to pay for installing a Fog device. The values of each parameter of the PPL are mentioned in Table 4.

3.2 Experimental Results

In this section, the results obtained using the WS-MOOM are presented and analyzed. Table 5 shows that the FDNDP is solved for 11 combinations of $w1$ and $w2$. The maximum and minimum value of set of minimum values are highlighted in bold.

Table 5. The pareto optimal set using WS-MOOM

w1	w2	R10	R20	R40	R80	R160
1.0	0.0	0.0000	0.0000	0.0000	0.0000	0.0000
0.9	0.1	0.0577	0.0631	0.0799	0.1000	0.0961
0.8	0.2	0.0799	0.0937	0.1026	0.1566	0.1350
0.7	0.3	**0.0885**	0.0963	0.1201	0.1972	0.1739
0.6	0.4	0.0866	**0.0980**	**0.1223**	0.2250	0.2127
0.5	0.5	0.0722	0.0967	0.1085	**0.2312**	0.2251
0.4	0.6	0.0577	0.0773	0.0868	0.2150	**0.2291**
0.3	0.7	0.0433	0.0580	0.0651	0.1620	0.2021
0.2	0.8	0.0289	0.0387	0.0434	0.1080	0.1347
0.1	0.9	**0.0144**	**0.0193**	**0.0217**	**0.0540**	**0.0674**
0.0	1.0	0.0000	0.0000	0.0000	0.0000	0.0000

The graphical representation of minimum, average, and maximum values of different data sets is presented in Fig. 3(a, b), and (c), respectively. In each scenario, the output values are increasing as the data set size increases.

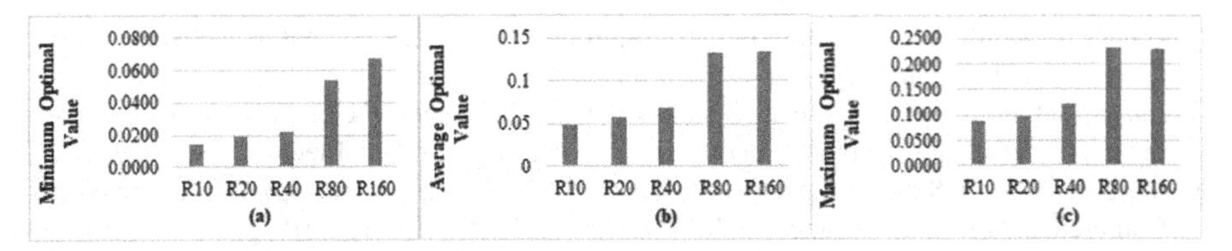

Fig. 3. (a) The minimum values of different data sets (b) The average values of different data sets (c) The maximum value of different data sets

3.3 Analysis of Results

The WS-MOOM is one of the most widely used methods due to its simplicity. However, it is usually difficult to generate a good set of points that are uniformly distributed on the Pareto front. Furthermore, proper scaling or normalization of the objectives are often needed so that the ranges/values of each objective should be comparable. Otherwise, the weight coefficients are not well distributed leading to biased sampling on the Pareto front. So, we have performed both the steps of choosing normalized value and optimal weight. In the context, we have tried and compared with various permutations of weights to explore the most optimal value to minimize deployment cost and traffic while considering the decision-making problems in Fog device network design problem.

4 Conclusion

This paper uses a linear mathematical formulation to solve a Fog Device Network Design Problem with an application to university campus digitalization. The proposed solution finds the optimal placement locations for Fog devices, suitable capacity, and the connection links. Two objectives have been considered, i.e., deployment cost and traffic. This multi-objective optimization problem is solved using Weighted Sum multi-objective optimization method. The IBM CPLEX optimization tool is used to evaluate the proposed multi-objective optimization problem. The experimental results exhibit optimal placement of Fog devices with minimum deployment cost.

References

1. Asghar, A., et al.: Fog based architecture and load balancing methodology for health monitoring systems. IEEE Access **9**, 96189–96200 (2021)
2. Awaisi, K.S., et al.: Towards a fog enabled efficient car parking architecture. IEEE Access **7**, 159100–159111 (2019)
3. Balfaqih, M., et al.: Design and development of smart parking system based on fog computing and internet of things. Electron. **10**(24), 1–18 (2021)
4. Celaya-Echarri, M., et al.: Building decentralized fog computing-based smart parking systems: from deterministic propagation modeling to practical deployment. IEEE Access **8**, 117666–117688 (2020)
5. Garach, P.V., Thakkar, R.: A survey on FOG computing for smart waste management system. In: ICCT 2017 - International Conference on Intelligent Computing and Communication Technologies, pp. 272–278 (2018)

6. Haider, F., et al.: On the planning and design problem of fog computing networks. IEEE Trans. Cloud Comput. **9**(2), 724–736 (2021)
7. Ijaz, M., et al.: Integration and applications of fog computing and cloud computing based on the internet of things for provision of healthcare services at home. Electron. **10**, 9 (2021)
8. Maiti, P., et al.: QoS-aware fog nodes placement. In: Proceedings of the 4th IEEE International Conference on Recent Advances in Information Technology RAIT 2018, pp. 1–6 (2018)
9. Marler, R.T., Arora, J.S.: The weighted sum method for multi-objective optimization: new insights. Struct. Multidiscip. Optim. **41**(6), 853–862 (2010)
10. OpenfogConsortium: OpenFog Reference Architecture for Fog Computing Produced. Reference Architecture, pp. 1–162 (2017)
11. Rahmani, A.M., et al.: Exploiting smart e-Health gateways at the edge of healthcare Internet-of-Things: a fog computing approach. Futur. Gener. Comput. Syst. **78**, 641–658 (2018)
12. Saroa, M.K., Aron, R.: Fog computing and its role in development of smart applications. In: Proceedings - 16th IEEE International Symposium on Parallel and Distributed Processing with Applications. 7th IEEE International Conference Ubiquitous Computing Communications. 8th IEEE International Conference on Big Data Cloud Computing 11t. pp. 1120–1127 (2019)
13. Shaheen, Q., et al.: A Lightweight Location-Aware Fog Framework (LAFF) for QoS in Internet of Things Paradigm. Mobile Information Systems 2020 (2020)
14. Sham, E.E., Vidyarthi, D.P.: Admission control and resource provisioning in fog-integrated cloud using modified fuzzy inference system. J. Supercomput. **78**, 1–41 (2022)
15. Sham, E.E., Vidyarthi, D.P.: CoFA for QoS based secure communication using adaptive chaos dynamical system in fog-integrated cloud. Digit. Signal Process. **126**, 103523 (2022)
16. Sham, E.E., Vidyarthi, D.P.: Intelligent admission control manager for fog-integrated cloud: a hybrid machine learning approach. Concurr. Comput. Pract. Exper. **34,** 1–27 (2021)
17. da Silva, R.A.C., da Fonseca, N.L.S.: On the location of fog nodes in fog-cloud infrastructures. Sensors (Switzerland). **19**, 11 (2019)
18. Da Silva, R.A.C., Da Fonseca, N.L.S.: Location of fog nodes for reduction of energy consumption of end-user devices. IEEE Trans. Green Commun. Netw. **4**(2), 593–605 (2020)
19. Sohag, M.U., Podder, A.K.: Smart garbage management system for a sustainable urban life: an IoT based application. Internet of Things. **11**, 100255 (2020)
20. Srikanth, C.S., et al.: Smart waste management using internet-of-things (IoT). Int. J. Innov. Technol. Explor. Eng. **8**(9), 2518–2522 (2019)
21. Tang, C., et al.: Towards smart parking based on fog computing. IEEE Access **6**, 70172–70185 (2018)
22. Tomovic, S., Yoshigoe, K., Maljevic, I., Radusinovic, I.: Software-defined fog network architecture for IoT. Wireless Pers. Commun. **92**(1), 181–196 (2016). https://doi.org/10.1007/s11277-016-3845-0
23. Vilela, P.H., et al.: Performance evaluation of a Fog-assisted IoT solution for e-Health applications. Futur. Gener. Comput. Syst. **97**, 379–386 (2019)
24. Zhang, D., et al.: Model and algorithms for the planning of fog computing networks. IEEE Internet Things J. **6**(2), 3873–3884 (2019)

Word Sense Disambiguation from English to Indic Language: Approaches and Opportunities

Binod Kumar Mishra(✉) and Suresh Jain

Computer Science and Engineering Department, Medi-Caps University, Indore, India
bkmishra21@gmail.com, suresh.jain@rediffmail.com

Abstract. Ambiguity is one of the major challenges in Natural Language Processing and the process to solve is known as Word Sense Disambiguation. It is useful to determine the appropriate meaning of polysemy words in a given context using computational methods. Generally, Knowledge, Supervised, and Unsupervised based approaches are the most common methods used to resolve ambiguity problems that occur in a sentence. The government of India has initiated many digital services for its citizen in the last decade. All these services require natural language processing to be easily accessed by web portals or any electronic gadget. Also, these services are provided by the government in Hindi or other Indian languages to better serve Indian citizens. Since English and other languages like Chinese, Japanese, and Korean have plenty of resources available to build applications based on natural language processing but due to low resources available for disambiguating polysemous words in Hindi and other Indian languages, it becomes a hindrance to building any application based on these languages. In this paper, the suggested method enables the assessment of the correct meaning in terms of sustaining data sequences. In order to automatically extract features, the proposed method uses an RNN neural network model. Additionally, it integrates glosses from IndoWordNet. The outcomes demonstrate that the suggested technique performs consistently and significantly better than the alternatives.

Keywords: Word sense disambiguation · RNN · WordNet · LESK · Naïve Bayes

1 Introduction

Language technologies are the basic instrument used by millions of people in everyday life. Various NLP applications which are based upon these technologies like Machine Translation or Web Search engines rely on linguistic knowledge. This development did not make a high impact on the majority of the people due to little awareness and not being available in his own language. Ambiguity is one of the most fundamental and intermediate problems found in every NLP application. It is also considered an AI-complete problem (Navigli 2009).

K. K. Patel et al. (Eds.): icSoftComp 2022, CCIS 1788, pp. 135–146, 2023.
https://doi.org/10.1007/978-3-031-27609-5_11

Words can have various meanings depending on the situation due to the ambiguity of natural language. For instance, the word "bank" can signify either "a bank as a financial entity" or "the bank of a river". As a person, it appears to be quite simple to determine the appropriate sense of the given context, but for a machine, it is highly challenging to find out exact sense. In order to determine the appropriate meaning, it is necessary to process an extremely large amount of data and to store that data in a specific location (Ng and Lee 1996; Nguyen et al. 2018; Sarika and Sharma 2015). Sometimes, part-of-speech (POS) tags can help to resolve ambiguity to some extent, but even for the same part-of-speech words that have the same POS tags, the word senses are still very unclear (Taghipour and Ng 2015).

Over the past decade, the government of India has introduced a wide variety of digital services for the country's population. The government makes these services available to Indian citizens in Hindi as well as in other Indian languages so that it can serve them more effectively. All of these digital services demand natural language processing, which will make it possible for them to be conveniently accessed through online portals or any other electronic device.

Polysemous words are those that can be comprehended in more than one way, and they are found in all natural languages. Word Sense Disambiguation is a tool that assists natural language processing applications in better understanding language and performing effectively.

This ambiguity problem can be solved using a variety of methodologies, including knowledge-based, supervised, and unsupervised methods. For disambiguation, each of these approaches required using at least one of two resources: wordnet or a corpus.

Since English and other languages like Chinese, Japanese, and Korean have plenty of resources available to perform word sense disambiguation. The limited resources that are now available to disambiguate polysemous terms in Hindi and other Indic languages create a barrier to the development of any application that is based on one of these languages. In order for users who are only competent in a regional language to be able to engage with or manage computer-based systems, it is necessary to have access to certain resources and tools that can convert natural language into a form that can be processed by computers. Therefore, the word sense disambiguation problem must be solved before processing input in any natural language application in order for it to produce better results.

The following sections of the paper are arranged as follows: The resources required for disambiguation are listed in Sect. 2 whereas variants of WSD describe in Sect. 3. Related work which compares existing methods for English and other languages is explained in Sect. 4 and Sect. 5 deals with proposed work for WSD. Section 6 elaborates on the result discussion. The conclusion and future directions are discussed in Sect. 7.

2 Resources Required for Disambiguation

Various resources such as WordNet and Corpus are required to disambiguate polysemous words. These knowledge sources contain information that is necessary for linking senses to a particular word. Corpus contains sense annotated or raw corpus. One or more resources are required for this task, the list of resources is as follows:

2.1 Machine Readable Dictionaries/Thesaurus

The advantage is taken by the research community when electronic forms of dictionaries are available during the 1970s and 1980s at that time it was quite popular. It offers a glossary of terms, definitions, and examples of how to use them. Similarly, in a thesaurus, instead of storing definitions, it includes word relationships such as synonyms, antonyms, and a variety of other lexical links (Wilks et al. 1996).

2.2 WordNet

It is a database of words in the English language. It was created by combining lexical relations like synonymy and antonymy with semantic relations like hyponyms, hypernymies, meronyms, entailments, and toponymies. WordNet 3.0 is currently available, with around 155,000 words and 117,000 synsets (Miller 1995). The same structure created by IIT Bombay's Centre for Indian Language Technology (CFILT) (Jha et al. 2001) for Hindi and other Indian languages known as IndoWordNet. It contains synsets, hypernymy, hyponymy, holonomy, meronymy, antonymy, and many more relationship between words for 19 Indian languages.

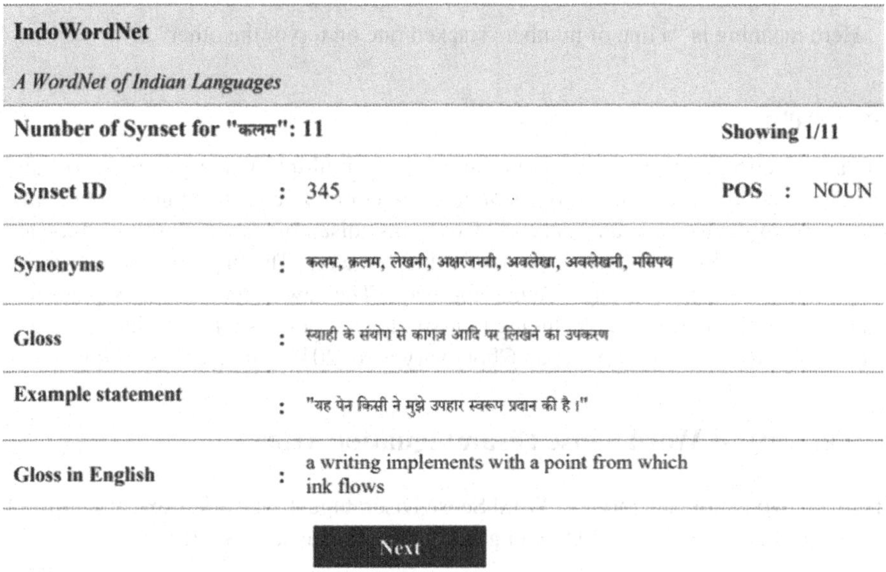

IndoWordNet

A WordNet of Indian Languages

Number of Synset for "कलम": 11		**Showing 1/11**
Synset ID	: 345	**POS** : NOUN
Synonyms	: कलम, क़लम, लेखनी, अक्षरजननी, अवलेखा, अवलेखनी, मसिपथ	
Gloss	: स्याही के संयोग से कागज़ आदि पर लिखने का उपकरण	
Example statement	: "यह पेन किसी ने मुझे उपहार स्वरूप प्रदान की है।"	
Gloss in English	: a writing implements with a point from which ink flows	

Next

Fig. 1. IndoWordNet (Bhattacharyya 2010; Jha et al. 2001)

A snapshot of IndoWordNet is shown in Fig. 1 taken from (Jha et al. 2001; Narayan et al. 2002) for the word "कलम". It gives 11 different meanings; all meaning is decided according to the context they used. These senses can be used to disambiguate polysemous words. Out of 11, five sense are as follow:

- स्याही के संयोग से कागज़ आदि पर लिखने का उपकरण

Here meaning is "an item used to write with that has an ink-flowing tip".

- पेड़ की वह टहनी जो दूसरी जगह बैठाने या दूसरे पेड़ में पैबंद लगाने के लिए काटी जाए

Here meaning is related to "a portion of a plant that is occasionally removed in order to root or graft a new plant".

- सिर के वे बाल जो कनपटी के पास होते हैं

Here meaning is "beard that has grown in front of a man's ears down the side of his face, especially when the rest of the beard has been shaven".

- चित्रकार के रंग भरने की कलम

Here meaning is related to "a brush used as an applicator (to apply paint)".

- बही-खाते आदि में लिखा जाने वाला कोई मद

Here meaning is "a line of numbers stacked one on top of the other".

2.3 Corpus

It is a collection of texts for a single language or multiple languages used to build language models. There are two types of corpora required i.e., sense annotated and raw corpus, for supervised and unsupervised word sense disambiguation. A sense annotated corpus is created by person who is expert in this language. The linguistic experts create sense annotation corpus using manual approach. The raw corpus contains plain text taken from different domains including tourism, health, news, sports, stories, literature, history, etc. (Ramamoorthy Narayan Choudhary et al. 2019; Singh and Siddiqui 2016).

3 Variants of Word Sense Disambiguation Work

There are two major categories of Word Sense Disambiguation work, Lexical sample or Target word and All-word WSD (Bhingardive and Bhattacharyya 2017):

3.1 Lexical Sample (or Target Word or One Word) WSD

It would apply when the system is required to disambiguate a single word in a particular sentence. In this case a machine learning approach are used for this dedicated word, to trained the model by using corpus. With the help of this it can determine the correct meaning for the target word in the given context.

3.2 All Word WSD

With the help of this it can predict more ambiguous word in the given sentence. In the provided context, it identifies every word that falls within the open-class category, including Nouns, Adjectives, Verbs, and Adverbs. Since Machine learning specially in supervised approach it required a large amount of tagged corpus also it cannot easily scale up. So, in this case knowledge-based or Graph based are most suitable methods for all-word WSD tasks.

4 Related Work

The problem of ambiguity can be solved by different way, and it depend upon the resources used to solve this problem. Generally, knowledge-based approach used WordNet or IndoWordNet, supervised based approach used sense annotated datasets and unsupervised based approach used raw corpus.

In knowledge-based approach generally we extract information from wordnet. Since lexical resources like WordNet playing important role to find out glosses of target word. Initially this approach developed by Lesk (Lesk 1986) which used overlap function between glosses of target word and context word. In this method it selects the word that most overlaps with words in context. Later this work is modified by Banerjee & Pedersen (Banerjee and Pedersen 2003; Banerjee et al. 2002) by using semantic relation of word with the help of WordNet. Basile et al. (2014) extend the LESK's work using distribution semantic model. The Distributional Semantics Models (DSM) realize the architectural metaphor of meanings, which have been represented as points in a space, where proximity is estimated by semantic similarity.

The majority of research on word sense disambiguation that has been published in the literature focuses on English as well as a number of other languages, including Arabic, Chinese, Japanese, and Korean. The first attempt to address the WSD problem in Hindi, however, was made by Sinha et al. (2004).

The supervised approach and Unsupervised technique require sense annotated corpus and raw corpus respectively (Bhingardive et al. 2015; Bhingardive and Bhattacharyya 2017). A sense annotated corpus was manually constructed by a language expert; as a result, it took a lot of time and labour and was occasionally improperly annotated. This method uses a word-specific classifier to determine the correct meaning of each word. It involves two steps. In the first step training the model is required and for this thing sense tagged corpus is used and this classifier capture the syntactic and semantic classifiers. The second step entails using classifiers to identify the meaning of an ambiguous word that better represents the surrounding context (Vaishnav and Sajja 2019).

Unsupervised approaches do not require sense-annotated corpora, saving a lot of time on corpus formation. With the help of context clustering, it discriminates different senses. Context and sense vectors are amalgamated to create clusters. By mapping the ambiguous word to a context vector in word space, the word is disambiguated in the given context. The closest sense vector receives the meaning reference. By mapping an ambiguous word to a context vector in Word Space, the context of the word is revealed. The sense with the closest sense vector is given the context. Another method is commonly used for disambiguation is known as Co-occurrences graphs. A graph for the target word is constructed with the use of corpora. In one paragraph, there are edges connecting two words that occur together. Each edge is given a weight in order to determine the relative frequency of the co-occurring words. Any nodes that represent a word's senses are connected to the target word when they are chosen. The distance is calculated by using Minimum Spanning Tree and result obtained from this is stored. All word WSD operations are performed using this spanning tree. Since supervised WSD approaches gives very good result in terms of accuracy, hence it overlap other approaches of Word Sense Disambiguation (Zhong and Ng 2010). Neural network technique also used corpus as well as it considers local context of the target word.

Through extensive literature review, it is observed that several shortcomings are associated with Hindi WSD. For this, we should not only depend upon WordNet but also Corpus. It is also noted down that word embedding technique is required for Hindi Text. To increase the accuracy for Hindi WSD, it is important to develop new techniques which will combine both senses and corpus to train the model.

The majority of the Indian language does not have rich resources like English who helps to solve the problem of WSD, so it requires more resources and efficient algorithms for better accuracy. (Sinha et al. 2004) developed a first-time statistical technique for Hindi WSD with the help of Indo-WordNet. Later many researchers developed models to solve Word Sense Disambiguation in Hindi and other Indian languages. Details of each approach are given in Table 1.

Table 1. Comparison of various Word Sense Disambiguation approaches

Approach	Type			
	Method used	Resources	Language	Accuracy
Knowledge based	LESK (Lesk 1986)	Machine readable dictionary	English	31.2%
	Extended LESK (Banerjee and Pedersen 2003; Banerjee et al. 2002)	WordNet	English	41.1%
	Extended LESK with TF-IDF (Basile et al. 2014)	BabelNet	English	71%
	LESK with Bi-gram and Tri-gram (Gautam and Sharma 2016)	Indo WordNet	Hindi	52.98%
	Map-Reduce function on Hadoop (Nair et al. 2019)	WordNet	English	51.68%
	Overlap based with Semantic relation (Sinha et al. 2004)	Indo WordNet	Hindi	40–70%

(continued)

Table 1. *(continued)*

Approach	Type			
	Method used	Resources	Language	Accuracy
Graph based	Score based LESK (Tripathi et al. 2020)	Indo WordNet	Hindi	61.2%
	Using Global and Local measure (Sheth et al. 2016)	Indo WordNet	Hindi	66.67%
Supervised based	SVM (Zhong and Ng 2010)	SemCor annotated corpus	English	65.5%
	Naïve Bayes (Singh et al. 2014)	Indo WordNet	Hindi	80.0%
	Cosine similarity (Sharma 2016)	Indo WordNet	Hindi	48.9%
	SVM with Embedding (Iacobacci et al. 2016)	WordNet	English	75.2%
	Random forest with Embedding(Agre et al. 2018)	SemCor	English	75.80%
	IMS with Massive Contest (Liu and Wei 2019)	SemCor	English	67.7%
	Embedding like Doc2Vec (Li et al. 2021)	WordNet	English	63.9%
Unsupervised	Expectation Maximization (Bhingardive and Bhattacharyya 2017; Khapra et al. 2011)	Indo WordNet	Hindi	54.98%
	Word2Vec (Kumari and Lobiyal 2020, 2021)	Indo WordNet	Hindi	52%
	Word2Vec with Cosine distance (Soni et al. 2021)	Indo WordNet	Hindi	57.21%
	ShotgunWSD 2.0 (Butnaru and Ionescu 2019)	WordNet	English	63.84%
	Train-O-Matric (Pasini and Navigli 2020)	WordNet	English	67.3%

5 Proposed Approach for WSD

Overview of the proposed WSD model explained in this section. It contains architecture, implementation details and evaluation.

5.1 Architecture of the Proposed WSD Model

Figure 2 depicts the suggested WSD model's general architecture, taken some ideas from (Kumari and Lobiyal 2020) There are three modules in it:

Distributed Word Representation: In this sub-module, numerical representation in terms vectors is generated. It requires corpus for Hindi text. After creating corpus, preprocessing is required which will remove punctuation, special symbols and stop words. Morphology is also required so that root words can be found and size of the corpus is reduced.

After preprocessing, each word or tokens is converted into vector representation using Bag of words, TF-IDF and Word2Vec approach.

Create Context and Sense Module: With the aid of the preceding phase, this sub-module generates vector representations of the input sentence. In each sentence at least one word represents as an ambiguous word or target word and remaining words represent as a context words. The context vector is a single representation of the context words and sense vector is a generated for each sense of target word. These senses are defined at Indo WordNet developed by (Jha et al. 2001).

Build Machine Learning WSD Model: Two inputs are necessary for creating a machine learning model: the context vector C and sense vectors for the ambiguous word those senses are defined WordNet. The sense vectors may be two or more, it depends upon the senses defined on Indo WordNet. In this model, memory module is also introduced, which will update memory to refine the sense.

5.2 Implementation Details

For building model, rule based, classical machine learning and neural network model was setup. Each model having own setup. To train model sense annotated corpus built, as well as to validate separate setup was built, it also includes same number of senses. These datasets keep them fixed for each model, so that comparable is easy.

For rule based WSD model simply, if-then-else statements are used. It includes LESK approach using overlap based to find out maximum score.

For classical machine leaning naïve bayes classifier is used. It contains groups of characteristics surrounding the target word within a specified timeframe, such as co-location and co-occurrence.

For Neural network, RNN model is used to maintain the sequences of the words. It shares parameters among all words. For hyperparameter tuning embedding size 100, dropout rate 0.2 and stochastic gradient descent optimizer is taken.

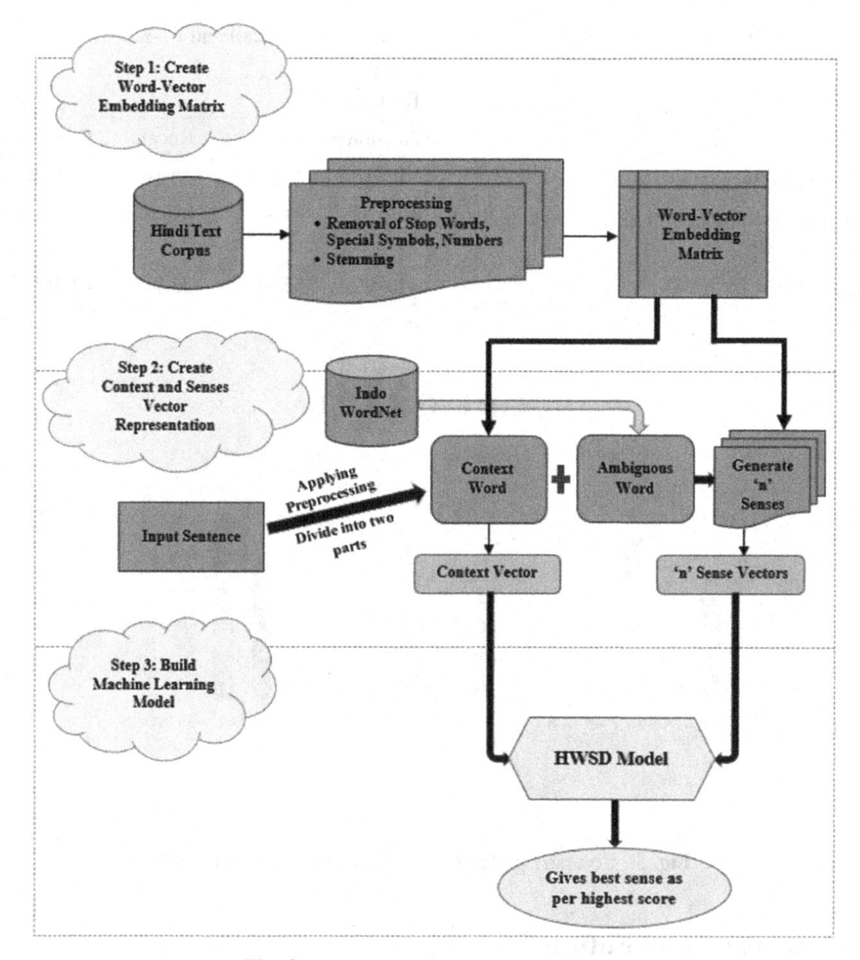

Fig. 2. Overview of the proposed model

6 Result Discussion

In this section, performance of each proposed model rule-based method, classical supervised method and neural-network based methods are describes. All method used same datasets for training and also for validation same datasets are used so that results can be comparable. Evaluation for each model, five different ambiguous sentences is used. Table 2 displays the outcomes for each model in terms of recall, accuracy, precision, and F1-score with our own dataset.

On observing Table 2, it shows that neural network based RNN model gives 41.61% accuracy, 55.9% Precision, 35.71% Recall and 43.58% F1-score, which are highest performance over others approach. This score has been found by applying concatenation of all ambiguous words.

Figure 3, show comparable scores for each approach, in this graph neural network based RNN model achieves highest performance in all measurement parameters.

Table 2. Result with respect to Accuracy, Precision, Recall and F1-score

HWSD Models	Results			
	Accuracy	Precision	Recall	F1-score
Rule based (Jha et al. 2001)	31.2%	25.2%	50.1%	33.53%
Classical Machine Learning (Singh et al. 2014) (Naïve Bayes)	33.67%	35.6%	38.71%	37.08%
*Neural Network (RNN)	**41.61%**	**55.9%**	**35.71%**	**43.58%**

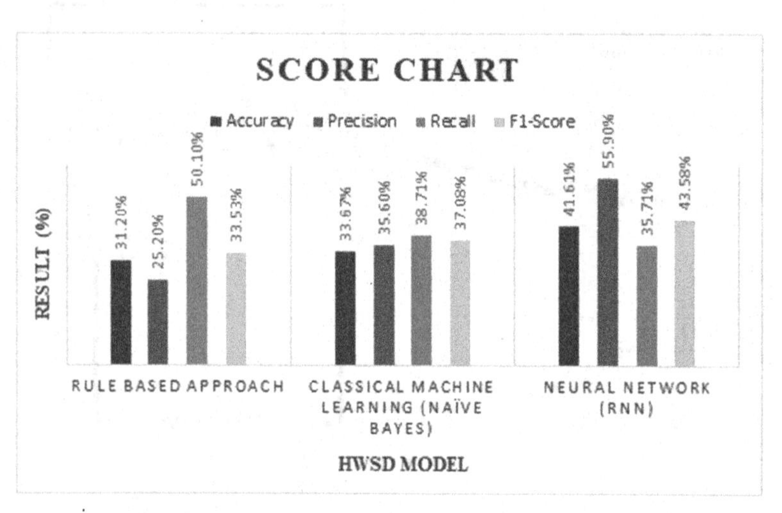

Fig. 3. Compare performance over existing approach

7 Conclusion and Future Directions

Word Sense Disambiguation is essential for any application of natural language processing. This work incorporates the IndoWordNet-defined concept of meaning into neural network algorithms. This article compares rule-based, statistical machine learning, and neural network-based RNN models on four very ambiguous terms, as well as the remaining seven ambiguous words combined. The results of the suggested model demonstrate that it outperforms the current techniques.

There remains one remaining challenge for future development. Since the knowledge-based technique does not include local context, it relies on WordNet, whereas the word-specific classifier employed in the supervised and neural-based approaches takes local context into account but disregards lexical information. For greater precision, disambiguation of every term requires both local context and lexical information. Combining knowledge-based and machine-learning methodologies in future work yields a superior outcome. Additionally, constructing a model utilizing the LSTM model may yield a superior outcome.

References

Agre, G., Petrov, D., Keskinova, S.: A new approach to the supervised word sense disambiguation. In: Agre, G., van Genabith, J., Declerck, T. (eds.) AIMSA 2018. LNCS (LNAI), vol. 11089, pp. 3–15. Springer, Cham (2018). https://doi.org/10.1007/978-3-319-99344-7_1

Banerjee, S., Pedersen, T.: Extended gloss overlaps as a measure of semantic relatedness. In: Ijcai, pp. 805–810 (2003)

Banerjee, S., Pedersen, T.: An adapted lesk algorithm for word sense disambiguation using wordnet. In: Gelbukh, A. (ed.) CICLing 2002. LNCS, vol. 2276, pp. 136–145. Springer, Heidelberg (2002). https://doi.org/10.1007/3-540-45715-1_11

Basile, P., Caputo, A., Semeraro, G.: An enhanced lesk word sense disambiguation algorithm through a distributional semantic model. In: Proceedings of COLING 2014, the 25th International Conference on Computational Linguistics: Technical Papers, pp. 1591–1600 (2014)

Bhattacharyya, P.: Indowordnet. Lexical Resources Engineering Conference 2010 (Lrec 2010). Malta (2010)

Bhingardive, S., et al.: Unsupervised most frequent sense detection using word embeddings. In: DENVER, Citeseer (2015)

Bhingardive, S., Bhattacharyya, P.: Word sense disambiguation using IndoWordNet. In: Dash, N.S., Bhattacharyya, P., Pawar, J.D. (eds.) The WordNet in Indian Languages, pp. 243–260. Springer, Singapore (2017). https://doi.org/10.1007/978-981-10-1909-8_15

Butnaru, A.M., Ionescu, R.T.R.: ShotgunWSD 2.0: an improved algorithm for global word sense disambiguation. IEEE Access **7**, 120961–120975 (2019). https://doi.org/10.1109/ACCESS.2019.2938058

Gautam, C.B.S., Sharma, D.K.: Hindi word sense disambiguation using lesk approach on bigram and trigram words. In: Proceedings of the International Conference on Advances in Information Communication Technology & Computing, pp. 1–5 (2016)

Iacobacci, I., Pilehvar, M.T., Navigli, R.: Embeddings for word sense disambiguation: an evaluation study. In: Proceedings of the 54th Annual Meeting of the Association for Computational Linguistics (Volume 1: Long Papers), pp. 897–907 (2016)

Jha, S., Dipak, N., Prabhakar, P., Pushpak, B.: "A Wordnet for Hindi. In: International Workshop on Lexical Resources in Natural Language Processing. Hyderabad, India (2001)

Khapra, M.M, Joshi, S., Bhattacharyya, P.: It takes two to tango: a bilingual unsupervised approach for estimating sense distributions using expectation maximization. In: Proceedings of 5th International Joint Conference on Natural Language Processing, pp. 695–704 (2011)

Kumari, A., Lobiyal, D.K.: Word2vec's distributed word representation for Hindi word sense disambiguation. In: Hung, D.V., D'Souza, M. (eds.) ICDCIT 2020. LNCS, vol. 11969, pp. 325–335. Springer, Cham (2020). https://doi.org/10.1007/978-3-030-36987-3_21

Kumari, A., Lobiyal, D.K.: Efficient estimation of Hindi WSD with distributed word representation in vector space. J. King Saud Univ. Comput. Inform. Sci. **34**, 6092–6103(2021)

Lesk, M.: Automatic sense disambiguation using machine readable dictionaries: how to tell a pine cone from an ice cream cone. In: Proceedings of the 5th Annual International Conference on Systems Documentation, pp. 24–26 (1986)

Li, X., You, S., Chen, W.: Enhancing accuracy of semantic relatedness measurement by word single-meaning embeddings. IEEE Access **9**, 117424–117433 (2021). https://doi.org/10.1109/ACCESS.2021.3107445

Liu, Y.-F., Wei, J.: Word sense disambiguation with massive contextual texts. In: Li, G., Yang, J., Gama, J., Natwichai, J., Tong, Y. (eds.) DASFAA 2019. LNCS, vol. 11448, pp. 430–433. Springer, Cham (2019). https://doi.org/10.1007/978-3-030-18590-9_60

Miller, G.A.: WordNet: a lexical database for English. Commun. ACM **38**(11), 39–41 (1995)

Nair, A., Kyada, K., Zadafiya, N.: Implementation of word sense disambiguation on hadoop using map-reduce, pp. 573–580. Springer, In Information and Communication Technology for Intelligent Systems (2019)

Narayan, D., Chakrabarti, D., Pande, P., Bhattacharyya, P.: An experience in building the indo wordnet-a wordnet for Hindi. In: First International Conference on Global WordNet. Mysore, India (2002)

Navigli, R.: Word sense disambiguation: a survey. ACM Comput. Surv. **41**(2), 1–69 (2009)

Ng, H.T., Lee. H.B.: Integrating multiple knowledge sources to disambiguate word sense: an exemplar-based approach. In: 34th Annual Meeting of the Association for Computational Linguistics, Santa Cruz, pp. 40–47. Association for Computational Linguistics, California, USA (1996). https://www.aclweb.org/anthology/P96-1006

Nguyen, Q.-P., Vo, A.-D., Shin, J.-C., Ock, C.-Y.: Effect of word sense disambiguation on neural machine translation: a case study in Korean. IEEE Access **6**, 38512–38523 (2018). https://doi.org/10.1109/ACCESS.2018.2851281

Pasini, T., Navigli, R.: Train-O-Matic: supervised word sense disambiguation with no (manual) effort. Artific. Intell. **279**, 103215 (2020). https://doi.org/10.1016/j.artint.2019.103215

Ramamoorthy, N.C., et al.: A Gold Standard Hindi Raw Text Corpus. Central Institute of Indian Languages, Mysore (2019)

Sharma, D.K.S.: A comparative analysis of hindi word sense disambiguation and its approaches. In: International Conference on Computing, Communication & Automation, pp. 314–321 (2015)

Sharma, D.K.: Hindi word sense disambiguation using cosine similarity. In: Satapathy, S., Joshi, A., Modi, N., Pathak, N. (eds.) Proceedings of International Conference on ICT for Sustainable Development. AISC, vol. 409, 801–808. Springer, Singapore (2016). https://doi.org/10.1007/978-981-10-0135-2_76

Sheth, M., Popat, S., Vyas, T.: Word sense disambiguation for indian languages. In: Shetty, N., Patnaik, L., Prasad, N., Nalini, N. (eds.) Emerging Research in Computing, Information, Communication and Applications. ERCICA 2016. Springer, Singapore (2018). https://doi.org/10.1007/978-981-10-4741-1_50

Singh, S., Siddiqui, T.J.: Sense annotated Hindi Corpus. In: 2016 International Conference on Asian Language Processing (IALP), pp. 22–25. IEEE (2016)

Singh, S., Siddiqui, T.J., Sharma, S.K.: Naïve bayes classifier for hindi word sense disambiguation. In: Proceedings of the 7th ACM India Computing Conference, pp. 1–8 (2014)

Sinha, M., et al.: Hindi word sense disambiguation. In: International Symposium on Machine Translation, Natural Language Processing and Translation Support Systems. Delhi, India (2004)

Soni, V.K., Gopalaniî, D., Govil, M.C.: An adaptive approach for word sense disambiguation for Hindi Language. In: IOP Conference Series: Materials Science and Engineering, p. 12022. IOP Publishing (2021)

Taghipour, K., Ng, H.T.: Semi-supervised word sense disambiguation using word embeddings in general and specific domains. In: Proceedings of the 2015 Conference of the North American Chapter of the Association for Computational Linguistics: Human Language Technologies, pp. 314–323 (2015)

Tripathi, P., et al.: Word sense disambiguation in hindi language using score based modified lesk algorithm. Int. J. Comput. Dig. Syst. **10**, 2–20 (2020)

Vaishnav, Z.B., Sajja, P.S.: Knowledge-Based Approach for Word Sense Disambiguation Using Genetic Algorithm for Gujarati, pp. 485–494. Springer, In Information and Communication Technology for Intelligent Systems (2019)

Wilks, Y.A., Slator, B.M., Guthrie, L.: Electric Words: Dictionaries, Computers, and Meanings. The MIT Press (1996). https://doi.org/10.7551/mitpress/2663.001.0001

Zhong, Z., Ng, H.T.: It makes sense: a wide-coverage word sense disambiguation system for free text. In: Proceedings of the ACL 2010 System Demonstrations, pp. 78–83 (2010)

Explainable AI for Predictive Analytics on Employee Attrition

Sandip Das[1], Sayan Chakraborty[2(✉)], Gairik Sajjan[1], Soumi Majumder[3], Nilanjan Dey[4], and João Manuel R. S. Tavares[5]

[1] Department of CSE, JIS University, Kolkata, India
[2] Department of CSE, Swami Vivekananda University, Kolkata, India
sayanc@svu.ac.in
[3] Dept. of Business Administration, Future Institute of Engineering and Management, Kolkata, India
[4] Department of CSE, Techno International New Town, Kolkata, India
[5] Departamento de Engenharia Mecânica, Faculdade de Engenharia, Instituto de Ciência e Inovação em Engenharia Mecânica e Engenharia Industrial, Universidade do Porto, Porto, Portugal

Abstract. Employees are the key to an organization's success. An employee can directly affect the productivity of any organization. Hence, retention of the employees becomes great challenge for every organization. Artificial Intelligence is extensively used in such cases to analyze the problem behind a company's low retention rate of valuable employees. The analysis can help organizations to build up strategies to deal with the problems of employees leaving an organization. The current work aims to detect key features or main reasons behind an employee's key decision to stay or leave the organization. Primarily in this work machine learning is used extensively to create a create a prediction model on the collected data from human resources of different organization. In addition, the proposed framework uses explainable Artificial Intelligence methods such as SHAPley Additive exPlanations and Local Interpretable Model-Agnostic Explanations to analyze the key factors behind employees leaving the organizations. The current work managed to accurately create a prediction model with a score of 96% and detect three key features behind employee's decision of staying or leaving any organization.

Keywords: Random Forest · Explainable AI · Employee retention · Machine learning · LIME · SHAP

1 Introduction

In the present era, organizations are looking to increase their productivity which has a direct effect on employee retention. It has been observed that employees [1] are often reluctant to work in any particular organization for a long period of time. As a result, it affects the productivity of the organization. To hire new employees and to make them as competent as their predecessors takes lot of time and effort. Meanwhile, in this evolving

K. K. Patel et al. (Eds.): icSoftComp 2022, CCIS 1788, pp. 147–157, 2023.
https://doi.org/10.1007/978-3-031-27609-5_12

era of technology, some organizations tend to take benefit by adapting to the changes. Hence, it is an endless cycle which affects the productivity of any organization.

It is hard to predict when an employee is going to leave the organization. Although it is possible to analyze the reason behind employees leaving [2] an organization if the data of the same can be collected and analyzed properly. This can lead to organization building up strategies to retain [3] their significant employees before they can think of leaving the organization. This will not only help to increase the productivity, but also will enhance the environment work culture [4] inside the organization. Analysis of the data will not be enough though, as the reason behind low employee retention should be analyzed in such cases also. Without knowing the actual problem, organizations will not be able to strategize employee retention.

Explainable AI (XAI) [5] is one of the rapidly growing domains of Artificial Intelligence. It provides the community the insights of black box machine learning algorithms. In the current work, XAI [6] is extensively used to analyze the data gathered from organizations regarding their employees leaving the company. Explainable AI is a set of tools and frameworks [7, 8] to help to understand and interpret predictions made by machine learning models, natively integrated with a number of Google's products and services. With it, a person can debug and improve the performance of the machine learning model, and help others understand the models' behavior. Thus, explainable AI is used to describe an AI model, expected impact and potential biases.

The rest of this article is structured in the following manner: Sect. 2 is focuses on the background of the current study. In Sect. 3, the dataset information along with preprocessing steps and the methods used are explained. Section 4 presents the analysis of the obtained results, and Sect. 5 concludes the article.

2 Related Works

In order to strategize the employment retention process, appraisal, etc., decision making plays a vital role for the management of any organization. Employee retention is a common problem faced by most of the organizations that demands proper decisions by the management to retain valuable employees. Various sub-domains of Artificial Intelligence play a vital role in these decision-making systems. Some of the previous works include similar frameworks where artificial intelligence has been extensively used to identify employee attrition. In 2012, Anand et al. presented an analysis on employee [1] attrition. In this work, the authors focused on business process outsourcing industry and presented a practical study on this kind of industry. A framework was built to firstly provide questionnaire, which helped to gather the data. Later, the authors analyzed the data using several statistical methods such as chi-square test, percentage analysis and analysis of variance.

In 2018 Shankar et al. analyzed and predicted [2] reasons behind employee attrition using data mining. In this work, Decision Tree (DT), Logistic Regression, Support Vector Machine (SVM), K-Nearest Neighbor (KNN), Random Forest (RF), and Naive Bayes methods were used to make analysis and prediction on employee attrition data. In 2019, Bhartiya et al. proposed a prediction model [4] for employee attrition in which they used classifiers to create the prediction model. The authors used DT, KNN, RF,

SVM and Naive Bayes algorithms to create the prediction model for employee attrition. In 2020, Jain et al. explained [4] and analyzed the reasons behind employee attrition using machine learning. In this work, the authors built up a prediction model [10] using machine learning and model-agnostic, local explanation approaches.

In 2021, Joseph et al. [11] used Machine Learning and Depression Analysis to find the reasons behind employee attrition. The authors pre-processed the data and used DT, RF, and SVM base classifiers on the used dataset, and obtained and accuracy score of 86.0%. In 2022, Krishna and Sidharth analyzed Employee attrition [12] data using artificial intelligence. This model used RF and Adaptive Boosting to create prediction models. The authors also made an analysis on the key-factors that directly has an effect on employee attrition of any organization to have a clear view of the situation in front of the management, so that key decisions can be taken regarding the employee retention.

The work of Sekaran et al. demonstrated the robustness of XAI models such as: Local Interpretable Model-Agnostic Explainer (LIME) and SHAPley Additive exPlanations (SHAP) on detecting [13] the key-reasons for employee attrition. These models help to provide logical insights and clear perspective on the data that could help the management.

The aforementioned works mostly were mainly focused either on the prediction model or on the factors behind employee attrition. Explainable Artificial Intelligence was used to understand and indicate important features behind Employee attrition. The previous existing methods mostly used machine learning to predict employee attrition's significant reasons. The research lacked the investigation behind the factors and their effect on employee attrition which was a major research gap in this domain. The current work aimed to combine both of these concepts to have a clear perception on the employee attrition situation in organizations.

3 Materials and Methods

3.1 Dataset Description

A human resources data [3] was used, which contains 15000 rows and 10 columns. The data contains 9 parameters of 15000 employees. The target data is labeled as per employee's current status (whether he/she is presently working). The dataset was collected from Kaggle.com.

3.2 Data Preprocessing

The dataset [3] used here contains both categorical and numerical values. It has 2 categorical and 8 numerical features. In order to pre-process [11] the data, the proposed model applied label [9] encoding to the categorical features. Using label encoding, it was assigned a number to each class in the categorical features [28].

3.3 Explainable AI

Almost every machine learning classifier (Support Vector Machine, XGBoost, KNN, etc.) available today is a black box model, i.e., a particular output is not known due to some given input. Hence, better accuracy from such models cannot guarantee a robust model. XAI [6] is one of the most interesting areas of interpreting black box machine learning algorithms. Regression algorithms are most commonly used for interpretability. But complex algorithms give better results, so we need to understand them. LIME [10] and SHAP [7, 8] are the two commonly used algorithms for explaining complex machine learning models. LIME and SHAP are chosen due to their black-box model which helps to achieve high accuracy and high interpretable while making decision. As the current work aims to find the key reasons behind employees leaving any organization, LIME and SHAP are very much relatable to finding out the exact reasons or key features from the database. Local Interpretable Model-agnostic Explanation.

LIME provides a local explanation [8, 9] for a particular prediction. It explains which feature how much has contributed and why for a particular row in a dataset. It perturbs the data samples and observes the impact of it on the original data and, based on that, shows the feature importance of that particular sample.

3.4 SHAPley Additive Explanations

SHAP is a game theory-based [7] approach, where it is explained why a particular prediction is different from the baseline [8] and which feature contributed much to pull or push that prediction value from or to the direction of the base value. Hence, it is capable of debugging a machine learning model and observes the reason behind its predictions which provides an insight to the prediction model.

4 Results and Discussion

The present work was executed on Google's Collaboratory notebooks with the help of an explainable Artifitial Intelligence (AI) framework. Different classifiers were used to create the prediction model, and later explainable AI was used to identify importance of features in the prediction model. Multiple classifiers were applied on the data such as (I) Random Forest, (II) Naive Bayes, (III) K Nearest Neighbor, and (IV) Support Vector Classifier. The accuracy scores obtained by the used classifiers are reported in Table 1, which clearly shows that the Random Forest classifier outperformed the others in terms of accuracy.

Table 1. Accuracy scores obtained by the used different classifiers.

Model	Accuracy (%)
Random Forest	98.88
XGBoost	96.44
KNN	93.54
Naive Bayes	78.36
SVC	77.12

Random Forest Classifier is an ensemble learning algorithm, which uses classification as its mechanism. A Receiver Operating Characteristic (AUC) value of 0.99 for the training data and of 0.988 for the testing data as obtained by the Random Forest classifier, Fig. 1.

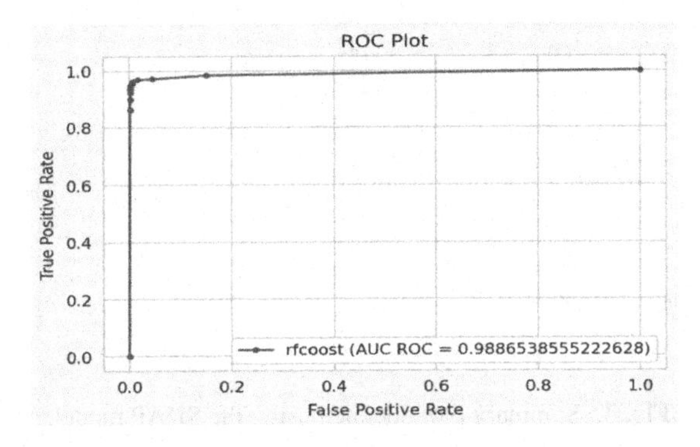

Fig. 1. Receiver Operating Characteristic (ROC) curve of the Random Forest classifier.

The SHAP model [7, 8] was trained with 66.67% of the total data. This game theoretic approach was applied in order to calculate and analyze the importance of each used feature. The approach shuffled the value of each feature and observed its effect on prediction. In Fig. 2, the built feature importance plot is presented, which indicates the important features as to employee retention.

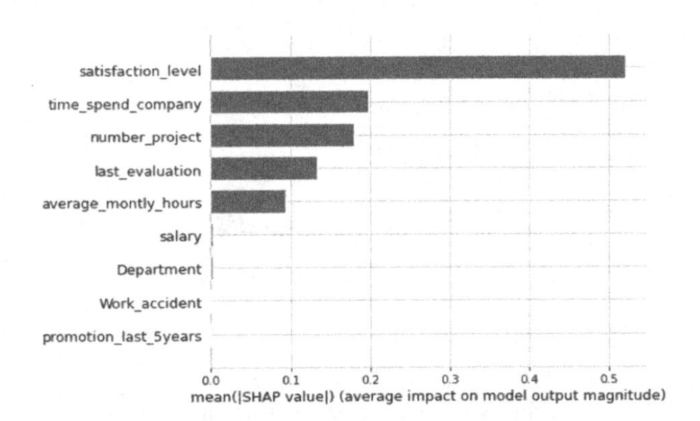

Fig. 2. Feature importance plot obtained using the SHAP model.

In Fig. 2, the global impact [14, 15] of every feature is shown in decreasing order. As per the figure, satisfaction level is the most important for retaining an employee [16, 17] by an organization, followed by the employee's tenure [18], number of projects worked on, evaluation, monthly hours, etc.

A summary plot is presented in Fig. 3 in order to make this analysis further clear. This plot shows the features arranged in descending order according to their importance level in employee [19, 20] retention.

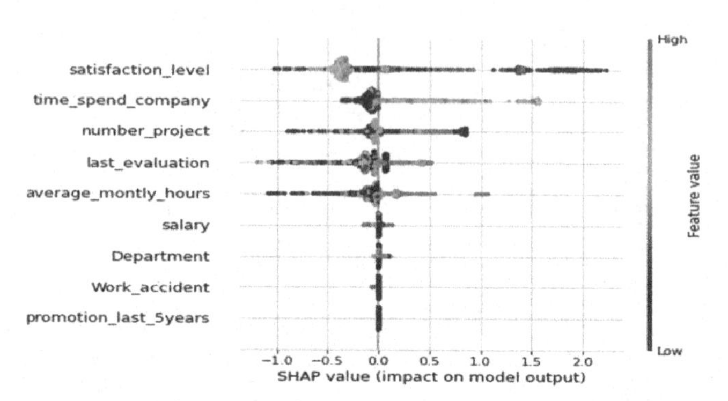

Fig. 3. Summary plot obtained using the SHAP model.

It can be observed from the summary plot (Fig. 4a) that if the 'satisfaction level' is 'low', then the employee has a higher chance of leaving the company. Similarly, if an employee's 'time spent' with the company is high he/she has a good chance of leaving the company (Fig. 4b).

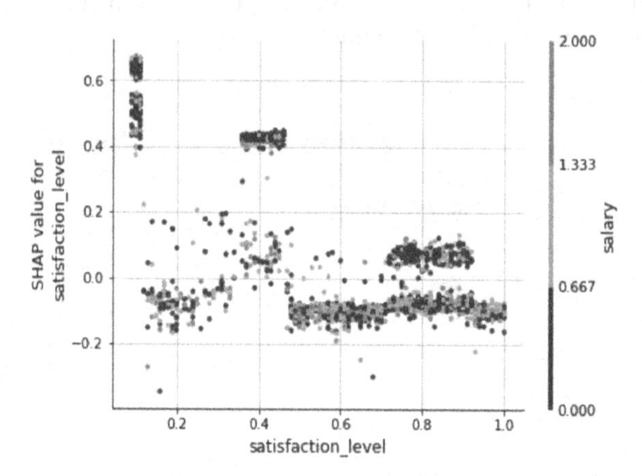

Fig. 4a. SHAP values for 'satisfaction_level'.

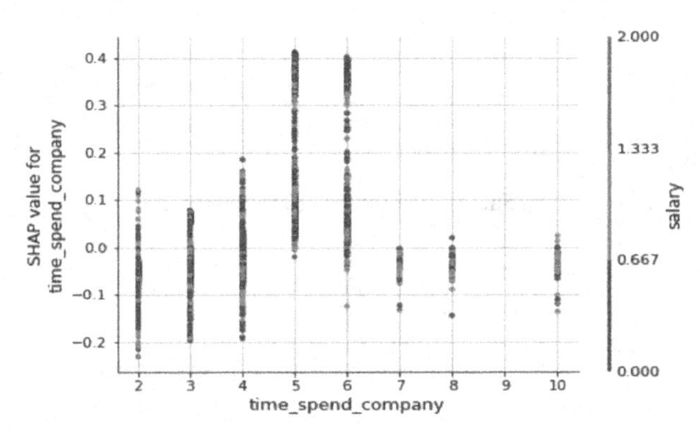

Fig. 4b. SHAP values for 'time_spend_company'.

In Fig. 4a and 4b, the top 3 important features as reported in Fig. 2, are analyzed through SHAP Values. The important features behind employee attrition as reported in Fig. 3 are 'satisfaction_level', 'time_spend_company' and 'number_project'. In Figs. 4a, 4b, 4c, color blue signifies the range 0.000 to 0.667, the violet and red colors represent the ranges 0.667–1.333 and 1.333–2.000, respectively. The SHAP values explained individual prediction with the help of base prediction and effect of each feature on the base value [21].

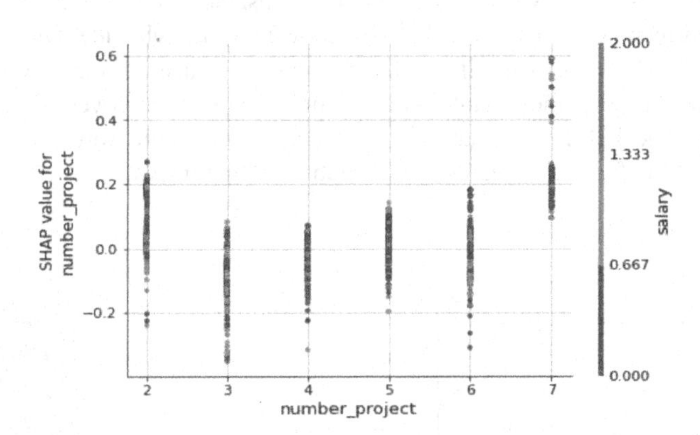

Fig. 4c. Shap values for 'number_project'.

In Figs. 5a, 5b, 5c, three different Force plots are shown for 3 individual instances, which are presented in order to provide in depth analysis of contribution of each feature on the prediction model. These values can change for different instances.

In Figs. 5a, 5b, 5c, the features with top contributions to the prediction model for 1575th, 2012th and 10463rd instance are shown. The local explanation of the prediction of any particular instance is shown using LIME, which provides explanations based upon the perturbed data samples.

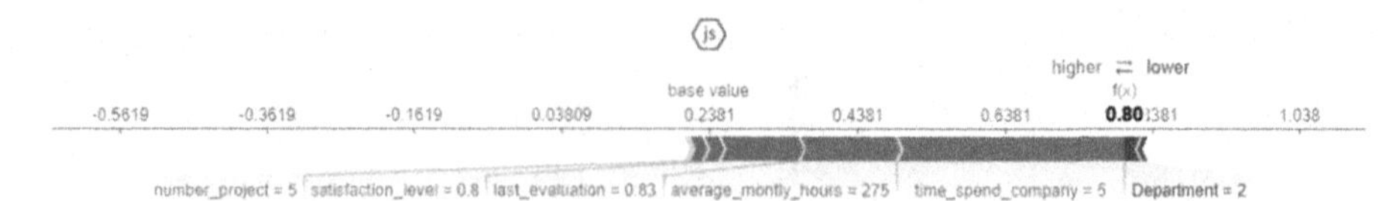

Fig. 5a. Force plot for 1575th instance.

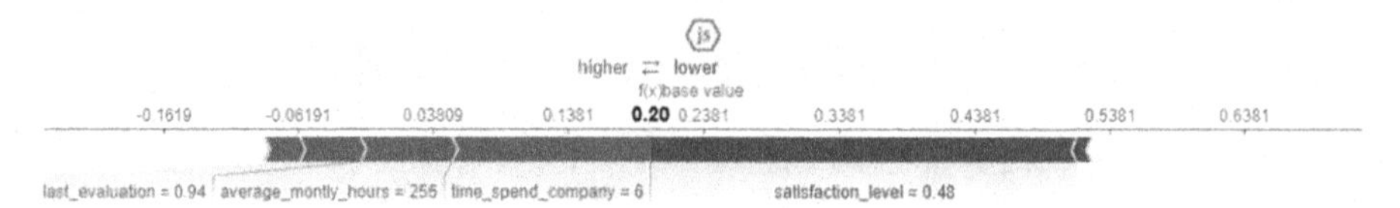

Fig. 5b. Force plot for 2012th instance.

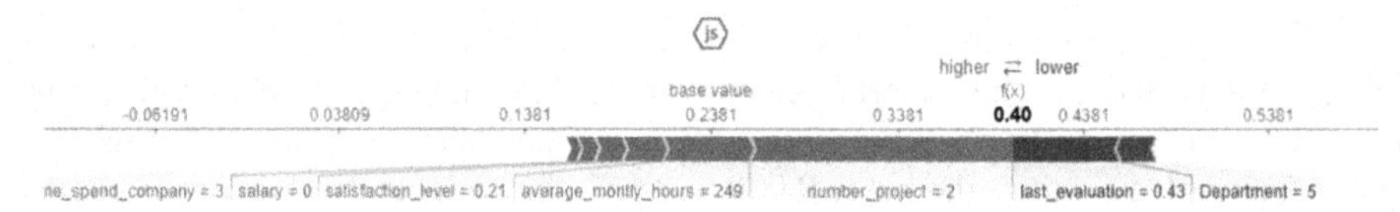

Fig. 5c. Force plot for 10463rd instance.

In Figs. 6a, 6b, the LIME explanations for 2 instances are shown. In both shown graphs, the corresponding feature values are shown on the right side. On the left, 'predicted value' are shown, that signifies if the particular instance was classified as 0 (zero). If it is classified as zero, then it automatically indicates that employee is likely to leave the company. The LIME explanation also shows each feature values as well as the contribution of the features towards the prediction of that instance.

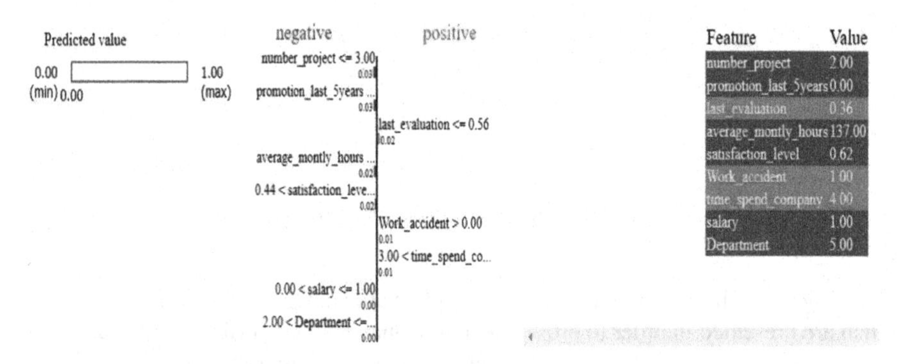

Fig. 6a. LIME explanation (left = 1).

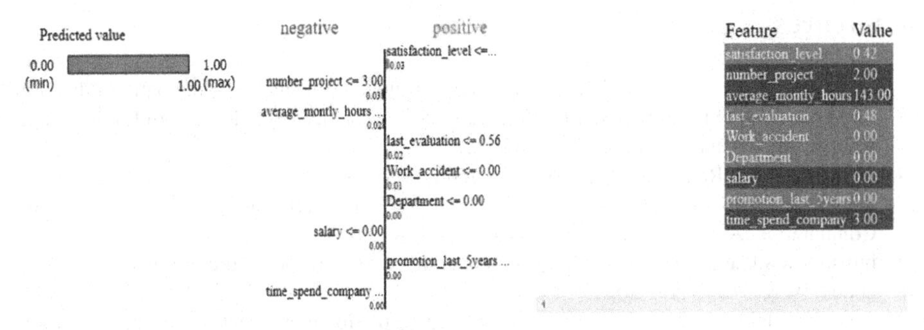

Fig. 6b. LIME explanation (left = 0).

The current work used different classification frameworks to create the prediction model, out of which Random Forest due to its superior accuracy score was chosen for further processing. The prediction model [22, 23] was further analyzed using ROC curve [22–24] and the AUC [25, 26] score. SHAP was introduced [27, 28] to work specifically on detection importance of the features involved. SHAP trained the dataset and provided an in-depth analysis on the features and their contribution in the used prediction model. Using SHAP, it was observed that 'satisfaction_level', 'time_spend_company' and 'number_project' are the three important reasons behind an employee's decision to stay or leave the organization. LIME was additionally used [9] to explain the features' contribution in the analysis. Data taken at different instances helped to explain the importance of the features in the prediction model as well.

5 Conclusions

The current work had analyzed the possible factors behind employees leaving the company. From the analysis and the discussion made, it was quite clear that from the given dataset, satisfaction level is the top priority of employees. Additionally, the employee's tenure at the current organization and workload is also other major factors behind employee's decision to stay or leave any organization. Only the gathered data was analyzed, which may not have included other factors that could result into different analysis and outcomes. This work was focused on results obtained from the Random Forest classifier-based model, as in the comparative study made, it was the one that led to the highest accuracy score among the tested frameworks. Then, based on the obtained results, the study was focused on in depth analysis of features using SHAP and LIME in order to explain the features and their importance level in details. Future work may consider other factors and use different explainable AI frameworks to make further analysis. Additionally, other techniques such as Deep Learning or Reinforced Learning may be used to create the prediction model.

References

1. Vijay Anand, V., Saravanasudhan, R., Vijesh, R.: Employee attrition - a pragmatic study with reference to BPO Industry. In: IEEE-International Conference On Advances In Engineering, Science And Management, pp. 42–48 (2012)
2. Shankar, R.S., Rajanikanth, J., Sivaramaraju, V.V., Murthy, K.V.S.S.R.: Prediction of employee attrition using data mining. In: 2018 IEEE International Conference on System, Computation, Automation and Networking, pp. 1–8 (2018)
3. https://www.kaggle.com/code/adepvenugopal/employee-attrition-prediction-using-ml/notebook. Accessed 15 Aug 2022
4. Jain, P.K., Jain, M., Pamula, R.: Explaining and predicting employees' attrition: a machine learning approach. SN Appl. Sci. **2**, 757–761 (2020)
5. Došilović, F.K., Brčić, M., Hlupić, N.: Explainable artificial intelligence: a survey. In: 2018 41st International Convention on Information and Communication Technology, Electronics and Microelectronics, pp. 0210–0215 (2018)
6. Ye, Q., Xia, J. Yang, G.: Explainable AI for COVID-19 CT classifiers: an initial comparison study. In: 2021 IEEE 34th International Symposium on Computer-Based Medical Systems (CBMS), pp. 521–526 (2021)
7. Marcflio, W.E., Eler, D.M.: From explanations to feature selection: assessing SHAP values as feature selection mechanism. In: 2020 33rd SIBGRAPI Conference on Graphics, Patterns and Images (SIBGRAPI), pp. 340–347 (2020)
8. Kumar, C.S., Choudary, M.N.S., Bommineni, V.B., Tarun, G., Anjali, T.: Dimensionality reduction based on SHAP Analysis: a simple and trustworthy approach. In: 2020 International Conference on Communication and Signal Processing (ICCSP), pp. 558–560 (2020)
9. Sahay, S., Omare, N., Shukla, K.K.: An approach to identify captioning keywords in an image using LIME. In: 2021 International Conference on Computing, Communication, and Intelligent Systems (ICCCIS), pp. 648–651 (2021)
10. Slack, D., Hilgard, S., Jia, E., Singh, S., Lakkaraju, H.: Fooling lime and shap: adversarial attacks on post hoc explanation methods. In: AIES '20: Proceedings of the AAAI/ACM Conference on AI, Ethics, and Society, pp. 180–186 (2019)
11. Joseph, R. Udupa, S., Jangale, S., Kotkar, K., Pawar, P.: Employee attrition using machine learning and depression analysis. In: 5th International Conference on Intelligent Computing and Control Systems (ICICCS), pp. 1000–1005 (2021)
12. Krishna, S., Sidharth, S.: Analyzing employee attrition using machine learning: the new AI approach. In: 2022 IEEE 7th International conference for Convergence in Technology (I2CT), pp. 1–14 (2022). https://doi.org/10.1109/I2CT54291.2022.9825342
13. Sekaran, K. Shanmugam. S.: Interpreting the factors of employee attrition using explainable AI. In: 2022 International Conference on Decision Aid Sciences and Applications (DASA), pp. 932–936 (2022). https://doi.org/10.1109/DASA54658.2022.9765067
14. Usha, P., Balaji, N.: Analyzing employee attrition using machine learning. Karpagam J. Comput. Sci. **13**, 277–282 (2019)
15. Ponnuru, S., Merugumala, G., Padigala, S., Vanga, R., Kantapalli, B.: Employee attrition prediction using logistic regression. Int. J. Res. Appl. Sci. Eng. Technol. **8**, 2871–2875 (2020)
16. Alao, D.A.B.A., Adeyemo, A.B.: Analyzing employee attrition using decision tree algorithms. Comput. Inform. Syst. Develop. Inform. Allied Res. J. **4**(1), 17–28 (2013)
17. Sarah, S., Alduay, J., Rajpoot, K.: Predicting employee attrition using machine learning. In: 2018 International Conference on Innovations in Information Technology, pp. 93–98 (2018)
18. Boomhower, C., Fabricant, S., Frye, A., Mumford, D., Smith, M., Vitovsky, L.: Employee attrition: what makes an employee quit. SMU Data Sci. Rev. **1**(1), 9–16 (2018)

19. Jantan, H., Hamdan, A.R., Othman, Z.A.: Towards applying data mining techniques for talent managements. In: 2009 International Conference on Computer Engineering and Applications IPCSIT, vol. 2, p. 476–581 (2011)
20. Srinivasan Nagadevara, V., Valk, R.: Establishing a link between employee turnover and withdrawal behaviours: application of data mining techniques. Res. Pract. Hum. Resour. Manag. **16**(2), 81–97 (2008)
21. Hong, W.C., Wei, S.Y., Chen, Y.F.: A comparative test of two employee turnover prediction models. Int. J. Manag. **24**(4), 808–813 (2007)
22. Kamal, M.S., Northcote, A., Chowdhury, L., Dey, N., Crespo, R.G., Herrera-Viedma, E.: Alzheimer's patient analysis using image and gene expression data and explainable-AI to present associated genes. IEEE Trans. Instrum. Meas. **70**, 1–7 (2021)
23. Kamal, M.S., Chowdhury, L., Dey, N., Fong, S.J., Santosh, K.: Explainable AI to analyze outcomes of spike neural network in Covid-19 chest X-rays. In: 2021 IEEE International Conference on Systems, Man, and Cybernetics (SMC), pp. 3408–3415 (2021)
24. Majumder, S., Dey, N.: Explainable Artificial Intelligence (XAI) for Knowledge Management (KM). In: Majumder, S., Dey, N. (eds.) AI-empowered Knowledge Management, pp. 101–104. Springer Singapore, Singapore (2022). https://doi.org/10.1007/978-981-19-0316-8_6
25. Singh, P.: A novel hybrid time series forecasting model based on neutrosophic-PSO approach. Int. J. Mach. Learn. Cybern. **11**(8), 1643–1658 (2020). https://doi.org/10.1007/s13042-020-01064-z
26. Singh, P.: FQTSFM: a fuzzy-quantum time series forecasting model. Inf. Sci. **566**, 57–79 (2021). https://doi.org/10.1016/j.ins.2021.02.024
27. Chou, Y.-L., Moreira, C., Bruza, P., Ouyang, C., Jorge, J.: Counterfactuals and causability in explainable artificial intelligence: theory, algorithms, and applications. Inform. Fus. **81**, 59–83 (2022). https://doi.org/10.1016/j.inffus.2021.11.003
28. Shinde, G.R., Majumder, S., Bhapkar, H.R., Mahalle, P.N.: Quality of Work-Life During Pandemic: Data Analysis and Mathematical Modeling, pp. 16–27. Springer, Singapore (2021)

Graph Convolutional Neural Networks for Nuclei Segmentation from Histopathology Images

Karishma Damania and J. Angel Arul Jothi(✉)

Department of Computer Science, Birla Institute of Technology and Science, Pilani, Dubai Campus, Dubai International Academic City, Dubai, UAE
f20180008d@alumni.bits-pilani.ac.in,
angeljothi@dubai.bits-pilani.ac.in

Abstract. The analysis of hematoxylin and eosin (H&E) stained images obtained from breast tissue biopsies is one of the most dependable ways to obtain an accurate diagnosis for breast cancer. Several deep learning methods have been explored to perform the analysis of breast histopathology images to automate and improve the efficiency of computer aided diagnoses (CAD) of breast cancer. Cell nuclei segmentation is one of the predominant tasks of a CAD system. Traditional convolutional neural networks (CNNs) have seen remarkable success when it comes to this task. However, there has been recent success suggested by several studies that made use of graph convolutional neural networks (GCNN) with computer vision tasks. This paper aimed to design an architecture called GraphSegNet, a GCNN that follows the encoder-decoder structure of the well-known network for image segmentation, SegNet to perform nuclei segmentation. The proposed model achieved an accuracy, Dice coefficient, Intersection over Union (IoU), precision and recall scores of 90.75%, 83.74%, 83.2%, 75.40% and 79.93% respectively on a UCSB bio-segmentation benchmark breast histopathology dataset.

Keywords: Deep learning · Graph convolutional neural networks · Image segmentation · Cell nuclei segmentation · GraphSegNet

1 Introduction

Breast cancer is one of the leading causes of mortality of women across the world. The most effective way to reduce this is through a timely diagnoses and treatment. The benchmark for diagnoses of breast cancer is through the analysis of breast tissue samples obtained from biopsies using a microscope. However, the process of manually going through a large number of histopathological images is time consuming as well as subject to human error due to the complexity of breast cells. Thus, several researches have been conducted over the years in order to make this process automated through the deployment of deep learning techniques to develop a computer aided diagnosis (CAD) system. The first part of examining histopathology images is to identify the cell nuclei; then, observe clusters and patterns that help determine whether the cell growth is benign

K. K. Patel et al. (Eds.): icSoftComp 2022, CCIS 1788, pp. 158–169, 2023.
https://doi.org/10.1007/978-3-031-27609-5_13

or malignant in nature. Thus, the process of efficiently segmenting the cell nuclei from these images is vital in the process of diagnosis and early detection of the disease. The CAD systems developed till date have used many different machine learning (ML) and deep learning (DL) techniques and their combinations for this purpose [1], however minimal research has been conducted to test the efficacy of using graph convolutional neural networks (GCNN) for cell nuclei segmentation and hence, this paper aims to design a GCNN called GraphSegNet based on the famous encoder decoder segmentation architecture, SegNet.

To realize the functioning of GCNNs and their advantages over traditional convolutional neural networks (CNNs), there is a need to understand the difference between the data that these two neural networks work with. CNNs work with Euclidean data but their performance deteriorates significantly when dealing with non-Euclidean i.e. graphical data [2]. However, several real-world data are actually expressed as graphical data. Graphs, are defined as $G(V, E)$ comprising of a set of vertices V and edges E. For those real-world problem that can be defined in terms of objects and the connections between them, GCNNs perform significantly better than traditional CNNs. This is evident through the several applications of GCNNs in fields of text classification to produce document labels, relations extraction to extract semantic relations from text, protein interface prediction for drug discovery, and computer vision tasks such as object detection, region classification, image segmentation, etc. Each of these applications follows the simple idea that data is often related or associated with one another and these dependencies can be learned and replicated to produce the necessary outputs with the help of GCNNs.

Thus, GCNNs are an optimizable transformation on the attributes of a graph (nodes, edges, global context) that preserve their symmetries i.e. the network is able to use the features represented by the graph, learn the associations and produce outputs as required [3]. Most commonly, for feature representations between data, an adjacency matrix is used. An adjacency matrix is simply an $N \times N$ matrix where N represents the number of nodes, and 0s and 1s are used to represent the edges between two nodes. Convolutions and operations in GCNNs are performed on the adjacency matrices for images such that each pixel becomes a node and the kernel sizes determine the number of nodes processed in one convolution. For a kernel size of 3 i.e. a 3×3 convolution while dealing with images, 9 pixels are processed at a time and since each of these pixels are associated with 8 other pixels, the problem becomes that of an 8-nearest neighbour graph on a 2D grid. Since each image is considered as a graph, this is a graph level task.

This study, focus to design and experiment with a GCNN to assess its performance to segment cell nuclei from a dataset of breast histopathology images. To do this, we derive motivation from the well-known architecture SegNet, to create a GCNN that draws on its structure and underlying ideas to create GraphSegNet.

This paper has been segregated into the following sections. Section 2 explores the literature review of recent research conducted using graph neural networks. Section 3 presents the dataset used for this study. Section 4 explains the techniques used during this research methodology and the GraphSegNet architecture that was implemented. Section 5 details the experimental set up and the evaluation metrics. Section 6 presents the experimental results achieved by the model followed by the Sect. 7, which summarizes the conclusion and highlights the merits of the model.

2 Literature Review

One of the very first works that explored extending neural networks to be able to work on graphical data was proposed by Franco Scarselli et al. in [4]. This model is able to work on several types of graphs: cyclic, acyclic, directed and undirected with the help of function that maps a graph and its nodes onto a dimensional Euclidean space and the enforces a supervised learning algorithm to solve the problem of graph and node classification. The work described in [5], focused on weakly supervised semantic segmentation in order to address the challenges of image-level annotations and pixelwise segmentation. They proposed a GNN for doing so where the images are converted into graph nodes and the associations between them are labelled through an attention mechanism. Another contribution of this work was the proposal of a graph dropout later that aided the model to form better object responses. Upon experimenting with the well-known PASCAL and COCO datasets, this methodology was able to outperform several other state-of-the-art models and achieve mean testing and validation IoU of 68.2% and 68.5% respectively.

In [6], Zhang, B et al. aimed at the training of annotations using bounding boxes. This work proposed the affinity attention GNN (A^2GNN) for this. Their methodology involved initially creating seeds that were pseudo semantically aware and then transformed them into graphs through their model. The model consists of an attention layer which helps to transfer the labels to the unlabeled pixels by extracting the necessary information from the edges of the graph. They also introduced a loss function to optimize their model performance and conducted experiment with the PASCAL VOC dataset to achieve validation and test accuracies of 76.5% and 75.2% respectively. Since making use of contextual information over a long range is vital when it comes to pixel-wise prediction tasks, the graph neural network proposed in [7] i.e. the Dual graph convolutional network (GCN) aims to achieve this by exploiting the global contexts of the features that are input into the model by creating a pair of orthogonal graphs. The first helps to replicate and learn the spatial relationships between the pixels where the second helps to associate the dependencies along the channels of feature map produced by the network. By creating a lower dimensional space where the features are defined as pairs before reprojecting them into the initial space. Upon experimenting with the Cityscapes and Pascal context datasets, this work was able to achieve mean IoUs of 82.0% and 53.7%, respectively. The paper [8] presents a representative graph (RepGraph) layer with the aim of using non-local operations to work with long-distance dependencies. It does this by initially sampling a set of characteristic and informative features instead of passing messages from all the parts of the graph which helps to significantly reduce repetition since it considers and represents the response of a single node with the help of only few other informative nodes and these nodes are derived from a spatial matrix. The flexibility of this layer means that it can be integrated for a variety of object detection tasks and thus, experiments were performed on three benchmark datasets ADE20K, Cityscapes and PASCALContext using ResNet50 as a baseline to achieve mean IoUs of 43.12%, 81.5% and 53.9% respectively. Since it is well established that long-distance pixel associations help to improve global context representations and consequently increase the efficiency of semantic segmentation. However due to challenges such as complex models with high computational cost required in order to do this, Qinghui et al. in [9] proposed the self-constructing graph model that learns these dependencies straight from the image and thus passes on the contextual information more effectively. This work proposed a novel enhancement method for graph reconstruction which when incorporated into a neural network was able to perform end-to-end semantic segmentation. Upon conducting experiments on the ISPRS Potsdam and Vaihingen datasets, this methodology was able to achieve a mean F-1 score of 92.0% while having a lesser number of parameters and reduced, more optimized computational cost.

In [10], the motivation is derived from the fact that spatial relationships are not well captured during the feature extraction process using traditional convolutions and pooling layers. Since spatial relationships are vital when it comes to differentiating objects between classes, this work proposes the AttResUNet which combines the capabilities of a deep semantic segmentation network and a graph convolution network in order to better the extraction capabilities along with superpixels being used to represent the graph nodes and assigning different weights to the spatial and spectral information. This model thus is able to conduct more efficient classification while reducing noise due to the pixels. The experimental results obtained on two datasets; UCM and DeepGlobe and acheieved an IoU of 73.99%. The paper [11] also addressed the problem of local information being overlooked during feature extraction by conventional deep learning neural networks which is actually of key importance when it comes to performing semantic segmentation. Thus, the proposed methodology utilizes a fully convolutional graph network to do this by converting the image into a graph structure which then becomes a graph node classification problem that is solved by a GNN. Being one of the first works that used this methodology to performance semantic segmentation with images, it achieved a mean IoU score of 65.91% when applied on the VOC Dataset. Taking inspiration from the well-known architecture when it comes to sematic segmentation, UNet, [12] recreated this encoder-decoder structure to work with graphical data. However, since images can be represented as nodes on 2D lattices to be fed into a graph neural network, the methodology proposed in this paper addresses the challenges of upsampling and pooling operations that would need to be performed on graphical data. Thus, by designing the novel gPool and gUnpool operations in order to recreate the encoder-decoder structure, this work proposed the Graph U-Net to do this. These gPool layers select a few nodes to create a smaller graph as the pooling operation whereas the gUnpool layer recreates the smaller graph back into the original with the help of the positions selected in the corresponding gPool layers. Upon working with the protein datasets D&D, PROTEINS and COLLAB, Graph U-Net was able to achieve graph classification accuracies of 82.43% 77.68% and 77.56% respectively. In order to improve the efficacy of semi-supervised learning of graph data, one of the first works [13] was able to produce a model that was a significant alternative to traditional CNNs as they were able to perform the same transformations on graphical data. By linear scaling of the graph edges and nodes, it is able to learn the hidden representations that help to understand the graph structure and features. With the help of first order approximations on spectral graph convolutions, this work was able to perform a number of experiments on the citation network datasets – Citeseer, Cora, Pubmed and NELL to achieve classification accuracies of 70.3, 81.5,

79.0 and 66.0 respectively hence demonstrating that their work was able to outperform several other state-of-art methods by a compelling margin.

3 Dataset Description

The dataset used for this study is taken from UCSB bio-segmentation benchmark which consisted of 58 hematoxylin and eosin (H&E) stained images obtained from breast tissue biopsies [14]. The H&E is a commonly used dye which bind particularly to some components of the cell, highlighting them and thus making it easier to identify cell structures. The dataset consists of the stained histopathology images and their corresponding masks that indicate the cell nuclei segments that are used to identify whether the cells are benign or malignant. Some samples and their masks from this dataset are as shown in Fig. 1.

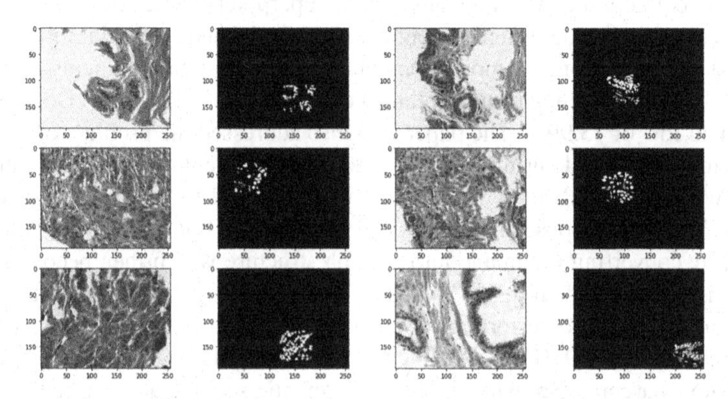

Fig. 1. Sample images and ground truth masks

4 Research Methodology

4.1 Data Augmentation

One of the prominent challenges of working with histopathology images is the lack of data. For training a neural network, the more the number of available training images, the better is the subsequent performance of the model. Thus, in order to create more training samples from the preexisting images, data augmentation is required [15]. While dealing with segmentation, any augmentation performed on the image needs to be simultaneously performed on the masks as well. In this paper, we applied horizontal and vertical flipping, as well as random rotation in order to increase the number of available images. The augmentations applied to a sample image and its mask is as shown in Fig. 2.

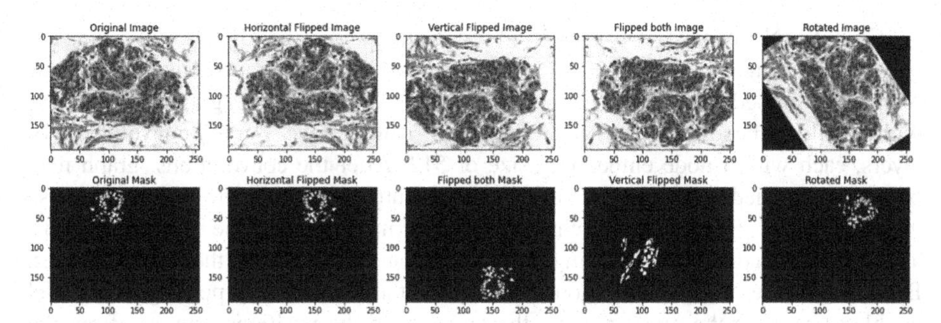

Fig. 2. Sample image and augmented images obtained from it

4.2 Image Normalization

Another lingering challenge when analyzing H&E stained images using deep learning techniques is the variations of color and intensity due to the nature and extent of the stain. Hence, it is essential to perform image normalization in order to rescale the pixel intensities of the image from a range of 0–255 to fall into a scale between 0 to 1 and this is achieved with the min-max normalization. Thus, for each pixel, the normalized value is calculated as the difference of the original pixel value and the minimum pixel value in the image divided by the difference of the maximum and minimum pixel values in the image as given by (1).

$$x_{norm} = \frac{x - x_{min}}{x_{max} - x_{min}} \tag{1}$$

where x_{max} and x_{min} refer to the maximum and minimum pixel intensity values of the images, respectively.

4.3 GraphSegNet Architecture

Like the SegNet architecture [16], the model proposed in this paper also follows an encoder and decoder structure. The underlying principle behind the success of SegNet is that while performing pooling during the encoding and down sampling stage, the pooling indices are retained. Since pooling helps to reduce feature map sizes by retaining the more representative features, storing their pooling indices and passing them onto the decoding layers helps the network to efficiently place and localize the most significant pixels after unpooling and up sampling.

In order to transform the images into an adjacency matrix, they were projected onto a 2D lattice and the consequent features were used to obtain the graph. Further, adjacency were used matrices for the representation of the features of this graph, the problem was reduced to a graph classification task which was then resolved with the help of the model proposed. GraphSegNet makes use of semi-supervised learning in its graph convolutions followed by graph pooling and unpooling to perform the up sampling and down sampling of the images.

GraphSegNet consists of stages 1–5 for encoding and stages 6–10 for decoding for encoding. During encoding, stages 1 and 2 have two layers of graph convolutions each with a node embedding size of 64 and 128 respectively, stage 3 has 3 graph convolutions with a node embedding size of 256 and stages 4 and 5 have 3 graph convolutions of layers, each with a node embedding size of 512. After the convolutions, graph max pooling is applied to reduce the size of the feature set by returning the batch-wise graph-level outputs by taking the channel-wise maximum through the node dimensions. These pooling indices are saved in order to recall and pass them to the decoder stages. By accurately recalling the previous locations of the nodes before pooling, the model is able to recreate the graph placing the nodes from the positions acquired from the corresponding pooling layers. For the decoding part of the network, the pooling indices passed on from the respective stages is used for the up sampling and similarly, stages 6 and 7 consist of 3 graph convolutions with node embeddings of 512, followed by stage 8 which consists of 3 graph convolutions of node embedding size 256 and lastly, stages 9 and 10 consist of 2 graph convolutions with node embedding sizes of 64 as depicted in Fig. 3. Each of the convolutions are followed by batch normalization and a rectified linear unit (ReLU).

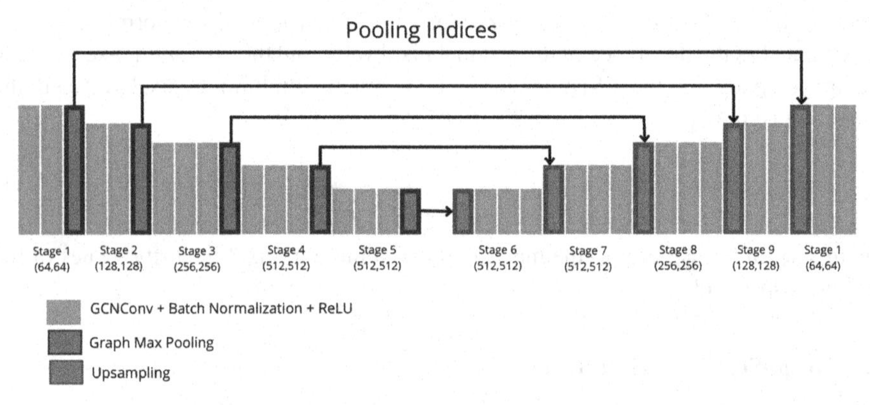

Fig. 3. Encoder decoder structure of GraphSegNet

5 Experimental Setup

5.1 Implementation Details

For the training and testing images in the dataset, the data was segregated with an 80–20 split. The PyTorch module was used in order to load, concatenate, normalize, and augment the images as part of the data cleaning and preprocessing. The GCNConv convolution layer offered by the torch_geometric module was used to build the convolutional blocks of the network. This is based on semi supervised learning with graph-learning convolutional networks which outperformed several state-of-the-art graph convolutional layers [17]. The success of these convolution layers is that it is able to learn the optimal graph structure which then is processed by the neural network for semi-supervised learning by using both graph convolutions and graph learning in a unified network architecture. The batch size was set to 32 and the model was trained for 100 epochs and a callback function was used in order to monitor the validation in order to reduce unnecessary training and overfitting by interrupting the training if sufficient improvement had not occurred for the patience value of 15 epochs. For training, the Adam optimizer with a learning rate of 0.001 was used along with binary cross entropy for the loss function was used. In order to carry out the experiments, a Python 3.7.6 environment was used with the help of a Nvidia Tesla P100 Graphics Processing Unit (GPU) and the visualizations were illustrated with the help of the Matplotlib library.

5.2 Evaluation Metrics

5.2.1 Intersection Over Union (IoU)

The Intersection Over Union (IoU) is the most highly used evaluation metric in order to assess model performance relating to image segmentation. The IoU metric given by Eq. (2) measures the number of pixels that overlap between the predicted output mask and the ground truth mask of the image divided by the total number of pixels across both masks.

$$IoU = \frac{Ground\ Truth \cap Prediction}{Ground\ Truth \cup Prediction} \tag{2}$$

5.2.2 Dice Coefficient

The Dice coefficient (DC) is another evaluation metric used to represent the efficacy of a model for image segmentation. It is given by (3) and is defined as the twice the intersection of the ground truth and prediction by the summation of the intersection and union. This metric provide an understanding on how similar the ground truth and the predicted masks are.

$$DC = \frac{2 * (Ground\ Truth \cap Prediction)}{(Ground\ Truth \cap Prediction) + (Ground\ Truth \cup Prediction)} \tag{3}$$

5.2.3 Pixelwise Accuracy

Pixelwise accuracy simply refers to measuring the percentage of pixels in the image that were correctly segmented and is given by (4). Since in performing semantic segmentation, the aim is to predict the mask, this is done by considering the pixels as binary where 0 or negatives represent the background class and 1 or positive represent the object of interest. Thus, the pixelwise accuracy is defined as the number of true positives (TP) and negatives (TN) upon the total number of pixels.

$$Accuracy = \frac{TP + TN}{TP + TN + FP + FN} \tag{4}$$

5.2.4 Precision

Precision quantifies how many pixels that were belonging to the object of interest were correctly predicted in the mask that was produced by the model. Relating to semantic segmentation, a true positive refers to when a prediction mask has an IoU score that exceeds a certain threshold, a false positive is when a pixel from the predicted object mask has no association with the ground truth object mask and a false negative is when a pixel from a ground truth object mask has no association with the predicted object mask. Precision is hence measured as the ratio of true positives to the sum of true positives and false positives (FP) and is given by (5).

$$Precision = \frac{TP}{TP + FP} \tag{5}$$

5.2.5 Recall

Recall helps to describe how effectively the positively predicted pixels relate to the ground truth i.e. from all the pixels relating to the object of interest, how many of them were accurately represented by the predicted mask. Hence, Recall is calculated as the ratio of the true positives to the sum of true positives and false negatives (FN) and is given by Eq. (6).

$$Recall = \frac{TP}{TP + FN} \tag{6}$$

6 Results

6.1 Model Performance

After training the model as per the implementation details explained above, the results of the model are obtained as depicted in Table 1. The training and validation accuracy, dice coefficient, pixelwise accuracy, IoU, precision and recall are as illustrated in Fig. 4.

Table 1. Experimental results

Model	IoU	Pixelwise accuracy	Dice coefficient	Precision	Recall
GraphSegNet	83.2%	84.75%	83.74%	75.40%	79.93%

Figure 5 shows the predicted masks through which it is evident that the semantic segmentation was performed well and the output mask predicted by the model is quite close to the ground truth mask, hence providing a visual confirmation of the success of this approach.

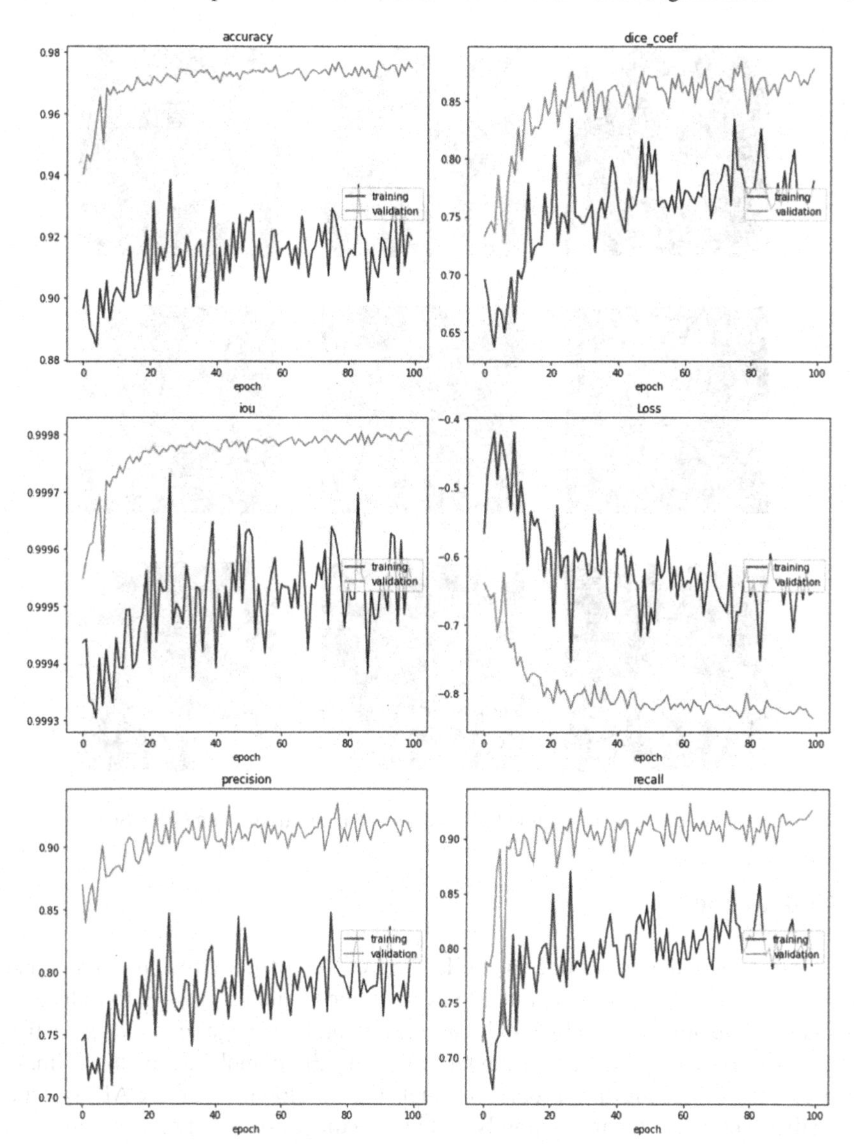

Fig. 4. Training and validation evaluation metrics

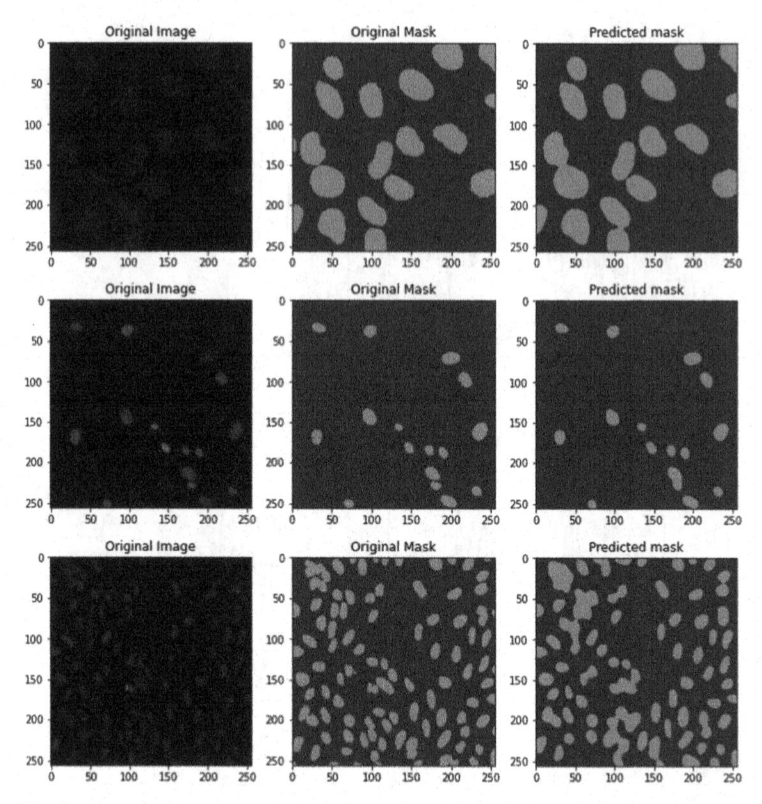

Fig. 5. Sample images, ground truth masks and predictions from GraphSegNet

7 Conclusion

Breast cancer remains to be one of the leading causes of mortality in women and a timely diagnosis followed by treatment has been one of the main ways to reduce this. The diagnosis requires the analysis of breast histopathology images obtained from tissue biopsies and due to the plausible human error and the complexity of these images, the need for automating this process through the development of a CAD system is imperative. They key step in accurately obtaining a diagnosis is to first perform the cell nuclei segmentation, followed by their analysis to classify them as benign or malignant. Thus, through this study, it was aimed to design a graph convolutional neural network with an encoder-decoder architecture called the GraphSegNet which was based on the well-known SegNet, to perform semantic segmentation of cell nuclei. The methodology proposed involved performing data preprocessing such as image normalization and data augmentation to obtain a more defined and larger set of images. These images were then translated to adjacency matrices for feature representation and fed into the GraphSegNet architecture which has an encoder-decoder structure that preserves and passes on the pooling indices obtained during encoding to the decoding stages thus allowing for the model to recreate the output graphs and consequently the images more accurately by placing the nodes according to the corresponding pooling indices. The experimental results demonstrate the efficacy of the model as it was able to outperform several other state-of-the-art methods. Thus, it can be concluded that the architecture performed well for the cell nuclei segmentation from breast histopathology images and the future scope of this model can be extended to simultaneously perform classification of the obtained segments to fully automate the process of detection of malignancies.

References

1. https://nanonets.com/blog/deep-learning-for-medical-imaging/. Accessed 02 Feb 2022
2. Daigavane, A., et al.: Understanding convolutions on graphs. Distill **6**, e32 (2021)
3. https://towardsdatascience.com/understanding-graph-convolutional-networks-for-node-classification-a2bfdb7aba7b#:~:text=The%20major%20difference%20between%20CNNs,non%2DEuclidean%20structured%20data)
4. Scarselli, F., Gori, M., Tsoi, A.C., Hagenbuchner, M., Monfardini, G.: The graph neural network model. Trans. Neur. Netw. **20**(1), 61–80 (2009). https://doi.org/10.1109/TNN.2008.2005605
5. Li, X., Zhou, T., Li, J., Zhou, Y., Zhang, Z.: Group-wise semantic mining for weakly supervised semantic segmentation. In: Proceedings of the AAAI Conference on Artificial Intelligence, vol. 35, no. 3, pp. 1984–1992 (2021). https://ojs.aaai.org/index.php/AAAI/article/view/16294
6. Zhang, B., Xiao, J., Jiao, J., Wei, Y., Zhao, Y.: Affinity attention graph neural network for weakly supervised semantic segmentation. IEEE Trans. Pattern Anal. Mach. Intell. **44**, 8082–8096 (2021)
7. Zhang, L., Li, X., Arnab, A., Yang, K., Tong, Y., Torr, P.H.: Dual graph convolutional network for semantic segmentation. arXiv, abs/1909.06121 (2019)
8. Yu, C., Liu, Y., Gao, C., Shen, C., Sang, N.: Representative graph neural network. In: Vedaldi, A., Bischof, H., Brox, T., Frahm, J.-M. (eds.) ECCV 2020. LNCS, vol. 12352, pp. 379–396. Springer, Cham (2020). https://doi.org/10.1007/978-3-030-58571-6_23
9. Liu, Q., Kampffmeyer, M., Jenssen, R., Salberg, A.-B.: SCG-Net: self-constructing graph neural networks for semantic segmentation (2020)
10. Ouyang, S., Li, Y.: Combining deep semantic segmentation network and graph convolutional neural network for semantic segmentation of remote sensing imagery. Remote Sens. **13**(1), 119 (2021). https://doi.org/10.3390/rs13010119
11. Lu, Y., Chen, Y., Zhao, D., Chen, J.: Graph-FCN for image semantic segmentation. In: Lu, H., Tang, H., Wang, Z. (eds.) ISNN 2019. LNCS, vol. 11554, pp. 97–105. Springer, Cham (2019). https://doi.org/10.1007/978-3-030-22796-8_11
12. Gao, H., Ji, S.: Graph U-Nets. IEEE Trans. Pattern Anal. Mach. Intell. https://doi.org/10.1109/TPAMI.2021.3081010
13. Kipf, T.N., Welling, M.: Semi-supervised classification with graph convolutional networks. arXiv:1609.02907 cs, stat (2017)
14. Drelie Gelasca, E., Byun, J., Obara, B., Manjunath, B.S.: Evaluation and benchmark for biological image segmentation. In: 2008 15th IEEE International Conference on Image Processing, San Diego, CA, pp. 1816–1819 (2008)
15. Aquino, N.R., Gutoski, M., Hattori, L.T., Lopes, H.S.: The effect of data augmentation on the performance of convolutional neural networks, 21528/CBIC2017-51 (2017)
16. Badrinarayanan, V., Kendall, A., Cipolla, R.: SegNet: a deep convolutional encoder-decoder architecture for image segmentation. IEEE Trans. Pattern Anal. Mach. Intell. **39**(12), 2481–2495 (2017). https://doi.org/10.1109/TPAMI.2016.2644615
17. Jiang, B., Zhang, Z., Lin, D., Tang, J., Luo, B.: Semi-supervised learning with graph learning-convolutional networks. In: 2019 IEEE/CVF Conference on Computer Vision and Pattern Recognition (CVPR), pp. 11305–11312 (2019). https://doi.org/10.1109/CVPR.2019.01157

Performance Analysis of Cache Memory in CPU

Viraj Mankad, Virag Shah, Sachin Gajjar, and Dhaval Shah(✉)

Department of Electronics and Communication Engineering, Institute of Technology, Nirma University, Ahmedabad, Gujarat, India
{sachin.gajjar,dhaval.shah}@nirmauni.ac.in
https://ec.nirmauni.ac.in/

Abstract. The architecture of a computer can be efficiently designed, by synchronizing the major sections of a computer namely the Central Processing Unit (CPU), memory, data path, pipeline structure, Arithmetic, and Logic Unit (ALU). The data storage and organization depend on the memory allotment and specific memory design carried out in accordance with the CPU specifications. A specific memory hierarchy is followed for the CPU, which starts with the units near the processor such as memory registers, cache memory, main memory, secondary memory, and finally the flash storage devices. This hierarchy is based on parameters like speed, response time, program complexity, and overall storage capacity for the computer. This work focuses on the cache memory which is a high-speed special memory, essential in the memory section of computer architecture. A modified version of the vector triad benchmark program is run on four computer systems with different specifications to check the performance and understand the cache levels. The problem size, i.e. the execution of the benchmark program is defined to get iterations and plotting comparison curves for time and throughput analysis for test cases of computers. As the level of cache changes from level 1 to level 3, the throughput curve decreases. Also, when more processes are active, while the benchmark code is running, then a significant rise in computation time, data access time, and total time values are observed which reflects on the computer performance.

Keywords: Cache memory · Memory hierarchy · Benchmark programs · CPU Performance · Vector triad

1 Introduction

The storage and execution of data and instructions is an integral part of computer architecture. The data transfer in a computer system involves parameters like speed, execution time, computer performance, hardware specifications, etc.

K. K. Patel et al. (Eds.): icSoftComp 2022, CCIS 1788, pp. 170–181, 2023.
https://doi.org/10.1007/978-3-031-27609-5_14

A specific hierarchy of memory can be developed to have a reliable and organized operation in the computer system. Exploring modern computer storage, a common pattern emerges, suggesting that any kind of processing or computing device with higher capacities may offer low performance. This trade-off between performance and capacity is termed as the memory hierarchy. In the memory hierarchy, a special memory called the cache memory has a major role in faster data access and storage [1].

In general, the cache is a more readily accessible place for data and operates with high speed for quicker computation [2]. The access time and response of a system for data instruction executed are minimized by a properly developed Memory Hierarchy. In the CPU, a specific Memory sequence or Memory Hierarchy is followed for proper organization and smooth transfer of data. After the Processor for the memory slots, there is a sequence of Registers, Cache Memory, Main Memory, and Secondary Memory. Storage locations like registers, CPU caches, main memory, and files on Flash disk, Traditional disk, etc. have varied access times, rates, and storage.

When a compute-intensive application is programmed, it is important to consider where data is stored and how frequently the program accesses data from which memory. For designing CPUs, knowing the memory hierarchy helps to trade-off between speed, cost, and storage capacity, Also, the cache memory, which is very near to the processor for data processing, has frequent use in the computer system. Primary storage devices are accessed directly by a program on the CPU.

The paper after the introduction is organized as follows: Sect. 2 describes the literature review and existing work for the memory hierarchy of CPU, then Sect. 3 deals with the methodology of the benchmark program, and the information about CPU Performance and the experiment parameters. The next Sect. 4 provides the computer specifications for all 4 computers, and it is followed by the benchmark logic. The results and graph analysis are included in the next section, followed by the last section as the conclusion and future scope.

2 Literature Review

Cache Memory is a very high-speed memory, used to speed up and synchronize with a high-speed CPU. Between the operation of RAM and CPU, this memory is a faster memory that can act as a buffer or intermediator. A special function of cache is that it is capable of data holding and supplying as and when needed and some important and frequently used data and instructions can be stored, as per the levels and memory capacity.

A cache is a fast semiconductor memory, where a frequently used data or instructions from secondary memory is stored [3]. For reduction of time required to locate and access data within the cache and provide it to the computer for processing, the smaller capacity of the cache can be helpful [4]. The working of cache depends on the principle of locality [3]. When the computer program accesses the same set of memory locations for a particular time interval, that particular phenomenon is called a locality of reference. The property of locality can be observed in subroutine calls or loops in a program. There are three types of cache: L1, L2, and L3.

- L1 cache or primary cache is extremely fast, and smaller in size. The access time of the L1 cache is comparable to the processor registers. It is very close to the ALU.
- L2 cache or the secondary cache is slower and has more storage capacity than the L1 cache. It is placed between the primary cache and the rest of the memory if the L3 cache is not present.
- L3 cache, a specialized cache that has the main task to improve the performance of L1 and L2. Generally, L1 and L2 caches are faster than the L3 cache. Many multi-core CPUs also share data between cores in a larger L3 cache.

Cache size is measured in bytes or words. More data can be stored in a larger cache but it will result in higher costs. Cache line width is defined as the unit of data transfer between the cache and main memory. Placement policies determine the memory location to store the value of the incoming cache line. To ensure optimal memory performance, changing any parameter like cache size often affects other parameters like speed, latency, etc. as they are interrelated [5]. The working of cache memory describes the memory hierarchy and data transfer between different cache levels, where fast and slow transfer takes place and the cache memory is found in configurations of single cache and multiple caches also [6]. Cache memory also points to the same set of instructions or data stored in memory executed multiple times for a fixed period refers to locality. In each level of the cache, there exists a separate sub-cache for data and instructions i.e. L1(d) and L1(i) [1].

The comparison of the performance of modern time processors with the various factors affecting cache memory performance is a major viewpoint carried out in [3], for the analysis of the memory hierarchy of the CPU. The parameters like hit rate, latency, speed, and energy consumption are taken into consideration. To improve the access time, the above parameters were studied using the SPEC2000 benchmark program. The conclusion through such conceptual work refers to the cache replacement policies that may gain more significance as caches are becoming more set-associated [3].

Another aspect of memory allocation which has been studied in [1], is the trade-off between reliability and performance. The accurate method to estimate the reliability of cache memories and avoid the soft errors in cache memory has been worked upon. The SPEC2000 benchmark program has been used to measure the meantime to failure ratio of the level 1 cache. For the analysis of the memory hierarchy, the Schoenauer vector triad benchmark program has been used for study and demonstration purposes [1].

The locality concept involved in the cache can prove handy for design specifications and analysis of the factors affecting computer performance. The literature review also links to the applications of cache memory have a greater impact in solving multi-core computing issues and providing a better transition of data [7]. Data and task scheduling related problems which involve multi-core processing are resolved by efficient usage of cache memory and the hierarchy. In case of real-time scheduler, if the device is to be used where there are physical conditions constraints like energy, temperature, pressure etc., then the memory configuration of the processing device, must have inclusion of cache as per its levels, so as to fetch data promptly [8]. A processor chip can have core, data and instructions part in-sync with the cache levels. Data efficiency and resource utilization improves in such systems where cache memory is interfaced for speed and time enhancements [9]. The benchmark laws and applications prove to be a precise way to verify the usage of cache memory in multi-core computing [10].

3 Methodology for Benchmark Program

Benchmark Programs and other verification codes are the main methodologies used for getting the computer accuracy or performance measurement based on a specific defined problem execution size. Here, it can be stated that the vector triad code is used as a problem to be executed, and taking that as a reference, the parameter of problem size can be further scaled to get a better view of the graph. The significance of memory to system performance motivates improving its efficiency [11]. There is a need to understand the memory hierarchy to simulate the results. For CPU memory hierarchy, cache memory is considered as primary storage along with Registers. Attempts have been made and methodologies have been developed in order to increase the speed and capability of data transfer in the cache. The cache is capable of holding a copy of such information which is frequently used and also accessed for specific operations [4].

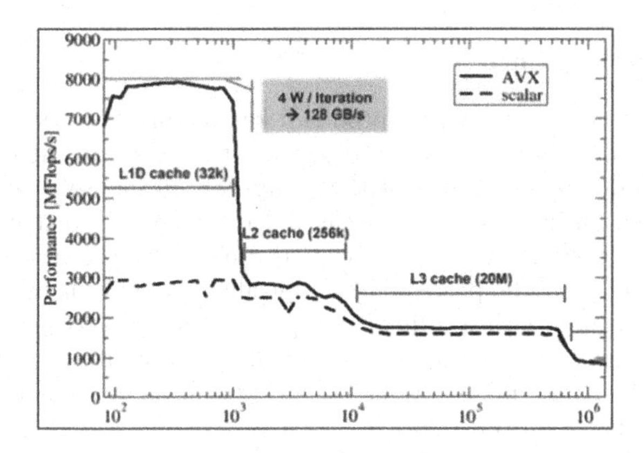

Fig. 1. Performance curve: cache Levels [12]

The measure of the number of units of information that can be processed by a system in a specific amount of time is called Throughput [13]. It is generally measured as bits/sec. So, the lower the data access time, the higher the throughput. Thus, the higher the value of throughput, the better the performance. In Fig. 1, the value of throughput can be observed to be decreasing significantly at ($x = 10^3$) and ($x = 10^4$). The decrease in throughput value indicates the increase in total runtime. If the problem size exceeds L1 memory ($x = 10^3$), then the L2 memory is used which results in higher data access time. The experimental curve for the L1 cache can be observed and the problem size relation with the actual CPU specification can be verified. The values of throughput for different problem sizes for a case explained in [12] are plotted as shown in Fig. 1.

Performance is reported in FLOPS (Floating Point Operations Per Second) for a wide range of problem sizes N and the value is chosen such that the benchmark runs for a sufficiently long time so that the wallclock time measurement is accurate. The main logic for the Vector Triad Code is shown below, as referred from the detailed explanation about memory hierarchy benchmarking in [14]. The vector triad function is used for the realization of the cache memory hierarchy. The vector triad benchmark is a simple program that measures memory bandwidth in terms of throughput. It performs arithmetic addition and multiplication operations for n number of times in a loop.

```
S = get_walltime()
do r = 1,NITER
    do i = 1,N
        A(i) = B(i) + C(i) * D(i)
    enddo
enddo
WT = get_walltime() - S
MFLOPS = 2.d0 * N * NITER / WT / 1.d6
```

3.1 CPU Performance

For computer performance, CPU speed is a key factor. The performance is the execution time for running a specific program. It depends on the response time, throughput, and execution time of the computer. Response Time is the time elapsed between an inquiry on a system and the response to that inquiry [15]. Throughput is a term used in computer performance that indicates the amount of information that can be processed in a specific amount of time. The throughput can be measured with a unit of bits per second or data per second. CPU execution time is the total time a CPU spends on computing a given task. So, the performance is determined by the execution time as performance is inversely related to the execution time [16].

The process of measuring the performance of the algorithm against other algorithms may be considered as better or as a reference, which is called Benchmarking. Benchmark refers to a process of running a specific computer program or a set of programs, or it can also be detailed operations for evaluating the relative performance by running some standard tests. System performance is determined by benchmark programs, and the programs can test stability, speed, effectiveness, etc. factors. The computer parameters considered for comparison are Problem Size (2^N), Throughput (FLOPS), Computation Time (ns), Data Access Time (ns), and Total Time (ns).

The performance and characteristics of computer hardware can be assessed by benchmark programs, e.g., CPU's floating-point operation performance: software testing. The hardware architecture of the given computer was studied using the lscpu command, which gathers CPU architecture information i.e., the clock speed, processor type, processor cores, etc. The impact of the memory hierarchy on the computation speed of a CPU was explored by Vector Triad Benchmarking.

4 Computer Specifications

Table 1 demonstrates the parameters and computer architecture specifications for all 4 computer systems, and the benchmark code experiment here is done, with multiple processes running, in Computer 4. For all the four computers, the Architecture is x86_64, the CPU op-mode is 32-bit, 64-bit and the Address Size is 36 bits physical, 48 bits virtual.

Table 1. Computer specifications

	Computer: 1	Computer: 2	Computer: 3	Computer: 4
Thread(s) per core	1	2	2	2
Operating systems	Linux	Windows	Windows	Windows
CPU MHz	1214.326	2000.000	1992.000	1190.000
Model name	Intel(R) Core(TM) 2 Duo	Intel(R) Core(TM) i3-5005U	Intel(R) Core(TM) i7-8565U	Intel(R) Core(TM) i5-1035G1
L1 cache	32 KB	128 KB	256 KB	320 KB
L2 cache	3.072 MB	512 KB	1.0 MB	2.0 MB
L3 cache	–	3 MB	8.0 MB	6.0 MB

- Problem Size: The number of times an instruction or benchmark program is executed. Here, it is the number of times the vector triad benchmark program is running.
- Throughput: The amount of data transferred or handled between two locations, within a fixed amount of time.
- Total Time = Data Access Time + Computation Time. It is time spent by the system executing that task, including the time spent executing run-time and accessing the data, calculated using the clockgettime function.
- Computation Time: It is the time required for only computational processes.
- Data Access Time: It is a time delay between a request to the system memory, and the requested data returned.

5 Benchmark Logic

The benchmark logic flow is about how the benchmark logic is used to analyze the computer performance. The variables for performance parameters are initialized after which array initialization would start. Calculation of time values and throughput is in the next step, which is done as per the vector triad logic functions. Runs of loop and total problem size are also considered and graphs are generated. The flow of the Benchmark Logic is shown in Fig. 2 and Fig. 3, as Part - 1 and 2 of the logic flow.

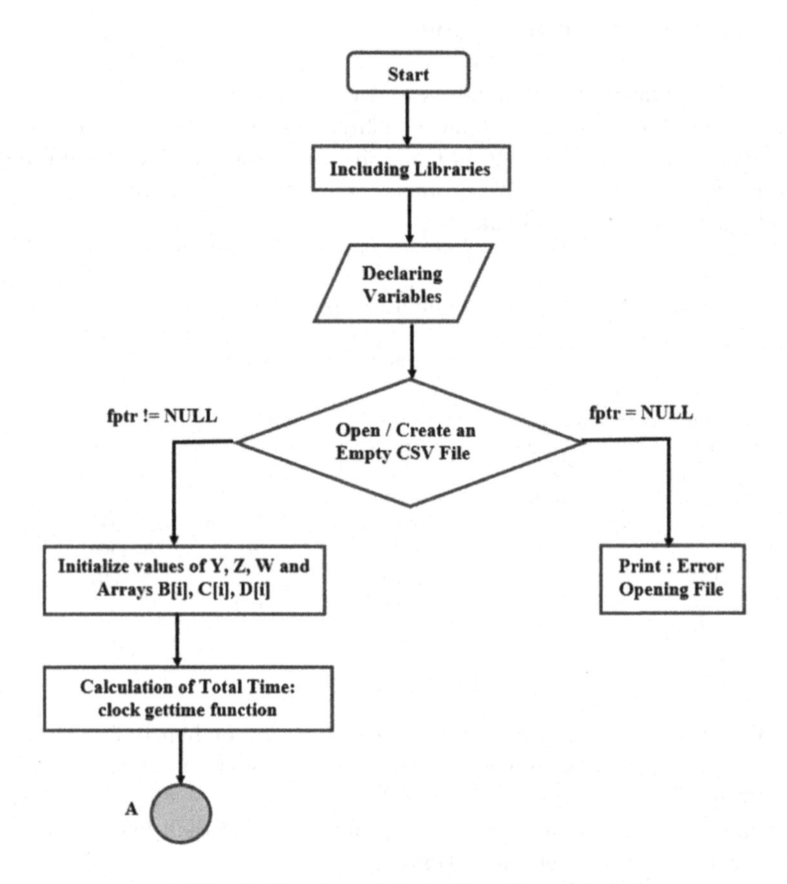

Fig. 2. Benchmark logic flow: Part - 1

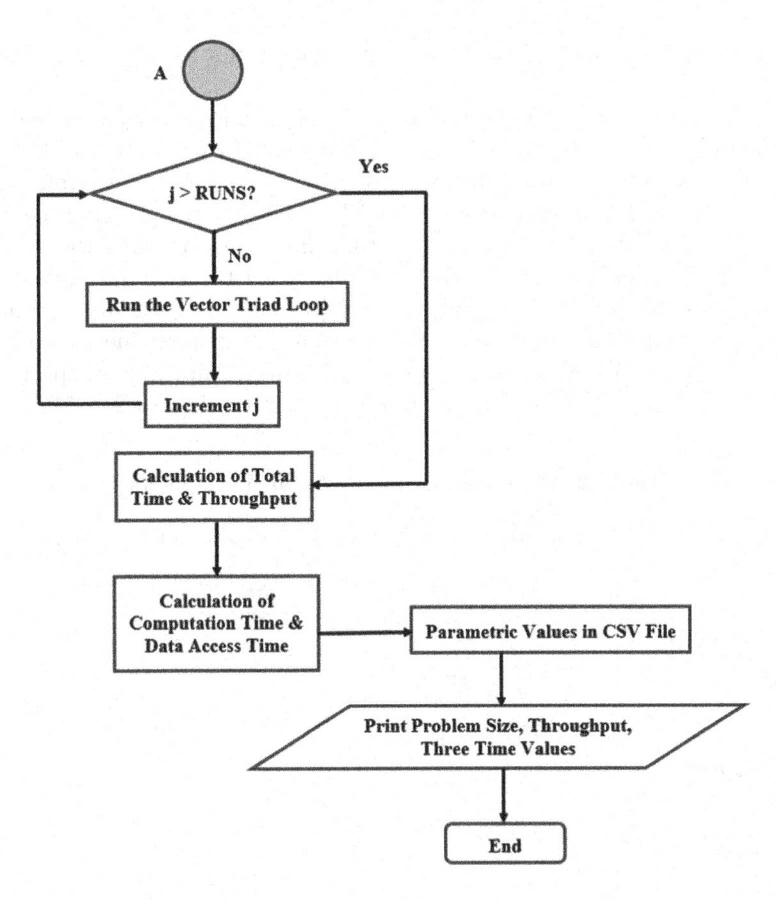

Fig. 3. Benchmark logic flow: Part - 2

6 Results and Graph Analysis

As ideal real-time cases are taken, when the benchmark program is running, the number of processes running in the computer, or processing done by the CPU would affect the parametric values. The problem size for each computer are pre-determined to be: Computer - 1 (2^5 to 2^{25}), Computer - 2 (2^5 to 2^{27}), Computer - 3 (2^5 to 2^{28}), Computer - 4 (2^5 to 2^{29}). If the N value, i.e. the initial instruction value for problem size, is taken more than the specified value, then the computer may stop working/malfunction/give garbage values.

The default storage values of the three cache levels are obtained and the final graphs are compared to those values, which show when the memory transition will take place from L1 to L2 to L3, which shows reduced computer performance. Here, the value of instruction for the start of the loop is taken as $N = 5$, as any value below that would not give proper distinguishable results, as the measurements are in sec and FLOPS, so for more precise and accurate graphs, the initial value of N is taken as 5. The defined problem size for all the computers and the theoretical and experiment-based values of cache limits for all four computers are shown in Table 2.

From general observation from Fig. 4(a) for throughput comparison, with the data in the L1 cache, we get throughput values close to 1 GFLOPS per second which shows the maximum performance as the data access time is least for L1. When the L1 limit exceeds, the CPU needs to use the L2 cache which is somewhat far. Thus, the data access time increases for the CPU resulting in a drop in the performance curve. Similarly, when the L2 limit exceeds, the CPU needs to use the L3 cache and data access time gets even larger, resulting in a drop in computer performance. And in some computers, the L3 cache has to be enabled or located or is not pre-integrated on-chip, due to specific PC requirements. For the levels of cache, in terms of speed, L1 > L2 > L3.

Table 2. Obtained cache levels for all 4 computers

Cache levels	CPU specifications	Actual problem size	Experiment-based problem size
Computer: 1			
L1	32 KB	2^{15}	2^{15}
L2	3 MB	$\approx 2^{22}$	2^{21}
Computer: 2			
L1	128 KB	2^{17}	2^{16}
L2	512 KB	2^{19}	2^{21}
L3	3 MB	$\approx 2^{22}$	2^{22}
Computer: 3			
L1	32 KB	2^{15}	2^{15}
L2	3 MB	$\approx 2^{22}$	2^{21}
Computer: 4			
L1	320 KB	2^{19}	2^{20}
L2	2 MB	2^{21}	2^{23}
L3	6 MB	2^{23}	–

Figure 4(a) shows the comparison of throughput for all 4 computers. The higher the throughput value, the higher will be the performance. The figure shows the performance of the test computers in terms of throughput. As the CPU performs a transition from L1 to L2 or L2 to L3, a drop in throughput value is observed. After graphical analysis, we can verify the theoretical cache limits of the computer by this method. The limits of L1, L2, and L3 are obtained from the task manager and final plots are used to compare the cache memory limits by analyzing the throughput performance of the computers.

Figure 4(b) shows the total time comparison for all four computers. The total time is similar for all cases, and there is a sudden rise in time value for 4th Computer (real-time multi-processing). Computation Time remains constant irrespective of problem size, which means the total time would increase only if the data access time increases.

Figure 5(a) and Fig. 5(b) show the computation time and data access time comparison respectively, for all the computers. The spike in both figures is due

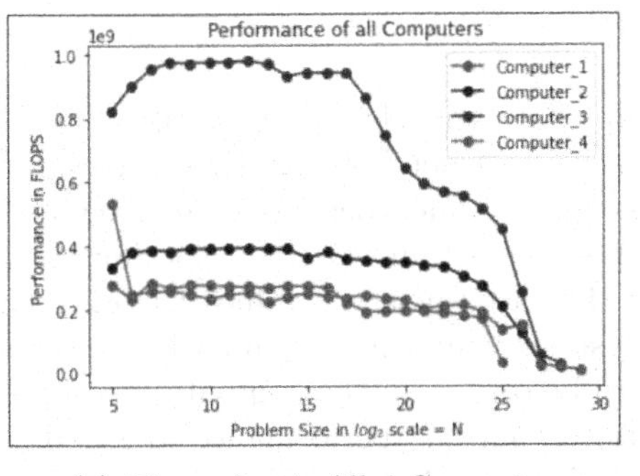

(a) Throughput: All 4 Computers

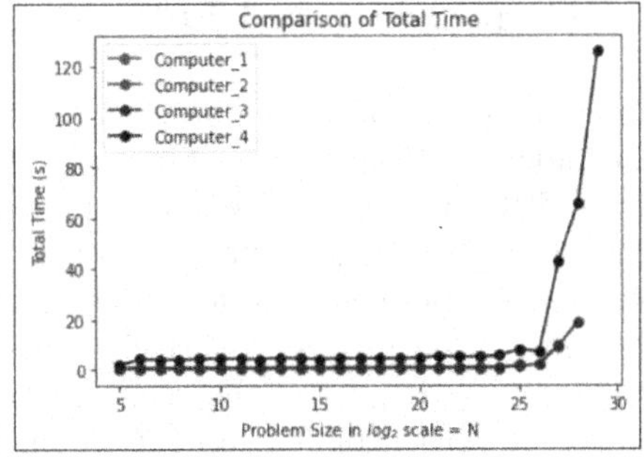

(b) Total Time: All 4 Computers

Fig. 4. Comparison of Throughput and Total Time

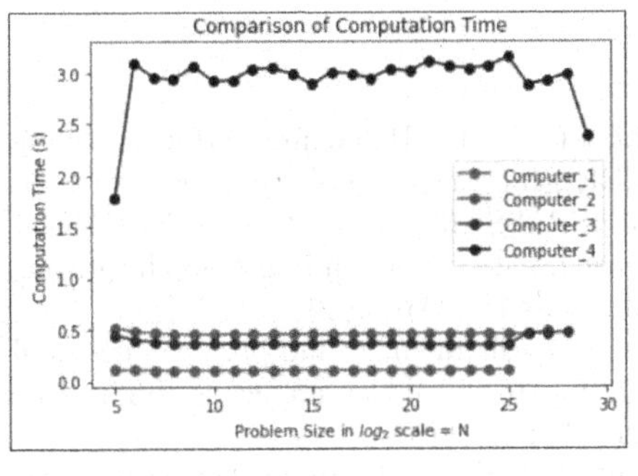

(a) Computation Time: All 4 Computers

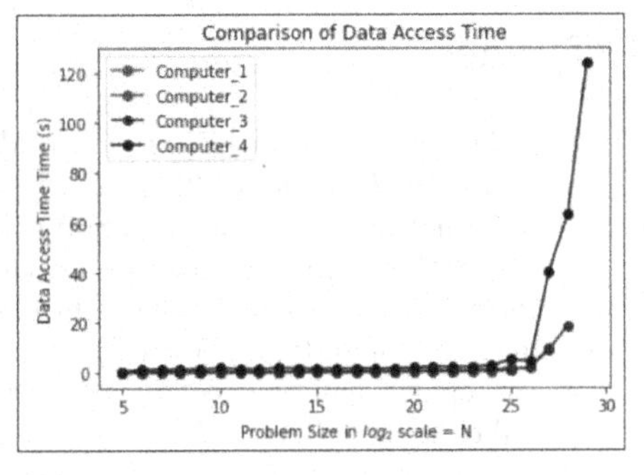

(b) Data Access Time: All 4 Computers

Fig. 5. Comparison of Computation Time and Data Access Time

to the number of active processes, while the benchmark runs for that particular Computer 4. As total time is the summation of data access time and computation time, the total time can be changed only by changing the data access time, as the computation time is the same for all the cases. A hike in the total time at L3 cache memory is observed when data access time increases and computation time remains constant. As the higher memory levels are needed to be accessed, more time is required. In the observation for total time, the curve of Computer 4 shows a significant rise at a point, which determines the change in the level of the memory hierarchy. Computation Time for Computer 4 is higher in the plot as many applications and processing softwares were kept open while running the program. The value for the computer problem size was chosen such that the benchmark ran for a sufficiently long time: the wall-clock time measurement was found to be accurate. The sizes of L1, L2, and L3 can be verified by observing the nature of the graph. Computer Performance is 1/CPU Execution Time, the base for analysis of memory hierarchy and computer output.

7 Conclusion and Future Scope

From the study of cache memory and implementation for the test cases, it was concluded that the time graph and performance depend on real-time values, i.e. if the CPU faces a high amount of processing at a time, then the time values and throughput would get affected. For computers, certain hardware specifications responsible for such a performance were the cache memory values L1, L2, and L3 and certain processes active in the system, while the benchmark code was running. The impact of the cache memory hierarchy on the computation speed of a CPU was explored, and it was noticed that the performance in FLOPS, depends on the range of problem sizes or N. The slope-based curve indicated changes in cache levels, for a particular value of N (Example: 2^5 to 2^{28}), and for the cache levels, as per speed, the obtained hierarchy was L1 $>$ L2 $>$ L3.

References

1. Asadi, G.H., Sridharan, V., Tahoori, M.B., Kaeli, D.: Balancing performance and reliability in the memory hierarchy. In: IEEE International Symposium on Performance Analysis of Systems and Software, ISPASS 2005, pp. 269–279. IEEE (2005)
2. Faris, S.: How important is a processor cache?. https://smallbusiness.chron.com/important-processor-cache-69692.html. Accessed 11 Apr 2021
3. Kumar, S., Singh, P.: An overview of modern cache memory and performance analysis of replacement policies. In: 2016 IEEE International Conference on Engineering and Technology (ICETECH), pp. 210–214. IEEE (2016)
4. Rodriguez E, Singh S.: Cache memory. https://www.britannica.com/technology/cache-memory. Accessed 8 Apr 2021
5. What is response time? - Definition from techopedia. https://www.techopedia.com/definition/9181/response-time. Accessed 6 Apr 2021
6. Cache memory. https://12myeducation.blogspot.com/2020/02/cache-memory-cach-memory.html. Accessed 11 Apr 2021
7. Moulik, S., Das, Z.: Tasor: A temperature-aware semi-partitioned real-time scheduler. In: TENCON 2019–2019 IEEE Region 10 Conference (TENCON), pp. 1578–1583. IEEE (2019)
8. Sharma, Y., Moulik, S., Chakraborty, S.: Restore: real-time task scheduling on a temperature aware finfet based multicore. In: 2022 Design, Automation & Test in Europe Conference & Exhibition (DATE), pp. 608–611. IEEE (2022)
9. Moulik, S.: Reset: a real-time scheduler for energy and temperature aware heterogeneous multi-core systems. Integration **77**, 59–69 (2021)
10. Moulik, S., Das, Z., Saikia, G.: Ceat: a cluster based energy aware scheduler for real-time heterogeneous systems. In: 2020 IEEE International Conference on Systems, Man, and Cybernetics (SMC), pp. 1815–1821. IEEE (2020)
11. Hallnor, E.G., Reinhardt, S.K.: A unified compressed memory hierarchy. In: 11th International Symposium on High-Performance Computer Architecture, pp. 201–212. IEEE (2005)
12. Microbenchmarking for Architectural Exploration (and more). https://moodle.rrze.uni-erlangen.de/pluginfile.php/15304/mod_resource/content/1/03_Microbenchmarking.pdf. Accessed 14 Apr 2021

13. Burke J.: What is throughput?. https://www.techtarget.com/searchnetworking/definition/throughput. Accessed 10 Apr 2021
14. Hager G.: Benchmarking the memory hierarchy of the new AMD Ryzen CPU using the vector triad. https://blogs.fau.de/hager/archives/7810. Accessed 7 Apr 2021
15. Jain R.: Memory hierarchy design and its characteristics. https://www.geeksforgeeks.org/memory-hierarchy-design-and-its-characteristics/. Accessed 12 Apr 2021
16. Parhami, B.: Computer Architecture: From Microprocessors to Supercomputers. Oxford University Press, New York (2005)

Extraction of Single Chemical Structure from Handwritten Complex Chemical Structure with Junction Point Based Segmentation

Shrikant Mapari(✉)

Symbiosis Institute of Computer Studies and Research (SICSR), Symbiosis International (Deemed University) (SIU), Model Colony, Pune, Maharashtra, India
shrikant.mapari@sicsr.ac.in

Abstract. In chemistry domain, lots of chemical compounds were represented by complex graphics. These complex graphics can be a combination of basic shapes from geometry like a circle, line, hexagon, etc. When such complex graphics have drawn with free handwriting, it was very difficult to recognize it for inputting it into a computer system. Hence the recognition of handwritten complex chemical structure (HCCS) has challenges in the segmentation of each chemical symbol or structure. Here in this paper, we had proposed an approach to segment the isolated chemical structures from HCCS. We had considered only the chemical structures made up with benzene rings. The proposed approach extracts isolated benzene structure from HCCS based on junction point base segmentation technique. Skeletonization of the input image has improved the accuracy of the junction point detection process. Here the skeletonization of an image has been achieved with morphological thinning. The proposed approach accepts the scanned image of HCCS as an input, and it produced the images of isolated chemical structures as an output.

Keywords: Segmentation · Connected symbol · Junction point · Thinning of the image · Morphological thinning · Complex chemical structures

1 Introduction

The chemical reactions, mostly from biochemistry and pharmacy domain have contents complex chemical compounds, which represented by complex structures when the reactions were written on paper. Recently, recognition of handwritten chemical expressions (HCE) has been a sizzling topic for researchers. Recognition of a chemical expression requires recognition of the chemical structures from that expression. Before recognition of HCE, these expressions have segmented in an isolated chemical symbol. There was some work has been reported by the

K. K. Patel et al. (Eds.): icSoftComp 2022, CCIS 1788, pp. 182–196, 2023.
https://doi.org/10.1007/978-3-031-27609-5_15

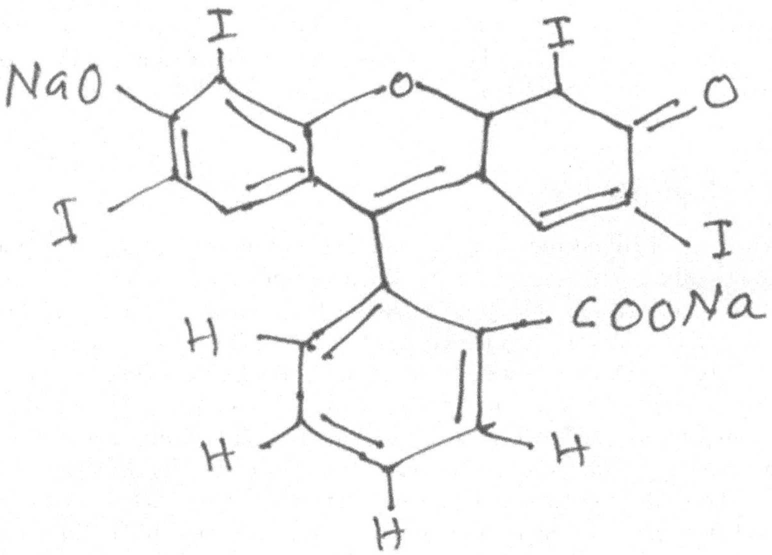

Fig. 1. Handwritten complex chemical structure

researchers to segment the handwritten chemical symbols form online handwritten chemical expression [2,14]. However, segmentation of a complex structure has a different kind of problem, where segmentation has done for the touching structures and symbols. There has been very little work has reported on this kind of segmentation problem in case of HCE. A complex chemical structure has represented by merging more structures into a single structure. Fig. 1 shows how a handwritten complex chemical structure has been looking. Here in this Fig. 1, there have been some benzene rings, which had merged to form a single complex chemical structure. Hence, to recognize this structure, it requires the separation of each benzene ring from it. This separation can achieve by a segmentation method. Segmentation is an essential phase for the recognition of handwritten chemical expression [17]. A free handwritten has no constraint on the writer, and it is writer-dependent. Hence recognition and segmentation of such complex handwritten chemical structures is a challenging problem. Here in this paper, we had developed algorithms to extract a single benzene structure from such HCCS. The proposed approach has accepted a scanned image of HCCS as an input. The inputted image has gone through the stages named as: pre-processing, skeletonization, Junction Point (JP) detection, and segmentation. The output of the proposed approach is the segmented benzene structure from the complex chemical structure. Once benzene structures had got isolated from HCCS, it makes easy to recognize the individual benzene structure, and the complete HCCS has been recognized by applying domain knowledge. The rest of the paper divided into four sections. The next Sect. 2 has described the existing solutions proposed by various researchers for the segmentation problem of connected symbols. The procedure of the proposed technique along with its stages has enlarged in Sect. 3. The explanation about experimental work and the results had discussed in Sect. 4. Section 5 summarizes the complete work along with feature expansion.

2 Critical Review

Segmentation of connected symbol, character and numerical has an importance in the recognition process of it. Fukushima K. and Imagawa T. has proposed a model for segmentation of English character from cursive handwriting [4]. This model also recognizes the segmented character with the help of a multilayer neural network. Recently there were some of the works has reported for segmenting character form regional languages [5,18]. While recognizing a printed mathematical expression Garain U. and Chaudhuri B.B. had dealt with the segmentation problem for touching mathematical symbols [6]. The proposed solution was based on the multi-factorial analysis. However, segmentation of handwritten chemical symbols from chemical expression has importance in HCE recognition. A novel approach has proposed by Zhao L. et al. to isolate chemical symbols from online handwritten chemical formulas [17]. The proposed approach has separated the connected inorganic symbols from handwritten chemical formulas. This approach has used freeman chain coding for segmentation. But the approach does not consider the complex chemical structures for segmentation. Fujiyoshi A. et al. proposed a robust method for segmentation and recognition of chemical structure [3]. The proposed method has considered optical chemical structures from Japanese language journal articles. This method has decomposed the single chemical structure into lines and curves by choosing cross points and bend points for decomposition. The proposed method had used printed chemical structure samples for the experiment. It has not verified the results with online or offline handwritten chemical structures. A method has been proposed by Yang J. et al. [16] to separate the handwritten benzene ring from chemical formulas. This proposed method has based on freshman chain code. This method has been used only to separate the single benzene ring from the other bonds and symbols. It cannot apply for segmentation of HCCS. Ouyang T. Y. and Davis R. [11] developed a stroke based segmentation algorithm to segment the chemical structures. This algorithm is unable to segment the HCCS. Recently a progressive approach has been implemented by Tang P. et al. [13] to split connected bond stroke of online handwritten chemical structures into several single bonds. The model has implemented for pen-based smart devices, where it used the online handwritten chemical structures as input. For the analysis of connected bonds of chemical structures, Chang M. et al. [2] proposed a modeling approach. In this approach, a structure analysis has done in recognition of connected bonds. This approach analyzes bonds and separates them from structures. The above investigation about existing segmentation techniques used in various domains shows that segmentation of connected symbols and character has been a challenging task. In case of handwritten chemical structure segmentation, very few researchers had addressed the segmentation of complex structures, and it has done for online handwritten chemical structures. Therefore, there has a scope to provide a solution for a segmentation problem in case of offline HCCS. While addressing this problem in this paper, we had proposed a junction point based technique for segmentation of the benzene structure from HCCS. The methodology proposed in this paper is different than the existing approaches for the segmentation of HCCS. The existing algorithms and systems are restricted to the online handwriting recognition as well as they were limited to the segmentation of single structures. The proposed approach, in this paper, has given the promising results for segmentation of HCCS. This approach uses a junction point based segmentation, which helps the separation of single benzene structures from the complex chemical structures. The proposed junction point based algorithm is different than the existing system because it separates the structures, from the connected handwritten complex structures. The details regarding the results of the proposed algorithm discussed in Sect. 4.3.

3 Segmentation Procedure

The procedure of our proposed approach for segmenting a single benzene structure from HCCS has divided into four steps. The steps had named as preprocessing, skeletonization, Junction Point (JP) detection, and segmentation. The functioning of each step has explained in the following subsequent sections.

3.1 Pre-processing

The input for this stage is a scanned image of HCCS. The inputted image has to scan through a standard scanner with 300dpi resolution. The scanned image has saved in Portable Network Graphics (PNG) format. These images were used as input for pre-processing. In pre-processing the coloured inputted image has to translate into the grayscale mode first. Then it has to convert to the binary image of two colours black and white. The binary conversion has done depending on threshold values of conversion. After the image has converted into binary a noise has removed to make the image sharp by applying a Laplacian filter. This sharpened image has given as input to the next step called as skeletonization.

3.2 Skeletonization

The process of skeletonization has done to reduce the thickness of handwritten drawing. This process has also known as thinning of the image. The hand-drawn shapes have multi-pixel thickness. This process of thinning converts such hand-drawn shapes with multi-pixel thickness to the single pixel thickness [12]. The thinning process of HCCS plays an important role in the detection of junction points. It helps to detect the accurate junction points and reduce the possibility of detecting false junction points. There are many approaches for skeletonization of the image. In this paper, we had applied mathematical morphological operator for thinning of HCCS. The morphological thinning process has discussed in the following section.

Morphological Thinning. Thinning is a mathematical morphology operation that has been used to remove some foreground pixels by applying erosion process on the image and covert image into single line pixel. The mathematical morphology is a theory of image transformation which derived its tools from set theory and geometric principle [7]. It applies the set-theoretical operations like union and intersections on binary image to perform the transformation of images. In the thinning process, in new thin image, each pixel value has been computed by performing some logical operations between the pixel itself and its neighbors. A structuring element (SE) has been described by defining a small set of neighbor pixels. An image translation has been done by efficiently implementing the erosion and dilation operations [8]. After that either ANDing or ORing operation has done with itself to obtain the thin image. The erosion and dilation operation has defined by following Eq. (1) and (2). Where 'I' denotes binary image, and 'S' denotes SE. The erosion and dilation have represented with symbol θ and ϕ respectively.

$$I\theta S = \bigcap_{z\epsilon S} I - Z \tag{1}$$

$$I\phi S = \bigcup_{z\epsilon S} Iz \tag{2}$$

where Iz is a translation of I, along the pixel vector z, and the set intersection and union operations represent bitwise AND and OR, respectively. Translation is always concerning the center of the SE. The morphological operation has done by positioning SE first at all possible locations in the image, and it is compared with the corresponding neighborhood of pixels. While comparing, the locations were marked where SE fits or hits. The image information about the structure of the image where such fits or hits occurred can obtained. The SE "fits" occurred when there were exists corresponding pixels with value 1 in the image for all pixels of SE with value 1. The pixels with value 0 (background) in SE for them corresponding image pixels values are irrelevant. The SE "hits" image when for any of the pixels from SE that has set to 1, the corresponding image pixel also has value 1. Image pixels had ignored for which the corresponding SE pixel has 0 values [15]. Here in this experiment, we had used a SE of size 3*3 to thin the binary image of HCCS and skeleton of an image has acquired. This image skeleton of HCCS has then inputted for the junction point detection stage.

3.3 Junction Point Detection

A point where multiple continues surfaces were connected has called as the junction point [1]. The different categories of junction points [10] have been defined depending on the geometric shape formed near to join. Some such categories of junction points have labelled as Y junction, T Junction, L junction, etc. These types of junction points can use for segmenting such connected surface from each other. A junction based segmentation algorithm has proposed by U. Jayarathna and G. Bandara [9] for segmenting the connected handwritten characters. This proposed algorithm shows that junction-based segmentation has produced isolated handwritten characters more accurately. A single pixel which has presences of pixels more than two directions has to denoted as a junction point. Such pixels had to be detected and used to create an image segment. In this paper, we had segmented a benzene structure from a scanned image of HCCS. The segmentation process based on the detected junction points. In this case, a junction point has represented by a pixel where two benzene structures are joined. Fig. 2 shows the HCCS with junction points marked as a small circle around it. According to a Fig. 2. possible joint of two benzene structure can represent by shape as Y, a mirror image of shape Y as λ and a shape X. Thus, we had categorized these junction points in three categories, namely Y junction point, λ junction point and X junction point. At this stage, a skeleton image of HCCS has accepted as input and detects the junction points of the above category from that image. To find out a junction point of a particular category, a satisfying criterion for each category of junction point has developed. If the pixel from the skeleton image of HCCS has satisfied the criterion for any one category, then that pixel has marked as a junction point. Consider X and Y are the coordinates of a current pixel for which the system has to check that either it falls into any one category of junction point. The following Table 1 had defined positions of other pixels as

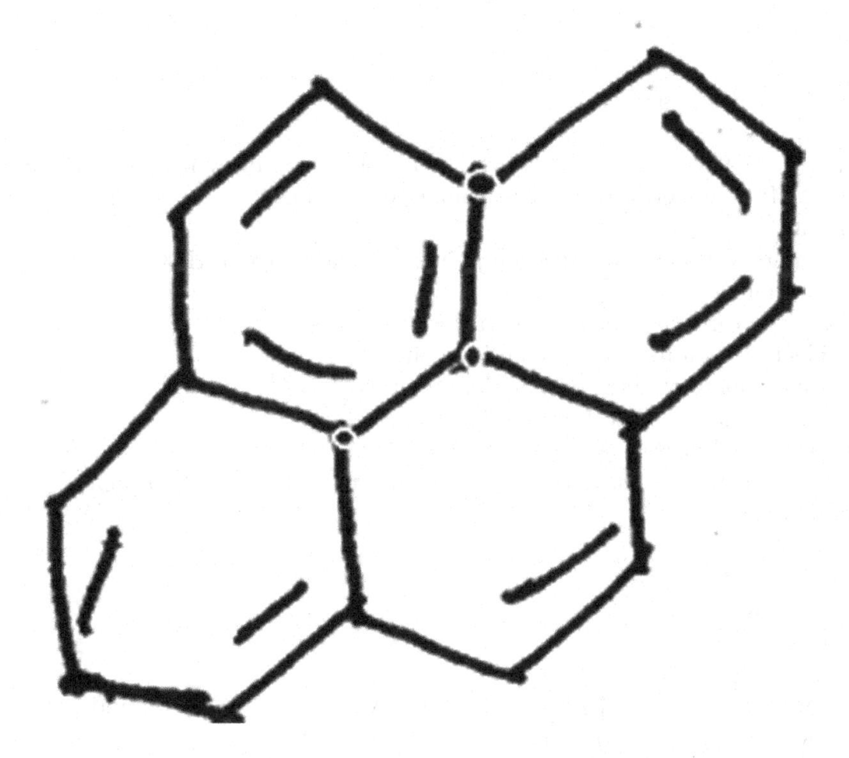

Fig. 2. HCCS showing junction point with small circles

a satisfying criterion for the respective categories. If there exist the pixels on all defined positions, then the current pixel has marked as a junction point of that category. The Table 1 shows the categories of the junction point with its receptive satisfying criterion. An inputted skeleton binary image of HCCS has scanned from its top left corner with a coordinate value at 0,0, and by applying the above criterion from Table 1, the junction points have been detected till the image get scanned up to right bottom corner. The algorithm has developed to find the junction points from an inputted binary image of HCCS. This developed algorithm has named as JP-Detect-Algo., which has shown as follows:

Table 1. Junction point categories with satisfaction criterion

Junction point category	Satisfaction criterion
Y Junction Point	X + 1 && Y + 1 , X + 1 && Y − 1 , X − 1 && Y
λ Junction Point	X − 1 && Y + 1 , X − 1 && Y − 1 , X + 1 && Y ,
X Junction Point	X + 1 && Y + 1 , X − 1 && Y − 1 , X − 1 && Y + 1 , X + 1 && Y − 1

Algorithm: JP-Detect-Algo
Input:
ImageM:- matrix of size M * N ; represents skeleton binary image of HCCS
Where M:- Number of rows , N:- Number of columns
Output:
JP: An array of detected junction points from an image of HCCS.
Assumptions:
JNC:- structure with the member as
id:- identification number of junction point.
a:- X coordinate of detected junction point.
b:- Y coordinate of detected junction point.
lbl:- Type of Junction point.
CJP:- An instance of structure JNC to indicate current junction point.
set Id = 0
set count = 0
for I = 0 to M
CJP = new JNC ( )
for J = 0 to N
if (ImageM[I][J] = 1) then
if ((ImageM[I+1] [J-1] =1) and (ImageM[I+1] [J+1] = 1) and (ImageM[I-1] [J] = 1)) then
CJP.id = Id
CJP.lbl = "Y Junction"

CJP.a = I
CJP.b = J
JP[count] = CJP
Id = Id +1
count = count + 1
end if
if ((ImageM[I-1] [J-1] = 1) and (ImageM[I-1] [J+1] = 1) and (ImageM[I+1] [J] = 1)) then
CJP.id = Id
CJP.lbl = "λ Junction"
CJP.a = I
CJP.b = J
JP[count] = CJP
Id = Id + 1
count = count + 1
end if
if ((ImageM[I-1] [J-1] = 1) and (ImageM[I+1] [J+1] =1) and (ImageM[I+1] [J-1] = 1) and (ImageM[I-1] [J+1] = 1)) then
CJP.id = Id
CJP.lbl = "X Junction"
CJP.a = I
CJP.b = J
JP[count] = CJP
Id = Id + 1
count = count + 1
end if
end if
end for
end for

The above JP-Detect-Algo. has produced an array of detected junction point named as JP, which can input to the next step of segmentation to segment the isolate benzene from HCCS.

3.4 Segmentation

After detection of the junction point, by implementing JP-Detect-Algo. an array of the junction points JP has produced, this has been used as an input for the segmentation process. The segmentation has done by traversing the junction point array JP. The image segments had created by selecting the image area which has a presence between two junction points. This process separates the isolated benzene from HCCS into a single image segment. For this process of segmentation, an algorithm has developed, which is named as JP-Seg-Algo. This algorithm has utilize to isolate the benzene structures from HCCS. The JP-Seg-Algo.is as follows: Algorithm: JP-Seg-Algo.
Input:
JP:- An array of Junction point; output of JP-Detect-Algo.

ImageM:- matrix of size M * N; represents a skeleton binary image of HCCS
Where M:- Number of rows , N:- Number of columns
Output:
SegImg:- An array of image segments, which represents isolated benzene structure from HCCS.
Assumptions:
TempImg:- matrix to store temporary image segment.
TempJP:- instance of structure JNC defined in JP-Detect-Algo.
r:- represents left most pixel of image or row element in image matrix
c :- represents top most pixel of image or column element in image matrix

```
set r = 0
set c = 0
set count = 0
for each jp ϵ JP do
TempJP = jp
L1 = 0
M1 = 0
for L = r to TempJP.a
for M = c to TempJP.b
TempImage[L1][M1]= ImageM[L][M]
M1 = M1 + 1
end for
L1 = L1 + 1
M1 = 0
end for
SegImg[count] = TempImage
count = count + 1
r = TempJP.a
c = TempJP.b
end for
```

The above algorithm produced the array of segmented images from HCCS. The single image from that array possible can be an isolated benzene structure which can further used for recognition of benzene structure or chemical symbols.

4 Experiment

Whereas when we had set up an experiment for segmentation of HCCS, we had considered the complex chemical structures which contents a benzene ring in it. The input, process, and output of the experiment have discussed in following subsequent sections

4.1 Input

For this experiment, the handwritten samples of HCCS have been collected from various peoples. For this, we had collected the sample of HCCS from the different professionals, including academician, students, scientists and others. We had collected the total 494 samples of complex structures. These samples have created by drawing the complex structures from the book of chemistry on plain paper with the help of a pen or pencil. We have collected samples of different complex chemical structures which has been written by various people. The collected samples have scanned with resolution at 300 dpi and converted to PNG image format. These images are used as an input for the experiment.

4.2 Process

The inputted image first has undergone through the pre-processing phase as discussed in Sect. 3.1 where the image has converted into a binary image. The converted binary image has to make thin using morphological thinning and the Skeletonization process as discussed in Sect. 3.2. This thinning of the image makes sure that at the time of junction point detection process an accurate junction point can detect. The Fig. 3 shows the original binary image of HCCS and Fig. 4 shows the skeleton of the original image of HCCS. Due to skeletonization, the image becomes thinner as compared to its original size as shown in Fig. 4.

The skeleton of an image of HCCS has inputted into the process of junction point detection. Junction point detection has carried out as per the JP-Detect-Algo. discussed in Sect. 3.3. The Fig. 5 shows the inputted image of HCCS and Fig. 6 shows the detected junction point it that image withdrawn circles. Once the junction point has detected and store in an array, using this information, segmentation has done. This segmentation process has done as per the JP-Seg-Aglo. discussed in Sect. 3.4. This segmentation process yields the isolated chemical structures in our case benzene structures from HCCS. The Fig. 7 shows HCCS, and Fig. 8 shows the segmented benzene structures with a drawn rectangle around it. Isolated benzenes have extracted by fragmenting the image using the rectangle area. The fragmented benzenes from the inputted HCCS have shown in Fig. 9.

4.3 Result and Discussion

The result of this setup experiment has isolated benzene structure or isolated chemical symbols which have extracted from HCCS as shown in Fig. 9 In this experiment, we had considered the chemical structures which have made up with benzene rings, hence the output of the experiment was the image segments array which mostly contents benzene structures. The total 494 samples of HCCS used to test the proposed algorithms on it. The following Table 2 has shown the statistical analysis of results The Table 2 has shown that the developed algorithm has successfully segmented approximately 93% of the sample images of HCCS. The sample images for which the algorithm fails to do segmentation has a very regular shape, drawn with noise and clumsy handwriting. The results of this experiment have further use for recognition of benzene structures.

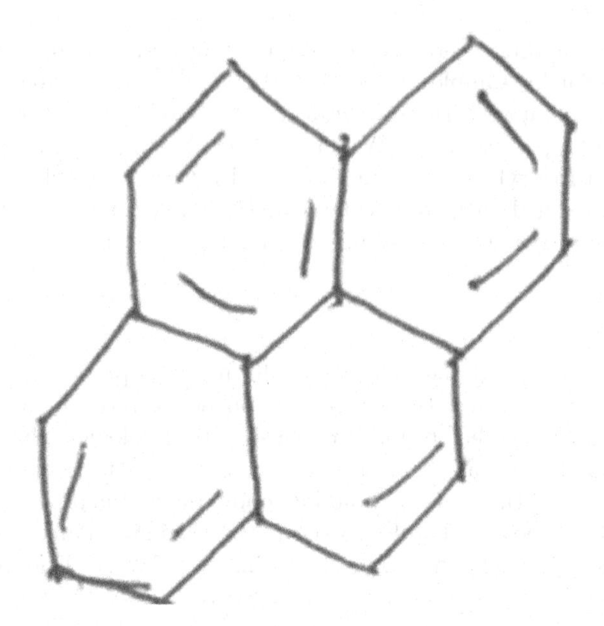

Fig. 3. Original image of HCCS

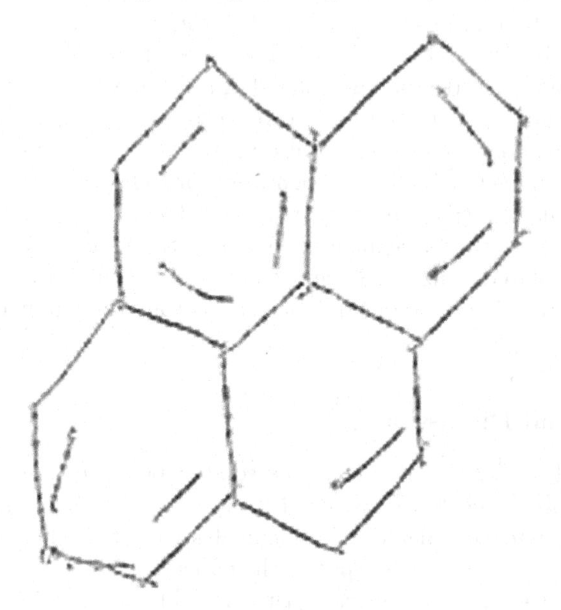

Fig. 4. Skeleton image of HCCS

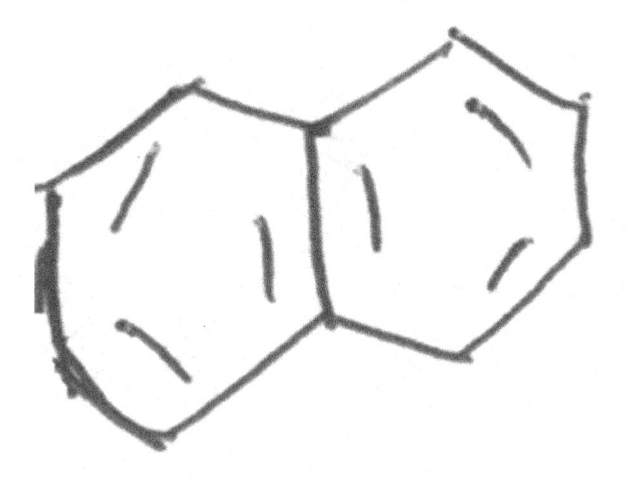

Fig. 5. Input image of HCCS

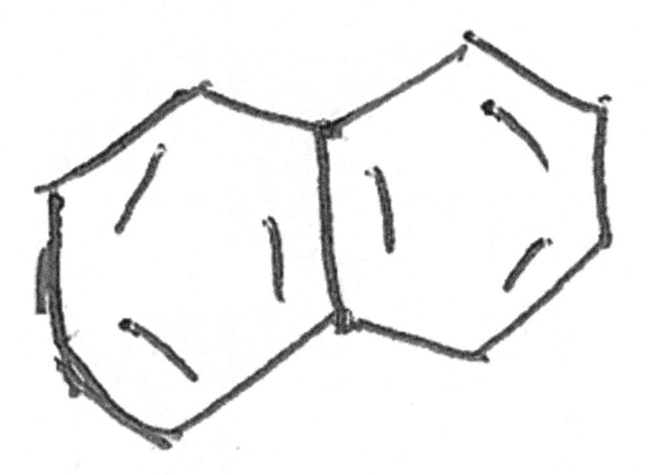

Fig. 6. Detected junction point drawn a small circle

Table 2. Statistical analysis of results data

Types of User	No. of samples collected	No. of samples segmented	No. of samples not segmented
Academician	125	120	5
Students	189	175	14
Scientist	112	108	4
Other	68	55	13
Total	494	458	36

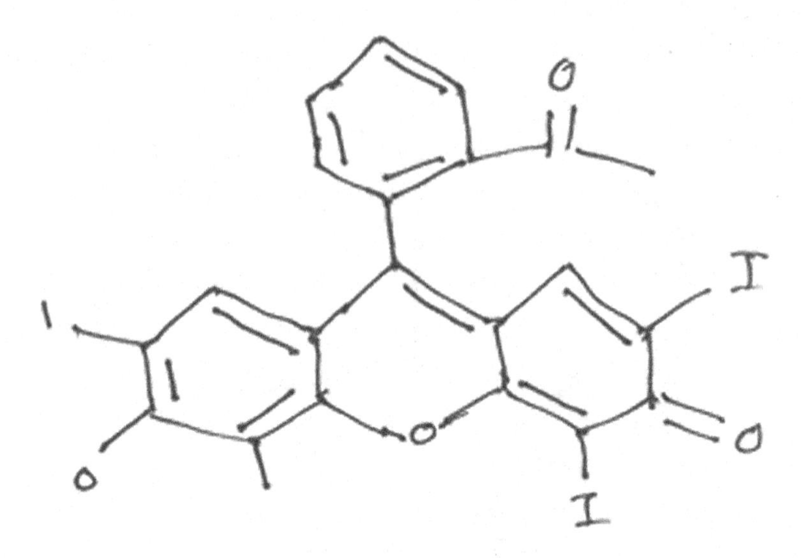

Fig. 7. Input image of HCCS

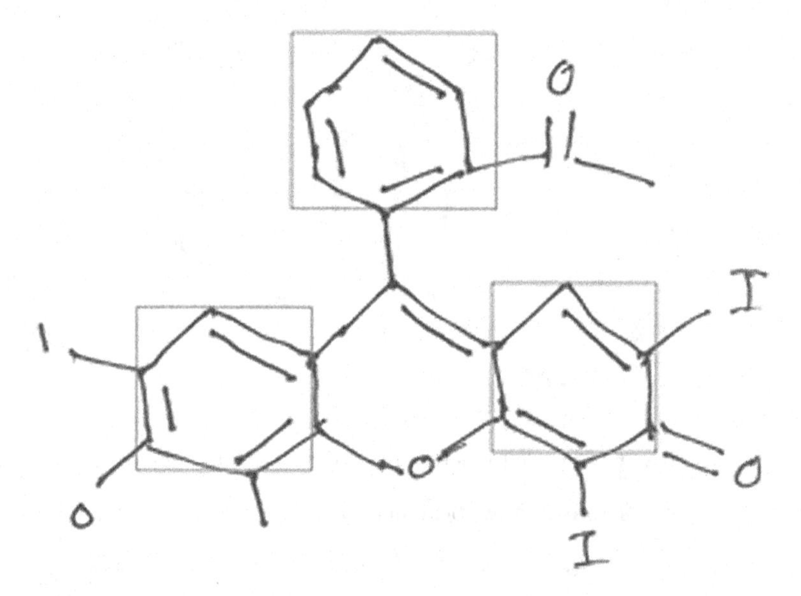

Fig. 8. Segmented benzene structures shown with drawn rectangle

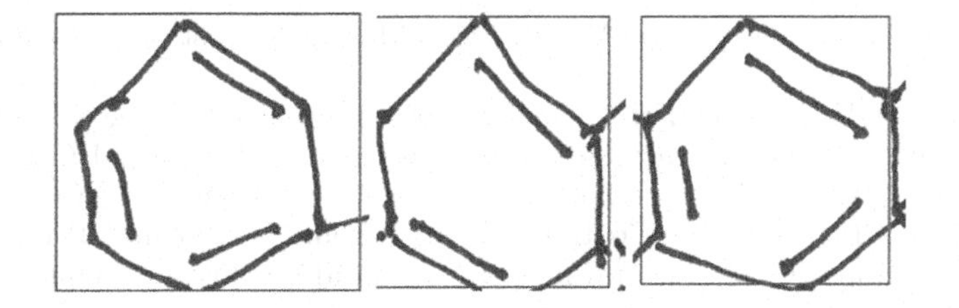

Fig. 9. Segmented benzene structures from HCCS shown in Fig. 8

5 Conclusion

This paper had proposed an algorithm to extract the single chemical structure from HCCS. The algorithm is based on junction point detection method. To verify this algorithm an experiment has set. This experiment is accepting scanned images of HCCS. The result of experiments has shown that the proposed approach has successfully separated the benzene rings from HCCS. The result and discussion section show that the success rate of this proposed algorithm is approximately 93%. The successes rate of experiment may decreased in the case of highly irregularly drawn shapes and shapes drawn with very clumsy handwriting. The algorithm has to refine for such cases in the feature. For this experiment, we had not considered samples of some other complex chemical structure which has made up of different chemical compounds. This can do as future expansion of this proposed work

References

1. Bergevin, R., Bubel, A.: Detection and characterization of junctions in a 2D image. Comput. Vis. Image Underst. **93**(3), 288–309 (2004)
2. Chang, M., Han, S., Zhang, D.: A unified framework for recognizing handwritten chemical expressions. In: 10th International Conference on Document Analysis and Recognition, pp. 1345–1349. IEEE (2009)
3. Fujiyoshi, A., Nakagawa, K., Suzuki, M.: Robust method of segmentation and recognition of chemical structure images in cheminfty. In: Pre-Proceedings of the 9th IAPR International Workshop on Graphics Recognition, GREC (2011)
4. Fukushima, K., Imagawa, T.: Recognition and segmentation of connected characters with selective attention. Neural Netw. **6**(1), 33–41 (1993)
5. Garain, U., Chaudhuri, B.B.: Segmentation of touching symbols for OCR of printed mathematical expressions: an approach based on multifactorial analysis. In: Eighth International Conference on Document Analysis and Recognition, pp. 177–181. IEEE (2005)
6. Garain, U., Chaudhuri, B.B.: Segmentation of touching characters in printed Devnagari and Bangla scripts using fuzzy multifactorial analysis. IEEE Trans. Syst. Man Cybern. Part C (Appl. Rev.) **32**(4), 449–459 (2002)
7. Heijmans, H.J., Ronse, C.: The algebraic basis of mathematical morphology I. Dilations and erosions. Comput. Vis. Graph. Image Process. **50**(3), 245–295 (1990)

8. Serra, J.: Image Analysis and Mathematical Morphology. 1st edn. Academic Press, London (1982)
9. Jayarathna, U.K.S., Bandara.: A junction based segmentation algorithm for offline handwritten connected character segmentation. In: International Conference on Intelligent Agents, Web Technologies and Internet Commerce, Computational Intelligence for Modelling, Control and Automation, pp. 147–147. IEEE (2006)
10. Malik, J.: Interpreting line drawings of curved objects. Int. J. Comput. Vis. **1**(1), 73–103 (1987)
11. Ouyang, T.Y., Davis, R.: Recognition of hand drawn chemical diagrams. Assoc. Adv. Artif. Intell. **7**(1), 846–851 (2007)
12. Prakash, R.P., Prakash, K.S., Binu, V.P.: Thinning algorithm using hypergraph based morphological operators. In: 2015 IEEE International Advance Computing Conference (IACC), pp. 1026–1029. IEEE (2015)
13. Tang, P., Hui, S.C., Fu, C.W.: Connected bond recognition for handwritten chemical skeletal structural formulas. In: 14th International Conference on Frontiers in Handwriting Recognition (ICFHR), pp. 122–127. IEEE (2014)
14. Tang, P., Hui, S.C., Fu, C.W.: Online structural analysis for handwritten chemical expression recognition. In: 8th International Conference on Information, Communications and Signal Processing (ICICS) pp. 1–5. IEEE (2011)
15. Tarabek, P.: Morphology image pre-processing for thinning algorithms. J. Inf. Control Manage. Syst. **5** (2007)
16. Yang, J., Wang, C., Yan, H.: A method for handling the connected areas in formulas. Phys. Procedia **33**, 279–286 (2012)
17. Zhao, L., Yan, H., Shi, G., Yang, J.: Segmentation of connected symbols in online handwritten chemical formulas. In: 2010 International Conference on System Science, Engineering Design and Manufacturing Informatization (ICSEM), pp. 278–281. IEEE (2010)
18. Zhao, S., Shi, P.: Segmentation of connected handwritten chinese characters based on stroke analysis and background thinning. In: Mizoguchi, R., Slaney, J. (eds.) PRICAI 2000. LNCS (LNAI), vol. 1886, pp. 608–616. Springer, Heidelberg (2000). https://doi.org/10.1007/3-540-44533-1_61

Automatic Mapping of Deciduous and Evergreen Forest by Using Machine Learning and Satellite Imagery

Rashmi Saini[1](✉), Suraj Singh[1], Shashi Kant Verma[1], and Sanjeevakumar M. Hatture[2]

[1] G. B. Pant Institute of Engineering and Technology, Pauri Garhwal 246196, India
2rashmisaini@gmail.com

[2] Basaveshwar Engineering College, Bagalkot 587103, India

Abstract. The existence of forests is crucial to the sustainability of life on earth. Automatic forest mapping is necessary to obtain accurate information about the deforestation rate, quantifying, monitoring and mapping. Such information is essential for various schemes to save forest. European satellite Sentinel-2 provide data at thirteen spectral band along with three different spatial resolution level. This satellite data is freely available having medium spatial resolution and faster revisit time, which makes it suitable choice for forest mapping. The objective of this study is automatic mapping of Deciduous and Evergreen Forest by using Sentinel-2 imagery (single date data) in district Dehradun, Uttarakhand, India. Two efficient Machine Learning (ML) approaches have been used for the classification i.e., Random Forest (RF) and *k*-Nearest Neighbor (*k*-NN). Sentinel-2 satellite spectral bands (10 m spatial resolution) namely Near-infrared (NIR), visible light band (Blue, Green and Red) are stacked for classification. In this study, overall classification accuracy attained by RF and k-NN is 81.52% (kappa value of 0.759) and 80.84% (kappa value of 0.751) respectively. Results indicates that both classifiers performed well, however, RF achieved slightly higher (+0.68%) accuracy as compared to k-NN classifier. It is found that RF obtained User Accuracy (UA) and Producer Accuracy (PA) 77.53% and 82.85% respectively for Deciduous Forest. Whereas, for Evergreen Forest UA and PA is 82.26% and 77.95% respectively. On the other hand, k-NN achieved UA of 77.33% and PA of 82.25%. For Evergreen Forest attained UA and PA of 81.81% and 76.95% respectively. Results demonstrated that Sentinel-2 multispectral satellite data is highly suitable for mapping of forests.

Keywords: Sentinel-2 · Forest mapping · k-Nearest Neighbor (k-NN) · Random Forest (RF) · Land Use Land Cover (LULC)

1 Introduction

Forests play a significant role in human lives as it affects the complete ecosystem. Forest are important for ecological balance, climate change, biodiversity, water conservation,

K. K. Patel et al. (Eds.): icSoftComp 2022, CCIS 1788, pp. 197–209, 2023.
https://doi.org/10.1007/978-3-031-27609-5_16

carbon balance and economic development of any country. Forests are a precious natural resource that maintains the environmental balance. The condition of a forest is the most reliable predictor of the region's ecological system. Uttarakhnad is the state in India, located in Northern region and most of the area is mountainous. In the past decade, migration increase at a rapid rate from Uttarakhand's hill areas to the urban or semi-urban areas. High density population leads to the physical expansion of land use, it exceeds cutting of trees in Terai bhabar and Shivalik zone and also affected vegetation area, crop land and Rajaji Reserve forest.

Automatic forest mapping through processing of satellite images provides quick and precise information. It can be applied to many assessments analysis like diversity of forest, density, volume, growth and deforestation rate, resource management and various decision-making process [1]. The selection of satellite for forest mapping depends upon many factors, availability of data, acquisition cost, spatial details, frequently updated forest structures and diversity of forest. Especially in developing nations with limited financial resources for remotely sensed data acquisition [2]. Many research utilized publicly available satellite data for forest mapping like MODIS, Landsat, PALSAR and ASTER, Sentinel-1 and Sentinel-2 [3–5]. However, due to low spatial resolution (250–500 m) mix-pixel problem was the major challenge by using MODIS dataset [6]. This data is suitable when study area is very large. Later on, with the availability of Landsat series satellites, mixed problem is resolved up to some extent. Sentinel-2A, European satellite that provides multispectral band, high spatial resolution and shorter revisit time (5 days), that can be utilized as alternative of low spatial resolution satellite imagery [7, 8]. The Multi-Spectral Instrument (MSI) on the Satellite Sentinel-2 has three spatial resolutions (10 m, 20 m, and 60 m) and thirteen spectral bands (Table 1). Since the establishment of a variety of remote sensing applications have utilized Sentinel-2 data. Scientific communities, government organizations and researchers for a different purpose, including forest monitoring, urban development, and agriculture mapping and monitoring [9–17].

Mondal et al. [18] compared Landsat-8 and Sentinel-2 satellite data for South Asian Forest degradation and used Random Forest classifier. This study's result demonstrated that the vegetation state in most countries has fluctuated over time. For the forest degradation assessment due to charcoal production authors applied multi-temporal Sentinel-2 data to monitor and assess the state of the forest. Sentinel-2 imagery was combined to generate a map with a resolution of 10 m [19]. The resulted map shown a rapid and intense deforestation as well as the temporal and spatial pattern of deforestation caused on by the production of charcoal. Hoscilo et al. [20] used multi-temporal data of Sentinel-2 satellite to map forest area and land cover by using Random Forest approach. This study demonstrated that the Sentinel-2 satellite has potential to classification for regional forest area cover, tree species and forest diversity. Another study utilized the worldview-2 imagery to map trees species and compared the performance of two algorithms [21]. The overall accuracy of the Support Vector Machine (SVM) classifier is 77%, which is slightly better as compared to Artificial Neural Network (75%). Saini and Ghosh [11] used Sentinel-2 data for crop classification and utilized two machine learning approaches (RF and SVM). The comparison result showed that the overall accuracy of the RF model was slightly better than the SVM model. Hawrylo et al. [22] used Sentinel-2 satellite imagery to assess and estimate scots pine stand defoliation in Poland. The outcome of the models proved that Sentinel-2 imagery data can provide crucial knowledge about forest cover, monitoring and forest defoliation.

As discussed above, the main focus of this research work is the mapping of Deciduous and Evergreen Forest in Dehradun, Uttarakhand, India. Further, this study also focuses on the investigate the suitability of satellite data (Sentinel-2) for mapping of Forest and successful generation of Land Use Land Cover (LULC) maps obtained by implementation of Random Forest (RF) and k-nearest neighbor (k-NN) classifier.

The paper is organized in the following way: Sect. 2 discuss about selected study locations and satellite dataset, Sect. 3 describe the methodology and selected algorithms, Sect. 4 Describe the results obtained from this study and last section i.e. Section 5 presents the conclusions of the study.

2 Study Area and Data

Sentinel-2 satellite dataset from the winter season, acquired on December 10, 2021 were used in this study. In the chosen study region forests play a crucial role in preserving the temperature (heat-waves) also ecosystem. This area is located near the Shivalik and some of area is Terai bhabar range of the Himalaya. Near Infrared (NIR) and visible light (Red, Green, and Blue) are the four bands, which have been used in research. Data collected from Sentinel-2 satellite at 10 m resolution utilized for the purpose of classification. The total area covered is 3088 km^2 and 78°1′55.8768″ E is the minimal bounding box coordinate at 30° 18′ 59.38567″ N and 30° 57′ 11.687″ E in the upper left corner, 30°03′47.051″ N in the lower right corner.

Table 1. Details of Sentinel-2 bands.

Band name	Resolution (m)	Wavelength (nm)	Band width (nm)
Coastal/aerosols	60	443	20
Blue (visible)	10	490	65
Green (visible)	10	560	35
Red (visible)	10	665	30
Red edge1	20	705	15
Red edge2	20	740	15
Red edge3	20	783	20
Near inferred	10	842	115
Narrow NIR	20	865	20
Water vapour	60	945	20
SWIR-cirrus	60	1375	30
SWIR1	20	1610	90
SWIR2	20	2190	180

There are 13 spectral bands of Sentenel-2 satellite, as a names, lists the characteristics of among bands, along with names, spatial resolution (in meter), and wavelengths are shown in Table 1. Study area in True Color Composite (FCC) created form Sentinel-2 imagery in ArcGIS software is represented in Fig. 1.

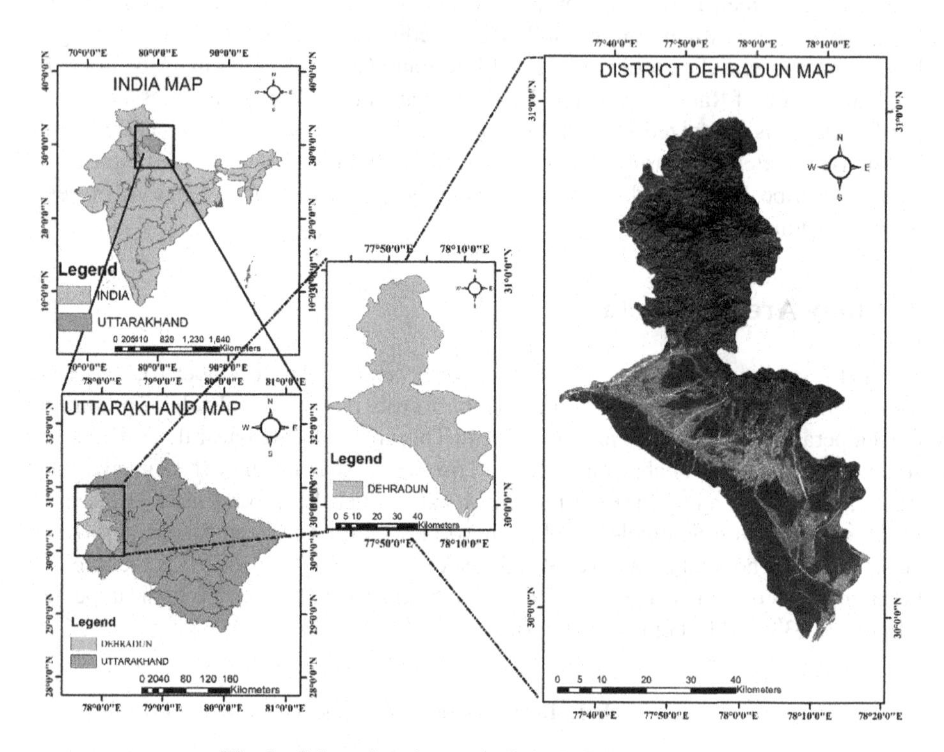

Fig. 1. Selected study area in True Color Composite.

3 Methodology

Figure 2 shows the proposed methodology and classification technique for automatic mapping of Deciduous and Evergreen Forest. Sentinel-2 imagery acquired on a single date has been used and then data preprocessing is performed using Sen2cor processor. After the atmospheric and radiometric correction, 10 m spatial resolution band of sentinel-2 satellite bands i.e. NIR, Blue, Green and Red bands have been stacked to generate a multispectral visible image tile. After the stacking operation, clipping of study area is done using shapefile, stacked image pixel holds spectral details and four-dimensional vector that are being considered. Ground truth data plays an important role for the classification using satellite imagery. Therefore, Bhuvan ISRO (thematic map) were used to construct the reference dataset, and some samples taken using Google Earth imagery (at high-resolution). The prepared reference dataset has been divided into two sets, one for training and one for testing. The ratio of partitioned dataset is 70% and

30% used for training and testing respectively. The chosen study area is mapped into eight LULC classes i.e. Water Body, River, Deciduous Forest, Evergreen Forest, Shrub land, Vegetation/ Crop land, Fallow Land, and Urban/Semi-urban. Forest and urban area cover up the majority of the land in the chosen region. Evergreen forests are particularly significant because they store water in their roots and provides water resources and oxygen, also help in maintaining the environment's temperature.

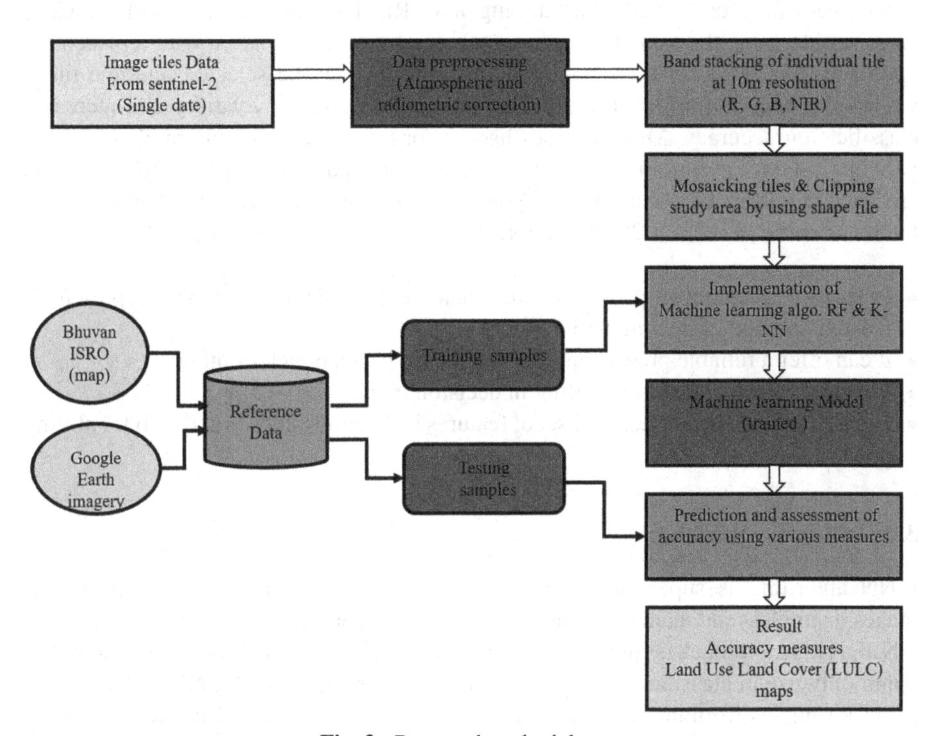

Fig. 2. Proposed methodology.

Machine Learning classifiers, RF and k-NN were trained using the training dataset. Both of the selected classifiers have been various classification applications to achieve high accuracy. Random Forest and k-NN have been trained using training samples. After the training of Machine Learning models, testing operation has been performed. Consequently, classification maps of the research areas are produced. Accuracy of classifiers is evaluated using Overall Accuracy or classification accuracy and kappa value. For LULC classes User Accuracy (UA) and Producer Accuracy (PA) has been used. In addition, LULC are also evaluated on the basis of F1-Score, which is a popular accuracy measure for classification. Therefore, F1-score was applied in this study to evaluate class-specific accuracy.

3.1 Random Forest

Random Forest is a widely known classifier, which has shown a high performance in various of remote sensing applications [12, 23]. RF is a non-parametric algorithm which can determine the applicability of variables and outcomes excellent result. [12, 23]. An ensemble classifier combines several classifiers to produce superior classification results. The ensemble learning technique combines of multiple decision trees algorithm to provide a final decision. The RF algorithm is a supervised machine learning method that builds a number of base models/learners and then uses a voting scheme to aggregate the responses of these models to make a final prediction. A decision tree was deployed as the basic model or learner in the building of the RF classifier. The RF classifier creates an ensemble using the same principle as bagging by using a random with replacement approach [23]. As a result, certain training samples may be selected multiple times while others are not selected at all. This approach decreases volatility and increases classification accuracy. Multiple trees had to construct the ensemble models and the variables used for partitioning at the nodes. Two tuning parameters for the RF algorithm are Mtry (number of predictor variable for node splitting) and ntree (number of trees to build ensemble [8, 12, 24, 25]. RF algorithm has the key characteristics [26]:

- In terms of accuracy, it performs better than the decision tree method and presents a feasible method of dealing with random errors.
- It can offer a reliable prediction even without hyper-parameter tuning.
- It addresses the issue of overfitting in decision trees.
- In each random forest tree, a subset of features is randomly chosen at the node splitting point.

3.2 k-Nearest Neighbor

k-NN algorithm is supervised non-parametric Machine Learning approach. Which makes it simplest and main advantage is the non-assumption on underlying data [27]. k-NN algorithm approach is suitable for both model regression and classification, however, commonly use in classification problem. The central point behind k-NN is discovers a set of k samples from the training data that, as per the function of Euclidean distance, are closest to the unidentified samples. Calculating average of variables from k samples yields the supervised class from unidentified sample of class attributes of the k-NN [28]. In this algorithm, k is the parameter and plays an important role in its performance. k-NN has Some key characteristics following are [29]:

- k-NN is simplest and common classifier.
- performance depend on k value and distance.
- distances tolerate noise to a certain degree.
- When compared to the other distances studied, the non-convex distance performed the best on most datasets.

4 Results and Discussion

The mapping of the Deciduous and Evergreen forest reserve (Rajaji national) has been performed using RF and k-NN approaches by utilizing satellite Sentinel-2 data. In this study, 10 m resolution resulting image used to map Deciduous, Evergreen forest and six other LULC classes. This study utilized stratified random sampling approach. Reference dataset has been collected with the help of Thematic map (Bhuvan ISRO) and Google earth engine. Partitioned reference data pixels are mutually exclusive for training and testing purpose. The R programming language is used to implement both Machine Learning approaches.

This study aims to maps two different types of forest (Deciduous and Evergreen Forest) in the selected study region. Therefore, the major focus of the analysis is on the target classes only i.e. forest types. In this study the confusion matrix obtained by a classifier's result is used to calculate Overall Accuracy, F1-score, User Accuracy (UA), Producer Accuracy (PA) and kappa coefficient. F1-score were utilized to determine class specific accuracy. The confusion matrices obtained by RF and k-NN lists producer's accuracy (Recall) and user's accuracy (Precision) with each geographical cover class, F1-score, Overall Accuracy, and kappa coefficient. According to the results (Table 2), RF achieved an Overall Accuracy of 81.52% and a kappa coefficient of 0.759. k-NN classifier attained an Overall Accuracy (OA) and kappa coefficient of 80.84% and 0.751 respectively. Both the classifiers performed well but RF achieve slightly better rise of + 0.681% over the k-NN. Many studies have demonstrated effectiveness of RF classifier for LULC classification [11, 12, 30].

Table 2. Overall accuracy and kappa coefficient value for RF and k-NN classifier.

Parameters	RF	k-NN
Overall accuracy%	81.52	80.84
Kappa	0.759	0.751

In the results section for confusion matrix following abbreviations are used all considered Land Use Land Cover (LULC) classes: Water body: WB, Rivers: RV, Deciduous forest: DF, Evergreen forest: EF, Shrub land: SL, Vegetation or Crop Land: CL, Fallow land: FL, Urban/semi-urban: UB. The resultant confusion matrix of RF classifiers is presented in Table 3. It is found that UA produce by RF for Deciduous forest and Evergreen forest is 77.53%, 82.26% and PA both forest class is 82.85%, 77.95% respectively shown in Table 3. It can be observed from the results that some of the misclassified sample of Deciduous forest is classified as Evergreen Forest. In the similar manner, some samples of Evergreen Forest are also misclassified as Deciduous forest. Some samples of both forest types are classified in other vegetation classes such as shrub land and crop land. It can be seen from the confusion matrix that other vegetation class i.e. shrubland is classified with lower User Accuracy value of 59.89% with a Producer value of 78.66%. On the other hand, Vegetation class is mapped with higher UA value of 92.12% and PA value of 86.60%. The major reason of such misclassification is the similar spectral signatures.

The resultant confusion matrix of k-NN classifier is shown in Table 4. It is found that UA produce by k-NN for Deciduous and Evergreen Forest is 77.33% and 81.81% respectively. On the other hand, obtained PA value for Deciduous forest classes are 82.85% and for Evergreen Forest is 76.95% (Table 4). It has been observed that UA and

PA value by k-NN for Deciduous forest class are nearly similar to RF classifier but for Evergreen Forest class there has difference (±1.1%) in UA and PA as compare to RF classifier Table 3.

Table 3. Confusion matrix obtained by RF classifier, Class Water body: WB, Rivers: RV, Deciduous forest: DF, Evergreen forest: EF, Shrub land: SL, Vegetation or Crop Land: VL, Fallow land: FL, Urban/semi-urban: UB.

RF	Classes	Truth/reference data									
		WB	RB	DF	EF	SL	VA	FL	UE	CO	UA%
Classified data	WB	**180**	10	0	0	0	0	0	0	**190**	94.73
	RB	11	**149**	0	0	0	0	26	75	**261**	57.08
	DF	0	0	**1657**	338	26	116	0	0	**2137**	77.53
	EF	0	0	244	**1559**	11	81	0	0	**1895**	82.26
	SL	0	0	53	34	**236**	71	0	0	**394**	59.89
	VA	0	0	46	69	27	**1732**	6	0	**1880**	92.12
	FL	0	4	0	0	0	0	**150**	19	**173**	86.70
	UE	9	38	0	0	0	0	16	**207**	**270**	76.66
TO		**200**	**201**	**2000**	**2000**	**300**	**2000**	**198**	**301**	7200	
PA%		90	74.12	82.85	77.95	78.66	86.6	75.75	68.77		
F1-S		**92.30**	**64.50**	**80.10**	**80.05**	**68.01**	**89.27**	**80.86**	**72.50**		
OA%		**81.528**									
Kappa		**0.759**									

It has been noted from the confusion matrix that there is misclassification among the vegetation classes i.e. Crop Land, Shrub Land, Evergreen Forest and Deciduous Forest. For the other vegetation class i.e. shrubland k-NN obtained UA of 58.04%, and PA value of 77.00%. Vegetation land is classified with UA of 91.31% and PA value of 86.65%. It has been observed that Vegetation land is classified with higher accuracy by both the classifiers (k-NN and RF). On the other hand, comparatively a lower accuracy has been reported for shrubland class by both the Machine Learning classifiers. It has been observed that similarity in the spectral signature leads to misclassification among the LULC classes.

Results of this study also revealed that for the other LULC classes such as Water body and Urban/Semi-urban there is no misclassification with vegetation classes. This observation is similar for the results obtained by both the classifiers. This is because such classes have different spectral signatures. For example, Water body has entirely different spectral signature from vegetation class. As a results there is no misclassification among such LULC classes.

In addition, F1-score is also computed for all LULC classes and results obtained by RF and k-NN are presented in Table 5. Both classifiers produced statistically close class-specific accuracy estimates (the difference is less than 2%), however, RF produce little bit high accuracy for both Deciduous and Evergreen forest class and other six LULC. Classification maps produced RF and k-NN classifiers are shown in Fig. 3, and Fig. 4 respectively.

Table 4. Confusion matrix obtained by RF classifier, following abbreviations are used: Water body: WB, Rivers: RV, Deciduous forest: DF, Evergreen forest: EF, Shrub land: SL, Vegetation or Crop Land: VL, Fallow land: FL, Urban/semi-urban: UB.

k-NN	Classes	Truth/reference data									
		WB	RB	DF	EF	SL	VA	FL	UE	CO	UA%
Classified data	WB	**179**	11	0	0	0	0	0	0	**190**	94.21
	RB	15	**148**	0	0	0	0	31	81	**275**	53.81
	DF	0	0	**1645**	338	32	112	0	0	**2127**	77.33
	EF	0	0	248	**1539**	12	82	0	0	**1881**	81.81
	SL	0	0	57	39	**231**	71	0	0	**398**	58.04
	VA	0	0	50	84	25	**1735**	6	0	**1900**	91.31
	FL	0	5	0	0	0	0	**143**	17	**165**	86.66
	UE	7	36	0	0	0	0	20	**201**	**264**	76.13
TO		**201**	**200**	**2000**	**2000**	**300**	**2000**	**200**	**299**	7200	
PA%		89.05	74	82.25	76.95	77	86.65	71.5	67.22		
F1-S		**91.56**	**62.31**	**79.71**	**79.30**	**66.18**	**88.97**	**78.35**	**71.40**		
OA%		**80.847**									
Kappa		**0.75**									

Table 5. F1-score value obtained by RF and k-NN classifiers.

Class name	RF%	k-NN%
Water body	92.30	91.56
Rivers	64.50	62.31
Deciduous forest	80.10	79.71
Evergreen forest	80.05	79.30
Shrubland	68.01	66.18
Vegetation	89.27	88.97
Fallow land	80.86	78.35
Urban/semi-urban	72.50	71.40

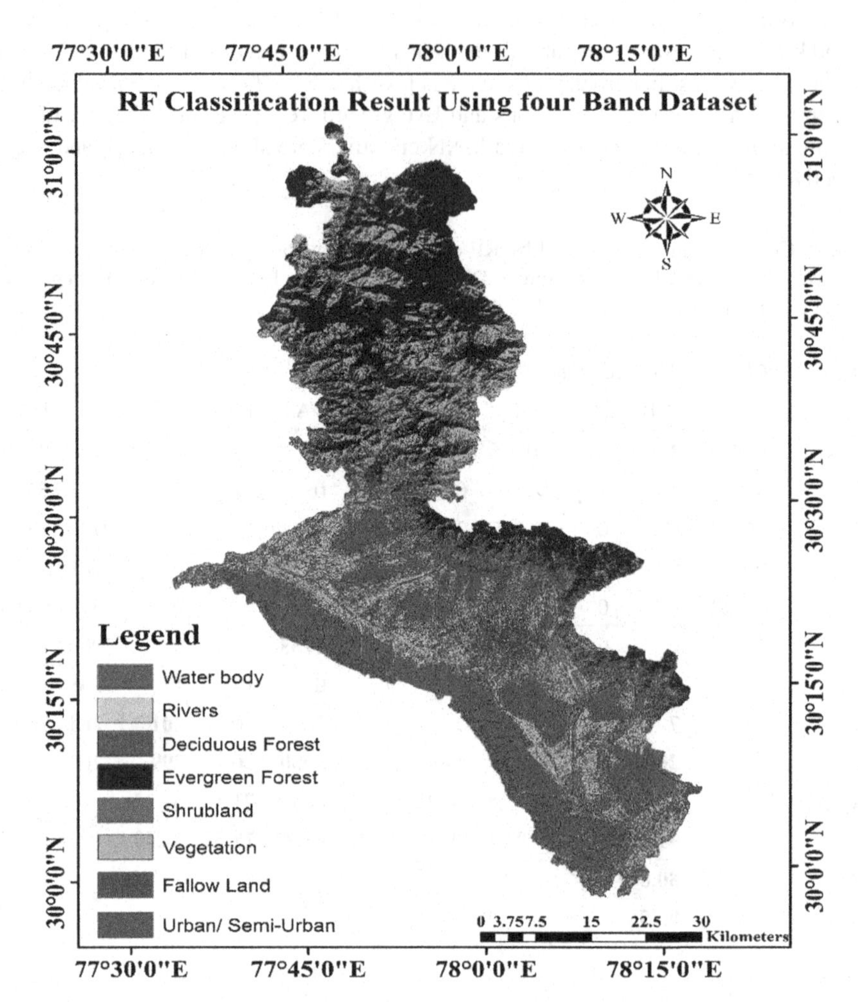

Fig. 3. Classification map obtained by RF classifier.

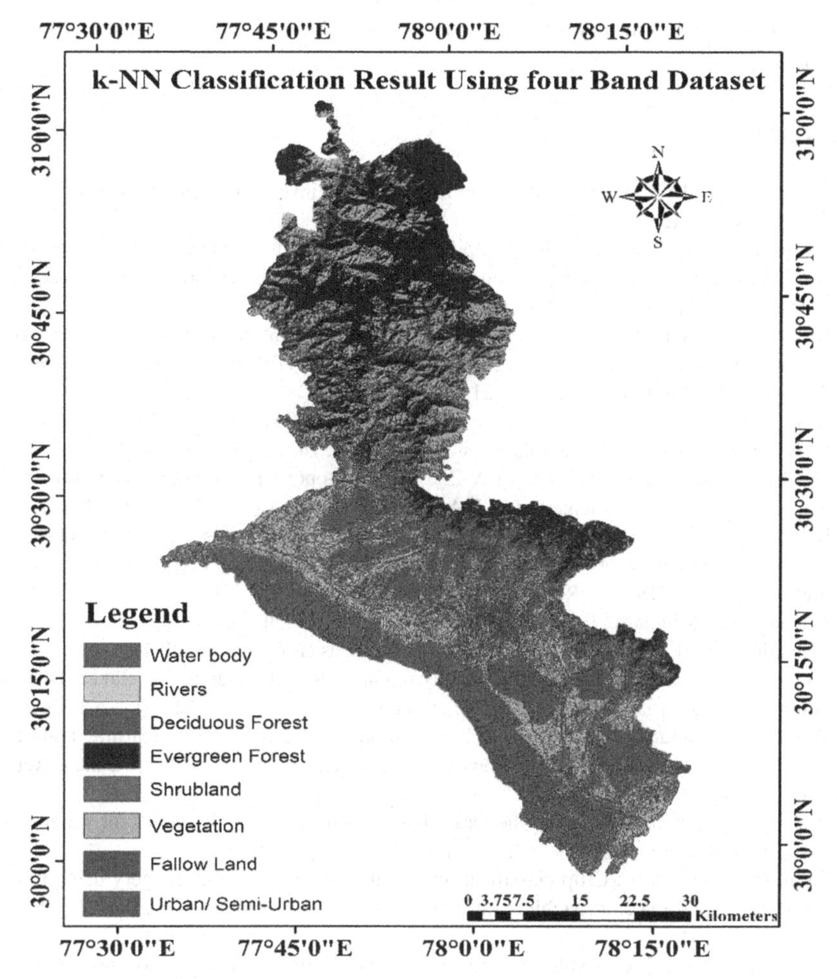

Fig. 4. Classification map obtained by k-NN classifier.

5 Conclusions

The objective of this study is automatic mapping of Deciduous and Evergreen Forest using Machine Learning algorithms namely RF and k-NN by utilizing single date Sentinel-2 imagery. In this study, four spectral bands (10 m spatial resolution) are considered for LULC classification for both classifiers. The RF classifier outperforms by obtaining an Overall Accuracy (OA) of 81.52% over the k-NN (80.84%). As per the results of the implementation, class specific accuracies of Deciduous and Evergreen Forest, it is concluded on the basis of various accuracy measures that both classifiers successfully extracted the forest type in the selected study region. However, some misclassification among the vegetation classes are also noted due to the similar properties of spectral signature. The findings of this study demonstrated that the Sentinel-2 has great of potential to classify forests cover types and LULC using Machine Learning classifiers.

References

1. Liang, X., et al.: Terrestrial laser scanning in forest inventories. ISPRS J. Photogramm. Remote. Sens. **115**, 63–77 (2016)
2. Morin, D., et al.: Estimation and mapping of forest structure parameters from open access satellite images: development of a generic method with a study case on coniferous plantation. Remote Sens. **11**(11), 1275 (2019)
3. Barakat, A., Khellouk, R., El Jazouli, A., Touhami, F., Nadem, S.: Monitoring of forest cover dynamics in eastern area of Béni-Mellal Province using ASTER and Sentinel-2A multispectral data. Geol. Ecol. Landsc. **2**(3), 203–215 (2018)
4. Zhang, Y., et al.: Mapping annual forest cover by fusing PALSAR/PALSAR-2 and MODIS NDVI during 2007–2016. Remote Sens. Environ. **224**, 74–91 (2019)
5. Yin, H., Tan, B., Frantz, D., Radeloff, V.C.: Integrated topographic corrections improve forest mapping using Landsat imagery. Int. J. Appl. Earth Obs. Geoinf. **108**, 102716 (2022)
6. Rahman, A.F., Dragoni, D., Didan, K., Barreto-Munoz, A., Hutabarat, J.A.: Detecting large scale conversion of mangroves to aquaculture with change point and mixed-pixel analyses of high-fidelity MODIS data. Remote Sens. Environ. **130**, 96–107 (2013)
7. Nomura, K., Mitchard, E.T.: More than meets the eye: using Sentinel-2 to map small plantations in complex forest landscapes. Remote Sens. **10**(11), 1693 (2018)
8. Phiri, D., Simwanda, M., Salekin, S., Nyirenda, V.R., Murayama, Y., Ranagalage, M.: Sentinel-2 data for land cover/use mapping: a review. Remote Sens. **12**(14), 2291 (2020)
9. Wessel, M., Brandmeier, M., Tiede, D.: Evaluation of different machine learning algorithms for scalable classification of tree types and tree species based on Sentinel-2 data. Remote Sens. **10**(9), 1419 (2018)
10. Bonansea, M., et al.: Evaluating the feasibility of using Sentinel-2 imagery for water clarity assessment in a reservoir. J. S. Am. Earth Sci. **95**, 102265 (2019)
11. Saini, R., Ghosh, S.K.: Crop classification on single date Sentinel-2 imagery using random forest and support vector machine. Int. Arch. Photogramm. Remote Sens. Spat. Inf. Sci. **42**, 683–688 (2018)
12. Saini, R., Ghosh, S.K.: Exploring capabilities of Sentinel-2 for vegetation mapping using random forest. Remote Sens. Spatial Inf. Sci. **XLII**, 247667 (2018)
13. Themistocleous, K., Papoutsa, C., Michaelides, S., Hadjimitsis, D.: Investigating detection of floating plastic litter from space using sentinel-2 imagery. Remote Sens. **12**(16), 2648 (2020)
14. Pageot, Y., Baup, F., Inglada, J., Baghdadi, N., Demarez, V.: Detection of irrigated and rainfed crops in temperate areas using Sentinel-1 and Sentinel-2 time series. Remote Sens. **12**(18), 3044 (2020)
15. Zheng, Q., Huang, W., Cui, X., Shi, Y., Liu, L.: New spectral index for detecting wheat yellow rust using Sentinel-2 multispectral imagery. Sensors **18**(3), 868 (2018)
16. Rawat, S., Saini, R., Kumar Hatture, S., Kumar Shukla, P.: Analysis of post-flood impacts on Sentinel-2 data using non-parametric machine learning classifiers: a case study from Bihar floods, Saharsa, India. In: Iyer, B., Crick, T., Peng, S.L. (eds.) ICCET 2022. SIST, vol. 303, pp. 152–160. Springer, Singapore (2022). https://doi.org/10.1007/978-981-19-2719-5_14
17. Saini, R., Verma, S.K., Gautam, A.: Implementation of machine learning classifiers for built-up extraction using textural features on Sentinel-2 data. In: 2021 7th International Conference on Advanced Computing and Communication Systems (ICACCS), vol. 1, pp. 1394–1399. IEEE (2021)

18. Mondal, P., McDermid, S.S., Qadir, A.: A reporting framework for Sustainable Development Goal 15: multi-scale monitoring of forest degradation using MODIS, Landsat and Sentinel data. Remote Sens. Environ. **237**, 111592 (2020)
19. Sedano, F., et al.: Monitoring intra and inter annual dynamics of forest degradation from charcoal production in Southern Africa with Sentinel–2 imagery. Int. J. Appl. Earth Obs. Geoinf. **92**, 102184 (2020)
20. Hościło, A., Lewandowska, A.: Mapping forest type and tree species on a regional scale using multi-temporal Sentinel-2 data. Remote Sens. **11**(8), 929 (2019)
21. Omer, G., Mutanga, O., Abdel-Rahman, E.M., Adam, E.: Performance of support vector machines and artificial neural network for mapping endangered tree species using WorldView-2 data in Dukuduku forest, South Africa. IEEE J. Sel. Top. Appl. Earth Observ. Remote Sens. **8**(10), 4825–4840 (2015)
22. Hawryło, P., Bednarz, B., Wężyk, P., Szostak, M.: Estimating defoliation of Scots pine stands using machine learning methods and vegetation indices of Sentinel-2. Eur. J. Remote Sens. **51**(1), 194–204 (2018)
23. Breiman, L.: Random forests. Mach. Learn. **45**(1), 5–32 (2001)
24. Son, N.T., Chen, C.F., Chen, C.R., Minh, V.Q.: Assessment of Sentinel-1A data for rice crop classification using random forests and support vector machines. Geocarto Int. **33**(6), 587–601 (2018)
25. Whyte, A., Ferentinos, K.P., Petropoulos, G.P.: A new synergistic approach for monitoring wetlands using Sentinels-1 and 2 data with object-based machine learning algorithms. Environ. Model. Softw. **104**, 40–54 (2018)
26. Rodriguez-Galiano, V.F., Ghimire, B., Rogan, J., Chica-Olmo, M., Rigol-Sanchez, J.P.: An assessment of the effectiveness of a random forest classifier for land-cover classification. ISPRS J. Photogramm. Remote. Sens. **67**, 93–104 (2012)
27. Duda, R.O., Hart, P.E.: Pattern Classification and Scene Analysis, vol. 3, pp. 731–739. Wiley, New York (1973)
28. Akbulut, Y., Sengur, A., Guo, Y., Smarandache, F.: NS-k-NN: neutrosophic set-based k-nearest neighbors classifier. Symmetry **9**(9), 179 (2017)
29. Prasath, V.B., et al.: Distance and similarity measures effect on the performance of k-nearest neighbor classifier–a review. arXiv preprint arXiv:1708.04321 (2017)
30. Shetty, S., Gupta, P.K., Belgiu, M., Srivastav, S.K.: Assessing the effect of training sampling design on the performance of machine learning classifiers for land cover mapping using multi-temporal remote sensing data and google earth engine. Remote Sens. **13**(8), 1433 (2021)

Systems and Applications

RIN: Towards a Semantic Rigorous Interpretable Artificial Immune System for Intrusion Detection

Qianru Zhou[1(✉)], Rongzhen Li[1], Lei Xu[1], Anmin Fu[1], Jian Yang[1], Alasdair J. G. Gray[2], and Stephen McLaughlin[3]

[1] School of Computer Science and Engineering, Nanjing University of Science and Technology, Nanjing, China
zhouqianru@njust.edu.cn
[2] School of Engineering and Physical Sciences, Heriot-Watt University, Edinburgh EH14 4AS, UK
[3] Department of Computer Science, Heriot-Watt University, Edinburgh EH14 4AS, UK

Abstract. The Internet is the most complex machine humankind has ever built, and how to defense it from intrusions is even more complex. With the ever increasing of new intrusions, intrusion detection task rely on Artificial Intelligence more and more. Interpretability and transparency of the machine learning model is the foundation of trust in AI-driven intrusion detection results. Current interpretation Artificial Intelligence technologies in intrusion detection are heuristic, which is neither accurate nor sufficient. This paper proposed a rigorous interpretable Artificial Intelligence driven intrusion detection approach, based on formal logic calculations. Details of rigorous interpretation calculation process for a decision tree model is presented. *Prime implicant* explanation for benign traffic flow are given in detail as rule for negative selection of the cyber immune system. Experiments are carried out in real-life traffic.

Keywords: Interpretable machine learning · Explainable artificial intelligence · Prime implicant · Artificial immune system · Intrusion detection

1 Introduction

The ultimate goal of cyber intrusion detection is an artificial immune system for the cyber that could identify intrusions automatically without any intervene from human or outsider software [1]. The performance of the cyber immune system depends entirely on the rule to identify benign traffic. Artificial intelligence based decision making has been used extensively in intrusion detection, trying to identify the patterns (or rules) hidden inside the traffic data [2]. However, in domains where security is of utmost importance, trust is the fundamental

K. K. Patel et al. (Eds.): icSoftComp 2022, CCIS 1788, pp. 213–224, 2023.
https://doi.org/10.1007/978-3-031-27609-5_17

basis and guarantee of the validity and prosperity for adopting of AI-based decision making strategies. People's trust on the decisions made is based on the interpretability and transparency of the machine learning models make them [2,3]. Unfortunately, most of the popular machine learning models, such as deep learning, neural networks, and even the tree-based models are uninterpretable (although the tree-based models are believed to be interpretable for they can provide the decision paths that lead to the decisions, many have point out that these explanations are "shallow" and contain potentially too many redundant features and rules, and thus actually unable to provide rigorous sufficient reasons, also known as *prime implicant explanations*, or *minimal sufficient reasons* [3]). Consequences of the decision made by uninterpretable machine learning models are occasionally catastrophic, for example the fatal car crushes by Google's autonomous car[1] and Tesla's autopilot system[2]; An automatic bail risk assessment algorithm is believed to be biased and keep many people in jail longer than they should without explicit reasons, and another machine learning based DNA trace analysis software accuses people with crimes they did not commit[3]; Millions of African-American could not get due medical care by a biased machine learning assessment algorithm[4]; In Scotland, a football game is ruined because the AI camera mistakes the judge's bald head as the ball and keep focusing on it rather than the goal scene[5]. The key reasons lay in that all machine learning models suffer from overfitting [4]. And overfitting could be seriously exacerbated by noisy data, and real-life data is, and almost always, noisy. These, among many other reasons (like GDPR requirements and judicial requirements), have driven the surge of research interest on the interpretation of machine learning models, analyzing the reasons for positive or negative decisions, interrogating them by human domain experts, and adjusting them if necessary. That gives rise to the surge of research interest in Explainable Artificial Intelligence (XAI) or Interpretable Machine Learning (IML)[6] [2].

In this paper, a rigorous XAI methodology is proposed to interpret the benign traffic pattern learned by a machine learning model with acceptable performance in detecting intrusions[7]. Use the rigorous explanations as rules for detecting benign traffic, an artificial immune system architecture is proposed.

[1] https://www.govtech.com/transportation/google-autonomous-car-experiences-another-crash.html.

[2] https://www.usatoday.com/story/money/cars/2021/08/16/tesla-autopilot-investigation-nhtsa-emergency-responder-crashes/8147397002/.

[3] See https://www.nytimes.com/2017/06/13/opinion/how-computers-are-harming-criminal-justice.html.

[4] See https://www.wsj.com/articles/researchers-find-racial-bias-in-hospital-algorithm-11571941096.

[5] see https://www.ndtv.com/offbeat/ai-camera-ruins-football-game-by-mistaking-referees-bald-head-for-ball-2319171.

[6] there is subtle difference between explainable and interpretable AI, but this is not within the focus of this paper, so we will use XAI to represent both methodologies throughout the paper.

[7] with accuracy in terms of AUC nearly 1.

The paper is organized as follows, Sect. 2 provide a overall literature review for XAI methodologies in intrusion detection; Sect. 3 provides details of rigorous XAI methodologies, including continuous features discretization, Map and Merge (M&M) algorithm, and the proposed RIN system; Sect. 4 present the details of rigorous explanation calculated from the target model, evaluated with the results on real-life traffic flow instances. Section 5 summarize the work.

2 Explainable Artificial Intelligent Driven Intrusion Detections

Various approaches trying to explain machine learning models for cybersecurity have been proposed [5–8]. Luca Vigano and his group proposed a new paradigm in security research called Explainable Security (XSec) in [9]. They propose the "Six Ws" of XSec (Who? What? Where? When? Why? and How?) as the standard perspectives of XAI in cybersecurity domain.

Marco Melis trys to explain malicious black-box android malware detections on any-type of models [5]. This work leverages a gradient-based approach to identify the most influential local features. It also enables use of nonlinear models to potentially increase accuracy without sacrificing interpretability of decisions on the DREBIN Dataset. Drebin as such explains its decisions by reporting, for any given application, the most influential features, i.e., those features present in a given application and assigned the highest absolute weights by the classifier. [8] tries to use adversarial machine learning to find the minimum modifications (of the input features) required to correctly classify a given set of misclassified samples, to be specific, it tries to find an adversarial sample that is classified as positive with the minimum distance between the real sample and the modified sample. Other works on explainable android malware detection do not specifically use explainable machine learning models but do make use of other feature analyses to reduce uncertainty of information. For example, [6] uses static analysis and probability statistics-based feature extraction analysis to detect and analyze malicious Android apps. [7] try to interpret the rules for malicious node identification by directly use paths in decision tree model trained by KDD dataset. However, many researchers believe the direct interpretation provided by decision tree model is "shallow" and redundancy, and thus is not necessarily the minimal prime implicant interpretation, in other words, it is not the radical reason.

A brief summary of current XAI applications on cybersecurity is presented in Table 1, with details of the methodology, datasets used, and target models. In the author's humble knowledge, almost all the state-of-the-art methodologies in intrusion detection are heuristic or simply direct "shallow" interpretations provided by tree-based models, which are neither accurate nor sufficient, thus cannot be considered really "interpretable" [10,11].

3 RIN– Rigorous XAI Driven Intrusion Detection

Table 1. Summary of XAI methodologies in intrusion detection.

Name	Description	Target models	Dataset	Ref.
Marino et.al.'s work	Use adversarial machine learning by finding an adversarial sample that is classified as positive while minimizing the distance between the real sample and the modified sample	DNN	NSL-KDD	[8]
Marco Melis et.al.'s work	Use a gradient-based approach to identify the most influential local features	Any	DREBIN	[5]
DeNNeS	An embedded, deep learning-based cybersecurity expert system extracting refined rules from a trained multilayer DNN	DNN	UCI's phishing websites dataset Android malware dataset	[12]
J. N. Paredes et.al.'s work	An vision of combining knowledge reasoning-driven and data-driven approaches	N/A	National Vulnerability Database MITRE CVE MITRE CWE MITRE ATT&CK	[11]
LEMNA	Explain the model by approximating a local area	DNN	Binary dataset generated in BYTEWEIGHT[a]	[13]
RIN	**Provide rigorous explanation with 100% accuracy**	**Any**	**CIC-AWS-2018**	**Proposed**

[a] https://security.ece.cmu.edu/byteweight/.

3.1 Rigorous XAI

While heuristic XAI methods compute approximations of real explanations, rigorous explanations are guaranteed to represent exactly the same behavior with the model. In rigorous XAI theory [14], a classifier is a *Boolean function* which can be represented by a propositional formula Δ. An *implicant* τ of a propositional formula Δ is a term that satisfies Δ, namely $\tau \models \Delta$. A *prime implicant* is an implicant that is not subsumed by any other implicants, that is there is no implicate τ' that contains a strict subset of the literals of τ. *Prime implicant* have been used to give rigorous explanations in XAI. Explanations given using *prime implicant* are also called *sufficient reasons*, which are defined formally below.

Definition 1 (Sufficient Reason [14]**).** *A sufficient reason for decision Δ_α is a property of instance α that is also a prime implicant of Δ_α (Δ_α is Δ if the decision is positive and $\neg\Delta$ otherwise).*

A *sufficient reason* (or *prime implicant explanation*) is also the *minimum explanation*. The major difference between *sufficient reason* and *prime implicant* is that *sufficient reason* disclose the reasons of a certain instance while *prime implicant* illustrate the essential characteristics of the model [14]. *Sufficient reason* explains the root cause of the decision for an instance, in terms of the *prime implicants* involved. The decision will stay unchanged no matter how the other characters change, and none of its strict subsets can justify the decision. Please be noted that a decision may have multiple sufficient reasons, sometimes many [14]. The Quine–McCluskey algorithm (QMC) (also known as the *method of prime implicants* or *tabulation method*) is used in this paper to calculate prime implicants of a Boolean expression[8].

3.2 M&M – Discretize Continuous Features

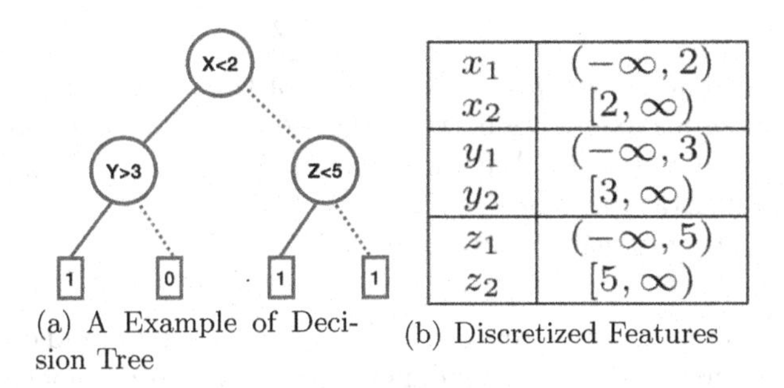

x_1	$(-\infty, 2)$
x_2	$[2, \infty)$
y_1	$(-\infty, 3)$
y_2	$[3, \infty)$
z_1	$(-\infty, 5)$
z_2	$[5, \infty)$

(a) A Example of Decision Tree (b) Discretized Features

Fig. 1. A simple decision tree with continuous values.

Rigorous logical reasoning methods work directly on boolean expressions. Machine learning models with boolean features can be immediately represented as boolean circuits [14]. Models with discrete features can be transformed into boolean expressions with atom variables [14]. However, most of the classifiers used in intrusion detection have continuous features. Darwiche et.al. proposed a mapping method to map continuous features into discrete ones [14], as presented below in Fig. 1. I enhanced the method by further simplify it with a merge process, hence map and merge (M&M) method. The detail of M&M discretize method is presented below.

[8] It is a classic boolean expression minimization method developed by Willard V. Quine in 1952 and extended by Edward J. McCluskey in 1956.

Map. Take the decision tree in Fig. 1(a) for example, the solid lines represent if the node is true, while the dashed lines represent false. Based on each decision node, features with continuous values are discretized into several variables, each represent an interval divided by decision nodes. As shown in Fig. 1(b), feature X in Fig. 1(a) are discretized into x_1, and x_2, representing the intervals $(-\infty, 2)$ and $[2, \infty)$ respectively, thus we have

$$x_1 \vee x_2 \models U \;\&\&\; x_1 \wedge x_2 \models \oslash \quad (1)$$

$$y_1 \vee y_2 \models U \;\&\&\; y_1 \wedge y_2 \models \oslash \quad (2)$$

$$z_1 \vee z_2 \models U \;\&\&\; z_1 \wedge z_2 \models \oslash \quad (3)$$

Thus, the decision rule of the decision tree in Fig. 1(a) can be represented by boolean expression

$$\Delta = (x_1 \wedge y_1) \vee (x_2 \wedge z_1) \vee (x_2 \wedge z_2) \quad (4)$$

According to Eq. 3, we have

$$z_1 = \bar{z_2}$$

With De Morgan's law, Eq. 4 can be further simplified to

$$\Delta = (x_1 \wedge y_1) \vee x_2 \quad (5)$$

In which $x_1 \wedge y_1$ and x_2 are *prime implicants* of the decision tree in Fig. 1(a), and thus the rigorous explanation of the decision tree are *"So long as $x_1 \wedge y_1$ or x_2, the decision will be 1"*. According to the discretization rule, the rigorous explanation can further be *"So long as $X < 2$ and $Y < 3$, or $X \geq 5$ in the instance, the decision is guarrenteed to be 1."*

Merge. The discrete features get from the map process may (and often) contain (potentially a large number of) redundancy. As the number of boolean expression is $N = 2^n$ where n is the number of discrete features, directly transform the discrete features into boolean circuits may experience huge waste of computing and storage expense, due to *Combinatorial Explosion*. Thus the merge process proposed in Algorithm 1 is used in RIN .

Algorithm 1. Merge Rule.

For $x_i = (v_1, v_2]$ and $x_{i+1} = (v_2, v_3]$, $(v_1 \leq v_2 \leq v_3)$
AND $s_i = \{set\ of\ features\ in\ rule\ r_i\}$
IF $\neg\exists((x_i \in s_i) \wedge (x_{i+1} \notin s_i)) \vee ((x_i \notin s_i) \wedge (x_{i+1} \in s_i))$
THEN DELETE x_{i+1} AND $x_i = x_i \cup x_{i+1} = (v_1, v_3]$

For example, let $\Delta = ((x_1 \vee x_2) \wedge y_1) \vee ((x_1 \vee x_2) \wedge z_2)$ be the boolean expression of a model after map process, according to Algorithm 1, $(x_1 \vee x_2)$ fits the requirement of merge rule, and thus the model can be simplified into $\Delta = (x_{1new} \wedge y_1) \vee (x_{1new} \wedge z_2)$, where $x_{1new} = x_{1old} \vee x_{2old}$.

3.3 Architecture of RIN

Based on the previous discussion, the architecture of the proposed rigorous XAI driven intrusion detection system RIN is shown in Fig. 2.

4 Evaluation Results of RIN

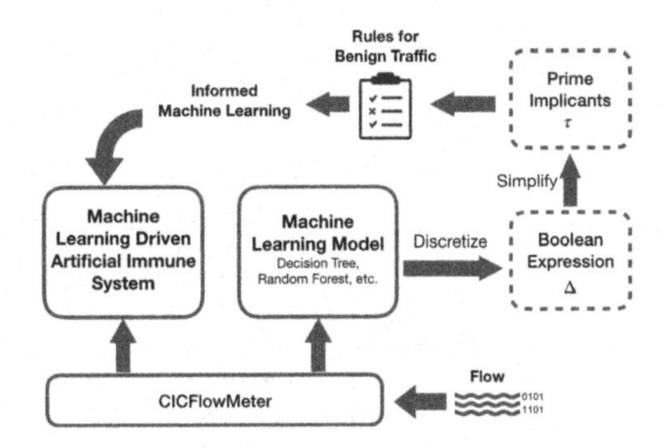

Fig. 2. The proposed architecture of RIN, an cyber-immune system driven by rigorous XAI technology.

The target model is selected from our previous work [15]. We have evaluated 8 kinds of common machine learning models on eleven different kinds of real-life intrusion traffic data, and decision tree has the best performance in both accuracy and time expense when detecting known intrusions (with AUC almost 1). The model has done learning when it tends toward stability. From Fig. 3 it is evident that the *number of leaves, maximum depth*, and *node count* of decision tree model trained per round fluctuate within a narrow range, do not show any pronounce trend (of increase or decline). It is reasonable to believe that the model is stable. A modest model (in terms of *number of leaves, maximum depth*, and *node count*) is selected as the target model for explanations computing, with 14 features, 19 leaves, and maximum depth of 11.

4.1 Feature Discretization

M&M discretization method in Sect. 3.2 is used to transform continuous features into discretized one. The Features before and after discretization are presented in Table 2. After discretization, 14 continuous features are transformed to 37 discrete variables. Thus, based on their feature, each instance can be mapped into a 37 bits binary expression

$$a_1a_2a_3b_1b_2b_3b_4c_1c_2d_1d_2d_3e_1e_2e_3f_1f_2f_3g_1g_2$$

$$h_1h_2h_3i_1i_2j_1j_2j_3k_1k_2l_1l_2m_1m_2n_1n_2$$

Table 2. Feature discretization for the target decision tree model.

Continuous	Discrete
A: Fwd_Pkt_Len_Min	a_1: $(-\infty, 49.5)$
	a_2: $[49.5, 110.5)$
	a_3: $[110.5, +\infty)$
B: Fwd_Pkts_s	b_1: $(-\infty, 0.01)$
	b_2: $[0.01, 315.7)$
	b_3: $[315.7, 3793)$
	b_4: $[3793, +\infty)$
C:Idle_Mean	c_1: $(-\infty, 15000095)$
	c_2: $[15000095, 48801184)$
	c_3: $[48801184, +\infty)$
D: Flow_Duration	d_1: $(-\infty, 194)$
	d_2: $[194, 119992424)$
	d_3: $[119992424, +\infty)$
E: Fwd_IAT_Std	e_1: $(-\infty, 14587155)$
	e_2: $[14587155, 15289483.5)$
	e_3: $[15289483.5, +\infty)$
F: Subflow_Fwd_Pkts	f_1: $(-\infty, 95)$
	f_2: $[95, 62410)$
	f_3: $[62410, +\infty)$
G: Fwd_Header_Len	g_1: $(-\infty, 12186)$
	g_2: $[12186, +\infty)$
H: Fwd_IAT_Max	h_1: $(-\infty, 25696)$
	h_2: $[25696, 41484286)$
	h_3: $[41484286, +\infty)$
I: Fwd_Pkt_Len_Max	i_1: $(-\infty, 685.5)$
	i_2: $[685.5, +\infty)$
J: Dst_Port	j_1: $(-\infty, 434)$
	j_2: $[434, 444)$
	j_3: $[444, +\infty)$
K: Flow_IAT_Min	k_1: $(-\infty, 190)$
	k_2: $[190, +\infty)$
L: Idle_Min	l_1: $(-\infty, 48801184)$
	l_2: $[48801184, +\infty)$
M: Fwd_IAT_Tot	m_1: $(-\infty, 1241282.5)$
	m_2: $[1241282.5, +\infty)$
N: Fwd_Seg_Size_Avg	n_1: $(-\infty, 486)$
	n_2: $[486, +\infty)$

Table 3. Feature discretization for the target decision tree model.

$Fwd\_Pkt\_Len\_Min = 0.0$	$Fwd\_Pkts\_s = 64.42$
$Idle\_Mean = 0.0$	$Flow\_Duration = 31045$
$Fwd\_IAT\_Std = 0.0$	$Subflow\_Fwd\_Pkts = 2.0$
$Fwd\_Header\_Len = 40$	$Fwd\_IAT\_Max = 31045$
$Fwd\_Pkt\_Len\_Max = 0.0$	$Dst\_Port = 80$
$Flow\_IAT\_Min = 31045$	$Idle\_Min = 0.0$
$Fwd\_IAT\_Tot = 31045$	$Fwd\_Seg\_Size\_Avg = 0.0$

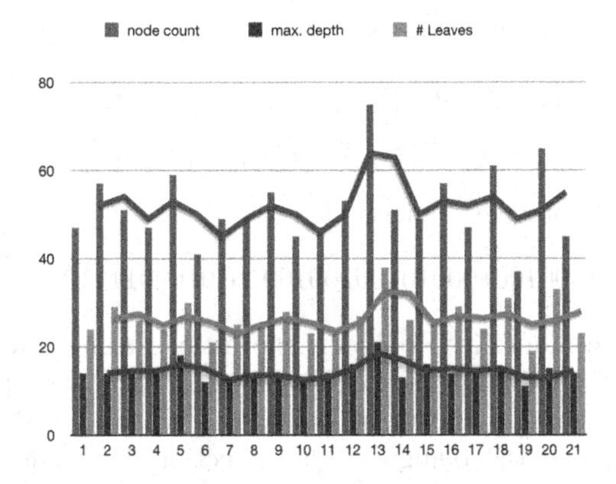

Fig. 3. The model features' (*number of leaves, maximum depth,* and *node count*) fluctuation and trend during the fitting process.

Table 4. Prime Implicants for benign traffic in the decision tree model.

#	Minterm	Boolean Expression
τ_1	$a_1\bar{a}_2\bar{a}_3b_1\bar{b}_2\bar{b}_3\bar{b}_4\bar{c}_1c_2\bar{c}_3$	1001000010---------------------------
τ_2	$\bar{a}_1a_2\bar{a}_3b_1\bar{b}_2\bar{b}_3\bar{b}_4\bar{c}_1c_2\bar{c}_3$	0101000010---------------------------
τ_3	$a_1\bar{a}_2\bar{a}_3b_1\bar{b}_2\bar{b}_3\bar{b}_4\bar{c}_1\bar{c}_2c_3$	1001000001---------------------------
τ_4	$\bar{a}_1a_2\bar{a}_3b_1\bar{b}_2\bar{b}_3\bar{b}_4\bar{c}_1\bar{c}_2c_3$	0101000001---------------------------
τ_5	$a_1\bar{a}_2\bar{a}_3\bar{b}_1\bar{b}_2b_3\bar{b}_4d_1\bar{d}_2\bar{d}_3e_1\bar{e}_2\bar{e}_3f_1\bar{f}_2\bar{f}_3\bar{i}_1i_2$	1000010---100100100-----01-----------
τ_6	$a_1\bar{a}_2\bar{a}_3\bar{b}_1\bar{b}_2b_3\bar{b}_4\bar{d}_1d_2\bar{d}_3e_1\bar{e}_2\bar{e}_3f_1\bar{f}_2\bar{f}_3\bar{i}_1i_2$	1000010---010100100-----01-----------
τ_7	$a_1\bar{a}_2\bar{a}_3\bar{b}_1b_2\bar{b}_3\bar{b}_4d_1\bar{d}_2\bar{d}_3\bar{e}_1e_2\bar{e}_3f_1\bar{f}_2\bar{f}_3\bar{h}_1\bar{h}_2h_3$	1000100---100010100--001-------------
τ_8	$a_1\bar{a}_2\bar{a}_3\bar{b}_1b_2\bar{b}_3\bar{b}_4d_1\bar{d}_2\bar{d}_3\bar{e}_1e_2\bar{e}_3f_1\bar{f}_2\bar{f}_3\bar{h}_1h_2\bar{h}_3$	1000100---100010100--010-------------
τ_9	$a_1\bar{a}_2\bar{a}_3\bar{b}_1b_2\bar{b}_3\bar{b}_4\bar{d}_1d_2\bar{d}_3\bar{e}_1e_2\bar{e}_3f_1\bar{f}_2\bar{f}_3h_1\bar{h}_2\bar{h}_3$	1000100---010010100--100-------------
τ_{10}	$a_1\bar{a}_2\bar{a}_3\bar{b}_1b_2\bar{b}_3\bar{b}_4\bar{d}_1d_2\bar{d}_3\bar{e}_1e_2\bar{e}_3f_1\bar{f}_2\bar{f}_3\bar{h}_1h_2\bar{h}_3$	1000100---010010100--010-------------
τ_{11}	$a_1\bar{a}_2\bar{a}_3\bar{b}_1\bar{b}_2b_3\bar{b}_4d_1\bar{d}_2\bar{d}_3\bar{e}_1e_2\bar{e}_3f_1\bar{f}_2\bar{f}_3h_1\bar{h}_2\bar{h}_3$	1000010---100010100--100-------------
τ_{12}	$a_1\bar{a}_2\bar{a}_3\bar{b}_1\bar{b}_2b_3\bar{b}_4d_1\bar{d}_2\bar{d}_3\bar{e}_1e_2\bar{e}_3f_1\bar{f}_2\bar{f}_3\bar{h}_1h_2\bar{h}_3$	1000010---100010100--010-------------
τ_{13}	$a_1\bar{a}_2\bar{a}_3\bar{b}_1\bar{b}_2b_3\bar{b}_4\bar{d}_1d_2\bar{d}_3\bar{e}_1e_2\bar{e}_3f_1\bar{f}_2\bar{f}_3h_1\bar{h}_2\bar{h}_3$	1000010---010010100--100-------------
τ_{14}	$a_1\bar{a}_2\bar{a}_3\bar{b}_1\bar{b}_2b_3\bar{b}_4\bar{d}_1d_2\bar{d}_3\bar{e}_1e_2\bar{e}_3f_1\bar{f}_2\bar{f}_3\bar{h}_1h_2\bar{h}_3$	1000010---010010100--010-------------
τ_{15}	$a_1\bar{a}_2\bar{a}_3\bar{b}_1b_2\bar{b}_3\bar{b}_4d_1\bar{d}_2\bar{d}_3e_1\bar{e}_2\bar{e}_3\bar{f}_1f_2\bar{f}_3g_1\bar{g}_2$	1000100---10010001010----------------
τ_{16}	$a_1\bar{a}_2\bar{a}_3\bar{b}_1b_2\bar{b}_3\bar{b}_4d_1\bar{d}_2\bar{d}_3e_1\bar{e}_2\bar{e}_3\bar{f}_1\bar{f}_2f_3g_1\bar{g}_2$	1000100---10010000110----------------
τ_{17}	$a_1\bar{a}_2\bar{a}_3\bar{b}_1b_2\bar{b}_3\bar{b}_4d_1\bar{d}_2\bar{d}_3\bar{e}_1e_2\bar{e}_3\bar{f}_1f_2\bar{f}_3g_1\bar{g}_2$	1000100---10001001010----------------
τ_{18}	$a_1\bar{a}_2\bar{a}_3\bar{b}_1b_2\bar{b}_3\bar{b}_4d_1\bar{d}_2\bar{d}_3\bar{e}_1e_2\bar{e}_3\bar{f}_1\bar{f}_2f_3g_1\bar{g}_2$	1000100---10001000110----------------
τ_{19}	$a_1\bar{a}_2\bar{a}_3\bar{b}_1b_2\bar{b}_3\bar{b}_4\bar{d}_1d_2\bar{d}_3e_1\bar{e}_2\bar{e}_3\bar{f}_1f_2\bar{f}_3g_1\bar{g}_2$	1000100---01010001010----------------
τ_{20}	$a_1\bar{a}_2\bar{a}_3\bar{b}_1b_2\bar{b}_3\bar{b}_4\bar{d}_1d_2\bar{d}_3e_1\bar{e}_2\bar{e}_3\bar{f}_1\bar{f}_2f_3g_1\bar{g}_2$	1000100---01010000110----------------
τ_{21}	$a_1\bar{a}_2\bar{a}_3\bar{b}_1b_2\bar{b}_3\bar{b}_4\bar{d}_1d_2\bar{d}_3\bar{e}_1e_2\bar{e}_3\bar{f}_1f_2\bar{f}_3g_1\bar{g}_2$	1000100---01001001010----------------
τ_{22}	$a_1\bar{a}_2\bar{a}_3\bar{b}_1b_2\bar{b}_3\bar{b}_4\bar{d}_1d_2\bar{d}_3\bar{e}_1e_2\bar{e}_3\bar{f}_1\bar{f}_2f_3g_1\bar{g}_2$	1000100---01001000110----------------
τ_{23}	$a_1\bar{a}_2\bar{a}_3\bar{b}_1\bar{b}_2b_3\bar{b}_4d_1\bar{d}_2\bar{d}_3e_1\bar{e}_2\bar{e}_3\bar{f}_1f_2\bar{f}_3g_1\bar{g}_2$	1000010---10010001010----------------
τ_{24}	$a_1\bar{a}_2\bar{a}_3\bar{b}_1\bar{b}_2b_3\bar{b}_4d_1\bar{d}_2\bar{d}_3e_1\bar{e}_2\bar{e}_3\bar{f}_1\bar{f}_2f_3g_1\bar{g}_2$	1000010---10010000110----------------
τ_{25}	$a_1\bar{a}_2\bar{a}_3\bar{b}_1\bar{b}_2b_3\bar{b}_4d_1\bar{d}_2\bar{d}_3\bar{e}_1e_2\bar{e}_3\bar{f}_1f_2\bar{f}_3g_1\bar{g}_2$	1000010---10001001010----------------
τ_{26}	$a_1\bar{a}_2\bar{a}_3\bar{b}_1\bar{b}_2b_3\bar{b}_4d_1\bar{d}_2\bar{d}_3\bar{e}_1e_2\bar{e}_3\bar{f}_1\bar{f}_2f_3g_1\bar{g}_2$	1000010---10001000110----------------
τ_{27}	$a_1\bar{a}_2\bar{a}_3\bar{b}_1\bar{b}_2b_3\bar{b}_4\bar{d}_1d_2\bar{d}_3e_1\bar{e}_2\bar{e}_3\bar{f}_1f_2\bar{f}_3g_1\bar{g}_2$	1000010---01010001010----------------
τ_{28}	$a_1\bar{a}_2\bar{a}_3\bar{b}_1\bar{b}_2b_3\bar{b}_4\bar{d}_1d_2\bar{d}_3e_1\bar{e}_2\bar{e}_3\bar{f}_1\bar{f}_2f_3g_1\bar{g}_2$	1000010---01010000110----------------
τ_{29}	$a_1\bar{a}_2\bar{a}_3\bar{b}_1\bar{b}_2b_3\bar{b}_4\bar{d}_1d_2\bar{d}_3\bar{e}_1e_2\bar{e}_3\bar{f}_1f_2\bar{f}_3g_1\bar{g}_2$	1000010---01001001010----------------
τ_{30}	$a_1\bar{a}_2\bar{a}_3\bar{b}_1\bar{b}_2b_3\bar{b}_4\bar{d}_1d_2\bar{d}_3\bar{e}_1e_2\bar{e}_3\bar{f}_1\bar{f}_2f_3g_1\bar{g}_2$	1000010---01001000110----------------
τ_{31}	$a_1\bar{a}_2\bar{a}_3\bar{b}_1b_2\bar{b}_3\bar{b}_4d_1\bar{d}_2\bar{d}_3\bar{e}_1\bar{e}_2e_3l_1\bar{l}_2$	1000100---100001---------------10----
τ_{32}	$a_1\bar{a}_2\bar{a}_3\bar{b}_1b_2\bar{b}_3\bar{b}_4\bar{d}_1d_2\bar{d}_3\bar{e}_1\bar{e}_2e_3l_1\bar{l}_2$	1000100---010001---------------10----
τ_{33}	$a_1\bar{a}_2\bar{a}_3\bar{b}_1\bar{b}_2b_3\bar{b}_4d_1\bar{d}_2\bar{d}_3\bar{e}_1\bar{e}_2e_3l_1\bar{l}_2$	1000010---100001---------------10----
τ_{34}	$a_1\bar{a}_2\bar{a}_3\bar{b}_1\bar{b}_2b_3\bar{b}_4\bar{d}_1d_2\bar{d}_3\bar{e}_1\bar{e}_2e_3l_1\bar{l}_2$	1000010---010001---------------10----
τ_{35}	$a_1\bar{a}_2\bar{a}_3\bar{b}_1b_2\bar{b}_3\bar{b}_4\bar{d}_1\bar{d}_2d_3f_1\bar{f}_2\bar{f}_3$	1000100---001---100------------------
τ_{36}	$a_1\bar{a}_2\bar{a}_3\bar{b}_1b_2\bar{b}_3\bar{b}_4\bar{d}_1\bar{d}_2d_3\bar{f}_1f_2\bar{f}_3$	1000100---001---010------------------
τ_{37}	$a_1\bar{a}_2\bar{a}_3\bar{b}_1\bar{b}_2b_3\bar{b}_4\bar{d}_1\bar{d}_2d_3f_1\bar{f}_2\bar{f}_3$	1000010---001---100------------------
τ_{38}	$a_1\bar{a}_2\bar{a}_3\bar{b}_1\bar{b}_2b_3\bar{b}_4\bar{d}_1\bar{d}_2d_3\bar{f}_1f_2\bar{f}_3$	1000010---001---010------------------
τ_{39}	$\bar{a}_1a_2\bar{a}_3\bar{b}_1b_2\bar{b}_3\bar{b}_4\bar{m}_1m_2$	0100100--------------------------01--
τ_{40}	$\bar{a}_1a_2\bar{a}_3\bar{b}_1\bar{b}_2b_3\bar{b}_4\bar{m}_1m_2$	0100010--------------------------01--
τ_{41}	$\bar{a}_1\bar{a}_2a_3n_1\bar{n}_2$	001--------------------------------10

For example, a flow instance with the features shown in Table 3 is mapped into

$$a_1\bar{a_2}\bar{a_3}\bar{b_1}b_2\bar{b_3}\bar{b_4}c_1\bar{c_2}\bar{c_3}\bar{d_1}d_2\bar{d_3}e_1\bar{e_2}\bar{e_3}f_1\bar{f_2}\bar{f_3}g_1\bar{g_2}$$
$$\bar{h_1}h_2\bar{h_3}i_1\bar{i_2}j_1\bar{j_2}\bar{j_3}\bar{k_1}k_2l_1\bar{l_2}m_1\bar{m_2}n_1\bar{n_2}$$

which can be represented in numeric:

1000100100010100100100101010001101010

After discretize the model into boolean expressions using M&M, prime Implicants are calculated with Quine-McCluskey algorithm. The prime implicants presented in minterms are shown in Table 4, in which a "–" means a "don't care". The behavior of the benign traffic detection of the decision tree model can be rigorously interpreted as $\Delta = \tau_1 \vee \tau_2 \vee \cdots \vee \tau_{41}$.

These *prime implicants* of the model works as *antibody* in cyber-immune system RIN, which form the rules of the negative selection process.

4.2 Semantic Rigorous Explanations

Prime implicants explanation examples for five real-life benign traffic flow instances explanation is presented in detail in Table 5. The original features, discretized features, and boolean expression of each instance are presented in detail. The reason for why it is classified as benign is provided and marked in red in its boolean expression, by the prime implicant that is used to make the decision, which is sufficient and rigorous to explain the decision. For example, in flow instance 1, its boolean expression is

0011000010010100100100101000101100110

after feature discretization and mapping, and it matches with *prime implicant* τ_{41}:

001--------------------------------10

which means *"if a flow's boolean expression start with* 001, *and end with* 10, *no matter what values is for the other bits, it is a benign flow."*. With the feature discretization mapping method, *prime implicant* τ_{41} is originally "$\bar{a_1}\bar{a_2}a_3n_1\bar{n_2}$" which can also be interpreted as *"if a flow's Fwd_Pkt_Len_Min is larger than* 110.5, *and Fwd_Seg_Size_Avg is smaller than* 486, *then it is benign."* Thus, the model decides this flow is benign. The whole decision process is formal and rigorous, and the reasons is sufficient, as proved in Sect. 3.1. In another example flow instance 5, whose feature is mapped into boolean expression

0100100100010100100100101010001100110

is classified into "Benign" because it matches with *prime implicant* τ_{39}:

0100100--------------------------01--

which is

$$\bar{a_1} a_2 \bar{a_3} \bar{b_1} b_2 \bar{b_3} \bar{b_4} \bar{m_1} m_2$$

says *"if a flow has Fwd_Pkt_Len_Min between* 49.5 *and* 110.5*, Fwd_Pkt_s between* 0.01 *and* 3793*, and Fwd_IAT_Tot larger than* 486*, then it is benign."* It is worth noting that one instance can match more than one (sometimes even many) *prime implicants* (although I did not have such experience in this experiment), which means that there are multiple explanations for the decision made on that instance, and each one of these reasons is sufficient and rigorous.

5 Conclusion and Future Research Challenges

Table 5. Examples of prime implicant explanations for benign flow instance.

	A:	B:	C:	D:	E:	F:	G:	H:	I:	J:	K:	L:	M:	N:
1	146	0.0	30043443.7	90130331	9627.8	4	32	30051116	146	17500	30032640	30032640	90130331	146
	$\bar{a_1}\bar{a_2}a_3b_1\bar{b_2}\bar{b_3}\bar{b_4}\bar{c_1}c_2\bar{c_3}\bar{d_1}d_2\bar{d_3}e_1\bar{e_2}\bar{e_3}f_1\bar{f_2}\bar{f_3}g_1\bar{g_2}\bar{h_1}h_2\bar{h_3}i_1\bar{i_2}\bar{j_1}\bar{j_2}j_3\bar{k_1}k_2l_1\bar{l_2}\bar{m_1}m_2n_1\bar{n_2}$													
	0011000010010100100100101000101100110													
PI τ_{41} :	001 −10													
	$\bar{a_1}\bar{a_2}a_3n_1\bar{n_2}$													
2	0.0	1	0.0	119999476	130.9	121	0	1000244	0.0	0.0	999705	0.0	119999476	0.0
	$a_1\bar{a_2}\bar{a_3}\bar{b_1}b_2\bar{b_3}\bar{b_4}c_1\bar{c_2}\bar{c_3}\bar{d_1}\bar{d_2}d_3e_1\bar{e_2}\bar{e_3}\bar{f_1}f_2\bar{f_3}g_1\bar{g_2}\bar{h_1}h_2\bar{h_3}i_1\bar{i_2}j_1\bar{j_2}\bar{j_3}\bar{k_1}k_2l_1\bar{l_2}\bar{m_1}m_2n_1\bar{n_2}$													
	1000100100001100010100101010001100110													
PI τ_{36} :	1000100 − − − 001 − − − 010 −													
	$a_1\bar{a_2}\bar{a_3}\bar{b_1}b_2\bar{b_3}\bar{b_4}\bar{d_1}\bar{d_2}d_3\bar{f_1}f_2\bar{f_3}$													
3	0	0.2	45007702.5	90808764	15355813.3	17	352	45015090	517	443	0	45000315	90803655	67.7
	$\bar{a_1}\bar{a_2}a_3\bar{b_1}b_2\bar{b_3}\bar{b_4}\bar{c_1}c_2\bar{c_3}\bar{d_1}d_2\bar{d_3}\bar{e_1}e_2\bar{e_3}f_1\bar{f_2}\bar{f_3}g_1\bar{g_2}\bar{h_1}\bar{h_2}h_3i_1\bar{i_2}\bar{j_1}j_2\bar{j_3}k_1\bar{k_2}l_1\bar{l_2}\bar{m_1}m_2n_1\bar{n_2}$													
	001010010010010100100011001010100110													
PI τ_{41} :	001 −10													
	$\bar{a_1}\bar{a_2}a_3n_1\bar{n_2}$													
4	40	0.0	38448540.3	115345621.0	358398.1	4	32	38687191	40	1947	38036412	38036412	115345621	40
	$a_1\bar{a_2}\bar{a_3}b_1\bar{b_2}\bar{b_3}\bar{b_4}\bar{c_1}c_2\bar{c_3}\bar{d_1}\bar{d_2}d_3e_1\bar{e_2}\bar{e_3}f_1\bar{f_2}\bar{f_3}g_1\bar{g_2}\bar{h_1}h_2\bar{h_3}i_1\bar{i_2}\bar{j_1}\bar{j_2}j_3\bar{k_1}k_2l_1\bar{l_2}\bar{m_1}m_2n_1\bar{n_2}$													
	1001000010010100100100101000110100110													
PI τ_1 :	1001000010 −													
	$a_1\bar{a_2}\bar{a_3}b_1\bar{b_2}\bar{b_3}\bar{b_4}\bar{c_1}c_2\bar{c_3}$													
5	50	2	0.0	1514340	7697.6	3.0	24	762613	50	137	751727	0.0	1514340	50
	$\bar{a_1}a_2\bar{a_3}\bar{b_1}b_2\bar{b_3}\bar{b_4}c_1\bar{c_2}\bar{c_3}\bar{d_1}d_2\bar{d_3}e_1\bar{e_2}\bar{e_3}f_1\bar{f_2}\bar{f_3}g_1\bar{g_2}\bar{h_1}h_2\bar{h_3}i_1\bar{i_2}\bar{j_1}\bar{j_2}\bar{j_3}\bar{k_1}k_2l_1\bar{l_2}\bar{m_1}m_2n_1\bar{n_2}$													
	0100100100010100100100101010001100110													
PI τ_{39} :	0100100 −01 − −													
	$\bar{a_1}a_2\bar{a_3}\bar{b_1}b_2\bar{b_3}\bar{b_4}\bar{m_1}m_2$													

This paper has calculated rigorous explanations from intrusion detection system driven by machine learning for the first time in our humble knowledge, rules for classify benign traffic flow are extracted and presented with formal logic methodology. The target model has achieved almost 100% accuracy on CIC-AWS-2018 dataset, it is evident to claim the rules extracted have the same accuracy for this dataset. Despite the progress, challenging work remain, for example, how to deal with dependency between features, how to discretize complex models in a scalable way, and how to leverage human experts' knowledge in it.

Acknowledgments. This work is supported by the Starting Program of Nanjing University of Science and Technology (No. AE89991/324), the Science and Technology Program of STATE GRID Corporation of China (5200-202140365A), the National Natural Science Foundation of China (No. 61973161, 61991404), and Jiangsu Science and technology planning project (No. be2021610).

References

1. Morel, B.: Anomaly based intrusion detection and artificial intelligence, In Tech Open Book Chapter, pp. 19–38 (2011)
2. Ribeiro, M.T., Singh, S., Guestrin, C.: Why should i trust you? explaining the predictions of any classifier. In: 22nd ACM SIGKDD, pp. 1135–1144 (2016)
3. Audemard, G., Bellart, S., Bounia, L., Koriche, F., Lagniez, J., Marquis, P.: On the explanatory power of decision trees, CoRR, vol. abs/2108.05266 (2021)
4. Domingos, P.: The master algorithm: how the quest for the ultimate learning machine will remake our world. Basic Books (2015)
5. Melis, M., Maiorca, D., Biggio, B., Giacinto, G., Roli, F.: Explaining black-box android malware detection. In: 26th EUSIPCO, pp. 524–528 (2018)
6. Grosse, K., Manoharan, P., Papernot, N., Backes, M., McDaniel, P.: On the (statistical) detection of adversarial examples. arXiv:1702.06280 (2017)
7. Mahbooba, B., Timilsina, M., Sahal, R., Serrano, M.: Explainable artificial intelligence (xai) to enhance trust management in intrusion detection systems using decision tree model. Complexity **2021** (2021)
8. Marino, D.L., Wickramasinghe, C.S., Manic, M.: An adversarial approach for explainable AI in intrusion detection systems. In: IEEE IECON, pp. 3237–3243. IEEE (2018)
9. Viganò, L., Magazzeni, D.: Explainable security. In: IEEE EuroS&PW, pp. 293–300 (2020)
10. Ignatiev, A.: Towards trustable explainable AI. In: IJCAI, pp. 5154–5158 (2020)
11. Paredes, J.N., Teze, J.C.L., Simari, G.I., Martinez, M.V.: On the importance of domain-specific explanations in AI-based cybersecurity systems (technical report), arXiv preprint arXiv:2108.02006 (2021)
12. Mahdavifar, S., Ghorbani, A.A.: DeNNeS: deep embedded neural network expert system for detecting cyber attacks. Neural Comput. Appl. **32**(18), 14753–14780 (2020). https://doi.org/10.1007/s00521-020-04830-w
13. Guo, W., Mu, D., Xu, J., Su, P., Wang, G., Xing, X.: Lemna: explaining deep learning based security applications. In: ACM SIGSAC, pp. 364–379 (2018)
14. Darwiche, A., Hirth, A.: On the reasons behind decisions, arXiv preprint arXiv:2002.09284 (2020)
15. Zhou, Q., Pezaros, D.: Evaluation of machine learning classifiers for zero-day intrusion detection-an analysis on CIC-AWS-2018 dataset, arXiv preprint arXiv:1905.03685 (2019)

SVRCI: An Approach for Semantically Driven Video Recommendation Incorporating Collective Intelligence

R. Ashvanth[1] and Gerard Deepak[2(✉)]

[1] Department of Computer Science and Engineering, National Institute of Technology, Tiruchirappalli, Tiruchirappalli, India

[2] Department of Computer Science and Engineering, Manipal Institute of Technology Bengaluru, Manipal Academy of Higher Education, Manipal, India

gerard.deepak.christuni@gmail.com

Abstract. In the modern era, the recommendation of educational resources, specifically in the form of videos, is a needful task. This paper proposes an SVRCI framework for education video recommendation. It is a query-centric knowledge-driven paradigm where the query terms are enriched by loading knowledge graphs from Google's knowledge graph (KG) API and subjecting it to structural topic modeling for aggregating relevant topics. Ontology alignment is also achieved to enrich much more relevant auxiliary knowledge. Cosine similarity, Twitter semantic similarity, and concept similarity are the three semantic similarity measures encompassed with differential thresholds for data point selection. The dataset is classified using a solid deep learning GRU classifier and the logistic regression feature control machine learning classifier. The education videos are recommended based on categories or the annotations appended to the videos in the dataset. Overall average precision of 95.66%, accuracy of 96.23%, F-measure of 96.23% and an nDCG of 0.99 has been achieved by this framework.

Keywords: Video recommendation · Collective intelligence · Semantic similarity · GRU · Logistic regression

1 Introduction

The evolution of the World Wide Web, combined with digitization, has drastically decreased the cost and time required to generate and distribute video content. On the Web, this has led to an information overload. There has been a huge increase in the amount of material that is accessible, indexed, and distributed as the cost and risk of creating and commercializing content have decreased substantially. As a huge amount of video content is regularly being made available online at an alarming rate, consumers are currently immersed in a practically unlimited supply of things to watch. The most popular video platform in the world, YouTube, receives one billion hours of video viewing per day. More than two billion people use it each month. The rise and expansion of OTT platforms has also resulted in the surge of digital content available today. Users

K. K. Patel et al. (Eds.): icSoftComp 2022, CCIS 1788, pp. 225–237, 2023.
https://doi.org/10.1007/978-3-031-27609-5_18

today are confronted with enormous amounts of data, making it difficult for average users to sort through the data in a productive and pleasant manner. Recommendation systems have been crucial in addressing this issue of information overload [1]. Users can save time and effort by letting video recommendations do the work of selecting suitable videos based on their past and present media consumption. Collaborative filtering is often utilized in the development of recommendation systems [2]. However, since various people typically possess diverse interests in videos and video platforms generating petabytes of data every second [3], the semantic relationship between objects should be crucial while building video recommendation systems, which is absent in conventional collaborative filtering frameworks [4]. It is necessary to develop knowledge-driven frameworks for video recommendation that are semantically oriented as the World Wide Web moves toward Web 3.0, which entails the integration of semantic technologies to make information more related thanks to semantic metadata.

Motivation: The absence of knowledge-centric frameworks for video recommendation requires frameworks, models, techniques, and paradigms for video recommendation, which incorporate and encompasses knowledge-centric frameworks for recommending videos. With the structure of the world wide web moving towards Web 3.0, a knowledge-centric video recommendation framework is required to scope the cohesive semantic structure of Web 3.0.

Contribution: The novel contributions of the SVRCI framework are as follows. Using the concept similarity, the recommendation of videos in the form of educational resources is achieved through query enrichment by Google's KG API, structural topic modeling, and ontology alignment with domain-centric educational ontology. Heterogeneous classifications are achieved by classifying the dataset using two distinct variational classifiers: a powerful auto-handcrafted feature-driven deep learning GRU classifier and a feature control machine learning logistic regression classifier to achieve variational heterogeneity in the classification. Semantic similarity computation using the cosine similarity and Twitter semantic similarity with differential thresholds and amalgamation of concept similarity into the model to rank and recommend ensures the best-in-class results. In the proposed methodology, as compared to the baseline models, precision, recall, F-Measure, accuracy, and the nDCG value are raised, while the FDR value is decreased.

Organization: A breakdown of the paper's structure is provided below. Sections 2 and 3 offer the related works and the proposed work, correspondingly. Section 4 contains the implementation and performance evaluation. Section 5 serves as the paper's conclusion.

2 Related Work

Deldjoo et al.'s [5] proposal for a novel content-based recommender system includes a method for autonomously evaluating videos and deriving a selection of relevant aesthetic elements based on already-accepted concepts of Applied Media Theory. Lee et al. [6] characterized recommendation as a similarity learning issue, and they developed deep video embeddings taught to forecast video associations. Yan et al. [7] developed a centralized YouTube video recommendation alternative: consumers' relevant data on Twitter is used to overcome the fundamental concerns in single network-based recommendation methods. Tripathi et al. [8] posited a customized, emotionally intelligent video recommendation engine. It measures the intensity of users' non-verbal emotional reactions to recommended videos through interactivity and facial expression detection for selection and video corpus development using authentic data streams. To address the cold-start issue, Li et al. [9] suggested a video recommendation method that benefited from deep convolutional neural networks. The suggested method performs admirably, particularly when there is significant data incoherence. Cui et al. [10] created a new recommendation system centered on a social network and video material. Huang et al. [11] created some novel methods to deliver reliable recommendations to consumers. Zhou et al. [12] created an innovative technique to recommendation in shared communities. By enabling batch recommendation to numerous new consumers and improving the subcommunity collection, a new approach is created. Duan et al. [13] presented JointRec, a video recommendation platform. JointRec enables pooled learning across dispersed cloud servers and links the JointCloud infrastructure into mobile IoT. In [14–19] several ontological and semantically driven frameworks in support of the literature of the proposed work have been depicted.

The gaps identified in the existing frameworks are that most current models are not compliant with Web 3.0, which is much more dense, cohesive, and has a high information density. Secondly, most models mainly focus on the cold-start problem or have proposed the content-based recommendation model only using semantic similarity measures along with some form of video embeddings. Some models use the actual contents, i.e., the static image features themselves, for recommendation. Some of the models use community ratings and collaborative filtering. Most of the approaches do not use auxiliary knowledge. Hybridized models are not present, and knowledge-centric frameworks are neglected. So, there is a need for a knowledge-centric framework that suits the highly dense and cohesive structure of the Web 3.0 or semantic web. Semantically compliant techniques are required, which should be a hybridization of machine intelligence with auxiliary knowledge and, if necessary, an optimization technique to yield the feasible solution set. Depending on the dataset and number of instances, the presence of a metaheuristic optimization model is optional based on the convergence rate of user satisfaction. Moreover, problems like personalization are only solved in the existing models; however, the problem of serendipity or diversification of results is not solved, and there is a need for an annotation-driven model in order to improve the cognitive capability by encompassing human thinking and reasoning and by lowering the semantic gap between the entities in the existing world wide web and the entities which are included in the localized recommendation framework.

3 Proposed Work

The proposed system architecture of the education recommendation framework is shown in Fig. 1. As with any recommendation system, the user query undergoes preprocessing, that comprises stopword removal, lemmatization, tokenization, and Named Entity Recognition (NER).

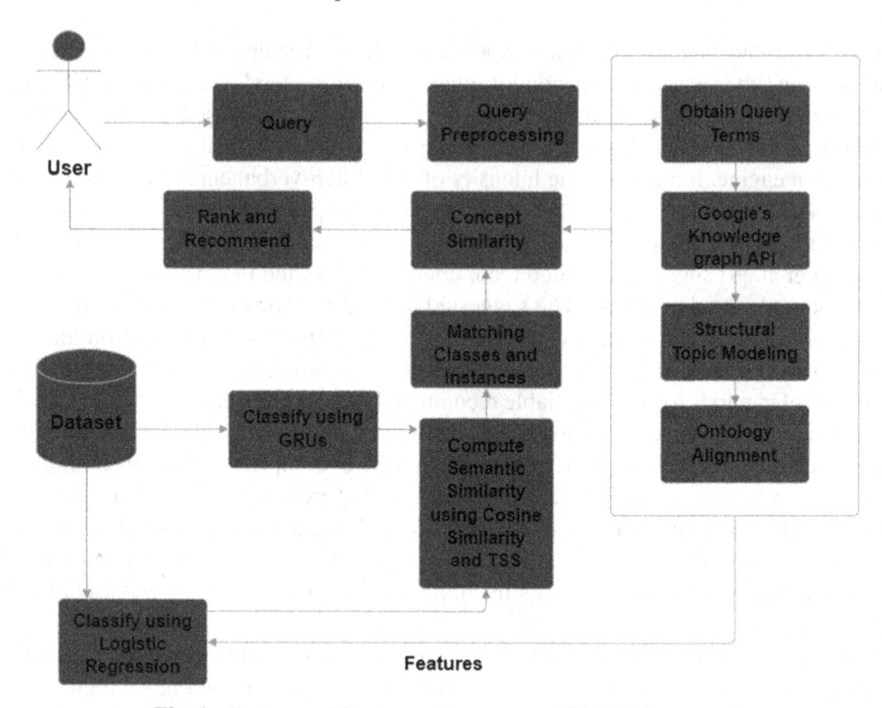

Fig. 1. System architecture of the proposed SVRCI framework

In the analog preprocessing phase, the individual Query Terms (QT) are obtained, which have to be enriched because query terms themselves are less informative. To do this knowledge enrichment, it is passed into Google's KG API. Google's KG API notes the individual subgraphs relevant to the QT, which is sent to the Structural Topic Modeling (STM) pipeline. STM is a topic modeling paradigm for aggregating uncovered but relevant topics from the external World Wide Web corpora. So topic discovery takes place using STM, which further enriches the topic categorization. In order to enhance much more instances in the model, domain-relevant educational ontology alignment is done. The ontologies are either automatically generated or manually modeled using Web Protégé. Ontology alignment takes place using concept similarity with a threshold of 0.5. The threshold is set to 0.5 to allow a more significant number of instances to be aligned into the model. In order to transform the query words that are less informative into much more informative words, entity enrichment takes place using Google's KG API, STM, and ontology alignment.

Subsequently, the categorical dataset for videos for the educational video repository is classified using two classifiers. One is a deep learning classifier, and the other is a machine learning classifier. The deep learning classifier is the Gated Recurrent Units (GRUs). It is applied because it works on the principle of auto-handcrafted feature selection where feature selection or extraction is implicit, and classification takes place as GRU is a high-power classifier with automatic feature selection. However, although deep learning algorithms may be pretty effective, they sometimes would overfit because there is no control of the features. So henceforth, the robust machine learning and logistic regression classifier are used by extracting features yielded by Google's KG API, STM, and ontology alignment to classify the dataset. A logistic regression classifier is used because there is control over the features sent into the framework. Finally, the classified instances yielded by the GRUs, and the logistic regression classifier are used to compute the semantic similarity using cosine similarity and Twitter semantic similarity (TSS). TSS and cosine similarity are again subjected to a threshold of 0.5 only because the classes and the instances under each class are parallelly computed for semantic similarity, so we do not require stringent matches for the final recommendation framework.

A Gated Recurrent Unit (GRU) is an advancement of the standard Recurrent Neural Network (RNN) architecture that employs connections among a number of nodes to perform machine learning operations. The vanishing gradient problem, a typical difficulty with RNNs, is resolved with the aid of GRUs by modifying neural network input weights. GRUs have an update gate and a reset gate that help to fine-tune the basic RNN topology. They decide which data should be transmitted to the output. They are remarkable in that it is possible to train them to remember past information without erasing it or deleting details which has nothing to do with the predictions. The update gate aids the model in determining the amount of past information from prior time periods that must be transmitted to the forecast. This is significant because the model has the option to duplicate all historical data, so removing the threat of the vanishing gradient problem. The model uses the reset gate to determine how much of the prior data to retain. The model can enhance its outputs by adjusting the flow of data through the system using these two vectors. Models containing GRUs retain information over time, making them "memory-centered" neural networks. On the other hand, other kinds of neural networks frequently lack the capacity to retain information since they lack GRUs.

Logistic regression is a robust machine learning approach for binary classification that may also be used for multiclass classification. Using multinomial logistic regression, events with more than two distinct possible outcomes can be modeled. The binary logistic model essentially divides outputs into two classes. The multinomial logistic model, in contrast, extends this to any number of classes without arranging them according to a logistic function. The input and output variables do not need to have a linear relationship for logistic regression to work. This is because the odds ratio underwent a nonlinear log change. Logistic regression is limited to values between 0 and 1. In logistic regression, a conditional probability loss function known as "maximum likelihood estimation (MLE)" is utilized, which is more suitable for data that are not perfectly correlated or when the samples have mismatched covariance matrices. Predictions are labeled as class 0 if the probability is greater than 0.5. If not, class 1 will be chosen. Logistic regression is quick, and by extending the fundamental ideas, it permits the evaluation of numerous explanatory variables as well.

All the matches and instances yielded from two distinct classifiers are used to compute the semantic similarity using the concept similarity between the initially yielded enriched entities from the query and the outcome of the matching classes, which are the outcome of the semantic similarity pipeline. Here the concept similarity is set to a threshold of 0.75 because only a relevant instance has to be yielded. Finally, the facets are ranked in the increasing order of concept similarity. Along with the facets, the videos containing these facets as metatags are recommended to the user under each category of facets. If the user is satisfied, the recommendation is stopped. If the user is unsatisfied, the recommendation is again fed into the pipeline as a preprocessed query as QT. This process continues until no further user clicks are recorded.

Cosine similarity of two vectors, X and Y is depicted by Eq. (1).

$$Sim(X, Y) = \cos(\theta) = \frac{X.Y}{|X||Y|} \tag{1}$$

For a term X, with time stamp series $\{\tau_i(X)\}$ of size N, the frequency $\Phi(X)$ is given by Eq. (2).

$$\Phi(X) = \left(\frac{\sum_{i=1}^{N-1}(\tau_{i+1}(X) - \tau_i(X))}{N-1}\right)^{-1} \tag{2}$$

Twitter semantic similarity of two terms, X and Y is depicted by Eq. (3).

$$Sim(X, Y) = \left(\frac{\Phi(X \wedge Y)}{\max(\Phi(X), \Phi(Y))}\right)^{\alpha} \tag{3}$$

where α is the scaling factor.

The information content similarity (*ics*) of two terms x_1, x_2 is given by Eqs. (4) and (5).

If $x_1 = x_2$,

$$ics(x_1, x_2) = 1 \tag{4}$$

Otherwise,

$$ics(x_1, x_2) = \frac{2\log p(x')}{\log p(x_1) + \log p(x_2)} \tag{5}$$

where x' is a provides the maximum information content shared by x_1 and x_2.

For concepts (U_1, V_1) and (U_2, V_2), sum of the *ics* is given by $M(V_1, V_2)$.

The concept similarity between (U_1, V_1) and (U_2, V_2) is depicted by Eq. (6).

$$Sim((U_1, V_1), (U_2, V_2)) = \frac{|(U_1 \bigcap U_2)|}{x} * w + \frac{M(V_1, V_2)}{y} * (1 - w) \tag{6}$$

where x and y have the highest value between the cardinalities of the sets U_1, U_2, and V_1, V_2, correspondingly. w is a weight lying between 0 and 1 to enrich flexibility.

4 Implementation and Performance Evaluation

To investigate the performance of the proposed education video recommendation framework, four independent datasets, namely the Video Recommendations Based on Visual Features Extracted with Deep Learning dataset by Kvifte et al. [20], Statista's Global online learning video viewership reach 2021, by region dataset [21], Panoramic video in an education-A systematic literature review from 2011 to 2021 dataset by Heng Luo [22], and Dataset for Instructor Use of Educational Streaming Video Resources by Horbal et al. [23] are all individually used to analyze the efficacy of the suggested framework. However, these datasets cannot be used as it is. Most of these base datasets were video-driven, but some were only literature driven. Since the video content is insufficient, the Collection of documents on the digitisation of higher education and research in Switzerland (1998–2020) by Sophie et al. [24] was also considered. The terms involved in the Panoramic video in education and Collection of documents on the digitisation of higher education and research in Switzerland were used. Several live educational video resources were crawled. Apart from these, indexes from standard textbooks in journalism, psychology, public policy, English literature, and digital humanities were considered. The indexes were alone parsed and crawled from these textbooks. These indexes were used to crawl educational video resources from a few other educational platforms like Coursera and the wide range of content on the world wide web. However, along with the crawled videos, the subsequent text was crawled, and tags were extracted. In total, 73,812 unique external videos were crawled, categorized, and tagged. Moreover, the videos or terms available in the independent datasets were used to crawl the videos, and another 56,418 categorized videos were integrated into the framework. However, not all the videos present in the datasets were considered. All these videos were categorized, and at least eight to ten annotations were included. The videos were rearranged such that the videos with similar annotations or categories were prioritized above.

Implementation was carried out using Google's Colaboratory IDE in a Intel Core i7 computer with a RAM of 32 GB and a clock speed of 3.6 Giga Hertz. Python's NLTK framework was employed for carrying out language processing tasks. Google's knowledge bases were used either by encompassing the API directly or by means of SPARQL query.

The performance of the proposed SVRCI, a semantically driven paradigm for video recommendation that incorporates collective intelligence, is compared using precision, accuracy, recall, and F-Measure percentages, False Discovery Rate (FDR), and Normalized Discounted Cumulative Gain (nDCG) as preferred parameters. To gauge how relevant the findings are, precision, recall, accuracy, and F-Measure are employed. The FDR recognises how many false positives the approach has acknowledged. The heterogeneity of the model's output is quantified by the nDCG.

$$Precision = \frac{Retrieved\ Videos \cap Relevant\ Videos}{Retrieved\ Videos} \tag{7}$$

$$Recall = \frac{Retrieved\ Videos \cap Relevant\ Videos}{Relevant\ Videos} \tag{8}$$

$$Accuracy = \frac{Precision + Recall}{2} \tag{9}$$

$$F - Measure = \frac{2 \times Precision \times Recall}{Precision + Recall} \tag{10}$$

$$FDR = 1 - Positive\ Predicted\ Value \tag{11}$$

Equations (7), (8), (9), (10) and (11) depict precision, recall, accuracy, F-Measure and FDR respectively which are used as standard metrics.

Table 1. Assessment of the proposed SVRCI's performance in relation to other options

Search technique	Average precision %	Average recall %	Average accuracy %	Average F-measure %	FDR	nDCG
DEMVR [1]	88.22	90.04	89.13	89.12	0.12	0.88
DVREB [2]	88.91	90.93	89.92	89.91	0.11	0.87
FLVR [3]	88.11	89.43	88.77	88.76	0.12	0.84
VRKGCF [4]	92.13	93.18	92.65	92.65	0.08	0.96
Proposed SVRCI	95.66	96.81	96.23	96.23	0.04	0.99

The proposed SVRCI model is baselined with DEMVR, DVREB, FLVR, and VRKGCF models. The same setting is used to test the baseline models for the same number of queries and the same dataset as the proposed framework to quantify and compare the results produced by the proposed framework. Table 1 indicates that the proposed SVRCI achieves the highest value of precision, recall, accuracy, and F-Measure of 95.66%, 96.81%, 96.23%, and 96.23% correspondingly, with the lowest FDR of 0.04 and the highest nDCG of 0.99. The DEMVR model obtains a precision of 88.22%, recall of 90.04%, accuracy of 89.13%, F-measure of 89.12%, and an FDR of 0.12 and nDCG of 0.88. The DVREB model yields a precision of 88.91%, recall of 90.93 accuracy of 89.92%, and an F-measure of 89.91% with FDR and nDCG of 0.11 and 0.87, respectively. The FLVR model furnishes a precision of 88.11%, recall of 89.43%, accuracy of 88.77%, F-measure of 88.76%, an FDR of 0.12, and an nDCG of 0.84. The VRKGCF model attains a precision of 92.13%, recall of 93.18%, accuracy of 92.65%, F-measure of 92.65%, an FDR of 0.08, and an nDCG of 0.96.

The reason why the proposed SVRCI performs the best is because it is a knowledge-centric semantically inclined framework for video recommendation which is driven by a query wherein the query is sequentially and strategically enriched using entity enrichment through the Google's KG API, Structured Topic Modeling, and ontology alignment. Entity enrichment through Google's KG API yields and loads lateral and relevant knowledge graphs and subgraphs. STM further discovers topics that are highly relevant to the framework. An ontology already being generated is aligned by using strategic ontology alignment techniques using cosine similarity with a specific threshold. This ontology alignment ensures a strategic growth of queries in a sequential manner.

Furthermore, the dataset is controlled by classification using logistic regression, a feature control classifier where the features are yielded from the sequential knowledge derived. The dataset is automatically classified GRUs. Differential classification based on a deep learning auto handcrafted feature selector GRU with a machine learning feature control logistic regression classifier ensures differential classification to increase the heterogeneity and further semantic similarity computation using cosine similarity and TSS with differential thresholds for matching instances and further incorporation of concept similarity with a certain specified threshold ensures that the proposed education video recommendation framework yields the highest precision, accuracy, recall, and F-measure. Robust relevance computation mechanism in terms of semantic similarities like cosine similarity, TSS, concept similarity, usage of strategic models like STM, alignment of lateral ontologies, and incorporation of auxiliary knowledge through knowledge graphs using knowledge graph API and having two differential classification systems in the framework ensures that the suggested SVRCI paradigm fares better than the baseline models.

The DEMVR is a video recommendation model with autoencoding and convolutional text networks. The text networks add a large amount of knowledge fed at a vast scale. The deep learning mechanism learns from this knowledge. However, when a surplus amount of knowledge without relevancy factoring or relevancy controlling is fed into a deep learning model, it results in an overfitting problem where the relevance of the results is lost. As a result, DEMVR does not perform as expected. The DVREB model, a video recommendation system using big data clustering and extracted words, is keyword driven, and big data cluster is used to add data. However, knowledge derivation reasoning is not present. Data is fed as it is channelized into keywords. The system overtrains data but lacks the reasoning capability in knowledge, and there are no learning methods, nor are there strong relevance computation mechanisms in the models. Henceforth the DVREB fails to perform as expected.

The FLVR model also does not perform as expected because fuzzy logic is applied for eLearning. Fuzzy logic, when applied to a system of recommendation, approximate yielding of computation takes place. Due to this approximate computation mechanism, the lack of appropriate scaling methods and lack of auxiliary knowledge addition makes this model lag to a greater extent. The VRKGCF is based on knowledge graph collaborative filtering for video recommendation. A knowledge graph yields a surplus amount of knowledge, but collaborative filtering requires rating every item. Item rating metrics have to be computed in the environment of collaborative filtering. Every video on the web cannot be rated. It is complicated to depend on a rating for recommendation because a single user need not see a relative rating. Also, the rating may not give good insights when a community is taken. Apart from this knowledge graph yields much auxiliary knowledge. However, the regulatory mechanism for the auxiliary knowledge and the learning mechanism to feed in the auxiliary knowledge is absent in this method. Henceforth, the VRKGCF model does not perform as expected.

The proposed SVRCI model yields the highest nDCG value because much auxiliary knowledge is fed based on staged auxiliary knowledge addition and a regulation mechanism is present. The VRKGCF model also yields a high nDCG value but not as high as the SVRCI because of knowledge graphs. In other models, knowledge is either absent or not appropriately regulated. Henceforth knowledge remains as data; therefore, the nDCG value remains low for the DEMVR, DVREB, and FLVR models.

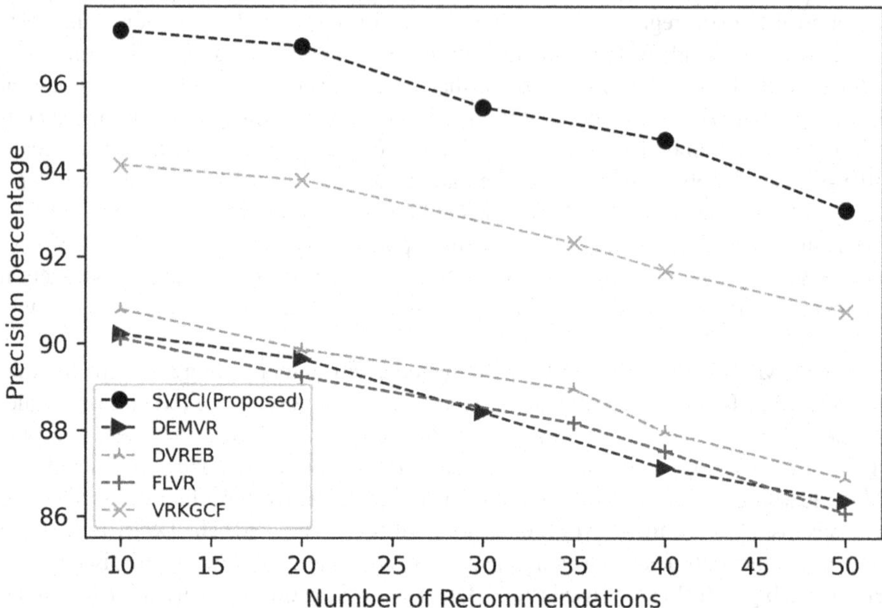

Fig. 2. Precision percentage vs number of recommendations distribution curve.

Figures 2 and 3 demonstrate the distribution curves for the precision percentage and the recall percentage in relation to the number of recommendations, respectively. The figures illustrate that the SVRCI occupies the uppermost position in the hierarchy for both the curves. Second, in the hierarchy is the VRKGCF. DVREB and DEMVR almost overlap the following two positions in the hierarchy. The FLVR occupies the lowermost position in the hierarchy in both the curves. The proposed SVRCI provides the best performance because of strategic models like STM, alignment of lateral ontologies, and incorporation of auxiliary knowledge by employing knowledge graphs using knowledge graph API and having two differential classifiers in the framework. GRUs and logistic regression classifier ensures differential classification to increase the heterogeneity and further semantic similarity computation using robust relevance computation mechanism in terms of semantic similarities like cosine similarity, TSS, and concept similarity, ensuring that the proposed SVRCI framework provides the best performance. When a surplus amount of knowledge is fed into the DEMVR model without relevancy factoring or relevancy controlling fed into a deep learning model, it results in an overfitting problem where the relevance of the results is lost.

The DVREB model overtrains data but lacks the reasoning capability in knowledge, and there are no learning methods, nor are there robust relevance computation mechanisms in the models. The FLVR model uses Fuzzy logic. Fuzzy logic, when applied to a system of recommendation, approximate yielding of computation takes place. Due to this approximate computation mechanism, the lack of appropriate scaling methods and lack of auxiliary knowledge addition makes this model lag to a greater extent. The VRKGCF

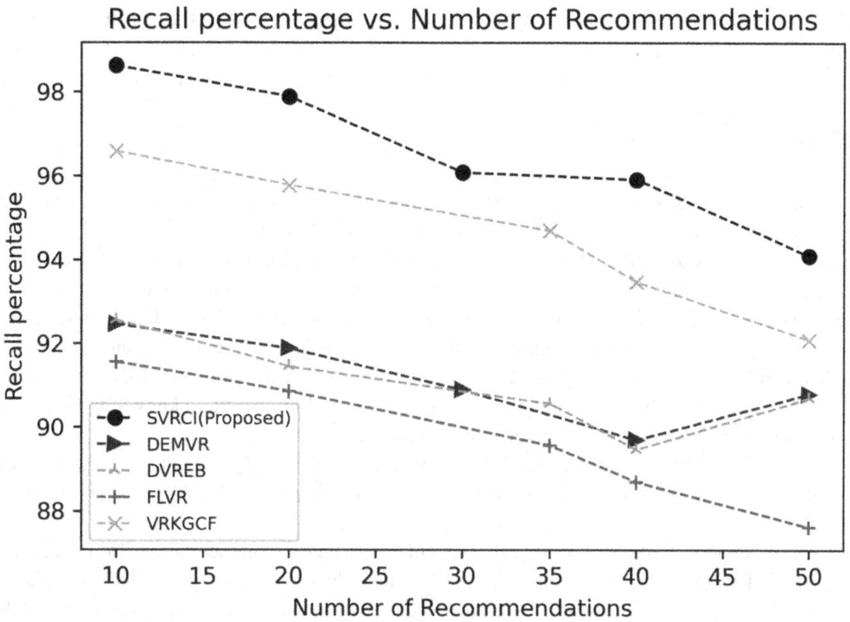

Fig. 3. Recall percentage vs number of recommendations distribution curve.

is based on knowledge graph collaborative filtering, which requires rating every item. It is complicated to depend on a rating for recommendation because a single user need not see a relative rating. Also, the rating may not give good insights when a community is taken. Apart from this knowledge graph yields much auxiliary knowledge. However, the regulatory mechanism for the auxiliary knowledge and the learning mechanism to feed in the auxiliary knowledge is absent in this method. These are the reasons why the suggested SVRCI framework performs better than the DEMVR, DVREB, FLVR, and VRKGCF models.

5 Conclusion

There is a need for knowledge-centric paradigms for recommending educational resources in the form of videos. This paper proposes the SVRCI model for recommending educational videos which is an ontology-driven framework for which the ontology alignment encompassing the concept similarity has been achieved along with structural topic modeling and loading knowledge graphs from Google's KG API for enriching the query terms. SVRCI integrates two distinct classifiers for classifying the dataset individually and independently by using the deep learning driven GRUs and the feature control logistic regression classifier to achieve variational heterogeneity in the classification of results and have variational perspectives in classification. Semantic similarity is computed using the cosine similarity, Twitter semantic similarity, and concept similarity with differential thresholds and several levels and stages to match classes and instances and recommend educational videos to the user in the increasing order of semantic similarity.

The proposed SVRCI framework achieved an overall average recall of 96.81%, with an FDR of 0.04 and an nDCG of 0.99.

References

1. Yan, W., Wang, D., Cao, M., Liu, J.: Deep auto encoder model with convolutional text networks for video recommendation. IEEE Access **7**, 40333–40346 (2019)
2. Lee, H.S., Kim, J.: A design of similar video recommendation system using extracted words in big data cluster. J. Korea Inst. Inf. Commun. Eng. **24**(2), 172–178 (2020)
3. Rishad, P., Saurav, N.S., Laiju, L., Jayaraj, J., Kumar, G.P., Sheela, C.: Application of fuzzy logic in video recommendation system for syllabus driven E-learning platform. In AIP Conference Proceedings, vol. 2336, no. 1, p. 040023. AIP Publishing LLC, March 2021
4. Yu, D., Chen, R., Chen, J.: Video recommendation algorithm based on knowledge graph and collaborative filtering. Int. J. Perform. Eng. **16**(12), 1933 (2020)
5. Deldjoo, Y., Elahi, M., Cremonesi, P., Garzotto, F., Piazzolla, P., Quadrana, M.: Content-based video recommendation system based on stylistic visual features. J. Data Semant. **5**(2), 99–113 (2016)
6. Lee, J., Abu-El-Haija, S.: Large-scale content-only video recommendation. In: Proceedings of the IEEE International Conference on Computer Vision Workshops, pp. 987–995 (2017)
7. Yan, M., Sang, J., Xu, C.: Unified YouTube video recommendation via cross-network collaboration. In: Proceedings of the 5th ACM on International Conference on Multimedia Retrieval, pp. 19–26, June 2015
8. Tripathi, A., Ashwin, T.S., Guddeti, R.M.R.: EmoWare: A context-aware framework for personalized video recommendation using affective video sequences. IEEE Access **7**, 51185–51200 (2019)
9. Li, Y., Wang, H., Liu, H., Chen, B.: A study on content-based video recommendation. In: 2017 IEEE International Conference on Image Processing (ICIP), pp. 4581–4585. IEEE, September 2017
10. Cui, L., Dong, L., Fu, X., Wen, Z., Lu, N., Zhang, G.: A video recommendation algorithm based on the combination of video content and social network. Concurr. Comput. Pract. Exp. **29**(14), e3900 (2017)
11. Huang, Y., Cui, B., Jiang, J., Hong, K., Zhang, W., Xie, Y.: Real-time video recommendation exploration. In: Proceedings of the 2016 International Conference on Management of Data, pp. 35–46, June 2016
12. Zhou, X., et al.: Enhancing online video recommendation using social user interactions. VLDB J. **26**(5), 637–656 (2017). https://doi.org/10.1007/s00778-017-0469-2
13. Duan, S., Zhang, D., Wang, Y., Li, L., Zhang, Y.: JointRec: A deep-learning-based joint cloud video recommendation framework for mobile IoT. IEEE Internet Things J. **7**(3), 1655–1666 (2019)
14. Roopak, N., Deepak, G.: OntoKnowNHS: ontology driven knowledge centric novel hybridised semantic scheme for image recommendation using knowledge graph. In: Villazón-Terrazas, B., Ortiz-Rodríguez, F., Tiwari, S., Goyal, A., Jabbar, M.A. (eds.) KGSWC 2021. CCIS, vol. 1459, pp. 138–152. Springer, Cham (2021). https://doi.org/10.1007/978-3-030-91305-2_11
15. Ojha, R., Deepak, G.: Metadata driven semantically aware medical query expansion. In: Villazón-Terrazas, B., Ortiz-Rodríguez, F., Tiwari, S., Goyal, A., Jabbar, M.A. (eds.) KGSWC 2021. CCIS, vol. 1459, pp. 223–233. Springer, Cham (2021). https://doi.org/10.1007/978-3-030-91305-2_17

16. Deepak, G., Surya, D., Trivedi, I., Kumar, A., Lingampalli, A.: An artificially intelligent approach for automatic speech processing based on triune ontology and adaptive tribonacci deep neural networks. Comput. Electr. Eng. **98**, 107736 (2022)
17. Krishnan, N., Deepak, G.: KnowSum: knowledge inclusive approach for text summarization using semantic allignment. In: 2021 7th International Conference on Web Research (ICWR), pp. 227–231. IEEE, May 2021
18. Arulmozhivarman, M., Deepak, G.: OWLW: ontology focused user centric architecture for web service recommendation based on LSTM and whale optimization. In: Musleh Al-Sartawi, A.M.A., Razzaque, A., Kamal, M.M. (eds.) EAMMIS 2021. LNNS, vol. 239, pp. 334–344. Springer, Cham (2021). https://doi.org/10.1007/978-3-030-77246-8_32
19. Surya, D., Deepak, G., Santhanavijayan: USWSBS: user-centric sensor and web service search for IoT application using bagging and sunflower optimization. In: Noor, A., Sen, A., Trivedi, G. (eds.) ETTIS 2021. AISC, vol. 1371, pp. 349–359. Springer, Singapore (2022). https://doi.org/10.1007/978-981-16-3097-2_29
20. Kvifte, T.: Video recommendations based on visual features extracted with deep learning (2021)
21. Statista: Global online learning video viewership reach 2021, by region (2022)
22. Luo, H.: Panoramic video in education-a systematic literature review from 2011 to 2021 (2022)
23. Horbal, A.: Dataset for instructor use of educational streaming video resources (2017)
24. Mützel, S., Saner, P.: Collection of documents on the digitisation of higher education and research in Switzerland (1998–2020) (2021)

Deep Learning Based Model for Fundus Retinal Image Classification

Rohit Thanki(✉)

IEEE, Gujarat Section, Rajkot, India
rohitthanki9@gmail.com

Abstract. Several types of eye diseases can lead to blindness, including glaucoma. For better diagnosis, retinal images should be assessed using advanced artificial intelligence techniques, which are used to identify a wide range of eye diseases. In this paper, support vector machine (SVM), random forest (RF), decision tree (DT), and convolutional neural network (CNN) methods are used to classify fundus retinal images of healthy and glaucomatous patients. This study tests various models on a small dataset of 30 high-resolution fundus retinal images. To classify these retinal images, the proposed CNN-based classifier achieved a classification accuracy of 80%. Furthermore, according to the confusion matrix, the proposed CNN model was 80% accurate for the healthy and glaucoma classes. In the glaucoma case, the CNN-based classifier proved superior to other classifiers based on the comparative analysis.

Keywords: Convolutional neural network · Fundus retinal image · Glaucoma · Deep learning · Image classification

1 Introduction

Glaucoma is characterized by increased pressure in the nerve structures of the eye and damage to their functionality. This disease causes blindness. According to the literature [1], this disease is estimated to cause blindness to 70.6 million people by 2020. The literature [2] indicates that glaucoma is India's third leading cause of blindness. Glaucoma affects around 12 million people in India, according to a survey conducted by the Glaucoma Society of India [2]. Many people suffer from this disease in India but are not identified or diagnosed. There is a process for processing retinal damage so that it is less evident to the patient. Blindness can be prevented by detecting this disease early and correctly diagnosing it. Glaucoma is diagnosed through various methods, including ophthalmoscopy, gonioscopy, and perimetry. Several factors can be used to diagnose glaucoma, including field vision tests, intraocular pressure, and optic nerve damage. A color fundus camera, optical coherence tomography (OCT), and visual test charts can be used to diagnose glaucoma. Different features can now be used to detect glaucoma, including cup-to-disk (CDR) and differences in disc rim thickness (ISNT). Recent research has focused on detecting and classifying glaucoma retinal images [3].

R. Thanki —Senior IEEE Member.

K. K. Patel et al. (Eds.): icSoftComp 2022, CCIS 1788, pp. 238–249, 2023.
https://doi.org/10.1007/978-3-031-27609-5_19

Modern glaucoma detection schemes classify and identify glaucomatous retinal images using image processing and machine learning algorithms [3]. The detection of glaucoma relies on both supervised and unsupervised machine learning algorithms. First, the enhanced retinal image is segmented by unsupervised machine learning. Then, a supervised machine learning algorithm classifies healthy and glaucomatous retinal images.

According to Acharya et al. [4], binary classifiers such as support vector machines (SVMs) and naive Bayes (NB) are helpful for the classification of retinal images. Gabor filters were used to extract Gabor features from retinal images. Gopalkrishan et al. [5] developed a least squares minimization-based classifier. A Gaussian smoothing filter was applied to extract retinal features. Chen et al. [6] proposed a classifier based on convolutional neural networks (CNNs). Singh et al. [7] developed the hypothesis thresholding method. The retinal features were extracted using morphological analysis and gray-level thresholding, two image processing methods. According to Isaac et al. [8], retinal images were classified using SVM and artificial neural networks (ANNs). This study classified retinal images based on their CDR features with these classifiers.

Claro et al. [9] developed a retinal image classification algorithm based on texture features. It was determined that texture features in this scheme were based on a gray-level co-occurrence matrix (GLCM). Various classifiers, such as multilayer perceptrons (MLPs), random committees, random forests, and support vector machines with radial basis functions, were used. Soman et al. [10] classified retinal images using wavelet features and supervised learning classifiers like SVM, random forest, and naive bayes. The K-nearest neighbor (KNN), the support vector machine (SVM), and the nave bayes algorithms can all be used for classifying retinal images, according to Singh et al. [11]. These approaches were evaluated using standard retinal datasets such as the Technische Fakultat. Sevastopolsky et al. [12] used a neural network (NN) to classify retinal images. Dey et al. [13] state that retinal images can be classified using SVM. To obtain statistical features of an image, this scheme combines color conversion, image resizing, Gaussian filter, and adaptive histogram equalization.

According to Septiarini et al. [14], retinal images can be identified using texture features and a supervised learning classifier. The texture features of retinal images were obtained using localization segmentation and histogram equalization. Texture features have been classified by naive Bayes, MLP, SVM, and K-NN classifiers. Using illumination correction, contrast enhancement, Radon transform, DWT, and PCA, Zou et al. [15] propose a glaucoma detection scheme based on hybrid features and supervised learning. To classify hybrid features of retinal images, SVM and random forest (RF) classifiers were applied. Manju et al. [2] developed a retinal image classification scheme using CDR features and thresholding. Devasia et al. [16] developed a retinal image classification scheme using hybrid features and various classifiers. CDR analysis, ISNT rules, neural networks, and cascade correlation neural networks (CCNNs) are all used in this scheme to classify data. Raghavendra et al. [17] developed a CNN-based classifier. A labeled retinal image classification scheme was evaluated and used. Recent studies have demonstrated the value of deep learning models, especially convolutional neural networks (CNNs) [18], which are considered the most effective models for image classification [19]. Medical images of various types have been classified using this model recently. However, it is essential to note that this model may only be effective for some medical image classification applications [19]. Based on traditional machine learning methods [6–16], several existing methods extract and analyze features using complex and time-consuming classifiers. Therefore, CNN classifiers are now widely used in many applications of medical image classification.

The paper proposes a CNN-based classification approach for retinal images of healthy patients and patients with glaucoma. Only a few previous studies have used deep learning to classify retinal images using existing literature [6, 17]. Based on existing studies, an innovative deep-learning model for retinal image classification is discussed. Support vector machines, random forests, and decision trees have been compared with the proposed CNN model's classification results to evaluate its performance. According to previous retinal image classification studies, glaucoma has been detected and classified based on retinal images [6–17]. However, some existing models could be more accurate [2, 5, 15]. Therefore, this proposed model improves the accuracy of the existing model for retinal image classification [2, 5, 15]. As a result of the proposed work, this has been achieved. This study's main limitation is the small number of retinal images in the dataset. The challenge of finding a dataset with large amounts of retinal images is also significant.

The rest of the paper is organized as follows: Sect. 2 discusses the methodology used in the proposed work. Section 3 discusses the details of the experimental setup with proposed scheme. Section 4 provides the results of the experiment and a discussion of them. Finally, Sect. 5 concludes with comments on the proposed work and future scope.

2 Used Methods

This section describes convolution neural networks with their working and traditional binary classifiers.

2.1 Basic of Convolutional Neural Network (CNN)

Convolutional neural networks (CNNs) are artificial neural networks with deep learning capabilities [18]. Several deep-learned layers are used in this model, including convolution layers, full-connection layers, and output layers. Additionally, many CNN models use softmax classifiers [20] as output layers. The CNN model shown in Fig. 1. This model has six layers: convolutional layer (C-layer), batch normalization layer, ReLU layer, pooling layer (P-layer), fully connected layer and softmax layer. Below you will find detailed information about each layer.

- **C Layer and ReLU Unit:** In the CNN model, this is the first step in extracting features from the input image. An image feature extraction process uses a small amount of information in an input image to determine the relationship between the pixels. Mathematically, the output is determined by two input values: the image pixel value and the filter mask value. For example, using the following relationship, the output can be obtained by the down relationship:

$$O = (M - f_M + 1) \times (N - f_N + 1) \quad (1)$$

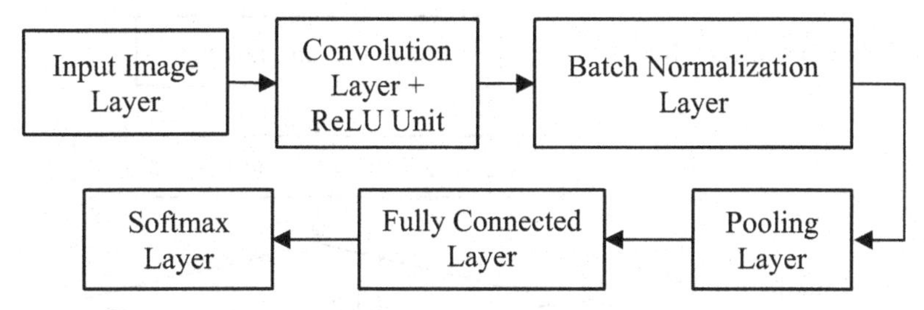

Fig. 1. Framework of basic CNN Model

where O is output, the dimension of an image pixel is $f_M \times f_N$ and dimension of a filter mask is M × N. After obtaining the value of output *O*, the obtained results are multiplied rectified linear unit (ReLU) [21] which removes negative values form the output before transmitting to the next C layer.

- **Batch Normalization Layer:** This layer improves model learning and performance [22]. As a result, adjacent layers are able to communicate with each other more efficiently.
- **Pooling Layer:** Using a pooling layer [22], each feature is upsampled or down sampled to reduce its dimensions. As part of CNN's dimension reduction process, max, sum, and average operations combine extracted features. For example, the max-pooling method calculates a feature's maximum value. This method effectively reduced the features derived from C layers.
- **Fully Connected Layer:** This layer [23] is composed of neurons connected by matrix multiplication. This layer converted each feature into a vector.
- **Softmax Layer:** This layer [17] applies a softmax function to the image features obtained. Mathematically, it is a normalized exponential function that converts features from 0 to 1. This layer also includes the class output label and loss function.

2.2 Working of Machine Learning Based Binary Classifiers

A binary classifier based on ML can be used to classify retinal images as shown in Fig. 2:

Step 1. Through image preprocessing, fundus retinal images are enhanced and reduced.
Step 2. A feature method is used to label retinal images and to extract retinal features.
Step 3. The whole dataset is used to create training and testing datasets.
Step 4. Using the training dataset to train classifiers.
Step 5. Classifiers are evaluated using a test dataset.

DT with different tree sizes, RF with different kernel functions, and SVM with different kernel functions are all used to classify the retinal image. These classifiers were chosen because they are widely used in image classification applications. This section does not cover RF, DT, and SVM details since they are covered in the literature [24, 25].

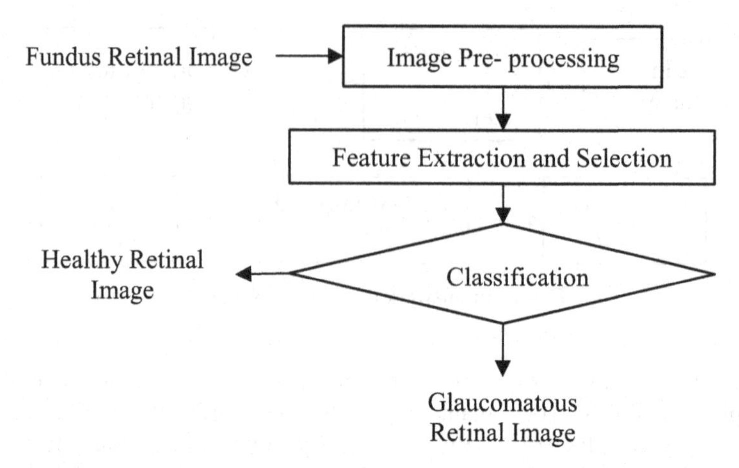

Fig. 2. Working of the ML based binary classifier

3 Experimental Setup

3.1 Dataset

This experiment labeled high-resolution color fundus retinal images as healthy or glaucomatous. The labeled dataset was developed by Budai et al. [26]. There are currently 15 images of healthy patients and 15 images of glaucoma patients in this dataset. Retinal image analysis experts and clinicians created this dataset from cooperating ophthalmology clinics. Sample images from the dataset are shown in Fig. 3.

3.2 Details of Proposed CNN Model

Figure 4 shows the proposed CNN model for retinal image classification. In the first step, the high-resolution color fundus retinal images (3504 × 2336) are labeled by a doctor before being resized (256 × 256) and divided into two classes. After assigning the training and test datasets to the CNN model, a cross-validation process is carried out. In the final step, we calculated the classification accuracy of our developed model using the testing dataset. The proposed architecture is inspired by Alexnet's architecture, which contains seven convolutional layers, seven pooling layers, three batch normalization layers, and three fully connected layers. Table 1 shows the output and parameters of the proposed CNN model.

In the proposed model, the kernels of each convolutional layer are 500, 61, 32, 32, 32, 32, and 62. Correspondingly, the output of features of each convolutional layer and pooling layer (C1–C7, P1–P7) are 256 × 256, 128 × 128, 128 × 128, 64 × 64, 64 × 64, 32 × 32, 32 × 32, 16 × 16, 16 × 16, 8 × 8, 8 × 8, 4 × 4, 4 × 4, and 2 × 2. There are three fully connected (FC) layers and one output layer in the proposed model. There are 2048 neurons in the first FC layer and 1024 neurons in the second. There are two neurons in the third FC layer, giving output as a two-class softmax output layer and input as a two-class softmax input layer.

Fig. 3. Sample retinal images (a) healthy (b) glaucomatous from the retinal dataset [26]

Table 1. Details of proposed CNN model with parameters

Layer	Output	Stride	Dropout ratio
Input	(3, 256, 256)	–	–
C1	(500, 256, 256)	0	–
P1	(500, 128, 128)	2	–
C2	(61, 128, 128)	0	–
P2	(61, 64, 64)	2	–
C3	(32, 64, 64)	0	–
P3	(32, 32, 32)	2	–
C4	(32, 32, 32)	0	–
P4	(32, 16, 16)	2	–
C5	(32, 16, 16)	0	–
P5	(32, 8, 8)	2	–
C6	(32, 8, 8)	0	–
P6	(32, 4, 4)	2	–
C7	(62, 4, 4)	0	–
P7	(62, 2, 2)	2	–
Flatten	248	–	–
FC1	2048	–	0.5
FC2	1024	–	0.5
FC3	2	–	–
Output	2	–	–

3.3 Parameters for Training of CNN Model

The learning rate, loss function, and learning method are hyperparameters used to determine the speed of each layer's learning when training a CNN model. Cross-entropy, learning rate, and epochs are other parameters used to train a model. For example, data complexity can be reduced using the cross-entropy loss function, which finds the entropy of the data. During training, the model is trained at a certain learning rate. A low rate

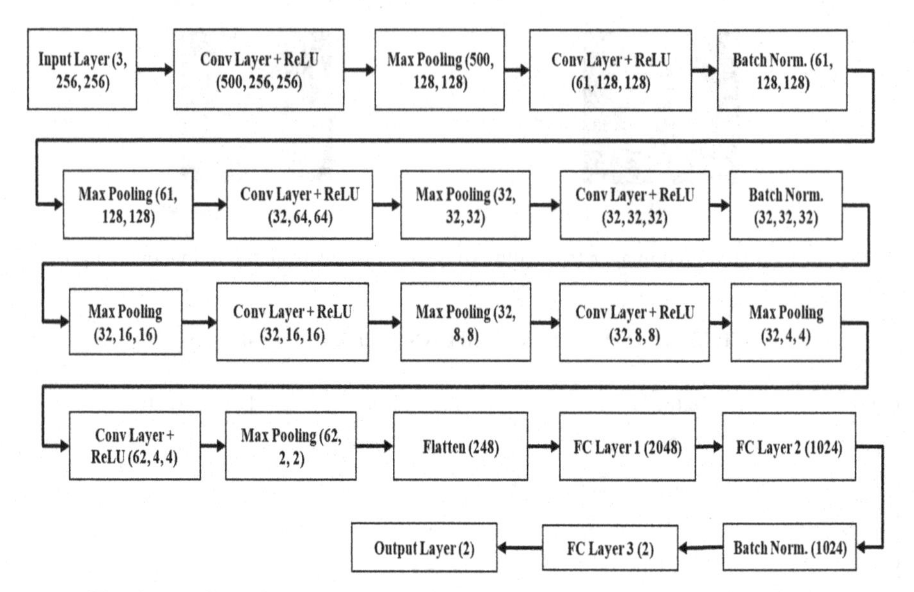

Fig. 4. Detail working of proposed CNN model for retinal image classification

of training will take longer. In this study, the learning rate is 0.1. Epochs are a way of validating a trained network against a testing dataset. The epochs used here are five.

3.4 Experimental Platform

The proposed work was conducted on an Intel Core i3-6006U 2 GHz processor with 8 GB of RAM. MATLAB 2016B and Deep Learning Studio were used for the experiment performance of the model.

4 Experimental Results and Discussion

A doctor labels each input color fundus retinal image (3504 × 2336), resize it (256 × 256), and divides it into two classes, such as healthy and glaucomatous, to evaluate the performance of the proposed CNN model. The output layer of the CNN model is softmax, which provides binary classification. The proposed model can be evaluated based on evaluation parameters such as confusion matrix [21], classifier accuracy, precision, sensitivity, specificity, and F1 score. In AI-based techniques for classifier performance evaluation, a confusion matrix [27] is used as a visualization tool. In binary classification, the confusion matrix gives the result for each class and compares it with the actual values predicted by the classifier. Table 2 shows a sample confusion matrix for the glaucomatous retinal class.

- **True Positive (TP):** In this case, a patient with a glaucomatous retinal image is predicted to have a glaucomatous retinal image.

Table 2. Sample confusion matrix for classification of glaucomatous retinal class

Predicted	Actual	
	Glaucomatous	Healthy
Glaucomatous	TP	FP
Healthy	FN	TN

- **True Negative (TN):** In this case, a patient with a healthy retinal image and a classified model will predict a healthy retinal image.
- **False Positive (FP):** The patient has a healthy retinal image, and the classified model predicts that the patient has a glaucomatous.
- **False Negative (FN):** A patient with glaucomatous retinal images and a classified model predicts a healthy retinal image.

Based on a confusion matrix, Table 3 shows the classification results for each class using the proposed CNN model. In 15 glaucomatous retinal labeled images, 12 retinal images are correctly predicted; 3 healthy retinal images are predicted as glaucomatous.

Table 3. Confusion matrix for classification results using proposed CNN-based classifier

Predicted value	Actual value	
	Glaucomatous	Health
Glaucomatous	12	3
Healthy	3	12

Based on Table 3, the proposed CNN model is evaluated based on accuracy, precision, sensitivity, specificity, and the F1 score for the glaucomatous retinal class (given in Table 4).

Table 4. Performance evaluation of proposed CNN-based classifier for glaucomatous retinal class

Accuracy	Precision	Sensitivity	Specificity	F1 score
0.8000	0.8000	0.8000	0.8000	0.8000

Tables 5, 6, 7 and 8 shows the classification results for the glaucomatous retinal class using traditional binary classifiers such as SVM with different kernel function, random forest, and decision tree with different size trees. Results in Tables 5, 6, 7 and 8 show that the maximum accuracy for classifying the glaucomatous retinal class is 0.7667 using a random forest classifier with 50 trees. The proposed CNN model has significant advantages compared to other classifiers according to the confusion matrix, such as equal classification of both classes by the model and a higher accuracy score of 80.0%, while its accuracy is inferior to that of SVM, RF, and DT-based classifiers.

According to Table 9, all classifiers achieve the same overall level of accuracy when classifying retinal images. Several recently published studies [2, 5, 15] have compared the proposed CNN model with their classification of glaucomatous retinal images. Table 9 summarizes the comparison based on the classifier, features, and maximum accuracy. Based on the comparison, this proposed CNN model outperformed existing work [2, 5, 15].

Table 5. Performance evaluation of SVM-based classifier for glaucomatous retinal class

Kernel function	Confusion matrix	Accuracy	Precision	Sensitivity	Specificity	F1 score
Linear	$\begin{bmatrix} 10 & 8 \\ 5 & 7 \end{bmatrix}$	0.5667	0.5556	0.6667	0.4667	0.6061
Quadratic	$\begin{bmatrix} 12 & 5 \\ 3 & 10 \end{bmatrix}$	0.7333	0.7059	0.8000	0.6667	0.7499
Polynomial	$\begin{bmatrix} 15 & 15 \\ 0 & 0 \end{bmatrix}$	0.5000	0.5000	1.0000	0.0000	0.6667
RBF	$\begin{bmatrix} 13 & 10 \\ 2 & 5 \end{bmatrix}$	0.6000	0.5652	0.8666	0.3333	0.6843
MLP	$\begin{bmatrix} 12 & 13 \\ 3 & 2 \end{bmatrix}$	0.4667	0.4800	0.8000	0.1333	0.6000

Table 6. Performance evaluation of RF based classifier for glaucomatous retinal class

Number of trees	Confusion matrix	Accuracy	Precision	Sensitivity	Specificity	F1 score
T = 5	$\begin{bmatrix} 8 & 8 \\ 7 & 7 \end{bmatrix}$	0.5000	0.5000	0.5333	0.4667	0.5161
T = 25	$\begin{bmatrix} 12 & 4 \\ 3 & 11 \end{bmatrix}$	0.7667	0.7500	0.8000	0.7333	0.7742
T = 50	$\begin{bmatrix} 12 & 5 \\ 3 & 10 \end{bmatrix}$	0.7591	0.7059	0.8000	0.6667	0.7499

Table 7. Performance evaluation of DT based classifier for glaucomatous retinal class

Number of trees	Confusion matrix	Accuracy	Precision	Sensitivity	Specificity	F1 score
T = 5	$\begin{bmatrix} 7 & 6 \\ 8 & 9 \end{bmatrix}$	0.5333	0.5385	0.4667	0.6000	0.5000
T = 25	$\begin{bmatrix} 6 & 7 \\ 8 & 9 \end{bmatrix}$	0.5000	0.4615	0.4000	0.5625	0.4286
T = 50	$\begin{bmatrix} 8 & 3 \\ 7 & 12 \end{bmatrix}$	0.6667	0.7273	0.5333	0.8000	0.6153

Table 8. Average accuracy obtained by each classifier

Classifiers	Accuracy (%)
SVM	57.33
RF	67.53
DT	56.67
CNN	80.00

Table 9. Comparison of prior work for classification of glaucomatous retinal images

Scheme	Features	Classifier	Maximum accuracy (%)
Gopalkrishan et al. (2015) [5]	CDR	Least square minimization	68
Manju's scheme (2018) [2]	CDR	Thresholding	73
Zou's scheme (2018) [15]	Hybrid	SVM and RF	74 and 78
Proposed	Deep	CNN	80

5 Conclusion

This study presented a CNN model for classifying high-resolution fundus retinal images into healthy and glaucomatous classes. Three traditional binary classifiers, SVM, RF, and DT, were used to evaluate the performance of CNN-based classifiers. Classifiers are evaluated using confusion matrices and other parameters such as accuracy, precision, sensitivity, etc. The results of the binary classifier are also satisfactory for the classification of glaucomatous retinal images. Based on comparisons with prior work, this proposed CNN-based classifier performed better than existing ones. A large retinal image dataset, such as ORIGA, can be used to evaluate the proposed CNN model in future work. Also, try some feature extraction methods, such as local binary patterns, co-occurrence matrix, etc., to obtain various features of retinal images before being fed to the CNN model and check the model's performance for this approach.

References

1. Kanse, S.S., Yadav, D.M.: Retinal fundus image for glaucoma detection: a review and study. J. Intell. Syst. **28**(1), 43–56 (2019)
2. Manju, K., Sabeenian, R.S.: Robust CDR calculation for glaucoma identification. Spec. Issue Biomed. Res. **2018**, S137–S144 (2018)
3. Das, S., Malathy, C.: Survey on diagnosis of diseases from retinal images. J. Phys. Conf. Ser. **1000**(1), 012053 (2018)
4. Acharya, U.R., et al.: Decision support system for glaucoma using Gabor transformation. Biomed. Signal Process. Control **15**, 18–26 (2015)
5. Gopalakrishnan, A., Almazroa, A., Raahemifar, K., Lakshminarayanan, V.: Optic disc segmentation using circular Hough transform and curve fitting. In: 2015 2nd International Conference on Opto-Electronics and Applied Optics (IEM OPTRONIX), pp. 1–4. IEEE, October 2015
6. Chen, X., Xu, Y., Yan, S., Wong, D.W.K., Wong, T.Y., Liu, J.: Automatic feature learning for glaucoma detection based on deep learning. In: Navab, N., Hornegger, J., Wells, W.M., Frangi, A.F. (eds.) MICCAI 2015. LNCS, vol. 9351, pp. 669–677. Springer, Cham (2015). https://doi.org/10.1007/978-3-319-24574-4_80
7. Singh, M., Singh, M., Virk, J.: A simple approach to Cup-to-Disk Ratio determination for Glaucoma Screening. Int. J. Comput. Sci. Commun. **6**(2), 77–82 (2015)
8. Issac, A., Sarathi, M.P., Dutta, M.K.: An adaptive threshold-based image processing technique for improved glaucoma detection and classification. Comput. Methods Programs Biomed. **122**(2), 229–244 (2015)
9. Claro, M., Santos, L., Silva, W., Araújo, F., Moura, N., Macedo, A.: Automatic glaucoma detection based on optic disc segmentation and texture feature extraction. CLEI Electron. J. **19**(2), 5 (2016)
10. Soman, A., Mathew, D.: Glaucoma detection and segmentation using retinal images. Int. J. Sci. Eng. Technol. Res. **5**(5), 1346–1350 (2016)
11. Singh, P., Marakarkandy, B.: Comparitive study of glaucoma detection using different classifiers. Int. J. Electron. Electr. Comput. Syst. **6**(7), 223–232 (2017)
12. Sevastopolsky, A.: Optic disc and cup segmentation methods for glaucoma detection with modification of U-Net convolutional neural network. Pattern Recogn. Image Anal. **27**(3), 618–624 (2017). https://doi.org/10.1134/S1054661817030269
13. Dey, A., Dey, K.N.: Automated glaucoma detection from fundus images of eye using statistical feature extraction methods and support vector machine classification. In: Bhattacharyya, S., Sen, S., Dutta, M., Biswas, P., Chattopadhyay, H. (eds.) Industry Interactive Innovations in Science, Engineering and Technology. LNNS, vol. 11, pp. 511–521. Springer, Singapore (2018). https://doi.org/10.1007/978-981-10-3953-9_49
14. Septiarini, A., Khairina, D.M., Kridalaksana, A.H., Hamdani, H.: Automatic glaucoma detection method applying a statistical approach to fundus images. Healthc. Inform. Res. **24**(1), 53–60 (2018)
15. Zou, B., Chen, Q., Zhao, R., Ouyang, P., Zhu, C., Duan, X.: An approach for glaucoma detection based on the features representation in radon domain. In: Huang, D.-S., Jo, K.-H., Zhang, X.-L. (eds.) ICIC 2018. LNCS, vol. 10955, pp. 259–264. Springer, Cham (2018). https://doi.org/10.1007/978-3-319-95933-7_32

16. Devasia, T., Jacob, K.P., Thomas, T.: Automatic early stage glaucoma detection using cascade correlation neural network. In: Satapathy, S.C., Bhateja, V., Das, S. (eds.) Smart Intelligent Computing and Applications. SIST, vol. 104, pp. 659–669. Springer, Singapore (2019). https://doi.org/10.1007/978-981-13-1921-1_64
17. Raghavendra, U., Fujita, H., Bhandary, S.V., Gudigar, A., Tan, J.H., Acharya, U.R.: Deep convolution neural network for accurate diagnosis of glaucoma using digital fundus images. Inf. Sci. **441**, 41–49 (2018)
18. Rawat, W., Wang, Z.: Deep convolutional neural networks for image classification: a comprehensive review. Neural Comput. **29**(9), 2352–2449 (2017)
19. Wang, Y., et al.: Classification of mice hepatic granuloma microscopic images based on a deep convolutional neural network. Appl. Soft Comput. **74**, 40–50 (2019)
20. Duan, K., Keerthi, S.S., Chu, W., Shevade, S.K., Poo, A.N.: Multi-category classification by soft-max combination of binary classifiers. In: Windeatt, T., Roli, F. (eds.) MCS 2003. LNCS, vol. 2709, pp. 125–134. Springer, Heidelberg (2003). https://doi.org/10.1007/3-540-44938-8_13
21. Jia, Y., et al.: Caffe: convolutional architecture for fast feature embedding. In: Proceedings of the 22nd ACM international conference on Multimedia, pp. 675–678. ACM, November 2014
22. Scherer, D., Müller, A., Behnke, S.: Evaluation of pooling operations in convolutional architectures for object recognition. In: Diamantaras, K., Duch, W., Iliadis, L.S. (eds.) ICANN 2010. LNCS, vol. 6354, pp. 92–101. Springer, Heidelberg (2010). https://doi.org/10.1007/978-3-642-15825-4_10
23. Serre, T., Wolf, L., Poggio, T.: Object recognition with features inspired by visual cortex. Department of Brain and Cognitive Sciences, Massachusetts Institute of Technology, Cambridge (2006)
24. Thrun, S., Pratt, L. (eds.): Learning to Learn. Springer, New York (2012)
25. Kotsiantis, S.B., Zaharakis, I., Pintelas, P.: Supervised machine learning: a review of classification techniques. In: Emerging Artificial Intelligence Applications in Computer Engineering, vol. 160, pp. 3–24 (2007)
26. Budai, A., Bock, R., Maier, A., Hornegger, J., Michelson, G.: Robust vessel segmentation in fundus images. Int. J. Biomed. Imaging **2013** (2013)
27. Ding, J., Hu, X.H., Gudivada, V.: A machine learning based framework for verification and validation of massive scale image data. IEEE Trans. Big Data **7**(2), 451–467 (2017)

Reliable Network-Packet Binary Classification

Raju Gudla(✉), Satyanarayana Vollala, and Ruhul Amin

Department of Computer Science and Engineering, Dr. SPM International Institute of Information Technology, Naya Raipur, Chhattisgarh 492 002, India
{raju.satya,ruhul}@iiitnr.edu.in
rajukits2008@gmail.com

Abstract. A network packet identification and classification is a fundamental requirement of network management to maintain the quality of service, quality of experience, efficient bandwidth utilization, etc. This becomes increasingly significant in light of the Internet's and online applications' rapid expansion. With the advent of secure applications, more and more encrypted traffic is proliferated on the internet. Specifically, peer-to-peer applications with user-defined protocols severely affect network management. So there is necessary to identify and classify encrypted traffic in a network. To overcome this, our proposed approach network-packet binary classification is implemented to classify the network traffic as encrypted or compressed packets, with better classification accuracy and with the usage of a limited amount of classification time. To achieve this, our model uses a Decision tree classifier with one of the efficient feature selection methods, Autoencoder. Our experimental results show that our model outperforms the most state-of-the-art methods in terms of classification accuracy. Our model achieved 100% classification accuracy within 0.009 s of processing time.

Keywords: Network packet classification · Machine learning algorithms · Dimensionality reduction techniques · Classification time

1 Introduction

As the amount of available information flows in computer networks grows, the problem of analyzing that information becomes more difficult, which can lead to delay in communication, especially in online and mission-critical systems [1,2]. The classification of network traffic is essential in network administration to improve the quality of service (QoS), quality of experience (QoE), bandwidth utilization, etc. In the early days of the internet, it was possible to perform traffic classification using port-based techniques and payload-based techniques or deep packet inspection (DPI) [3,4]. But, due to the proliferation of advanced

K. K. Patel et al. (Eds.): icSoftComp 2022, CCIS 1788, pp. 250–262, 2023.
https://doi.org/10.1007/978-3-031-27609-5_20

technologies in internet communication, like peer-to-peer applications with user-defined protocols, unregistered or dynamic port number usage, secure socket layer (SSL), secure shell (SSH), transport layer security (TLS), a virtual private network (VPN), the onion routing (Tor), etc., these lead to diminished network performance besides providing security and privacy to the network communications. Due to the application of these advanced encryption techniques to internet traffic, it is not possible to perform traffic classification using port-number-based and payload-based techniques because all internet traffic is encrypted. Also with DPI, data privacy preservation is omitted, which leads to legal issues in the case of confidential information [5].

A method for identifying and classifying network packets, ICLSTM is implemented. It uses neural networks to retrieve relevant traffic features, to avoid feature selection. Also mentioned is the approach to maintain the balanced information in the dataset [6]. In another approach, instead of using information from manually collected traffic features, it provides a technique to identify network traffic; it utilizes data related to temporal and spatial characteristics available in a network flow [7]. A reliable proposed encrypted traffic classification technique is implemented using path signature characteristics. Also, it consists of a traffic route based on the characteristics of bidirectional client-server communication to retrieve characteristics [8]. An approach discusses flow spatio-temporal features, which differentiate Tor and VPN network traffic. This set of features is a collection of properties that consists of timing components and packet length, which are related to identifying the traffic over the Tor and VPN communication networks [9]. The alternate solutions for encrypted traffic classification are statistical and machine learning-based methods. In these types of methods, statistical and time-related features of packets are considered for classification. Each packet has approximately 84 features apart from the label, i.e., the class of the packet. The set of 84 features consists of flow, source internet protocol address, source port number, destination internet protocol address, destination protocol number, protocol name, packet size, etc.

As far as network traffic classification is concerned, most internet traffic is encrypted. Our view is witnessed by [10] stating, "Google estimates that 95% of its internet traffic uses the encrypted HTTPS protocol, and most industry analyst firms conclude that between 80 to 90% of network traffic is encrypted today. This is a significant step forward for data integrity and consumer privacy". So, according to this statement, there is a necessity for encrypted network traffic identification and classification. And it leads to grabbing our motivation to develop a network-packet binary classification model, i.e., the identification and classification of internet traffic into encrypted or compressed (or unencrypted).

The major contributions of our proposed approach are as follows :

- In our approach, network packet binary classification is used to accurately distinguish encrypted and compressed packets because we achieved 100% accuracy in the case of Decision Tree with Autoencoder with 512 KB packets.

- With the help of machine learning algorithms, our model outperforms statistical based approaches like the High entropy distinguisher(HEDGE) [1].
- In our model evaluation apart from accuracy, we considered classification time to witness efficient timeliness.
- We evaluated five machine learning algorithms with the help of two efficient feature selection methods; we achieved promising results in significantly less time.

The rest of this paper is organized into the following sections. Section 2 reviews some important and recent work on network traffic binary classification. Section 3 presents our proposed approach for network-packet binary classification. The results of the proposed method for different ways of feature selection, analysis based on the classification time, and comparison with state-of-the-art methods are described in Sect. 4. Section 5 discusses the conclusion and possible future scope of the work.

2 Related Work

This section provides an review of the essential network packet categorization approaches. We can divide these methods into three ways, such as i) Port-number based techniques, ii) Deep packet inspection(DPI) or payload-based techniques, and iii) Machine learning-based techniques.

Here is a detailed review of the above-listed categories.

2.1 Port-number Based Techniques

The basic method in traffic classification by checking with the port number in an unencrypted packet header. The transport layer's 16-bit port numbers, which contain data about source and destination ports, are the requirements of the port-based classification system. This port number consists, of one of the simplest and quickest methods for identifying network traffic. In this method, the port number found in a packet is compared with the list of standard ports from Internet Assigned Numbers Authority (IANA) [11] to identify the type of data available in the packet payload. This technique is frequently used in networking devices like firewalls due to the quick extraction procedure [12]. The accuracy of this method has been greatly diminished by port forwarding, random port assignments, peer-to-peer user-defined protocols, and the usage of dynamic port numbers in communications.

2.2 Deep Packet Inspection(DPI) or Payload Based Techniques

The deep packet inspection technique for traffic classification looks for the specific pattern or signature in the payload of each packet flow. We can perform this DPI unless traffic is encrypted. The DPI is based on the data available in the payload. This technique is inefficient because most communication protocols may change the signature or pattern in the information over time, so it is impossible to update all the signature patterns. Also, it requires a large amount of computation resources, and classification accuracy depends on the underlined implementation. User privacy concerns are some of the most crucial points of consideration in terms of DPI. A technique has been developed to identify encrypted HTTPS traffic without performing decryption [13].

2.3 Machine Learning-Based Techniques

In this approach, packet classification identifies similar patterns in traffic flow packets. In most applications, there is a probability that similar patterns exist in the different packets of the same application data. These patterns are used to identify the application with a similar cluster of patterns. Machine learning-based classification approaches overcome the problem of depending on the payload. It uses payload-independent features such as packet inter-arrival time, packet length, flow duration, etc [14]. A subset of National Institute of Standards and Technology (NIST) tests and two flavors of Entropy test for finding the randomness of information to classify as compressed or encrypted packet [1]. And they achieved 68%, and 94% accuracy in the case of 4 KB and 64 KB packets, respectively. In machine learning-based classification approaches, the set of features is extracted from each packet and forms an array of values followed by a label. These feature sets are used for training and testing using different classifiers. An approach to identify encrypted VoIP traffic using NIST test suite and ENT test program and achieved 100% accuracy for distinguishing compressed versus encrypted VoIP [15]. A model for Encryption/Compression Distinguisher (ENCoD) to differentiate even smaller-sized packets as low as 512 bytes. It performs the binary and multi-class classification of the information using pattern learning [16].

A method Flow-Based Relation Network (RBRN) proposes an application of a meta-learning approach to encrypted traffic classification. This end-to-end classification model learns features from raw traffic flows and classifies them in a unified framework. Also, they designed "hallucinator" to convert an imbalanced to a balanced dataset by producing synthetic sample space [17]. A traffic classification method with Gaussian mixture models and hidden Markov models. This approach uses the packet's inter-packet arrival time and size to classify and analyze flow feature distribution and timing patterns. A novel technique for choosing model parameters that can improve classification outcomes while using less processing power [18]. An approach is discussed to identify and differentiate the type of protocol and application identification in Encrypted Two-Label Classification using CNN (ETCC) method. The first part identifies the protocol used for encrypted traffic and then uses the classifiers to classify as a type of application. ETCC achieved an accuracy of 97.65% [19]. Encrypted mobile communication traffic is identified in three ways i) Application identification, ii) In-application activity, and iii) Application activity. This approach uses 1D-CNN to process the first packets to extract spatial features and also bidirectional LSTM is used to acquire the temporal features from packet length and the flow direction [20].

3 Proposed Approach

In this work, we developed a framework using five machine learning algorithms, namely Support Vector Machine(SVM), Decision Tree(DT), Random-Forest(RF), Logistic Regression (LR), and K-Nearest Neighbors(K-NN), along with dimensionality reduction techniques such as Principal Component Analysis (PCA) and Autoencoder(AE).

3.1 Dimensionality Reduction Techniques

The dimensionality reduction technique consists of feature extraction and feature selection to convert higher-dimensional sample space to lower-dimensional sample space. Our proposed approach uses the following :

Principal Component Analysis (PCA): PCA [21] is a method of reducing the feature space of an information base by dividing a large group of components into a smaller one that still contains the essential data. In general, accuracy suffers when a dataset has fewer components. However, the dimensionality reduction only gives up a small amount of accuracy in exchange for ease. When data from a lower-dimensional space is projected onto input from higher-dimensional space, the lower-dimensional space input variance should increase more. Machine learning models can process quickly by reducing the feature space without information loss, which also helps us save classification time. It makes an effort to minimize the characteristics while retaining more information. The variance tells us how much of the total variation is attributable to each eigenvalue and significant component.

By applying the following formula, we can find the variation of each component:

$$Variance(PC_k) = \left(\frac{eigenvalueofPC_k}{\sum_{i=1}^{p} eigenvalueofPC_i} \right) \tag{1}$$

where p is the number of dimensions in primary feature set.

Autoencoder: Autoencoder [22] is a feed-forward neural network for the compression and decompression of feature space. It can change the data into a reduced state, which is used to reconstruct the information. It consists of three modules: Encoder, Code, and Decoder. The encoder module compresses the input and generates the code, which the decoder uses to recreate the information. An encoder mainly consists of a sequence of long-short-term-memory components. The decompression process reduces the loss rate from primary data to compressed data. This autoencoder removes the noise in the input data and collects only essential features. An autoencoder is used to encode linear as well as non-linear data.

3.2 Dataset

For this work, we use a dataset [23], that consists of five different sets, each with different-sized files of various file types (encrypted and compressed files of type Text, Image, PDF, MP3, Binary, Video, ZIP, GZIP, BZIP, and RAR), to simulate real-time network packets like files, with the file sizes of 64 KB, 128 KB, 256 KB, 512 KB, and 1024 KB. Each set contains an equal number of encrypted and compressed files of corresponding size and type. Figure 1 illustrate the dataset overview.

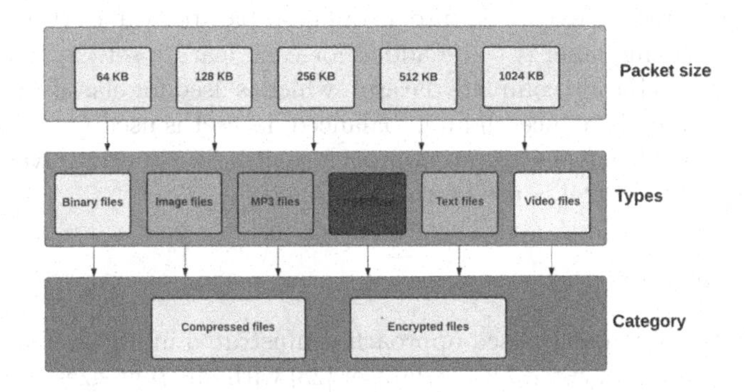

Fig. 1. Dataset Overview

Table 1. Number of files available in each class of dataset (packet size in KB)

Type of data	Class type	64 KB	128 KB	256 KB	512 KB	1024 KB
Binary file	Comp	1600	800	400	200	100
	Enc	1600	800	400	200	100
Image file	Comp	1600	800	400	200	100
	Enc	1600	800	400	200	100
MP3 file	Comp	1600	800	400	200	100
	Enc	1600	800	400	200	100
Text file	Comp	1600	800	400	200	100
	Enc	1600	800	400	200	100
PDF file	Comp	1600	800	400	200	100
	Enc	1600	800	400	200	100
Video file	Comp	1600	800	400	200	100
	Enc	1600	800	400	200	100

*Compressed file class (comp), Encrypted file class (enc).

All the category sets of files of different sizes are segregated into two types, viz., encrypted and compressed files, i.e., each type of data of different file sizes is combined to form half of the dataset. For example in Table 1 for Image category, dataset consists of separately (i.e. Encrypted or Compressed) 1600, 800, 400, 200, 100 files of 64 KB, 128 KB, 256 KB, 512 KB, 1024 KB files respectively.

3.3 Pre-processing and Labelling

This dataset's contents are captured in the TCP-Client-Server environment on a Local Area Network(LAN). On the client side, using Wireshark, all the data of a dataset is captured in the form of packet capture (PCAP) files. Then, from each PCAP file, features are extracted into a comma-separated values (CSV) file, and finally, each CSV file label is to be added for each feature set. At last, all CSV files are merged to form a complete dataset, which is used for classification. But a classifier's accuracy is reduced if an unbalanced dataset is used for classification. To overcome this disadvantage, an approach Synthetic Minority Over-sampling Technique (SMOTE) [24] is used for data.

3.4 Architecture

The architecture of the proposed approach is illustrated in Fig. 2. The collection of compressed and encrypted files of dataset [23] with different sizes is transferred from a source computer to a destination computer in a Local Area Network (LAN) using Client-Server socket programming modules. On the destination side, traffic must be collected as PCAP files for each category (compressed and encrypted).

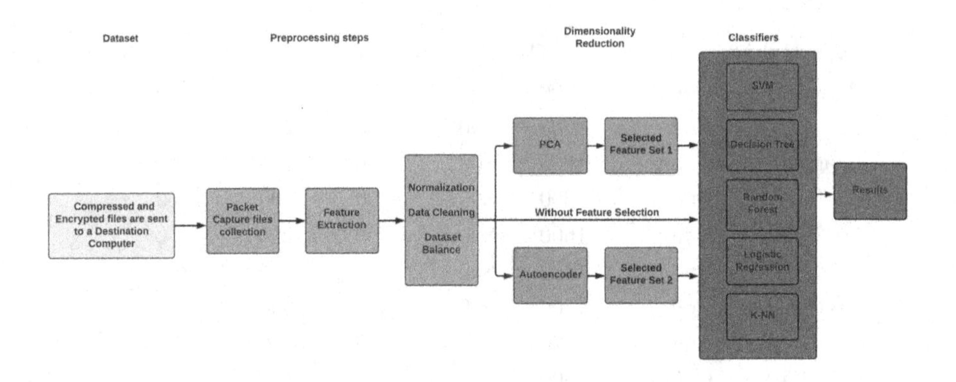

Fig. 2. Proposed approach Architecture

Then CICFlowmeter [25] is used to convert each PCAP file into a CSV of different traffic flows to extract the features of each network packet. Each packet label is to be added; the compressed packet label is '0', and for the encrypted packet, the label is '1'. Next, data cleaning is to be performed to remove packets with features of duplicate values, non-numerical values, NULL values, etc. After the data cleaning steps, check for an equal number of packets in each category to maintain a balanced dataset to achieve unbiased classification results. Due to data cleaning, a few packets are removed to maintain consistency within the dataset.

In our proposed approach to feature selection, dimensionality reduction methods like Principal Component Analysis (PCA) and Autoencoder are used to reduce the classification time in different classification models. And without feature selection, a complete feature set is used to identify and classify the packets. With and without feature selection, it is used to compare the performance of different classification approaches.

Finally, in each case of feature selection, all machine learning classifiers achieved optimal classification accuracy results in comparable classification times to those without feature selection.

4 Results and Analysis

We used the Ubuntu 16.04 operating system, Python 3.7, and 16 GB of memory to implement our proposed approach to network-packet binary classification. We use only one of the performance analysis metrics to evaluate the experimental results, i.e., accuracy, because we are performing binary classification of network packets. To evaluate the performance of our proposed approach, five machine learning algorithms: support vector machine(SVM), random forest(RF), decision tree(DT), K-nearest neighbors (K-NN), and logistic regression(LR) are used.

In our proposed model, classification accuracy without using any feature selection is illustrated in Fig. 3. SVM achieved 65% classification accuracy with 64 KB packets, and Decision Tree, K-NN achieved 83% classification accuracy with 128 KB packets. In the case of SVM, all 84 features' hyperplanes need to be created; therefore, SVM is not performing well without feature selection. In the case of a decision tree, all the features are organized in a tree data structure to identify the high information gain of each data point and split. In the case of K-NN, a maximum of 10 (i.e., $K = 10$) nearest neighbor data points are considered.

In our proposed model, classification accuracy with feature selection using PCA is illustrated in Fig. 4. Logistic regression achieved 67% classification accuracy with 128 KB packets; the Decision tree achieved 100% classification accuracy with 512 KB packets. In the case of logistic regression, the number of observations is less than the number of features in our model, leading to overfitting. And in the case of a decision tree, it outperforms the other classification models.

In our network packet binary classification model, classification accuracy with feature selection using Autoencoder is illustrated in Fig. 5. Logistic regression achieved 68% classification accuracy with 64 KB packets; the Decision tree achieved 100% classification accuracy with 512 KB packets. In the case of logistic regression, even though input space is compressed to a great extent due to the demerits of logistic regression, it is underperforming.

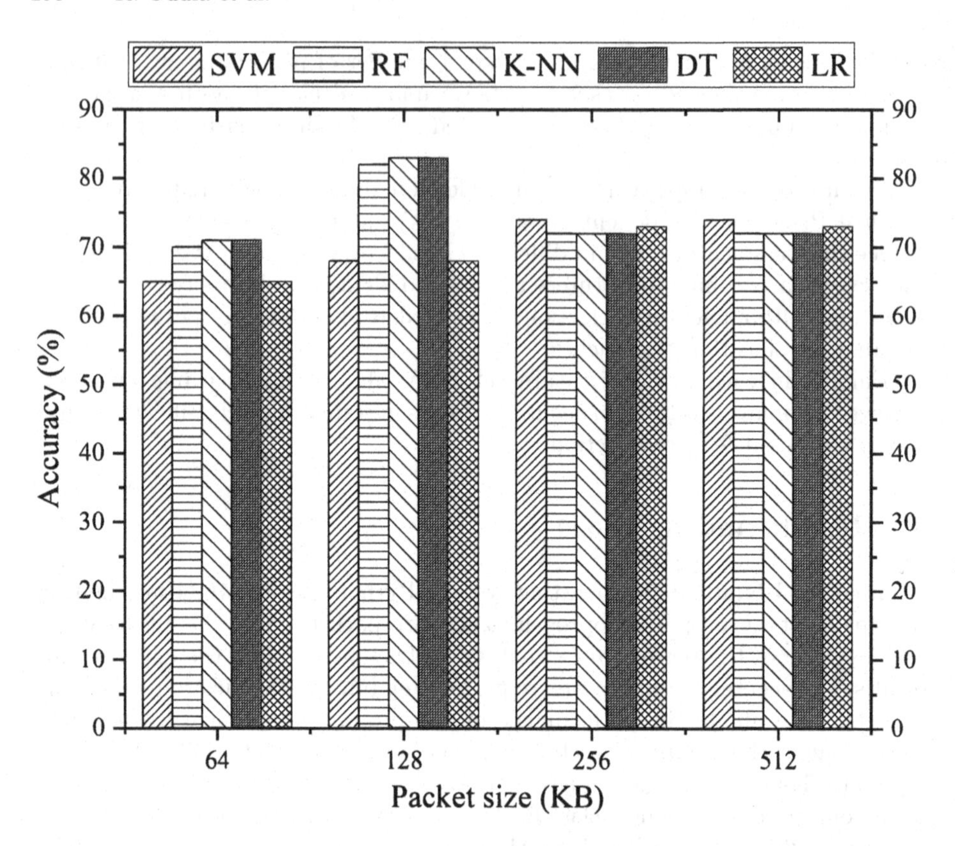

Fig. 3. Classification accuracy without feature selection

Table 2. Classification Time (in Seconds)

Classifier	Without FS				With PCA				With AE			
	64	128	256	512	64	128	256	512	64	128	256	512
SVM	16.127	6.734	0.980	0.507	16.000	13.350	1.350	0.020	15.000	6.750	0.400	0.970
DT	0.701	0.110	0.050	0.042	0.240	0.097	0.035	0.009	0.330	0.130	0.040	0.009
RT	0.742	0.350	0.117	0.115	0.840	0.404	0.180	0.053	1.230	0.280	0.090	0.028
LR	0.134	0.075	0.044	0.054	0.110	0.087	0.034	0.04	0.180	0.130	0.040	0.018
K-NN	0.127	0.109	0.023	0.036	0.270	0.273	0.082	0.08	0.250	0.230	0.040	0.022

*64, 128, 256, 512 packet size in KB.

In our network packet binary classification model, classification time analysis using different feature selection methods is illustrated in Table 2. The SVM classifies 64 KB packets in 16.127 s without feature selection; the decision tree classifies 512 KB packets with PCA or autoencoder in 0.009 s. The autoencoder outperforms in classification time. In the case without feature selection, the classification model needs to process all the data points in the input sample space and, based on the underlying architecture, directly affect the classification accuracy, such as by creating a hyperplane for each feature set. In the case of a

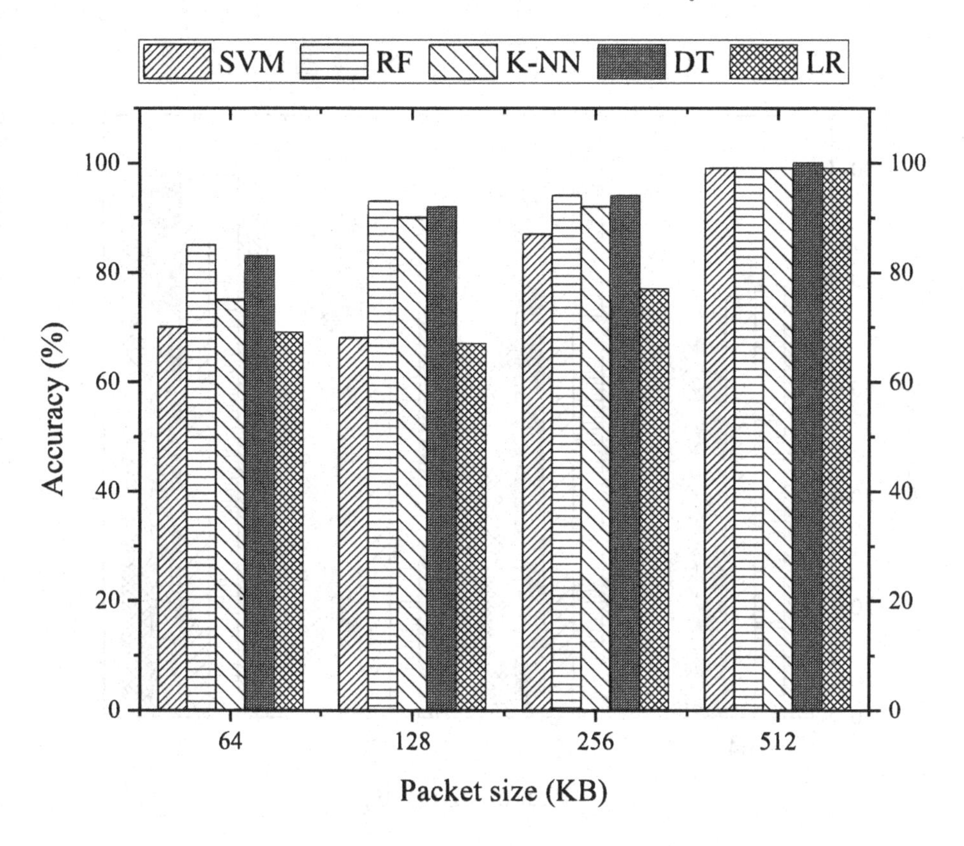

Fig. 4. Classification accuracy with PCA

decision tree, tree organization of data points, identification of information gain, and splitting each data point into leaf nodes play a vital role in achieving accurate and promising results with a limited set of computational resources like classification time and required memory.

4.1 Comparison

We compare our proposed approach's experimental results with state-of-the-art methods. In [1], a model named HEDGE was proposed to classify encrypted and compressed packets. It achieved 68% classification accuracy with 1 KB packets and 94 % classification accuracy with 64 KB packets by applying randomness tests like a subset of the NIST test suite. Those are the frequency within block test, the cumulative sums test, the approximate entropy test, the runs test, and two flavors of the ENT test program, such as the Chi-square test with absolute value and the Chi-square test with a percentage of confidence. Even though they achieved high classification accuracy with 64 KB packets, they used only statistical-based approaches instead of machine learning-based methods, which can handle large datasets. Also, with existing methods, it is impossible to classify

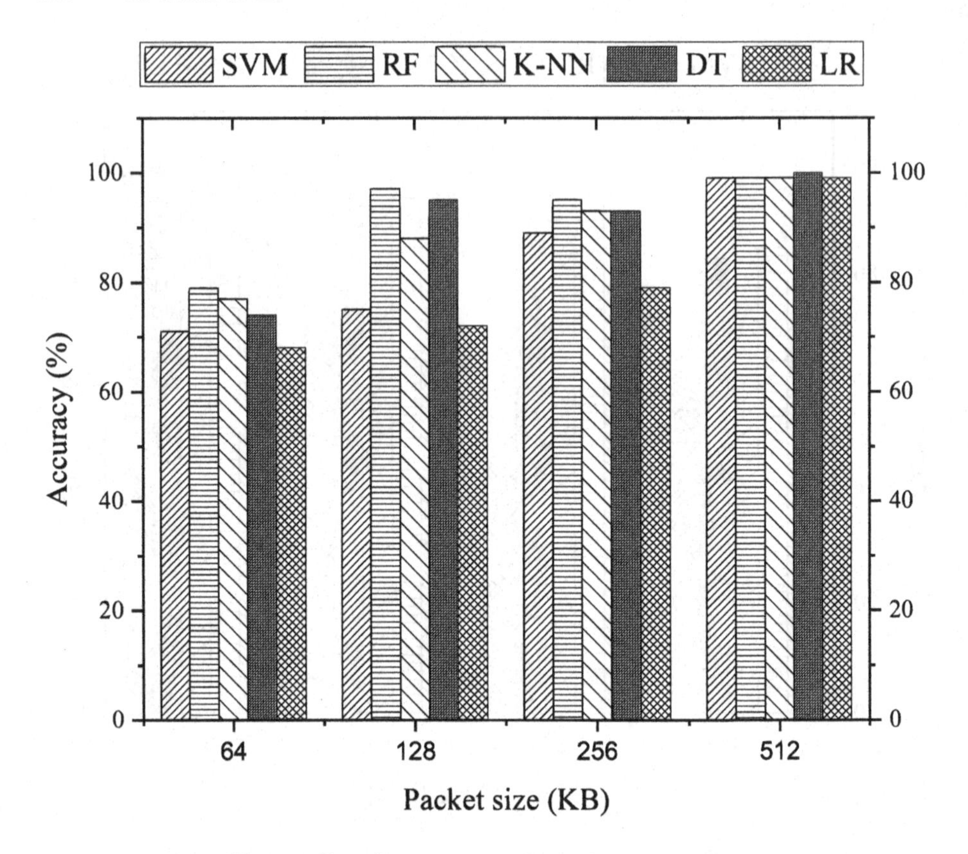

Fig. 5. Classification accuracy with autoencoder

a single message if a part of the message is encrypted and another part of the message is unencrypted, which is not a demerit in the case of machine learning-based classification techniques. And also, they have not used any feature selection methods for considering the classification time. As per our knowledge, none of the researchers considered the classification time as a parameter, as in our work.

5 Conclusion and Future Work

We developed a machine-learning model for network packet binary classification. It efficiently identifies and classifies encrypted and compressed packets from network traffic with better classification accuracy. In our approach, we used five machine learning methods: SVM, Random Forest, K-NN, Decision Tree, and Logistic regression, with three ways of feature selection and two different dimensionality reduction techniques i) Without feature selection, ii) With PCA, iii) With Autoencoder. These different ways of feature selection methods are used to extract and identify the most suitable set of features from the input sample space. Also used to achieve the results with a limited set of computational resources like processing time. We achieved 100% classification accuracy with the Decision tree with the Autoencoder feature selection method in the case of 512 KB packets of a dataset with a classification time of 0.009 s only. Our model's feature selection method, Autoencoder outperforms most of the cases of PCA.

Our model can classify network packets as encrypted or compressed. Still, in online network communications, a network can experience various content types like audio, video, text, binary, etc., for which different kinds of applications' traffic needs to be identified and classified. Online network traffic classification and application identification need to be precisely accomplished at line speed.

References

1. Casino, F., Choo, K.K.R., Patsakis, C.: HEDGE: efficient traffic classification of encrypted and compressed packets. IEEE Trans. Inf. Forensics Secur. **14**, 2916–2926, (2019)
2. Telikani, A., Gandomi, A.H., Choo, K.K.R., Shen, J.: A cost-sensitive deep learning-based approach for network traffic classification. IEEE Trans. Netw. Serv. Manage. **19**(1), 661–670 (2021)
3. Xu, C., Chen, S., Su, J., Yiu, S.M., Hui, L.C.: A survey on regular expression matching for deep packet inspection: applications, algorithms, and hardware platforms. IEEE Commun. Surv. Tutorials **18**(4), 2991–3029 (2016)
4. Xu, L., Zhou, X., Ren, Y., Qin, Y.: A traffic classification method based on packet transport layer payload by ensemble learning. In: IEEE Symposium on Computers and Communications (ISCC), pp. 1–6 (2019)
5. Song, W., et al.: A software deep packet inspection system for network traffic analysis and anomaly detection. Sens. J. **20**, 16–37 (2020)
6. Lu, B., Luktarhan, N., Ding, C., Zhang, W.: ICLSTM: encrypted traffic service identification based on inception-LSTM neural network. Symmetry J. MDPI **13**(6), 1080 (2021)
7. Roy, S., Shapira, T., Shavitt, Y.: Fast and lean encrypted Internet traffic classification. Comput. Commun. J. **186**, 166–173 (2022)
8. Xu, S.-J., Geng, G.-G., Jin, X.-B., Liu, D.-J., Weng, J.: Seeing traffic paths: encrypted traffic classification with path signature features. IEEE Trans. Inf. Forensics Secur. **17**, 2166–2181 (2022). https://doi.org/10.1109/TIFS.2022.3179955
9. Islam, F.U., Liu, G., Zhai, J., Liu, W.: VoIP traffic detection in tunneled and anonymous networks using deep learning. IEEE Access **9**, 59783–59799 (2021)
10. https://threatpost.com/decryption-improve-security/176613/ . Accessed 30 Jul 2022
11. Cotton, M., Eggert, L., Touch, J., Westerlund, M., Cheshire, S.: Internet assigned numbers authority (IANA) procedures for the management of the service name and transport protocol port number registry, RFC, pp. 1–33 (2011)
12. Qi, Y., Xu, L., Yang, B., Xue, Y., Li, J.: Packet classication algorithms: from theory to practice. In: INFOCOM, pp. 648–656 (2009)
13. Sherry, J., Lan, C., Popa, R.A., Ratnasamy, S.: Blindbox: deep packet inspection over encrypted traffic. In: ACM SIGCOMM Computer Communication Review, pp. 213–226 (2015)
14. Nguyen, T.T., Armitage, G.: A survey of techniques for internet traffic classification using machine learning. IEEE Commun. Surv. Tutorials **10**, 56–76 (2008)

15. Choudhury, P., Kumar, K.R.P., Nandi, S., Athithan, G.: An empirical approach towards characterization of encrypted and unencrypted VoIP traffic. Multimedia Tools Appl. **79**(1), 603–631 (2019). https://doi.org/10.1007/s11042-019-08088-w
16. De Gaspari, F., Hitaj, D., Pagnotta, G., De Carli, L., Mancini, L.V.: ENCOD: distinguishing compressed and encrypted file fragments. In: Kutyłowski, M., Zhang, J., Chen, C. (eds) Network and System Security. NSS 2020. Lecture Notes in Computer Science, vol. 12570, pp. 42–62. Springer, Cham (2020). https://doi.org/10.1007/978-3-030-65745-1_3
17. Zheng, W., Gou, C., Yan, L., Mo, S.: Learning to classify : a flow-based relation network for encrypted traffic classification. In: Proceedings of The Web Conference, pp. 13–22 (2020)
18. Yao, Z., Ge, J., Wu, Y., Lin, X., He, R., Ma, Y.: Encrypted traffic classification based on Gaussian mixture models and Hidden Markov Models. J. Netw. Comput. Appl. **166**, 102711 (2020). https://doi.org/10.1016/j.jnca.2020.102711
19. Li, Y., Lu, Y.: ETCC: encrypted two-label classification using CNN. Secur. Commun. Netw. **2021**, 1–11 (2021)
20. Zhang, H., Gou, G., Xiong, G., Liu, C., Tan, Y., Ye, K.: Multi-granularity mobile encrypted traffic classification based on fusion features. In: Lu, W., Sun, K., Yung, M., Liu, F. (eds.) SciSec 2021. LNCS, vol. 13005, pp. 154–170. Springer, Cham (2021). https://doi.org/10.1007/978-3-030-89137-4_11
21. Abdi, H., Williams, L.: Principal component analysis. Wiley Interdisc. Rev. Comput. Stat. **2**(4), 433–459 (2010)
22. Sakurada, M., Yairi, T.: Anomaly detection using autoencoders with nonlinear dimensionality reduction. In: Proceedings of the MLSDA 2014 2nd Workshop on Machine Learning for Sensory Data Analysis, pp. 4–11 (2014)
23. Casino, F., Hurley-Smith, D., Hernandez-Castro, J., Patsakis, C.: Dataset?: distinguishing between high entropy bit streams. Zenodo (2021). https://doi.org/10.5281/zenodo.5148542
24. Chawla, N.V., Bowyer, K.W., Hall, L.O., Kegelmeyer, W.P.: SMOTE: synthetic minority over-sampling technique. J. Artif. Intell. Res. **16**, 321–357 (2002)
25. Lashkari, A.H., Seo, A., Gil, G.D., Ghorbani, A.: CIC-AB: online ad blocker for browsers. In: International Carnahan Conference on Security Technology (ICCST), pp. 1–7. IEEE (2017)

SemKnowNews: A Semantically Inclined Knowledge Driven Approach for Multi-source Aggregation and Recommendation of News with a Focus on Personalization

Gurunameh Singh Chhatwal[1], Gerard Deepak[2(✉)], J. Sheeba Priyadarshini[3], and A. Santhanavijayan[4]

[1] Department of Electronics and Communication Engineering, University Institute of Engineering and Technology, Panjab University, Chandigarh, India
[2] Department of Computer Science and Engineering, Manipal Institute of Technology Bengaluru, Manipal Academy of Higher Education, Manipal, India
gerard.deepak.christuni@gmail.com
[3] Department of Data Science, CHRIST (Deemed to be University), Bangalore, India
[4] Department of Computer Science and Engineering, National Institute of Technology, Tiruchirappalli, Tiruchirappalli, India

Abstract. The availability of digital devices has increased throughout the world exponentially owing to which the average reader has shifted from offline media to online sources. There are a lot of online sources which aggregate and provide news from various outlets but due to the abundance of content there is an overload to the user. Personalization is therefore necessary to deliver interesting content to the user and alleviate excessive information. In this paper, we propose a novel semantically inclined knowledge driven approach for multi-source aggregation and recommendation of news with a focus on personalization to address the aforementioned issues. The proposed approach surpasses the existing work and yields an accuracy of 96.62%

Keywords: News recommendation · Ontology · Personalization · Transformers

1 Introduction

The volume of content on the internet has been on an exponential growth since the advent of the information age, with access to affordable internet majority of the content consumption has been shifted from traditional sources to online. News reading being a subset of this paradigm has also seen a similar substitution of newspapers and TV with online articles and websites. Many popular search engines such as Google, Yahoo, Bing etc., aggregate articles from the web and present it to the user. However, due to the sheer amount of data it is not possible for the user to go through each article to find the relevant news. Therefore, news aggregation with a personalized recommendation system is the necessary to improve user experience and increase engagement.

K. K. Patel et al. (Eds.): icSoftComp 2022, CCIS 1788, pp. 263–274, 2023.
https://doi.org/10.1007/978-3-031-27609-5_21

Motivation: Precise coordination between user interests and the available news content is critical for creating recommendation systems that help users alleviate the excess information overload that is present online. Existing models are either based on current user clicks or just the past interests in independent ways. Therefore, to create a user personalization aware recommendation network we have incorporated auxiliary knowledge with ontologies and multiple inputs are taken from heterogenous sources. Increasing user dependence on the digital online sources for news consumption requires intuitive recommendation systems to increase relatedness to user interest and improve the overall satisfaction.

Contribution: A semantically inclined knowledge driven framework for multi-source news aggregation and recommendation has been proposed with an emphasized focus on personalisation. The method is a sign of the integration of ontology, semantics, and artificial intelligence, which is achieved by employing bagging and transformers at various levels in the suggested framework. The network is given input from the user's clicks, queries, and prior interests, this helps network to not only stay relevant to the current interests but also incorporate past relevance. Further transformers have been used for ontology alignment and news input has been taken from various heterogenous sources. Improved precision, accuracy, recall, and f-measure has been observed with a lower false negative rate (FNR) and normalised discounted cumulative gain (nDCG).

Organization: The remainder of the paper is divided into the following sections. The relevant prior research on the subject is presented in Sect. 2. In Sect. 3, the proposed architecture is presented. The implementation is covered in Sect. 4. Performance evaluations and observed results are included in Sect. 5. The paper is concluded, Sect. 6.

2 Related Works

Bai et al. [1] have proposed an architecture for news recommendation which exploits the user search history for personalization. To address this challenge, they have created two profiles namely a user search profile and a news profile. Further, score aggregation and rank aggregation have been used to present recommendations to the user. Zhu et al. [2] in 2018 created a news recommendation model that refers user profile-based preferences to provide recommendations. It combines both long-term and short-term preferences selected from diversified sources. Further a preferential weight calculation is used with profile and news popularity as the input. He et al. [3] have proposed an algorithm for news recommendation where they are combining both content based and collaborative based filtering techniques. It amalgamates tag-space and vector space by utilising UILDA and map-reduce techniques.

Ferdous et al. [4] have create a semantic content-based approach that employs ontology translation for cross-lingual news recommendation utilising ontology matching. Bengali to English translation is used in this process. Zheng et al. [5] have developed a news recommendation algorithm using reinforcement learning based deep neural networks. It is based on Deep Q-Learning and have defined the future reward explicitly to inculcate diversity in results. An et al. [6] used a new strategy for user representations, they utilise both long-term and short-term dependencies via an encoder model based on attention. Further, multiple recombination strategies are used. An embedding based news recommendation was created by Okura et al. [7] They used a variety of methods, such as autoencoder representations, recurrent neural networks (RNN) with user internet history as input, and inner product operations, to create embeddings.

Wang et al. [8] have proposed a network based on amalgamation of knowledge graph representations and deep learning. Input entities are subject to knowledge distillation and fed to convolutional neural network to create embeddings for the candidate news. Attention has also been utilised for user interest extraction. Wu et al. [9] created an embedding based neural news recommendation framework. Multi-head self-attention model is utilised to make embeddings form news titles. Many techniques involving a combination of convolutional neural networks (CNN) and attention models are being used, Zhu et al. [10] have used CNN for aggregating user interests and RNN for the long hidden sequential features. Wu et al. [11] have proposed a personalised attention based neural news recommendation model. They have used CNN to learn latent representations of articles from titles and user representations are from user clicks and attention. In [12–17] several knowledge driven semantic strategies in support of the proposed literature have been depicted.

3 Proposed Architecture

The proposed architecture for the multi-source news recommendation and personalization is shown in Fig. 1. The approach is an indication towards the amalgamation of ontology, semantics with machine intelligence which is obtained by using bagging as well as transformers at different stages in the proposed framework. The current user clicks, the past interests as well as the user queries is subject to pre-processing. The user query is not a mandatory input it will be only considered if the user wishes to otherwise the past interests from the web usage data and current user cases will serve as the mandatory inputs.

Tokenization, lemmatization, stop word elimination, and named entity recognition are all part of the pre-processing. The python's natural language toolkit has been encompassed into the framework. As a result of pre-processing the terms with user interests are yielded which are further subject to alignment with the upper ontologies of the news categories. The upper ontology of news categories is a very shallow ontology based on several categories which appear in the news articles, it does not have any detailed sub-categories related to it. The reason for the separation of subcategories and upper ontologies is because it reduces the complexity of alignment at an initial stage.

Further, the detailed aggregation of the subcategories based on the category matching the previous step is incorporated. This clearly reduces the load on ontology alignment. For instance, if the category is 'crime' then all the subcategories under crime will be linked directly from the subcategory news ontology without any matching or ontology alignment operations. Next step is the topic modelling which is done using the Latent Dirichlet Allocation (LDA). Topic modelling enhances the robustness and the diversity of topics which are hidden and incorporates into the proposed framework. Entity enrichment procedure is performed with the help of ontology realignment. Index generation is done from a standard news API, in this case newsapi.org, irrespective of the source the listing categories are same for all news providers. Minute categories from these indexes are visualised as ontologies and ontology realignment is performed between the topics yielded in the previous stages until the topic modelling and index generated to formulate a large semantic network.

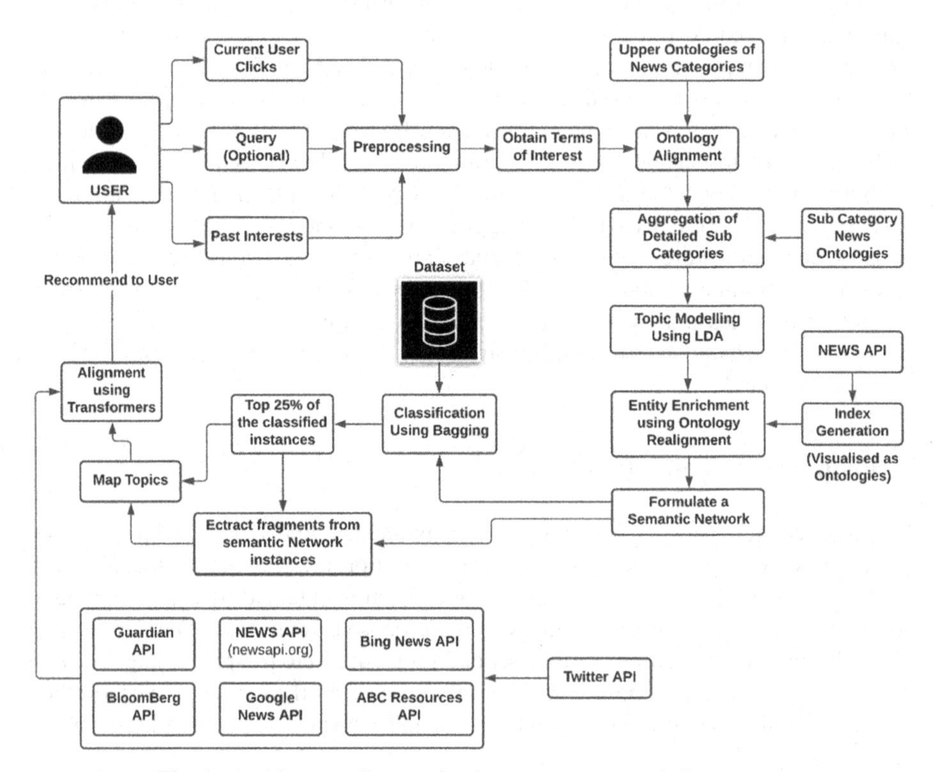

Fig. 1. Architecture diagram for the news recommendation system

The dataset is classified using the Bagging approach, which combines the SVMs and Random Forest classification algorithms, using the characteristics from the semantic network. The reason for the combination of these algorithms is, since SVM is a weak classifier and Random Forest is a strong classifier, the combination of two results in a better intermediate classification accuracy than any of them applied individually. More classes predicted better will be the final recommendation which is specific to this framework as it amalgamates both current and historical news from a combination of APIs. Top 25% of the classified instances from the bagging are considered to extract the fragments of the semantic network based on the vicinity of these classes in the network. To obtain the fragments, the depth of inheritance will be less than or equal to 5 for the relevant instances in the semantic network. The topics are mapped into a hash set based on these fragments as well as the upper 25% of the classified instances. Furthermore, based on these topics in the hash set alignment is done with input from categories in the news

APIs, contents are fetched from the twitter API and other news API source including guardian, Bing, google, Bloomberg, newsapi.org and ABC resource dynamically.

The alignment is further carried out using transformers based on the keywords and categories in the news fetched from the APIs. The reason for using transformers in this case and not ontology alignment is due to the fact that the amount of news contents generated from each of the API for the mapped topics is going to be extensively large and ontology alignment can become highly complex at this phase. Finally, all the aligned and relevant news containing these topics is mediated and suggested to the user. Up until there are no more user clicks or user-triggered inquiries, the process keeps going. The use of various APIs makes the network updated and socially aware for news recommendation. Moreover, using different APIs give the framework a multi-faceted approach towards acquisition of news stories, this not only gives diverse results but also encompasses news at all geographical levels. Usage of transformers ensures that the alignment is proper, and classifiers make a relation between the semantic intelligence and machine intelligence which is one of the most important characteristics of this framework.

The ontology alignment and the ontology realignment process take place using the combination of cosine similarity, Jaccard similarity, SemantoSim measure, concept similarity along with two diversity indexes, Kullback-Leibler divergence (KL divergence) and Bray-Curtis dissimilarity. The dissimilar points are rejected while the similar points are incorporated. Cosine similarity measures the resemblance between two corpuses utilising the property between two vectors i.e., The cosine of the angle between two vectors can be used to gauge their proximity, Eq. (1) illustrates the formal mathematical equation. Jaccard similarity measures the likeness between two sets using a comparison technique, it calculates the percentage of occurrence of members between the two sets which gives an insight into the exact similarity of two corpuses. Formal mathematical equation for Jaccard similarity has been given in Eq. (2). Concept similarity uses the semantic distance property to calculate relatedness of two instances.

$$\text{Cosine Similarity (X, Y)} = \frac{X \cdot Y}{\|X\| \|Y\|} = \frac{\sum_{i=1}^{n} X_i Y_i}{\sqrt{\sum_{i=1}^{n} X_i^2} \sqrt{\sum_{i=1}^{n} Y_i^2}} \tag{1}$$

$$\text{Jaccard Similarity (X, Y)} = \frac{|X \cap Y|}{|X \cup Y|} \tag{2}$$

The SemantoSim measure is an enhanced semantic measure between two entities. It has been further standardised and derivates from the pointwise mutual information (PMI) metric. The no. of input terms defines the formula as the total number of permutations are considered, the mathematical formula for two input terms is however given in Eq. (3). For the case of a single term the most closely related and relevant and itself is considered for calculating the semantic similarity. The probability of terms is considered in coexistence with the other permutations. The KL Divergence Eq. (4) is a method to measure the difference in probability distribution and quantify the mutual information gain. Another index used for measuring the dissimilarity between the entities is the Bray-Curtis dissimilarity, it calculates the structural dissimilarity between the two sets of specimens.

$$\text{SemantoSim (X, Y)} = \frac{PMI(X, Y) + P(X, Y) \cdot \log[P(X, Y)]}{[P(X) \cdot P(Y)] + \log(P(X, Y))} \tag{3}$$

$$\text{KL Divergence }(\mathrm{P}\|\mathrm{Q}) = \sum\nolimits_{n\in N} P(x)log\left(\frac{P(n)}{Q(n)}\right) \tag{4}$$

Finally, the curated set of selective news are recommended to the user based on the re-alignment of these ontologies and the similarity scores. The recommended news is personalised and relevant to the user.

4 Implementation

Utilizing Google Collaboratory as the chosen IDE, the proposed SemKnowNews framework was implemented using Python 3.9. For deep learning tasks, TensorFlow was employed, and the Python nltk library was used for pre-processing tasks. The IDE is enabled using an Intel i7 7th generation processor running at a maximum frequency of 4.20 GHz without the need of any external GPUs. The Microsoft News Recommendation Dataset (MIND), a high-quality benchmark dataset for news recommendation, was used for the experiment.

The dataset has been curated from anonymized behaviour logs of Microsoft News website. Randomly sampled users having multiple news clicks are taken whose behaviour is recorded and formatted into impression logs. There are a million cases on it, and there are more than 156 000 English news pieces, each of which has a rich textual description, title, and synopsis. Each label of the article is manually tagged by the editors into categories such as "Technology".

Table 1 depicts the proposed SemKnowNews algorithm, in the first step the proposed inputs undergo pre-processing and terms of interest are curated. Ontology alignment with the shallow upper ontologies of the news categories take with the curated terms of interest. Further, the aggregation of detailed subcategories based on the matching the previous step takes place. Topic modelling is done using LDA and index generation takes place using the standard news API service. The generated index is represented visually as the development of a semantic network employing realigned ontologies. Thereafter the dataset is classified with features of the semantic network based on the vicinity of these classes, using bagging method and top 25% of the classified instances are filtered.

Next, we collect a set of fragments from the network using the depth of inheritance. These set of fragments and the classified instances are curated into a set and these are subject to alignment using transformers with the inputs of news categories from the set of diverse news APIs. Finally, the semantic similarity score is calculated using the SemantoSim measure and the set of personalised selective news is recommended to the user.

5 Performance Evaluation and Results

The proposed SemKnowNews for news recommendation and personalisation is baselined with ESHNP [1], DPNRS [2], UPTreeRec [3], SCBCN [4] and SVM + Cosine Similarity + K-Means Clustering. With the use of the false discovery rate (FDR), normalised discounted cumulative gain (nDCG), precision, recall, accuracy, and f-measure,

Table 1. Proposed SemKnowNews algorithm

Input: Current user clicks, user input query (optional), user past interests, data from standard news APIs, Microsoft News Recommendation dataset **Output:** Curated Set of News Recommendations for the user

Step 1: Tokenization, lemmatization, stop word removal, and named entity recognition are performed on the suggested inputs, which include the Query (optional), user intent from current clicks, and user profile.
Step 2: Terms of interest T_i are curated from the described inputs.
Step 3: For each term in T_i
Ontology alignment with shallow upper ontologies of news categories
Aggregation of detailed sub-categories
Step 4: Topic modelling using LDA
Step 5: ***Perform entity enrichment and formulate a semantic network:***
5.1 Index generation with the help of standard news API
5.2 Visualization of generated index as ontologies
5.3 Formulation of semantic network using ontology realignment
Step 6: Classify the dataset with features of the semantic network based on the vicinity of these classes, using bagging method and filter top 25% of the classified instances into a set C.
Step 7: Compute depth of inheritance D_i for the network instances
If ($D_i > 5$)
Append fragment F_i to the set of fragments F
Step 8: **while** (C.next and F.next != NULL)
Map individual C_i to HashMap H
GenerateHashValue()
$H \leftarrow c, C_i$
$H \leftarrow f, F_i$
end while
Step 9: Alignment using transformers for the curated hash set along with the input from the categories of news APIs from a diversified set of sources.
Step 10: Calculation of semantic similarity scores with the help of SemantoSim measure.
Step 11: Present the set of selective news to the user.
End

the performance is compared. The relevance of outcomes is computed using the precision, recall, accuracy, and f-measure. The FDR shows how many false positives were detected by the system, while the nDCG gauges how diverse the SemKnowNews results were.

It is inferable from the Table 2 that With a precision of 95.81%, a recall of 97.43%, an accuracy of 96.62%, an f-measure of 96.61%, the lowest FDR value of 0.04, and an extremely high nDCG value of 0.96, the suggested SemKnowNews results in the best findings. ESHNP yields the lowest precision of 55.48%, lowest recall of 58.63%, lowest accuracy of 57.05%, lowest F-measure of 57.01% and 0.44 FDR with a low nDCG of 0.59. This approach utilises a very naïve methodology of cosine similarity for

Table 2. Comparison of the performance of the developed SemKnowNews to the existing techniques

Search technique	Average precision %	Average recall %	Accuracy %	F-Measure %	FDR	nDCG
ESHNP [1]	55.48	58.63	57.05	57.01	0.44	0.59
DPNRS [2]	68.12	73.28	70.70	70.60	0.31	0.73
UPTreeRec [3]	90.18	93.69	91.93	91.90	0.09	0.95
SCBCN	83.18	85.69	84.43	84.14	0.16	0.88
SVM + Cosine Similarity + K-Means Clustering	76.69	80.18	78.43	78.39	0.23	0.71
Proposed SemKnowNews	**95.81**	**97.43**	**96.62**	**96.61**	**0.04**	**0.96**

computing the relevance which affects the quantitative results of the recommendation algorithm. However, the compilation of the use of user profiles strategically ensures a very high degree of personalisation. Moreover, the incorporation of score aggregation, rank aggregation and user votes-based ranking helps yielding satisfactory recommendations qualitatively. The aggregation of user profiles does not assure knowledge which results in a low nDCG score.

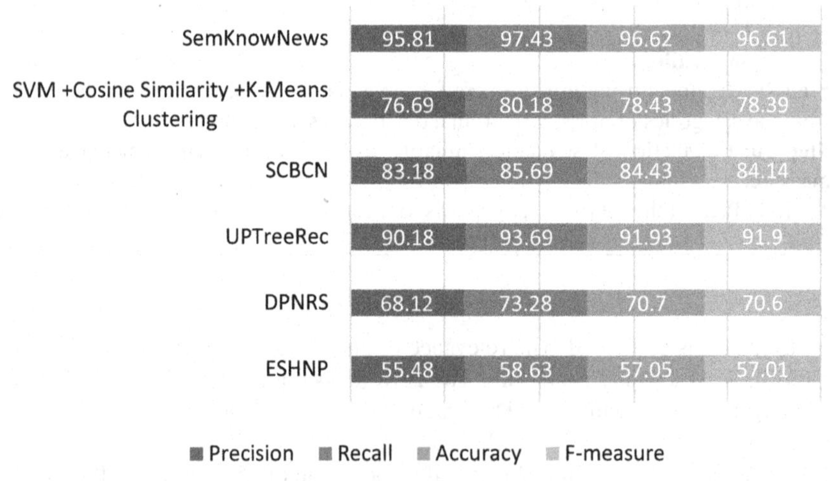

Fig. 2. Performance indicators of the proposed SemKnowNews are compared to those of competing methods.

The DPNRS combines the long-term and short-term user preferences i.e., selection of user preferences from various facets and perspectives is calculated. The approach follows a preferential weight calculation method using user behaviour and popularity of the news. Precision is 68.13%, recall is 73.28%, accuracy is 70.70%, F-measure is 70.60%, FDR is 0.31, and nDCG is 0.73% as a result of the method. The reason for the lag of DPNRS is the lag of classification or clustering mechanism, also the preferential weight calculation method is very weak to rely on it for relevance of topic and news recommendation computation. The approach also integrates new keyword extraction, named entity extraction, topic distribution analysis and news similarity computation. However, the similarity computation approach used is based on cosine similarity which does not yield best results for precision, recall, accuracy for this dataset. The several perspectives of the users being incorporated follows with addition of new knowledge which results in higher nDCG values compared to the relevance measures.

The UPTreeRec method uses both content based and collaborative based filtering techniques. This approach utilises UILDA method along with map-reduce, it unifies both tag-space and vector-space. Precision is 90.18%, recall is 93.69%, accuracy is 91.93%, F-measure is 91.90%, FDR is 0.09, and nDCG is 0.95% for the method. The reason for a high value of nDCG is mainly because of the LDA which uses the topic modelling. The map-reduce helps in mapping based on effective alignment of entities. However, the collaborative filtering method here requires external ratings which are inconsistent and a method which combines these two fails when the relevance of results is to be computed. The SCBCN is a semantic content-based method which uses ontology translation where Bengali to English translation is used for a cross-lingual news recommendation using ontology matching. Precision is 90.18%, recall is 93.69%, accuracy is 83.18%, F-measure is 85.69%, FDR is 0.16, and nDCG is 0.88% for the approach. The use of ontologies ensures that the nDCG value is higher however the ontology field is very static and sparse. As a result, the sparsity of static ontologies is a major lag, also the additional ontology matching algorithm i.e., the alignment extract algorithm is a traditional method which alone cannot be relied for relevance computation. SVM + Cosine Similarity + K-means clustering is an experimental approach where a classification and clustering mechanism is strategically combined along with a semantic similarity method. Increases in FDR are accompanied by decreases in precision, recall, accuracy, and f-measure. The lack of lateral knowledge fed into the system accounts for the low nDCG. Precision is 76.69%, recall is 80.18%, accuracy is 78.43%, F-measure is 78.39%, FDR is 0.23, and nDCG is 0.71% for the method.

The proposed SemKnowNews framework is a hybrid semantic intelligence knowledge driven framework. Here the upper ontologies of news categories are used, and these upper ontologies are aligned with the terms of interest which is extracted from the user profile as well as the current user clicks. The use of query is optional however, if used it enhances the quality of results. The ontology alignment takes place using the combination of cosine similarity, Jaccard similarity, SemantoSim measure, concept similarity along with two diversity indexes, Kullback-Leibler divergence (KL divergence) and Bray-Curtis dissimilarity. The detailed aggregation of news categories and sub-categories has been done after ontology alignment along with topic modelling done using LDA. News APIs have also been included in the framework to increase the density of real-world knowledge in the proposed framework. Utilizing the Twitter API, an index

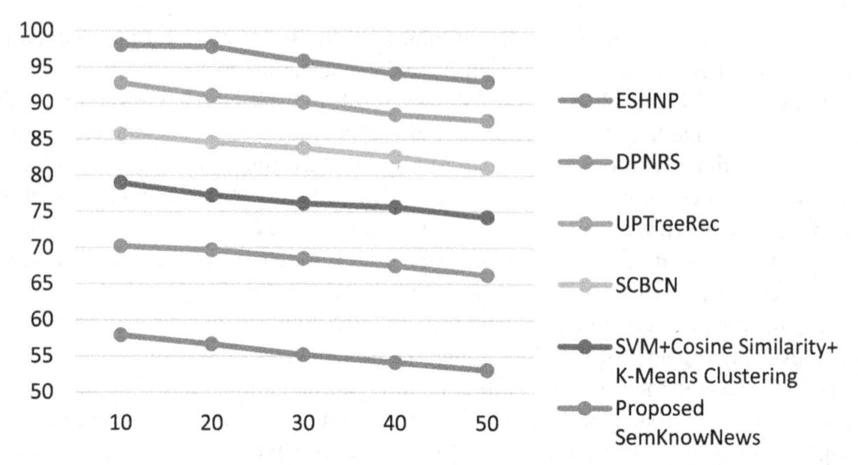

Fig. 3. Comparison of proposed SemKnowNews's precision distribution for 'N' random instances in a sample with existing methods

that is visualised as an ontology is created and social awareness is instilled. These collections of APIs include the incidence of the major news both present and archived which ensures the elimination of fake news implicitly since the incidences alone are considered here the unverified news will be eliminated. Topic mapping has also been incorporated along with alignment using transformers. The amalgamation of these numerous techniques in the framework results in maintenance of the relevance. Moreover, the dataset used is classified based on the semantic network which is initially formulated using the bagging technique which ensures that the most relevant instances based on the semantic network formation from the initial user preferences reduces the computation load by learning as well as classifying a certain number of entitiesA comparison of the precision versus distribution curve for the randomly chosen "N" instances in a sample is shown in Fig. 3. The graph clearly demonstrates that, when compared to the baseline models, the suggested SemKnowNews has a higher percentage of precision despite the volume of suggestions. The rationale is that it is a learning-infused semantic method, one of the finest in class when compared to other approaches. Figure 2 shows that the suggested strategy has a high degree of precision, recall, accuracy, and f-measure in addition to having a very low FDR and a high nDCG. The high nDCG value is a by-product of LDA, upper ontology supplication, index generation from the news API, social infusion, and knowledge influx into the framework.

6 Conclusion

In this research, we offer a personalised news recommendation system that is semantically inclined and knowledge aware. This approach aims to provide an intuitive model for accurate interest matching of the user. The approach takes into consideration the current user clicks, query and past interests of the user to yield terms of interest which are further aligned and modelled to formulate a semantic network. The news data input was gathered from a stack of disparate sources, and the dataset was classified using the bagging technique. Further the top classified instances and elements from semantic network are mapped into a frame-based data structure and aligned with the content using transformers. The SemantoSim metric is used to generate the semantic similarity score, and the user is then given recommendations for the top instances. We conduct exhaustive experiments on real word MIND dataset. The reported F-measure was 96.61% with an extraordinarily low FDR of 0.04.

References

1. Bai, X., Cambazoglu, B.B., Gullo, F., Mantrach, A., Silvestri, F.: Exploiting search history of users for news personalization. Inf. Sci. **385**, 125–137 (2017)
2. Zhu, Z., Li, D., Liang, J., Liu, G., Yu, H.: A dynamic personalized news recommendation system based on BAP user profiling method. IEEE Access **6**, 41068–41078 (2018)
3. He, M., Wu, X., Zhang, J., Dong, R.: UP-TreeRec: building dynamic user profiles tree for news recommendation. China Commun. **16**(4), 219–233 (2019)
4. Ferdous, S.N., Ali, M.M.: A semantic content based recommendation system for cross-lingual news. In: 2017 IEEE International Conference on Imaging, Vision and Pattern Recognition (icIVPR), pp. 1–6. IEEE, February 2017
5. Zheng, G., et al.: DRN: a deep reinforcement learning framework for news recommendation. In: Proceedings of the 2018 WWW Conference, pp. 167–176, April 2018
6. An, M., Wu, F., Wu, C., Zhang, K., Liu, Z., Xie, X.:. Neural news recommendation with long-and short-term user representations. In: Proceedings of the 57th Annual Meeting of the Association for Computational Linguistics, pp. 336–345, July 2019
7. Okura, S., Tagami, Y., Ono, S., Tajima, A.: Embedding-based news recommendation for millions of users. In: Proceedings of the 23rd ACM SIGKDD International Conference on Knowledge Discovery and Data Mining, pp. 1933–1942, August 2017
8. Wang, H., Zhang, F., Xie, X., Guo, M.: DKN: deep knowledge-aware network for news recommendation. In: Proceedings of the 2018 World Wide Web Conference, pp. 1835–1844, April 2018
9. Wu, C., Wu, F., Ge, S., Qi, T., Huang, Y., Xie, X.: Neural news recommendation with multi-head self-attention. In: Proceedings of the 2019 Conference on Empirical Methods in Natural Language Processing and the 9th International Joint Conference on Natural Language Processing (EMNLP-IJCNLP), pp. 6389–6394, November 2019
10. Zhu, Q., Zhou, X., Song, Z., Tan, J., Guo, L.: DAN: deep attention neural network for news recommendation. In: Proceedings of the AAAI Conference on Artificial Intelligence, vol. 33, no. 01, pp. 5973–5980, July 2019
11. Wu, C., Wu, F., An, M., Huang, J., Huang, Y., Xie, X.: NPA: neural news recommendation with personalized attention. In: Proceedings of the 25th ACM SIGKDD International Conference on Knowledge Discovery and Data Mining, pp. 2576–2584, July 2019
12. Deepak, G., Surya, D., Trivedi, I., Kumar, A., Lingampalli, A.: An artificially intelligent approach for automatic speech processing based on triune ontology and adaptive tribonacci deep neural networks. Comput. Electr. Eng. **98**, 107736 (2022)
13. Srivastava, R.A., Deepak, G.: Semantically driven machine learning-infused approach for tracing evolution on software requirements. In: Shukla, S., Gao, X.Z., Kureethara, J.V., Mishra, D. (eds.) Data Science and Security. LNNS, vol. 462, pp. 31–41. Springer, Singapore (2022). https://doi.org/10.1007/978-981-19-2211-4_3

14. Manoj, N., Deepak, G., Santhanavijayan, A.: OntoINT: a framework for ontology integration based on entity linking from heterogeneous knowledge sources. In: Saraswat, M., Sharma, H., Balachandran, K., Kim, J.H., Bansal, J.C. (eds.) Congress on Intelligent Systems. LNDECT, vol. 111, pp. 27–35. Springer, Singapore (2022). https://doi.org/10.1007/978-981-16-9113-3_3
15. Nachiappan, R., Deepak, G.: OSIBR: ontology focused semantic intelligence approach for book recommendation. In: Motahhir, S., Bossoufi, B. (eds.) ICDTA 2022. LNNS, vol. 454, pp. 397–406. Springer, Cham (2022). https://doi.org/10.1007/978-3-031-01942-5_40
16. Deepak, G., Teja, V., Santhanavijayan, A.: A novel firefly driven scheme for resume parsing and matching based on entity linking paradigm. J. Discret. Math. Sci. Cryptogr. **23**(1), 157–165 (2020)
17. Varghese, L., Deepak, G., Santhanavijayan, A.: A fuzzy ontology driven integrated IoT approach for home automation. In: Motahhir, S., Bossoufi, B. (eds.) ICDTA 2021. LNNS, vol. 211, pp. 271–277. Springer, Cham (2021). https://doi.org/10.1007/978-3-030-73882-2_25

Extraction and Analysis of Speech Emotion Features Using Hybrid Punjabi Audio Dataset

Kamaldeep Kaur[1(✉)] and Parminder Singh[2]

[1] IKGPTU, Kapurthala and Department of CSE, GNDEC, Ludhiana, India
kamal.gndec@gmail.com

[2] Department of CSE, GNDEC, Ludhiana, India

Abstract. This paper describes a Punjabi audio emotional dataset that was collected and developed for the process of Speech Emotion Recognition (SER), specifically done for Punjabi language. The database is designed, keeping in view the various categories of speech emotional databases. This database involves six emotions, namely surprise, fear, happy, sad, neutral and anger. A total of 900 sentences are put into as dataset, where 300 recordings are collected from TV shows, interviews, radio, movies, etc. Another 300 pre-defined sentences are narrated by professionals and remaining 300 by non-professionals. This dataset is further used for SER, in which various speech features are extracted and then selected for best features by feature selection algorithm LASSO. 1-D CNN is further used to classify the emotions. The system performs well with average values of accuracy, f1-score, recall and precision as 72.98% for each performance metric.

Keywords: Speech emotions · Punjabi speech dataset · Feature extraction · Feature selection · Emotion recognition

1 Introduction

Speech is one of the most natural and prevalent form of communication among humans. The researchers were inspired to create approaches for human-machine interaction utilizing speech signals as a case study. Human emotions have an important part in revealing the speaker's inner turmoil. They are reactions to events that occur both within and outwardly. As the mood changes, so does the motivation for opposing the message. In some cases, it is also deceiving for humans to gauge a person's feelings [1–4]. SER has been the focus of study since the 1970s, with a considerable variety of real-world applications. Researchers are looking into using human audio speech to judge emotions in a variety of medical and other applications [5, 6].

The creation of database is the first and most crucial phase in the SER process. A speech corpus is made up of audio recordings of spoken language samples. Differences in need, recording environment, speakers, equipment, and other factors can all contribute to the discrepancies [7].

Actor (simulated), elicited (induced), and natural databases are the three basic database collection methods that categorize databases into three kinds. The simulated

K. K. Patel et al. (Eds.): icSoftComp 2022, CCIS 1788, pp. 275–287, 2023.
https://doi.org/10.1007/978-3-031-27609-5_22

are obtained from experienced experts like as actors in the theatre or on the radio, who express pre-defined lines in desired emotions. Simulated emotions are regarded to be an appropriate method for efficiently conveying emotions. The evoked or induced are created without the speaker's knowledge by producing a fake emotion circumstance. The speaker is placed in a scenario or environment that allows him or her to exhibit the emotions required for the purpose of recording. In a mimicked or artificial environment, natural emotions cannot be documented, however contact center records, airline cockpit recordings, patient-doctor conversations, and other sources can be used to construct this type of database [8–10].

When the database is produced, a vital component of SER is the extraction of features representing speech, that appropriately reflect the speech emotions. The dataset, the standard of emotional characteristics, and the classifiers used to train the system all have an impact on the system's success. The recovered features may contain unnecessary, duplicated, or irrelevant data, lowering the model's performance or potentially lowering its accuracy. As a result, it has become vital to choose appropriate features, using some feature selection algorithm. The feature selection improves classification accuracy while simultaneously reducing the complexity of the algorithms [11–13].

In this paper, a database is presented, which has been specifically designed, recorded and verified, for Punjabi SER system. This database is designed keeping in view different types of databases, which is a mixture of recordings by professionals and non-professionals. This database is significant as there is no hybrid dataset found for Punjabi language SER system. A new hybrid dataset is designed, recorded, verified and prepared, then SER for Punjabi is developed. This database is further processed and features are extracted out of it, namely pitch, Mel Frequency Cepstral Coefficients (MFCC), Mel spectrogram, zero crossing rate, contrast, tonnetz, chroma, Linear Predictive Cepstral Coefficients (LPCC), shimmer, entropy, jitter, formant, Perceptual Linear Prediction coefficients (PLP), harmonic, duration and energy. After that, there's a feature selecting mechanism. To remove the unnecessary features, LASSO is used. The emotions are then recognized by using 1-D Convolutional Neural Network (CNN).

The rest of the paper is formulated as follows: The relevant work is detailed in Sect. 2, and the suggested SER process is discussed in Sect. 3, which includes the corpus dataset creation, extraction of features, selection of features, and classification. Section 4 details the experiment, the results, and the analysis. The work, as well as its future scope, are discussed and finished in the next section.

2 Related Work

In literature, various foreign languages are explored by researchers, including Chinese SER with Deep Belief Network [14], Mandarin [15], Persian [16] using Hidden Markov Model with 79.50% performance accuracy, for Polish [17] using k-nearest neighbor, Linear Discriminant Analysis, with performance of 80%. Emotion recognition has been improved by combining feature selection approaches, ranking models, and Fourier parameter models, as well as validating the models against standardized existing speech datasets including CASIA, EMODB, EESDB, FAU Aibo and LDC [18–20]. On Berlin EmoDB of speaker-dependent and speaker-independent tests, 2D CNN LSTM network achieves recognition accuracies of 95.33% and 95.89%, respectively. This contrasts favorably with the accuracy of 91.6% and 92.9% attained by conventional methods [21]. Eight emotional classes from the Ryerson Recordings-Visual Database of Emotional Speech and Song audio are used to train three proposed models (RAVDESS). The proposed models outperformed state-of-the-art models utilizing the same dataset, achieving overall precision between 81.5% and 85.5%.[22].

For Indian languages, [23] has shown work for Assamese using Gaussian Mixture Model, with performance of 74% and highest mean classification score as 76.5 and [24] has shown work for Odia language using prosodic features and Support Vector Machine, with 75% performance. Tamil has been explored by [25] with 71.3%, Bengali by [26] with 74.26%, Malayalam by [27] with Support Vector Machine and Artificial Neural Network with 88.4% performance, Telugu by [25, 28] with performance of 81% and Hindi with 74% [29]. [30] has completed the job for Marathi using the Gaussian Mixture Model, with an average accuracy rate of 83.33% and an average confusion rate of 16.67%. Using an adaptive artificial neural network, emotion recognition from a Marathi speech database has also been accomplished. The experimental research showed that the proposed modal is 10.85% more accurate than the standard models. [31].The work is also done for Punjabi with accuracy of 81% [32], using a Punjabi speech database collected only from non-professional speakers [33].

Although work on emotion recognition in a variety of foreign languages has been done, there has been less effort documented for Indian languages. For most of Indian languages, there are no standardized databases. Choosing which features to extract and which classifier to employ are still significant issues. In this field, no research on the Punjabi language has been recorded.

3 SER Process

The SER system involves various steps, including preparation of Audio Emotional Dataset, Feature Extraction, followed by Feature Selection, and then Classification.

3.1 Audio Emotional Dataset

The authors planned the design of this Punjabi Audio Emotional Dataset, keeping in view different types of standard speech database available. The audio dataset was designed, recorded and verified, by mixing different types. Six basic emotions, namely, anger, happy, fear, neutral, sad and surprise were chosen. A total of 900 speech files were constructed, out of which 300 recordings were done with 5 non-professional speakers, including 3 females and 2 male speakers, and 300 recordings were done with 5 professional speakers, including 2 females and 3 male speakers. And for this recording purpose, 10 Punjabi sentences were chosen that were neutral in character. So, for each emotion, 10 words were recorded per speaker, for a total of 600 utterances, with 100 utterances per emotion. The remaining 300 sentences of the dataset were collected from online resources such as plays, TV shows, radio, movies, vlogs, news, stories, interviews, political speeches, etc. with 50 sentences per emotion, yielding a total of 300 sentences of this type.

The Table 1 shows detailed characteristics of this dataset:

Table 1. Dataset specifications.

Specification	Description
Speakers	5 (2 males and 3 females) non-professional, 5 (3 males and 2 females) professional, 300 from online resources
Age group	20–45 years
Emotions	6 (fear, surprise, happy, angry, neutral, sad)
Sentences	900 (300 from non-professional, 300 from professional, 300 from online resources)
Environment	Studio
Hardware	Sennheiser e835 microphone
Software	Audicity 2.2.2
Sampling rate	16 kHz
Channel	Mono
Bit-rate	16 bit
Audio format	.wav

3.2 Extraction of Features

Pre-processing the signal is always required to remove noise and echo as well as to smooth the signal. The audio files in the database are converted to '.wav' format with mono-channel, 16 kHz sampling rate, and 16-bit rate for this study, and then pre-processed with several standard filters such as the Kalman filter, Normalized Mean Square (NLMS) with FIR filter, and Wiener filter. The Kalman filter is used to reduce noise, the NLMS is used to eliminate echo, and speech signal smoothening is done through Wiener filter.

After signal pre-processing, the signal is analyzed to extract number of features, which are now used to represent the human speech produced. These qualities best depict the emotions that are inherent in the input signal. MFCC, Mel Spectrogram and Pitch, tonnetz, zero crossing rate, jitter, shimmer, contrast, chroma, entropy, formant, harmonic, duration, PLP, LPCC, energy are extracted for this work.

In case of MFCC, the following equation is used to translate the power spectrum from the linear scale with frequency in Hz units onto the Mel frequency scale:

$$f_{mel}(f) = 2595\log_{10}(1 + \frac{f_{HZ}}{700})$$

For PLP, the bark frequency bank is produced through the fusion of frequency warping, smoothing, and sampling. Using a 256-point Fast Fourier Transform and a Hamming window with a window size of 20 ms, the quantized sampled signal is weighed (FFT). 200 voice signal samples are divided by the FFT into 56 zero-valued samples.

$$P(\omega) = \mathrm{Re}[S(\omega)]^2 + \mathrm{Im}[S(\omega)]^2$$

The power spectrum P(Ȧ) is warped for barking frequency by

$$\Lambda(\omega) = 6\ln\{\omega/1200\pi + [(\omega/1200\pi)^2 + 1]^{0.5}\}$$

A number of attributes on the either side of a feature is considered, yielding a total of 418 features. For this research work, mean, max, standard deviation, min, variance, median values are extracted for the features. The detailed features are shown in Table 2.

Table 2. Extracted feature attributes and their count

Feature	Count
Mfcc_mean	120
Chroma_mean	12
Mel_mean	128
Contrast_mean	7
Tonnetz_mean	6
Pitch: max, mean, tuning offset, std	76
Formant	9
Energy: max, mean, std, min	4
Jitter: Local, Localabsolute, Rap, ppq5, ddp	5
Shimmer: Local, db, apq3, aqpq5, apq11, dda	6
zcr_mean	1
Entropy: mean, max, std, var	4
Duration	1
Harmonic	1
Lpcc: coefficients, mean, std, var, max, min, median	19
Plp coefficients, mean, std, var, max, min, median	19
Total	**418**

3.3 Selection of Features

Feature selection methods take a subset of the total extracted features and pick a subset of them. It might get rid of redundant features or features that aren't important and don't add to the system's performance. The analysis of a high number of features may cause issues and slow down the SER system's learning. This problem is solved using feature selection approaches. The feature selection techniques enhance classification accuracy while simultaneously reducing the system's complexity. It also checks the problem of overfitting of the model by reducing the number of redundant and irrelevant features.

As per literature, there are three types of feature selection methods, namely-

i) the filter methods,
ii) the wrapper methods and
iii) the embedded methods.

The filter methods are based upon statistical tests to find the correlation of features with the expected outcome variable. The wrapper methods work better than the filter methods and solve the problem by reducing it to a search problem, by using subset of features. The properties of filter and wrapper methods are combined in the embedded methods, and they are better than the both types.

So, for this research work, an embedded method of feature selection, LASSO (Least Absolute Shrinkage and Selection Operator) is used to select relevant features. The statistical methods may have some prediction errors, which are significantly minimized by LASSO. LASSO selects every non-zero feature, then works on regularizing the model parameters, shrinks the coefficients of regression, which may lead to some of the coefficients as zero. Then it proceeds towards the feature selection phase. LASSO gives an upper bound to the sum of a model, to act as a constraint, to include and exclude some specific parameters.

It provides highly accurate prediction models. Since the method involves coefficient shrinking, which lowers variance and minimises bias, accuracy increases. It operates most effectively when there are many features.

The linear model is trained with L1 prior as regularizer. The optimization objective of Lasso is kept as

$$(1/(2 * \mathrm{n_{samples}})) * ||y - \mathrm{Xw}||_2^2 + \mathrm{alpha} * ||\mathrm{w}||_1$$

The constant alpha is kept as default float value of 1.0, which multiplies the L1 term, and controls the regularization strength.

In this experimental work, LASSO selected 288 features, out of 418 total features.

3.4 Recognition of Emotions

In this experiment, we use a 1-D Convolutional Neural Network (CNN). A 1D-CNN layer follows the input layer, and four more 1D-CNN layers with Batch Normalization, Max Pooling, and ReLU activation functions follow that. Then there are two dense layers that are totally coupled and have the Softmax activation function, followed by a flatten layer. The Batch Normalization layer is used to prevent overfitting. When batch normalization is applied, the input feature of each batch is always combined with other batch features. As a result, the Batch Normalization layer's output for each input feature is no longer a deterministic value. By allowing deep networks to generalize, this effect boosts up the training speed and as a result, overfitting is reduced in trials. The features are protected from noise and distortion by the Max Pooling layer. It divides the data into non-overlapping chunks and extracts the most value from each one. The Softmax classifier is the top layer of this architecture, and it is used to determine emotion based on learning features. The model takes training data as input and outputs predicted emotion with test data. The architecture of the model is shown in Fig. 1.

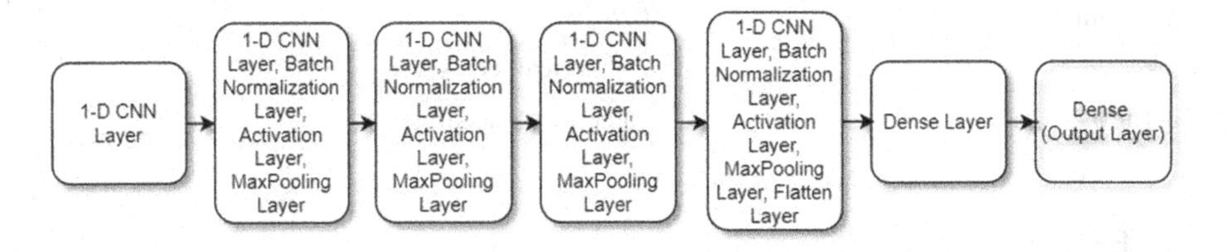

Fig. 1. CNN model for Punjabi SER

4 The Performance Evaluation

We detail the experiment environment in this section, as well as the recognition accuracy of the suggested CNN structures on the Punjabi emotional speech database. The database was utilized to train and test the network.

The total dataset of 900 speech signals is divided into two modules, that is, training and testing. 720 speech signals are utilized to train the system and 180 speech signals are used to test it, in this proposed methodology. The suggested model's performance is evaluated using statistical variables such as accuracy, precision, recall, and F1-score. This SER was created to identify the six emotion classes: anger, happiness, sadness, surprise, fear, and neutral.

4.1 Neural Network Configuration

Due to minor class imbalance problem in the training network, we resampled the data at 16000 Hz and oversampled selected datasets at random. The hyper parameters and settings used in our studies are presented in Table 3.

Table 3. The hyperparameters & network tuning

Parameter	Value
Number of convolution layers	5
Convolution filter size	256 for 1–3 layers, 128 for 4–5 layer
Activation function	relu
Maxpooling layer	Pool size = 4, Strides = 4
Dense layer	Units = 128, Activation: Softmax
Optimizer	SGD
Learning rate	0.0001

(continued)

Table 3. (*continued*)

Parameter	Value
Decay	1e-6
Momentum	0.9
Loss	categorical_crossentropy
Metrics	categorical_accuracy

4.2 Experimental Results

The results of our experimental study are shown emotion wise, in Table 4.

Table 4. Results of experiments

Emotion	Precision	Recall	f1-score	Accuracy
Happy	68.75	75.86	72.13	76
Sad	68.75	75.86	72.13	76
Neutral	81.48	75.86	78.57	76
Surprise	62.5	51.72	56.60	52
Fear	82.75	82.75	82.75	83
Anger	73.33	75.86	74.57	76
Average value	72.92	72.98	72.98	72.98

The results are also shown in the form of graphs and confusion matrix. Figure 2, 3, 4, 5 and 6 show precision, recall, f1-score, accuracy and confusion matrix for different emotions.

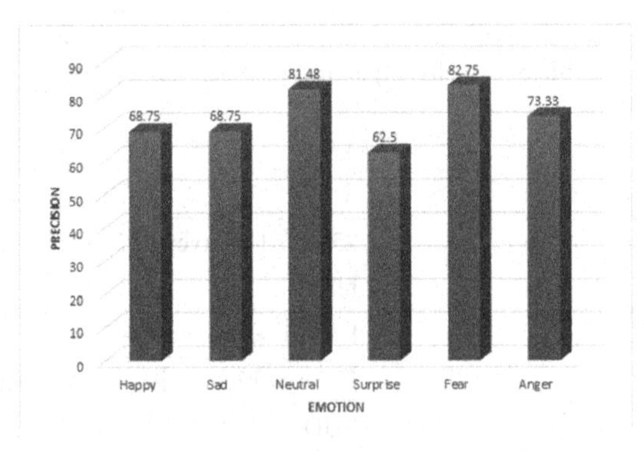

Fig. 2. Precision result for each emotion

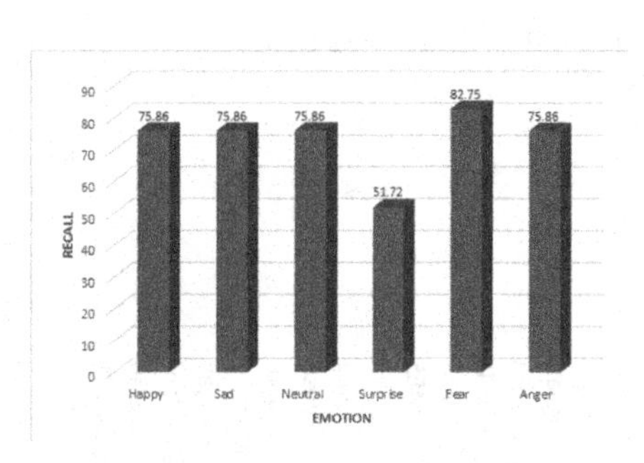

Fig. 3. Recall result for each emotion

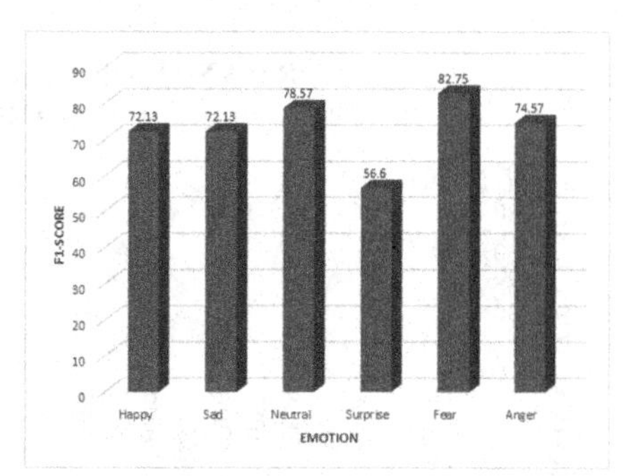

Fig. 4. F1-score result for each emotion

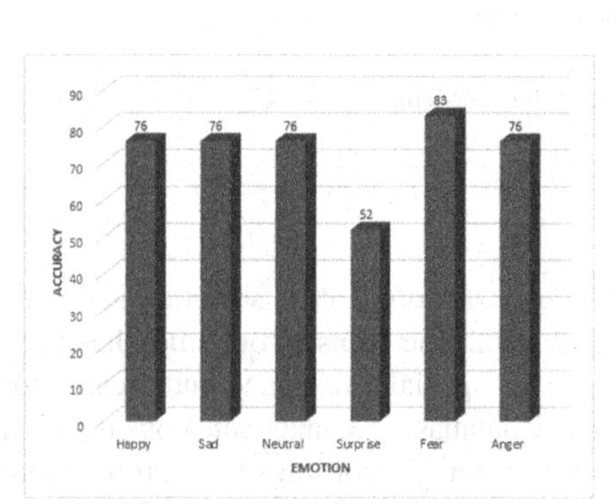

Fig. 5. Accuracy result for each emotion

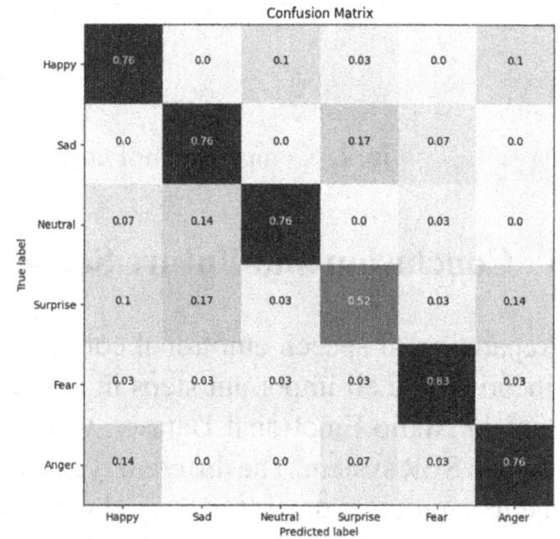

Fig. 6. Confusion Matrix

4.3 Comparison with other SER Systems

An extensive study has been done on SER system in a variety of languages, including English, Chinese, German, Japanese, Mandarin, Spanish, Swedish, Italian, Russian and others. Throughout the years, researchers designed speech datasets, exploited datasets that already exist, devised and analyzed different algorithms for feature selection process, and examined a variety of classification systems. Assamese, Hindi, Telugu, Marathi, Malayalam, Odia, Bengali, Odia, Tamil and other Indian languages were used in the study.

The results for Punjabi SER are comparable with other languages SER, which is shown graphically in Fig. 7.

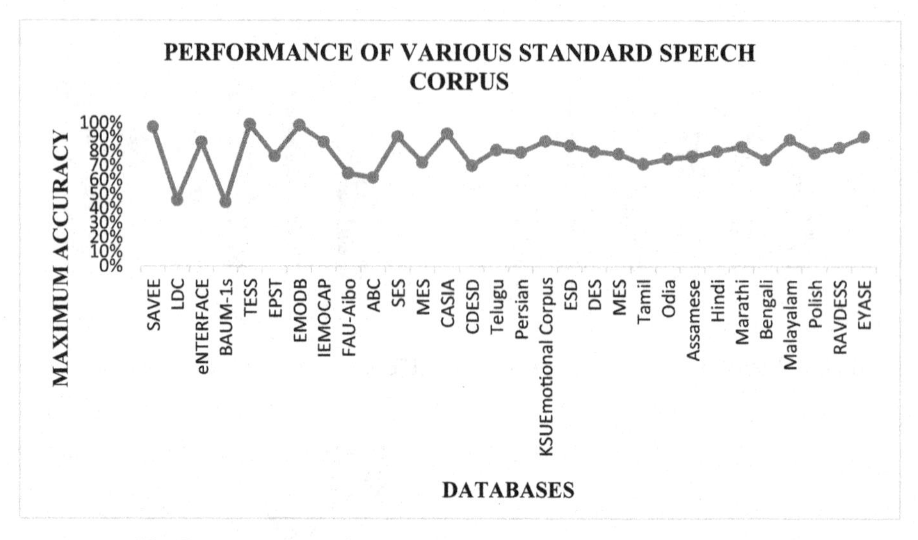

Fig. 7. Comparison of Punjabi SER with other language SER system

5 Conclusion and Future Scope

Preparation of speech emotional corpus, feature extraction, feature selection and classification, are all important steps in any SER system. The focus of our study has been Punjabi Audio Emotional Dataset, which has been specially designed and created for Punjabi SER system. The different types of speech databases are taken into consideration and then required dataset is created with a combination of recordings from professional speakers, non-professional speakers and from online resources such as plays, TV shows, radio, movies, vlogs, news, stories, interviews, political speeches, etc. The dataset after pre-processing, under gone through feature extraction phase where 418 features were extracted, out of which 288 features were kept and rest of the redundant features were removed by the feature selection algorithm LASSO. Finally, the classification was done through 1-D CNN. The system has shown good performance, which is shown in the form of precision, recall, f1-score, accuracy and confusion matrix.

The study can be extended to compare this work with other specific types of speech datasets, such as only natural or acted. More number of features can be added to improve the performance of the system further. The model can also be enhanced by adding more layers, utilizing 2-D CNN, or combining LSTM and CNN in some other way. More feature selection algorithms, such as ANOVA, RFE, SFFS, and t-statistics, can be investigated to increase system performance.

Acknowledgement. Guru Nanak Dev Engineering College, Ludhiana, and IKG Punjab Technical University, Kapurthala have supported this research. The authors are grateful to these organizations for assisting in this work.

References

1. Farooq, M., Hussain, F., Baloch, N.K., Raja, F.R., Yu, H., Zikria, Y.B.: Impact of feature selection algorithm on speech emotion recognition using deep convolutional neural network. Sensors (Switzerland) **20**(21), 6008 (2020). https://doi.org/10.3390/s20216008
2. El Ayadi, M., Kamel, M.S., Karray, F.: Survey on speech emotion recognition: features, classification schemes, and databases. Pattern Recogn. **44**(3), 572–587 (2011). https://doi.org/10.1016/j.patcog.2010.09.020
3. Luengo, I., Navas, E., Hernáez, I.: Feature analysis and evaluation for automatic emotion identification in speech. IEEE Trans. Multimedia **12**(6), 490–501 (2010). https://doi.org/10.1109/TMM.2010.2051872
4. Kuchibhotla, S., Vankayalapati, H.D., Vaddi, R.S., Anne, K.R.: A comparative analysis of classifiers in emotion recognition through acoustic features. Int. J. Speech Technol. **17**(4), 401–408 (2014). https://doi.org/10.1007/s10772-014-9239-3
5. Nicholson, J., Takahashi, K., Nakatsu, R.: Emotion recognition in speech using neural networks. Neural Comput. Appl. **9**(4), 290–296 (2000). https://doi.org/10.1007/s005210070006
6. Chandrasekar, P., Chapaneri, S., Jayaswal, D.: Automatic speech emotion recognition: a survey. In: 2014 International Conference on Circuits, Systems, Communication and Information Technology Applications, CSCITA 2014, pp. 341–346 (2014). https://doi.org/10.1109/CSCITA.2014.6839284
7. Bansal, S., Dev, A.: Emotional Hindi speech database. In: 2013 International Conference Oriental COCOSDA Held Jointly with 2013 Conference on Asian Spoken Language Research and Evaluation, O-COCOSDA/CASLRE 2013, pp. 1–4 (2013). https://doi.org/10.1109/ICSDA.2013.6709867
8. Koolagudi, S.G., Rao, K.S.: Emotion recognition from speech: a review. Int. J. Speech Technol. **15**(2), 99–117 (2012). https://doi.org/10.1007/s10772-011-9125-1
9. Gomes, J., El-Sharkawy, M.: i-Vector Algorithm with Gaussian Mixture Model for Efficient Speech Emotion Recognition. In: 2015 International Conference on Computational Science and Computational Intelligence (CSCI), pp. 476–480 (2015). https://doi.org/10.1109/CSCI.2015.17
10. Zhang, Z., Coutinho, E., Deng, J., Schuller, B.: Cooperative learning and its application to emotion recognition from speech. IEEE/ACM Trans. Audio Speech Lang. Process. **23**(1), 115–126 (2015). https://doi.org/10.1109/TASLP.2014.2375558
11. Özseven, T.: A novel feature selection method for speech emotion recognition. Appl. Acoust. **146**, 320–326 (2019). https://doi.org/10.1016/j.apacoust.2018.11.028
12. Kerkeni, L., Serrestou, Y., Raoof, K., Mbarki, M., Mahjoub, M.A., Cleder, C.: Automatic speech emotion recognition using an optimal combination of features based on EMD-TKEO. Speech Commun. **114**, 22–35 (2019). https://doi.org/10.1016/j.specom.2019.09.002
13. Kuchibhotla, S., Vankayalapati, H.D., Anne, K.R.: An optimal two stage feature selection for speech emotion recognition using acoustic features. Int. J. Speech Technol. **19**(4), 657–667 (2016). https://doi.org/10.1007/s10772-016-9358-0
14. Chen, B., Yin, Q., Guo, P.: A study of deep belief network based Chinese speech emotion recognition. In: Proceedings of the 2014 10th International Conference on Computational Intelligence and Security, CIS 2014, pp. 180–184 (2014). https://doi.org/10.1109/CIS.2014.148
15. Milton, A., Tamil Selvi, S.: Class-specific multiple classifiers scheme to recognize emotions from speech signals. Comput. Speech Lang. **28**(3), 727–742 (2014). https://doi.org/10.1016/j.csl.2013.08.004

16. Savargiv, M., Bastanfard, A.: Persian speech emotion recognition. In: 2015 7th Conference on Information and Knowledge Technology (IKT), pp. 1–5 (2015).https://doi.org/10.1109/IKT.2015.7288756
17. Majkowski, A., Kołodziej, M., Rak, R.J., Korczynski, R.: Classification of emotions from speech signal. In: Signal Processing - Algorithms, Architectures, Arrangements, and Applications Conference Proceedings, SPA, pp. 276–281 (2016). https://doi.org/10.1109/SPA.2016.7763627
18. Cao, H., Verma, R., Nenkova, A.: Speaker-sensitive emotion recognition via ranking: studies on acted and spontaneous speech. Comput. Speech Lang. **29**(1), 186–202 (2015). https://doi.org/10.1016/j.csl.2014.01.003
19. Wang, K., An, N., Li, B.N., Zhang, Y., Li, L.: Speech emotion recognition using Fourier parameters. IEEE Trans. Affect. Comput. **6**(1), 69–75 (2015)
20. Palo, H.K., Mohanty, M.N., Chandra, M.: Efficient feature combination techniques for emotional speech classification. Int. J. Speech Technol. **19**(1), 135–150 (2016). https://doi.org/10.1007/s10772-016-9333-9
21. Zhao, J., Mao, X., Chen, L.: Speech emotion recognition using deep 1D & 2D CNN LSTM networks. Biomed. Signal Process. Control **47**, 312–323 (2019). https://doi.org/10.1016/j.bspc.2018.08.035
22. Ezz-Eldin, M., Khalaf, A.A.M., Hamed, H.F.A., Hussein, A.I.: Efficient feature-aware hybrid model of deep learning architectures for speech emotion recognition. IEEE Access **9**, 19999–20011 (2021). https://doi.org/10.1109/access.2021.3054345
23. Kandali, A.B., Routray, A., Basu, T.K.: Emotion recognition from Assamese speeches using MFCC features and GMM classifier. In: IEEE Region 10 Annual International Conference Proceedings/TENCON (2008). https://doi.org/10.1109/TENCON.2008.4766487
24. Swain, M., Routray, A., Kabisatpathy, P., Kundu, J.N.: Study of prosodic feature extraction for multidialectal Odia speech emotion recognition. In: Proceedings/TENCON of IEEE Region 10 Annual International Conference, pp. 1644–1649 (2017). https://doi.org/10.1109/TENCON.2016.7848296
25. Krothapalli, S.R., Koolagudi, S.G.: Characterization and recognition of emotions from speech using excitation source information. Int. J. Speech Technol. **16**(2), 181–201 (2013). https://doi.org/10.1007/s10772-012-9175-z
26. Mohanta, A., Sharma, U.: Bengali speech emotion recognition. In: 2016 3rd International Conference on Computing for Sustainable Global Development (INDIACom), pp. 2812–2814 (2016)
27. Rajisha, T.M., Sunija, A.P., Riyas, K.S.: Performance analysis of Malayalam Language speech emotion recognition system using ANN/SVM. Procedia Technol. **24**, 1097–1104 (2016). https://doi.org/10.1016/j.protcy.2016.05.242
28. Koolagudi, S.G., Rao, K.S.: Emotion recognition from speech using source, system, and prosodic features. Int. J. Speech Technol. **15**(2), 265–289 (2012). https://doi.org/10.1007/s10772-012-9139-3
29. Bansal, S., Dev, A.: Emotional Hindi speech: feature extraction and classification. In: 2015 2nd International Conference on Computing for Sustainable Global Development (INDIACom), vol. 03, pp. 1865–1868 (2015)
30. Kamble, V.V., Gaikwad, B.P., Rana, D.M.: Spontaneous emotion recognition for Marathi Spoken Words. In: Proceedings of the International Conference on Communication and Signal Processing, ICCSP 2014, pp. 1984–1990 (2014). https://doi.org/10.1109/ICCSP.2014.6950191

31. Darekar, R.V., Dhande, A.P.: Emotion recognition from Marathi speech database using adaptive artificial neural network. Biologically Inspired Cognitive Architectures **23**, 35–42 (2018). https://doi.org/10.1016/j.bica.2018.01.002
32. Kaur, K., Singh, P.: Impact of feature extraction and feature selection algorithms on Punjabi speech. ACM Trans. Asian and Low-Resource Lang. Inf. Process. (2022). https://doi.org/10.1145/3511888
33. Kaur, K., Singh, P.: Punjabi emotional speech database: design, recording and verification. Int. J. Intell. Syst. Appl. Eng. **9**(4), 205–208 (2021). https://doi.org/10.18201/ijisae.2021473641

Human Activity Recognition in Videos Using Deep Learning

Mohit Kumar(✉), Adarsh Rana, Ankita, Arun Kumar Yadav, and Divakar Yadav

Department of Computer Science and Engineering, NIT Hamirpur, Hamirpur 177005, HP, India
mohit@nith.ac.in

Abstract. Human Activity Recognition (HAR) is a challenging classification task. In the past, it traditionally involved the identification of the movement and activities of a person based on sensor inputs, apply signal processing to receive features and fit the features into a machine learning model. In recent times, deep learning methods have shown good results in automatic Human Activity Recognition. In this paper, we propose a pre-trained CNN (Inception-v3) and LSTM based methodology for Human Activity Recognition. The proposed methodology is evaluated on the publicly available UCF-101 dataset. The results show that the proposed methodology outperforms recent state-of-art methods in terms of accuracy (79.21%) and top-5 accuracy (92.92%) on the HAR task.

Keywords: Human Activity Recognition · HAR · LSTM · CNN · Inception V3 · UCF-101

1 Introduction

Human activity recognition (HAR) is a broad field of research to identify the movement and activities of a person. In the past, sensor data for activity recognition was challenging to collect and it required specialized hardware. Nowadays, smartphones and other similar tracking devices used for fitness and health monitoring are easy to purchase. The ubiquity of these devices is useful to enable the collection of data easily. The identification of human activity is a challenging issue in current scenario because of its variations in types of actions, and meaning of actions. Human activity may be recognized in different categories of actions in videos, like blowing candles, body weight squats, handstand push-ups, rope climbing, and swing. Daily human actions, such as jogging and sleeping, are relatively simpler to recognize. However, sophisticated acts, such as peeling a potato, are difficult to detect. According to the authors of the paper [24], complex activities may be accurately identified if they are broken down into simpler activities. Group activities, gestures, human-object interaction, human-human interaction, and their behaviours are some examples of human activities. The manner in which humans conduct an activity is determined by their

K. K. Patel et al. (Eds.): icSoftComp 2022, CCIS 1788, pp. 288–299, 2023.
https://doi.org/10.1007/978-3-031-27609-5_23

habits, which makes determining the underlying activity challenging [2,3]. In the paper [11,16], the authors discuss the practical importance of activity recognition, particularly using cost-effective approaches. They further highlight HAR applications in behavioural biometrics, elderly care, robotics, gesture and posture analysis, health care privacy to monitor the patient activities and their behaviour, workplaces to detect how effectively their employees are functioning, in schools, and institutions where surveillance is important. For example, if a patient is in a coma and is brought to a hospital, a human nurse should monitor the patient under the traditional method. Using the HAR approach, the entire procedure may be automated. Valuable human efforts may be utilized in other meaningful pursuits while the automated model recognizes the patient activities without human intervention.

To address these issues, addressing the following three components is required: (i) background subtraction to separate the parts of the videos into images that are dynamic over time; (ii) tracking to locate the human body movement over time; and (iii) human activity recognition to localise a person's activity.

In the paper [18,25], the authors discuss the challenges in HAR. The mentioned challenges include 1.) a person doing the same activity may appear at different scales in different images 2.) the person holding the camera may shake it and the action may not be entirely seen in the frame 3) there may be background clutter, such as another person's activity in the video, human variation, or variation in their behaviours. These challenges clearly show that understanding people's activities and interactions with their surroundings is not a straightforward task.

As shown in Fig. 1, HAR is an important domain with extensive applications in various fields. Figure 1a depicts the representation of survey based on general HAR and specific HAR. It clearly shows that most of the research is going towards specific HAR. In the same way, Fig. 1b represents the study on HAR in the past 10 years based on various types of HAR subjects in various applications. It shows that 11% of research is done for HAR using deep learning. It motivates us to identify the scope of deep learning in HAR. The survey also indicates that applications of HAR may extend in various fields like biometrics, human-computer interaction (HCI), child monitoring, medical monitoring systems and human's activities in the entertainment field, such as holding the ball, kicking the ball, standing, or remaining silently.

This study focuses on HAR from input videos. Deep learning based approaches are applied on a publicly available dataset and their results are compared on accuracy metric. The major contributions of the paper may be summarized as follows:

- We have considered 3 highly effective different deep learning models, and evaluated them on publicly available UCF-101 dataset. The results of all the experiments are discussed in Sect. 4.
- Due to the use of highly efficient set of methods in experiments, the authors report the best performance on HAR on the UCF-101 dataset as compared to the recent state of art methods. The state of art comparison is shown in Sect. 4.

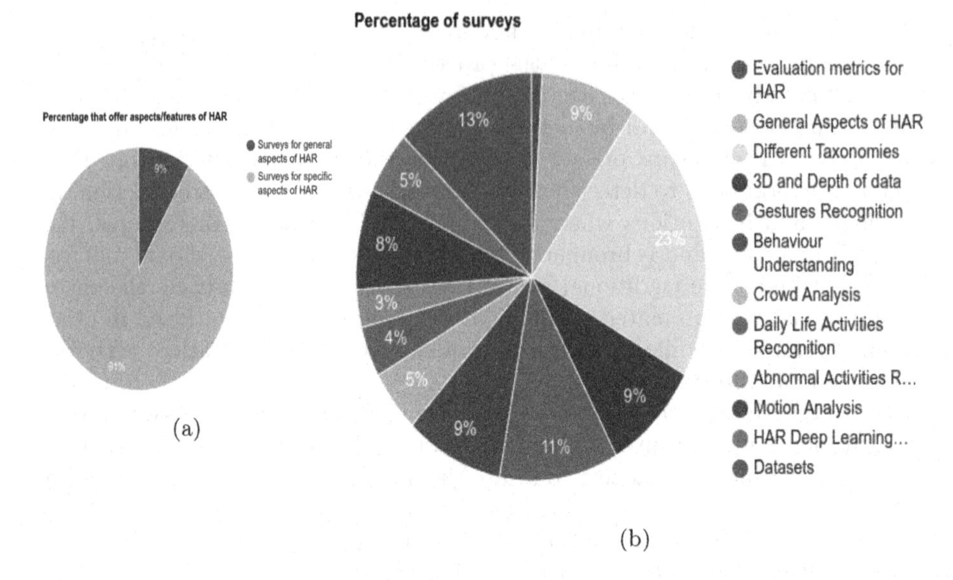

Fig. 1. (a) Percentage of past 10 year surveys showing general framework of HAR and specific taxonomies/domain specific of HAR (b) Most commonly discussed HAR subjects and percentage of surveys covering each of it for past 10 years [14] .

The rest of the paper is organised into the following sections: Sect. 2 discusses the recent contributions in the literature on HAR, along with the applications of deep learning in HAR. Section 3 describes the pre-processing, methodology, and details of the models used in the paper. Section 4 discusses and analyzes the results obtained and compares the results of the proposed approach with recent state of art methods. Section 5 concludes the paper with final thoughts and some directions for future work.

2 Literature Survey

This section discusses the literature on the application of deep learning on HAR. As discussed in the previous section, it is a challenging issue to recognise the activities that look similar, such as walking, jogging and running. HAR also plays an important role in biometric signatures, anti-terrorist and security solutions [20]. In the paper [23], the authors show the use of different features gained with the help of an overlapping windowing approach and random forest as classifiers. Their approach produced results with 92.71% accuracy in recognising activities.

In the paper [9], the authors used two pre-trained models, DenseNet201 and InceptionV3, for feature mapping. They extracted deep features using the Serial based Extended (SbE) approach. Results show that the proposed model achieved accuracies of 99.3%, 97.4%, 99.8%, and 99.9% on four different publicly available datasets, KTH, IXMAS, WVU, and Hollywood, respectively.

Similarly, in the paper [10], the authors worked on a deep learning model with HMDB-51 dataset with 82.55%, and Hollywood2 dataset with 91.99% accuracy. In the paper [19], the authors proposed a CNN based deep learning model for detection of motion and predicting activity like sitting, walking, standing, dancing, running, and stair climbing. The authors report more than 94% accuracy on WISDM dataset within the first 500 epochs of training.

In the paper [26], the authors proposed a deep neural network that combines a long-short term memory (LSTM) and conventional neural network. They evaluate the results on UCI dataset with 95.78% accuracy, WISDM dataset with 95.85% accuracy and Opportunity dataset with 92.63% accuracy. In the paper [15], the authors used the CNN-LSTM architecture and show the importance of LSTM units. They evaluate the results on KTH dataset with 93% accuracy, UCF-11 dataset with 91% accuracy and HMDB-51 dataset with 47% accuracy. According to research conducted in 2019 in the paper [5], the authors proposed two-level fusion strategies to combine features from different cues to address the problem of a large variety of activities. To solve the problem of diverse actions, they proposed machine learning techniques paired with two-level fusion features. This approach helped to increase the performance and improve upon state of art methods. They validated the proposed model with results on CAD-60 (98.52%), CAD-120 (94.40%), Activity3D (98.71%), and NTU-RGB+D (92.20%) datasets.

In the paper [6], the authors proposed a deep learning approach to identify human movements with the collection of body attitude, shape, and motion with 3-D skeletons. The proposed model (multi-modal CNN + LSTM + VGG16 pre-trained on ImageNet) was evaluated with two datasets, UTkinect Action3D and SBU Interaction.

The literature review shows HAR can play an important part in society, and a lot of work has been done in activity recognition. Also, the literature informs that HAR is extremely useful in the medical field to identify and diagnose various types of mental ailments of patents. It was also observed that there are considerable gaps in HAR research. Three major issues are identified and addressed as described here. First, research is required in deep learning to recognise Human activity as only 11% of research has explored deep learning approaches. Second, the challenges of large datasets with a variety of posture in videos are required to address the issues using deep learning. Finally, the chosen methods of investigation are evaluated comprehensively and reported using all popularly used performance metrics.

3 Methodology

This section describes the implementation of the proposed methodology based on issues identified in literature. It describes the dataset used, prepossessing of dataset, model evaluation and training. Also, the reasons for selecting specific deep learning model and their detailed classification are discussed.

3.1 Dataset

This study uses the publicly available UCF-101 dataset [21] for HAR. It contains 101 activity categories such as rope climbing, brushing teeth, playing the guitar, cliff diving, pushups, drumming, playing dhols, and so forth. The UCF-101 dataset is an expansion of the UCF-50 dataset. UCF-101 contains 13,320 clips from 101 activity categories. Because of the enormous number of classes, clips, camera movements, and crowded background, it is one of the most challenging dataset for HAR. The entire length of the video segments is 27 h, with a fixed frame rate of 25 frames per second and a resolution of 320×240. The sample frames of the UCF-101 dataset are shown in Fig. 2. The features of UCF-101 are shown in the Table 1 [22].

Fig. 2. Sample frames of UCF101 Dataset.

Table 1. UCF101 features

Action/clips	101/13320
Audio	Yes (51 actions)
Resolution	320X240
Frame rate	25fps
Total duration	1600 mins
Clips per action	4–7
Min Clip Length	1.06 s
Max Clip Length	71.04 s
Groups per action	25
Mean clip length	7.21 s

The footage from the various action categories is grouped into 25 groups of 4–7 clips each. Every video shares aspects such as the background. Figure 3 displays the number of clips, and the colours represent the various clips in that class. Figure 4 represents the length of clips using blue colors, while green represents the average clip length [1,21].

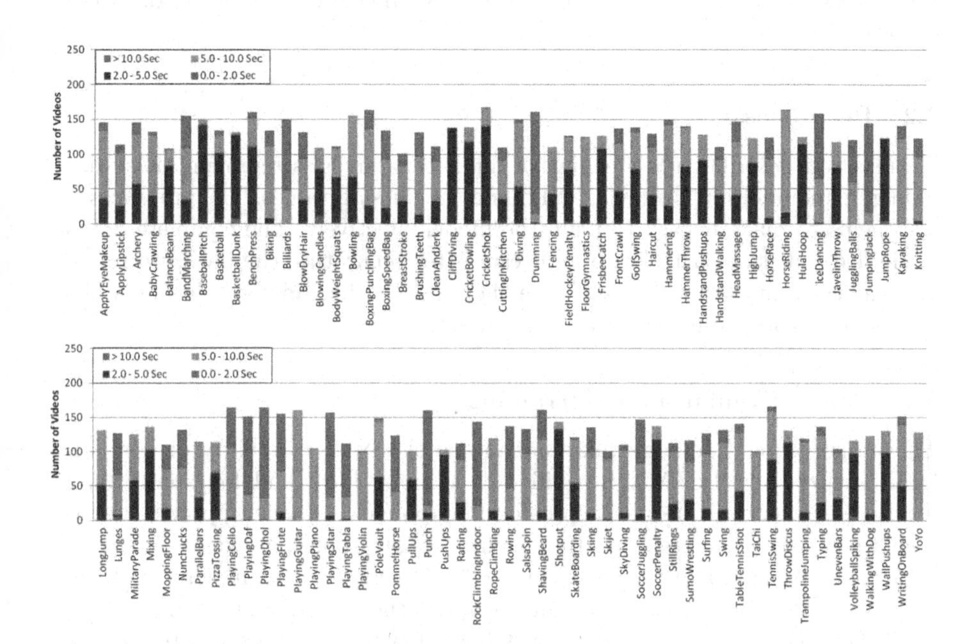

Fig. 3. Displays the number of clips, and the colours represent the various clips in that class.

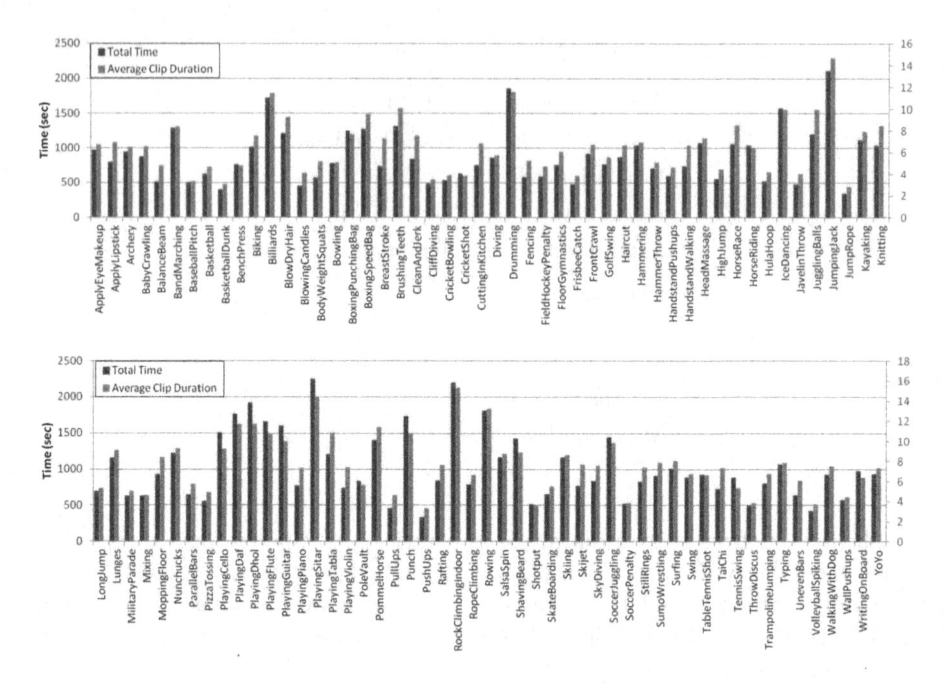

Fig. 4. Blue colours represent the length of clips, while green represents the average clip length. (Color figure online)

3.2 Preprocessing

This section describes the cleaning of input data such as removing noisy data, reducing anomalies in the data, and filling the missing values. The preprocessing step helps to make it easy to train the model effectively. The complete dataset is divided into two parts, train and test. After separating the data, we retrieve the frames from each video and save them as .jpg files in the relevant folders. We use **FFmpeg** to extract frames from each video.

3.3 Model Evaluation and Training

The proposed methodology takes advantage of CNN's capability of extracting local features and LSTM's capability of learning local and long-term dependencies, especially in sequence prediction problems. In this approach, we use InceptionV3 to extract features from frames of corresponding videos [7]. Figure 5 depicts the basic pipeline used in this work. The models are implemented using a Jupyter notebook (local environment) and the experiment code is written in Python. For intensive computing, Google Colab is a cloud environment for Jupyter notebooks that incorporate GPUs (Graphical Processing Units) and TPUs (Tensor Processing Units).

LSTM Model: It is a special RNN (Recurrent Neural Network) and is provided by Keras deep learning Python library. The Sequential model is composed of

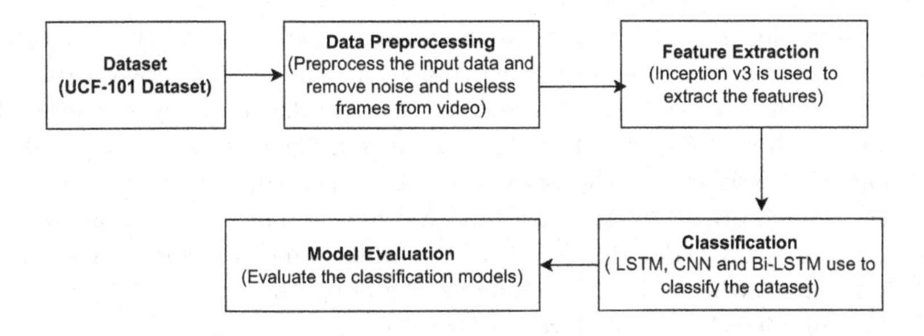

Fig. 5. Basic pipeline used in this work

four layers. The first layer of the neural network is the LSTM layer with 2048 as the number of units. It is the input layer in which features extracted via the use of pre-trained model are used by giving the matrix with the dimensions of 40×2048. Here 40 represents the number of frames for each video and the number 2048 represents the corresponding features. The second layer of the neural network is a dense layer with 512 units. This layer receives all outputs from the layer below it and distributes them to all of its neurons. Each neuron in this layer sends one output to the layer below it. The third layer is a Dropout layer, and its rate is set at 0.5 to reduce the impact of over-fitting. In the end, we use a Softmax activation function. This function's primary purpose is to supply a class as output expressed in terms of probability. As a result, it generates a probability distribution for all 101 classes. The Adam optimizer is used in the training process for the model, with categorical cross entropy serving as the loss function.

MLP (Multilayer Perceptron) Model: It is a feed forward Artificial Neural Network that generates a set of outputs from a set of inputs. The Sequential model of the Keras library is used to implement the suggested deep learning model with six layers. The first layer of the neural network is the flat layer, that flattens the sequence and passes the (2048×40) input vector into a fully connected network. The second layer of the neural network is the dense layer with several units as 512. This layer receives all outputs from the layer below it and distributes them to all of its neurons. The third layer is a Dropout layer, and its rate is set at 0.5 to reduce the impact of over-fitting. The fourth layer of the neural network is the dense layer with the number of units as 512. The fifth layer is a dropout layer of rate 0.5. In the end, we used a Softmax activation function, which outputs the probability got for the corresponding 101 classes.

CNN Model: It is an Artificial Neural Network used for processing structures arrays. The InceptionV3 pre-trained model is used for classifying the features got from the CNN model with four layers. In the first layer, we create the base pre-trained model. The second layer is the average pooling layer, that generalizes features extracted from the pre-trained model and helps the network recognize features. The third layer is the dense layer, which takes input from the pooling layer and distributes them to all of its neurons. The final layer uses the Softmax activation for classifying all the 101 classes using probability.

All the models are evaluated on standard evaluation metrics such as precision, recall, F1-score and accuracy. Also, we compute a metric that represents the number of times where the correct label is among the top k labels predicted (ranked by predicted scores). Here, the value of k is 5.

4 Results and Analysis

This section describes the three different deep learning models applied and evaluated on the UCF-101 dataset. As shown in Table 2, the CNN-LSTM model achieved an accuracy of **79.21%** with a validation loss of less than **1**. Furthermore, the top 5 categorical accuracy is observed as **92.92%**. The superior performance of the CNN-LSTM model may be attributed to the positive effects of both individual models. Whereas CNN helps to extract relevant features from the images, LSTM then uses those features to model the sequential relationship in frames. Together, this combined model effectively improves the HAR accuracy.

Table 2. Results of deep learning models on UCF-101 dataset

S.No	Model	Dataset	Accuracy (%)	Top 5 categorical Accuracy (%)
1	CNN	UCF101	64.56	87.82
2	CNN-MLP	UCF101	70.04	88.36
3	**CNN-LSTM**	**UCF101**	**79.21**	**92.92**

Comparison with state of the art In this section, we compare the results of the current study to previous research on human activity recognition. Table 3 provides comparison of the proposed model on other deep learning model on the same dataset. As can be seen in the table, the CNN-LSTM model used in this work outperforms the baseline and performs slightly better than the previous best method on the UCF-101 dataset.

Table 3. Comparison of proposed work with state-of-art methods

S. No	Model	Year	Dataset	Accuracy (%)
1	Slow Fusion + Finetune top 3 layers [8]	2014	UCF101	65.40
2	Two-Stream+LSTM [27]	2015	UCF101	88.60
3	I3D + PoTion [4]	2018	UCF101	29.30
4	CD-UAR [28]	2019	UCF101	42.50
5	3D-MobileNetV2 0.2x [12]	2019	UCF101	55.56
6	3D-ShuffleNetV2 0.25x [12]	2019	UCF101	56.52
7	MLGCN [13]	2019	UCF101	63.27
8	3D-SqueezeNet [12]	2019	UCF101	74.94
9	3D-ResNet-18 (VideoMoCo) [17]	2021	UCF101	74.10
10	R[2+1]D [17]	2021	UCF101	78.70
11	**This study(CNN-LSTM)**	**2022**	**UCF101**	**79.21**

5 Conclusion and Future Work

In the paper, the authors have presented a novel methodology to perform HAR. The main focus of the paper is to classify the type of activity performed by humans in videos. After the preprocessing steps, three deep learning based classification models have been used. In the best performing model, a pretrained CNN (Inception-V3) model for feature extraction and an LSTM model as the classifier, has been utilized. This model provides the best accuracy of 79.21% and best top-5 accuracy of 92.92% on HAR task.

In future, researchers can make improved and customized CNN model to improve accuracy, especially for extraction of features from the videos including spatio-temporal features. Spatio-temporal understanding may provide a better way to human activity recognition. Furthermore, it is observed that the size of kernel should be kept in mind while designing the models. This is because the kernel size depends on the size of images (frames) extracted from the videos.

References

1. Ahmad, Z., Illanko, K., Khan, N., Androutsos, D.: Human action recognition using convolutional neural network and depth sensor data. In: Proceedings of the 2019 International Conference on Information Technology and Computer Communications, pp. 1–5 (2019)
2. Avilés-Cruz, C., Ferreyra-Ramírez, A., Zúñiga-López, A., Villegas-Cortéz, J.: Coarse-fine convolutional deep-learning strategy for human activity recognition. Sensors **19**(7), 1556 (2019)
3. Banjarey, K., Sahu, S.P., Dewangan, D.K.: A survey on human activity recognition using sensors and deep learning methods. In: 2021 5th International Conference on Computing Methodologies and Communication (ICCMC), pp. 1610–1617. IEEE (2021)

4. Choutas, V., Weinzaepfel, P., Revaud, J., Schmid, C.: Potion: Pose motion representation for action recognition. In: Proceedings of the IEEE conference on computer vision and pattern recognition, pp. 7024–7033 (2018)
5. Das, S., Thonnat, M., Sakhalkar, K., Koperski, M., Bremond, F., Francesca, G.: A new hybrid architecture for human activity recognition from RGB-D videos. In: Kompatsiaris, I., Huet, B., Mezaris, V., Gurrin, C., Cheng, W.-H., Vrochidis, S. (eds.) MMM 2019. LNCS, vol. 11296, pp. 493–505. Springer, Cham (2019). https://doi.org/10.1007/978-3-030-05716-9_40
6. El-Ghaish, H., Hussien, M.E., Shoukry, A., Onai, R.: Human action recognition based on integrating body pose, part shape, and motion. IEEE Access **6**, 49040–49055 (2018)
7. Geng, C., Song, J.: Human action recognition based on convolutional neural networks with a convolutional auto-encoder. In: 2015 5th International Conference on Computer Sciences and Automation Engineering (ICCSAE 2015), pp. 933–938. Atlantis Press (2016)
8. Karpathy, A., Toderici, G., Shetty, S., Leung, T., Sukthankar, R., Fei-Fei, L.: Large-scale video classification with convolutional neural networks. In: Proceedings of the IEEE conference on Computer Vision and Pattern Recognition, pp. 1725–1732 (2014)
9. Khan, S., et al.: Human action recognition: a paradigm of best deep learning features selection and serial based extended fusion. Sensors **21**(23), 7941 (2021)
10. Khattar, L., Kapoor, C., Aggarwal, G.: Analysis of human activity recognition using deep learning. In: 2021 11th International Conference on Cloud Computing, Data Science & Engineering (Confluence), pp. 100–104. IEEE (2021)
11. Kong, Y., Fu, Y.: Human action recognition and prediction: A survey. arXiv preprint arXiv:1806.11230 (2018)
12. Kopuklu, O., Kose, N., Gunduz, A., Rigoll, G.: Resource efficient 3d convolutional neural networks. In: Proceedings of the IEEE/CVF International Conference on Computer Vision Workshops (2019)
13. Mazari, A., Sahbi, H.: Mlgcn: Multi-laplacian graph convolutional networks for human action recognition. In: The British Machine Vision Conference (BMVC) (2019)
14. Moussa, M.M., Hamayed, E., Fayek, M.B., El Nemr, H.A.: An enhanced method for human action recognition. J. Adv. Res. **6**(2), 163–169 (2015)
15. Orozco, C.I., Xamena, E., Buemi, M.E., Berlles, J.J.: Human action recognition in videos using a robust cnn lstm approach. Ciencia y Tecnologí 23–36 (2020)
16. Özyer, T., Ak, D.S., Alhajj, R.: Human action recognition approaches with video datasets-a survey. Knowledge-Based Systems **222**, 106995 (2021)
17. Pan, T., Song, Y., Yang, T., Jiang, W., Liu, W.: Videomoco: Contrastive video representation learning with temporally adversarial examples. In: Proceedings of the IEEE/CVF Conference on Computer Vision and Pattern Recognition, pp. 11205–11214 (2021)
18. Pareek, P., Thakkar, A.: A survey on video-based human action recognition: recent updates, datasets, challenges, and applications. Artif. Intell. Rev. **54**(3), 2259–2322 (2021)
19. Pienaar, S.W., Malekian, R.: Human activity recognition using lstm-rnn deep neural network architecture. In: 2019 IEEE 2nd Wireless Africa Conference (WAC), pp. 1–5. IEEE (2019)
20. Singh, R., Kushwaha, A.K.S., Srivastava, R., et al.: Recent trends in human activity recognition-a comparative study. Cognitive Syst. Res. (2022)

21. Soomro, K., Zamir, A.R., Shah, M.: Ucf101: A dataset of 101 human actions classes from videos in the wild. arXiv preprint arXiv:1212.0402 (2012)
22. Sultani, W., Shah, M.: Human action recognition in drone videos using a few aerial training examples. Comput. Vis. Image Underst. **206**, 103186 (2021)
23. Vijayvargiya, A., Kumari, N., Gupta, P., Kumar, R.: Implementation of machine learning algorithms for human activity recognition. In: 2021 3rd International Conference on Signal Processing and Communication (ICPSC), pp. 440–444. IEEE (2021)
24. Vrigkas, M., Nikou, C., Kakadiaris, I.A.: A review of human activity recognition methods. Front. Robot. AI **2**, 28 (2015)
25. Wang, L., Yangyang, X., Cheng, J., Xia, H., Yin, J., Jiaji, W.: Human action recognition by learning spatio-temporal features with deep neural networks. IEEE Access **6**, 17913–17922 (2018)
26. Xia, K., Huang, J., Wang, H.: Lstm-cnn architecture for human activity recognition. IEEE Access **8**, 56855–56866 (2020)
27. Ng, J.Y.-H., Hausknecht, M., Vijayanarasimhan, S., Vinyals, O., Monga, R., Toderici, G.: Beyond short snippets: Deep networks for video classification. In: Proceedings of the IEEE Conference On Computer Vision And Pattern Recognition, pp. 4694–4702 (2015)
28. Zhu, Y., Long, Y., Guan, Y., Newsam, S., Shao, L.: Towards universal representation for unseen action recognition. In: Proceedings of the IEEE Conference On Computer Vision And Pattern Recognition, pp. 9436–9445 (2018)

Brain Tumor Classification Using VGG-16 and MobileNetV2 Deep Learning Techniques on Magnetic Resonance Images (MRI)

Rashmi Saini[1](✉), Prabhakar Semwal[1], and Tushar Hrishikesh Jaware[2]

[1] G. B. Pant Institute of Engineering and Technology, Pauri Garhwal 246196, India
2rashmisaini@gmail.com

[2] R. C. Patel Institute of Technology, Shirpur, India

Abstract. Early diagnosis of brain tumor using Magnetic Resonance Image (MRI) is crucial and significantly effective in treatment planning and patient care. Recent advancement in deep learning techniques have shown excellent results for various medical imaging application. This paper aims to classify different brain tumors by implementing deep learning techniques and applied on MRI images. Two deep learning models namely VGG-16 and MobileNetV2 are employed to classify MRI images in the four categories: (i) Glioma (ii) Meningioma (iii) Non-tumor (Normal) (iv) Pituitary. In this study, complete data samples are partitioned into training (70%), validation (15%) and testing (15%) datasets. Both VGG16 and MobileNetV2 deep learning models are trained up to 50 epochs. Results of this study indicates that VGG16 model obtained an Overall Accuracy of 91.46%. The highest value of F1-score is 96.87% obtained for Non-tumor class using VGG-16 model. Whereas, MobileNetV2 model outperformed for brain tumor classification and reported accuracy of 97.46%, with maximum F1-score value of 98.87% for Non_tumor and F1-score of 98.72% for Pituitary Tumor. During the implementation phase, it has been observed that the parameter such as epoch affect the accuracy of model significantly. Therefore, it is recommended to train the model with sufficient number of epochs to obtain accurate classification results. Outcome of the study indicates that deep learning models have great potential to successfully classify brain tumor classes using MRI images.

Keywords: Deep learning · Brain Tumor · Classification · VGG-16 · MobileNet · MRI

1 Introduction

Medical Resonance Image (MRI) images play a significant role for effective detection of tumors in the domain of medical image processing. To determine the patient's state or disease, doctors evaluate the patient's signs and symptoms. Manual classification of different types of tumor is a challenging task, as well as depends on the expertise of the available human resources. Recent advancement in the learning based techniques have

K. K. Patel et al. (Eds.): icSoftComp 2022, CCIS 1788, pp. 300–313, 2023.
https://doi.org/10.1007/978-3-031-27609-5_24

shown promising results for tumor detection and classification [1–3]. In-depth evaluation using MRI by integrating advance learning based techniques such as Machine Learning (ML), Deep Learning (DL), Ensemble Learning (EL), Convolution Neural Network (CNN) may contribute in the detection and classification of tumors up to great extent.

A brain tumor is an abnormal growth of cells in the brain tissue. Earlier tumor identification is preferable, since, it can be beneficial to reduce the higher risk and save human life [2]. A brain tumor can have adverse effects in certain patients, including headaches, loss of hearing, difficulties in thinking, speaking, finding words and behavioral changes, etc. [4]. This research aimed to look the significance of MRI technique to quickly identify the tumor and its classification. Gliomas are a prominent type of tumor that is made up of neoplastic glial cells [5]. Nearly 30% of all primary brain tumors and 80% of all malignant ones are gliomas, which are also the predominant cause of mortality from primary brain tumors [6]. Patients with high-grade grade gliomas have survival periods of less than a year to three years after their first diagnosis [7]. Meningioma's are benign tumors that develop from arachnoid cap cells, which are non-neuroepithelial progenitor cells [8]. Meningioma are often detected incidentally (hence the term "incidentaloma") and exhibit little or very little development, especially in older patients with tumor calcification [9]. Pituitary adenomas (PAs), the majority of which are often benign tumors that are derived from anterior pituitary cells, are the most common kind of pituitary tumors. That govern bodily processes including hormone production. Pituitary tumor complications may result in permanent loss of eyesight and hormone insufficiency. Pituitary carcinomas (PC), which make up the remaining 0.1–0.2% of tumors, are tumors that have craniospinal or systemic metastases [10, 11].

Magnetic Resonance Images (MRI) images provide detailed information about the brain. It proves the tumor information like size, location, and shape [12, 13]. Based on the visual description, MR Images scans may be used to identify the presence of brain tumors and other abnormal diseases in the internal organs. To create a 3D image of the inside organs being scanned, the MRI equipment uses strong magnetic fields and radio waves. MR Images have a stronger contrast in soft tissue, they perform better in the field of Medical Detection Systems (MDS) than other imaging methods like Computed Tomography (CT) [14]. Therefore, accurate brain tumor MR images play a key role in clinical diagnosis and help to make decisions for the patient's treatment [15].

Recently, a huge growth of deep learning algorithms has been reported in the domain of medical imaging. The fundamental building block behind the deep learning techniques is the Artificial Neural Network (ANN). The neural network is made up of many interconnected neurons. A perceptron is considered as the fundamental neural network. Broadly, we can say that Input Layer, Output Layer, and Hidden Layer are the three primary blocks that make up all Deep Neural Networks. A large number of studies trying to attempt and analyze MRI for brain tumor identification or classification using deep learning algorithms [1–3, 12, 15–19]. Mehrotra et. al. [3] performed a study to classify brain tumor into two different categories i.e. malignant and benign using deep learning based approach on MRI dataset. Results of this study indicated that the classification accuracy of 99.04% has been obtained by AlexNet model. A study performed by Emrah

Irmak [2] utilized several deep learning techniques such as GoogleNet, AlexNet, VGG-16, ResNet-50 and Inceptionv3. In this study, author used four different datasets for the classification purpose. Three different deep learning models were proposed and brain tumor classification task achieved an accuracy of 92.66% for multiclass classification. Swati et. al. [1] proposed a deep learning based approach for brain tumor classification (three different types of brain tumor) using MRI images. This study demonstrated that an Overall Accuracy of 94.82% has been reported by the proposed method using 5-fold cross-validation scheme. A study performed by Almadhoun and Abu-Naser [16] to identify brain tumors in MRI scans. Authors employed the four pre-trained CNN models namely VGG-16, Inception, MobileNet, and ResNet50. Result of the study indicates that the maximum Overall Accuracy of 98.28%, has been achieved.

A deep convolutional neural network was employed to categorize brain cancers into three different categories (meningioma, glioma, and pituitary tumor) and to further categorize gliomas into different grades (grade II, III, and IV) using MRI scans. The maximum accuracy rates for the suggested architecture was reported as 98.70% [17]. In an another study, using MRI scans, authors suggested a novel CNN-based algorithm to categorize and segment the tumor, and brain lesions in the early stage. Eight distinct datasets (BRATS 2012 - 2015, and ISLES 2015, ISLES 2017) and five MRI modalities are used in this study to test the results (flair, DW1, T1, T2, and T1 contrast). The highest accuracy of the proposed DNN-based architecture was 100% for ISLES 2015 dataset, and the lowest accuracy was 93.1% for the BRATS 2014 Dataset [18]. Using brain MRI scans, Deepak and Ameer [19] classified brain malignancies (glioma, meningioma, and pituitary tumors) using a CNN model (GoogleNet). The experiment uses 5 Fold cross-validation and obtained nearly 98% classification accuracy.

This paper aims to classify MRI imagery into four different classes. Out of these four classes, three classes represent brain tumor, whereas, one class represents Non-tumor that signifies absence of any tumor or normal MRI image. Two advance deep learning approaches (VGG-16 and MobileNetV2) has been used for the classification purpose.

The paper is organized in the following manner: Sect. 2 presents the methodology or workflow of the study as well as the sample data representation. Section 3 presents results and analysis of this study along with the accuracy and loss graph obtained by both deep learning models. Section 4 presents the conclusions drawn from the study.

2 Methodology

The workflow of this study is demonstrated in Fig. 1. In this study, publically available kaggle input MRI images have been used. This dataset is available on the following link- https://www.kaggle.com/datasets/iashiqul/mri-image-based-brain-tumor-classification. Firstly, the selected input dataset is downloaded, which contain of a total number of 7445 MRI imagery. The used datasets consist four different types of MRI images: (i) Glioma (ii) Meningioma (iii) Non- tumor (Normal) (iv) Pituitary. Few samples of MRI images of each brain tumor type class and normal (Non-tumor) are shown in Fig. 2. Thereafter, the pre-processing is performed in order to obtain the required size of MRI images. The pre-processing involves the resize operation to convert the input imagery into a size of 224 × 224. The specified size is required in order to train the deep learning models. Next step is the partitioning of input data into training, testing and validation. Here, 70% of input MRI images have been used for training, whereas, rest of the 30% is equally partitioned as testing and validation datasets. Two deep learning methods i.e. VGG-16 and MobileNetV2 have been used for classification. A brief discussion about both the deep learning techniques is provided in subsequent section. Deep learning models are trained up to 50 epochs. After training and validation, testing is performed on unseen MRI images. Finally, the evaluation is performed based on various parameters for classification results obtained by using VGG-16 and MobileNetV2 model. The evaluation parameters are discussed in the subsequent section.

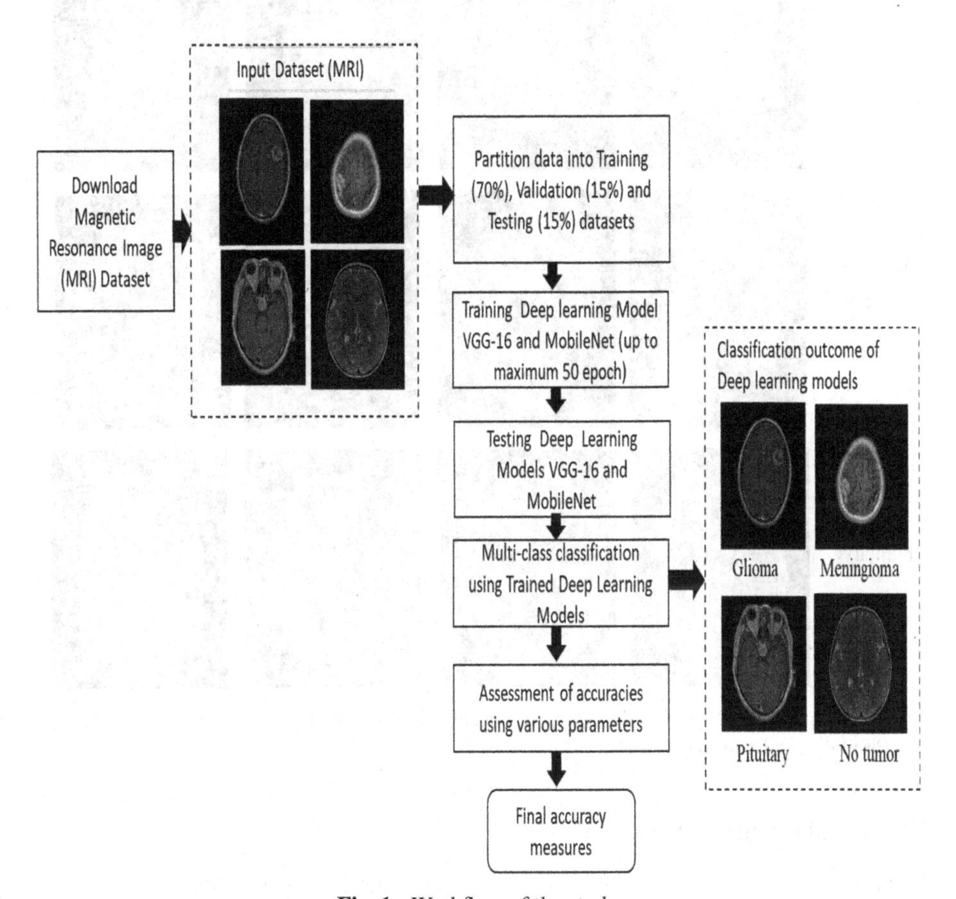

Fig. 1. Workflow of the study.

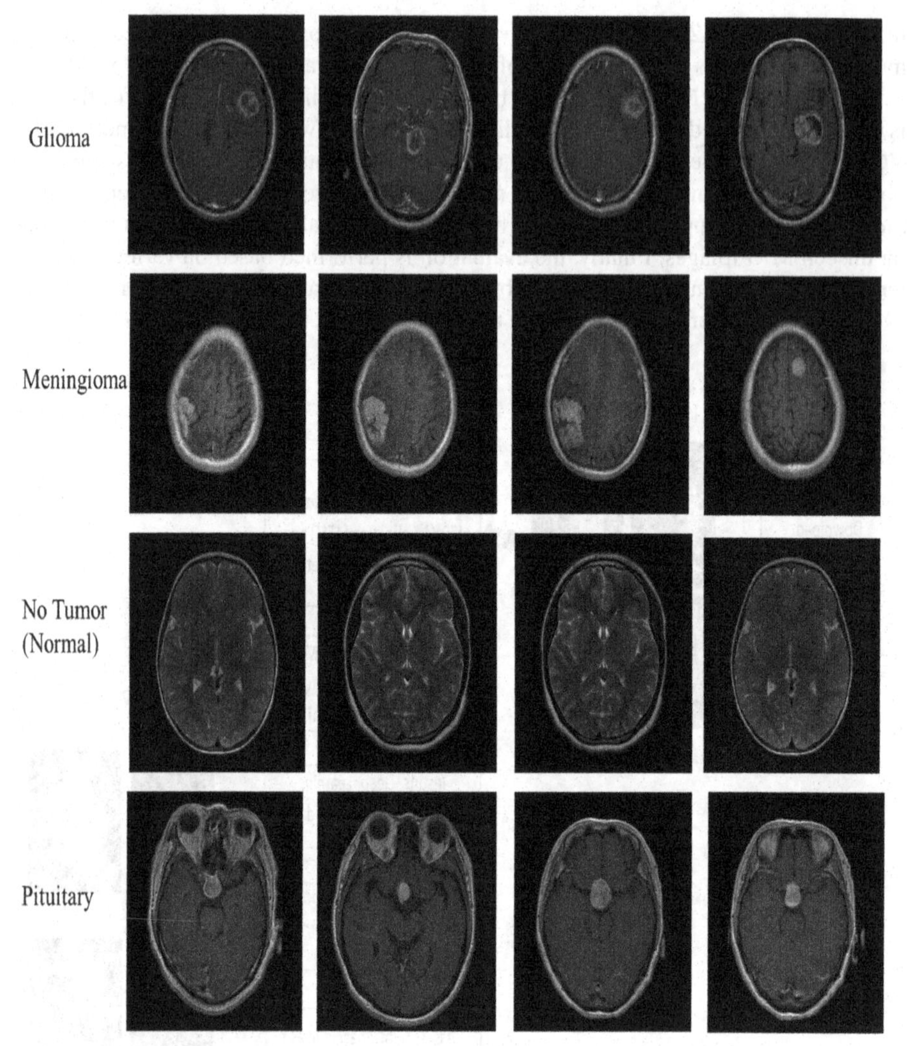

Fig. 2. Sample images used in this study represents various tumor ((i) Glioma (ii) Meningioma (iii) Pituitary) and non-tumor MRI.

2.1 Visual Geometry Group (VGG)

VGG-16 is a Convolution Neural Network (CNN) model developed at Oxford University by the Visual Geometry Group. AlexNet, the network's successor, was launched in 2012. The ILSVRC dataset was used to create and train the VGGNet. VGG-16 is made up of 13 convolution layers, three fully connected layers, and five max-pooling layers. To preserve the image's spatial resolution, spatial padding was applied. Similar operations may now be carried out using the VGG-16 network, which has been made open-source. Because certain frameworks, like Keras, provide pre-trained weights that can be used to build bespoke models with small changes, the model may also be used for transfer learning [19].

2.2 MobileNet

MobileNetV2's main structure is based on the previous version, MobileNetV1. MobilenetV2 uses the Depthwise Separable Convolution (DSC) technique for portability, and it not only improved the problem of information loss in non-linear layers in the convolutional block by employing Linear Bottlenecks, but it also introduced a new structure known as Inverted residuals to preserve the information. There are two kinds of blocks in MobileNetV2. One is a one-stride residual block. Another alternative for reducing is a block with two strides. Each block type has three levels. The first layer this time is a 1*1 convolution using ReLU. The depth wise convolution is the second layer. Whereas, third layer includes 1*1 additional convolutions.

2.3 Evaluation Parameters for Deep Learning Models

This paper aims to classify MRI images into four different categories namely Glioma, Meningioma, Pituitary tumor and Non-tumor (normal). Both the Deep learning models (VGG-16 and MobileNetV2) are trained up to maximum 50 epochs. The details of tuning parameters (learning rate, batch size, number of epoch etc.) for VGG-16 and MobileNetV2 are provided in Table 1. For the evaluation of Deep Learning models following accuracy measures are used: (i) Overall Accuracy, (ii) Precision, (iii) Recall, (iv) Specificity and (v) F1-score. These accuracy measures are computed using True Negative (TN), False Positive (FP), False Negative (FN) and True Positive (TP) values obtained by confusion matrix. The above said accuracy measures are computed using the following equations:

$$Overall\ Accuracy = \frac{TP + TN}{TP + TN + FP + FN} \quad (1)$$

$$Precision = \frac{TP}{TP + FP} \quad (2)$$

$$Recall = \frac{TP}{TP + FN} \quad (3)$$

$$Specificity = \frac{TN}{TN + FP} \quad (4)$$

$$F1 - score = 2 \times \frac{Precision \times Recall}{Precision + Recall} \quad (5)$$

3 Results and Analysis

In this study, two deep Learning models namely VGG-16 and MobileNetV2 have been developed to classify MRI images into four different categories (Glioma, Meningioma, Non_tumor, Pituitary). All the implementation is performed in Python programming language. Here, 5172 images are used for training purpose and 866 and 867 images are used for validation and testing purpose. Both the models are trained up to 50 epochs. The details about other parameters such as batch size, learning rate etc. is provided in Table 1.

Table 1. Details of parameters for both VGG-16 and MobileNetV2 deep learning models.

Parameters	VGG-16	MobileNetV2
Batch Size	32	32
Learning Rate	0.00003	0.00003
Epoch	50	50
No of Training image	5712	5712
No of Testing Image	867	867
No of Validation Image	866	866
Image Size	224*224	224*224
No of Classes	4 [Glioma, Meningioma, Non_tumor, Pituitary]	4 [Glioma, Meningioma, Non_tumor, Pituitary]
Total Parameters	134,309,700	5,146,180
Trainable Parameters	119,578,628	5,112,068
Non-Trainable Parameters	14,731,072	34,112

Figure 3 shows training and validation accuracy with respect to number of epochs for VGG-16 model. It can be seen from the obtained graph that as the number of epochs was very less (epochs < 10) training and validation accuracy was also very low. As the VGG-16 model trained with higher number of epochs there is a significant improvement observed in the accuracy measure. In the similar manner, the training loss and validation loss obtained up to 50 epochs is depicted in Fig. 4. It can be observed from the VGG-16 model loss curve (Fig. 4) that a rapid decrease in training and validation loss value up to 30 epochs. Thereafter, a slight decrease in loss is observed and model stable at nearly after 45 epochs.

The training and validation accuracy graph obtained for MobileNetV2 model is shown in Fig. 5. It can be seen that there is an abrupt change in training and validation accuracy as the number of epoch reaches to 20. Thereafter, a slight improvement is observed as the number of epoch increases. It can be observed that the MobileNetV2 model stable early as compared to VGG-16 model. On the other hand, training loss and validation loss graph have shown the similar pattern of reduction the loss value as the number of epochs increases (Fig. 6).

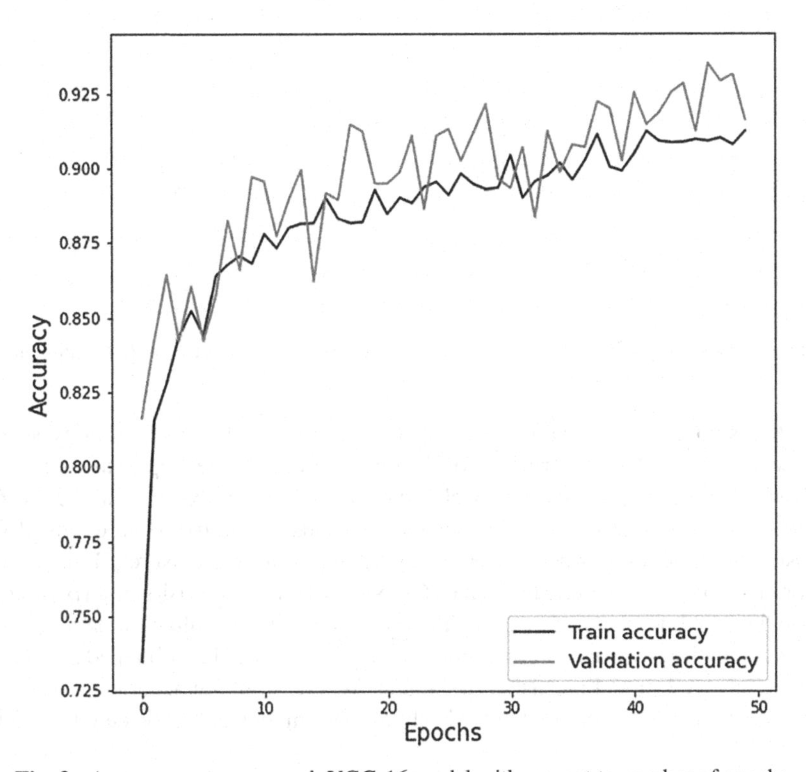

Fig. 3. Accuracy progress graph VGG-16 model with respect to number of epochs.

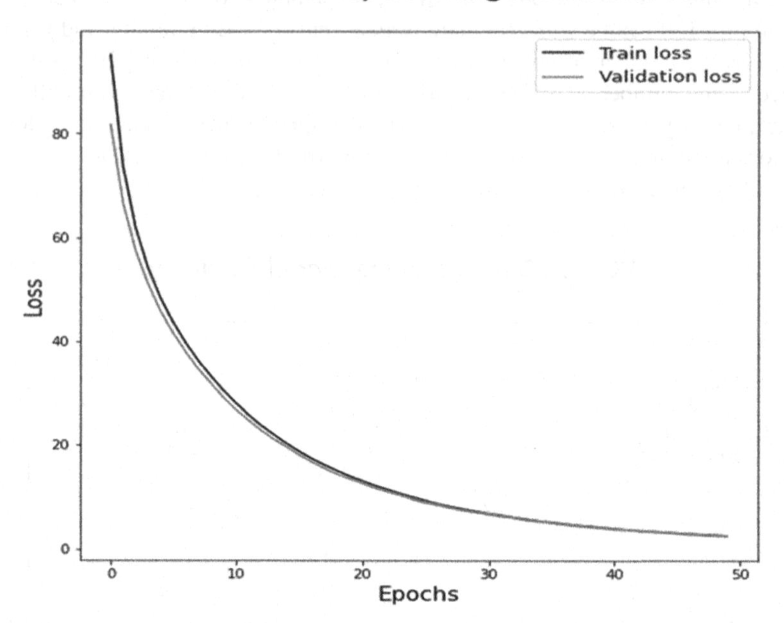

Fig. 4. Loss graph of VGG-16 deep learning model with respect to number of epochs.

In this study, several evaluation measures (as discussed in methodology section) are used to evaluate the performance of the deep learning models. Confusion matrix is table, which represents the accurately classified samples (True Positive) as well as false classified samples in terms of False positive and False Negative. Data obtained from confusion matrix is further used to calculate various accuracy measures. Table 2 shows the confusion matrix obtained by VGG-16 model, whereas, confusion matrix produced by MobileNetV2 model is depicted in Table 3. The resultant evaluation parameters i.e. precision, recall, F1-score and specificity for each individual class obtained by VGG-16 and MobileNetV2 model are listed in Table 4. Whereas, all the accuracy measures as discussed above are also computed for MobileNetV2 model and represented in Table 5.

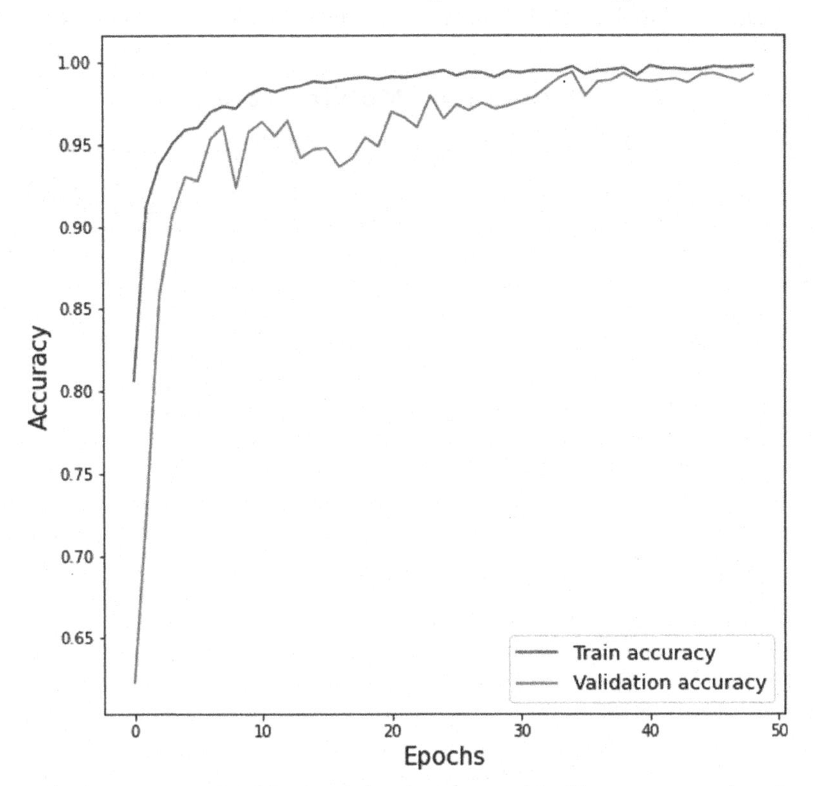

Fig. 5. Training progress of MobileNetV2 deep learning model with respect to number of epochs.

The outcomes of this study have shown that MobileNetV2 outperformed and obtained an accuracy of 97.46% and VGG-16 model obtained an accuracy of 91.46%. It is found that MobileNetV2 achieved 6.00% higher accuracy as compared to VGG-16 model. As far as the class-specific accuracy is concerned VGG-16 model obtained highest F1-score value of 96.87% for Non-Tumor class with precision and recall value of 94.28% and 99.62% respectively. It is found that minimum F1-score of 82.25% observed for Meningioma tumor with the precision value of 92.16% and recall value of 74.27%. It can be observed that all the other brain tumor categories are classified with a high F1-score value of greater than 90%.

Results obtained by using MobileNetV2 deep learning model indicates Non-tumor and Pituitary tumor classified with nearly same F1-score value (~99%). Resultant precision and recall values obtained for pituitary tumor are 98.98% and 98.48% respectively. In the similar way, Non-tumor class is also classified with higher F1-score of 98.87%. Whereas, Glioma and Meningioma classes are classified with nearly similar F1-score value of ~96.00%. It has been observed that MobileNetV2 model produced F1-score of greater than 95% for all the tumor classes, which is significantly higher than VGG-16 Model. Therefore, results revealed that MobileNetV2 deep learning model performed significantly better as compared to VGG-16 model. The findings of this study recommend the application of MobileNetV2 model for other tumor type detection studies.

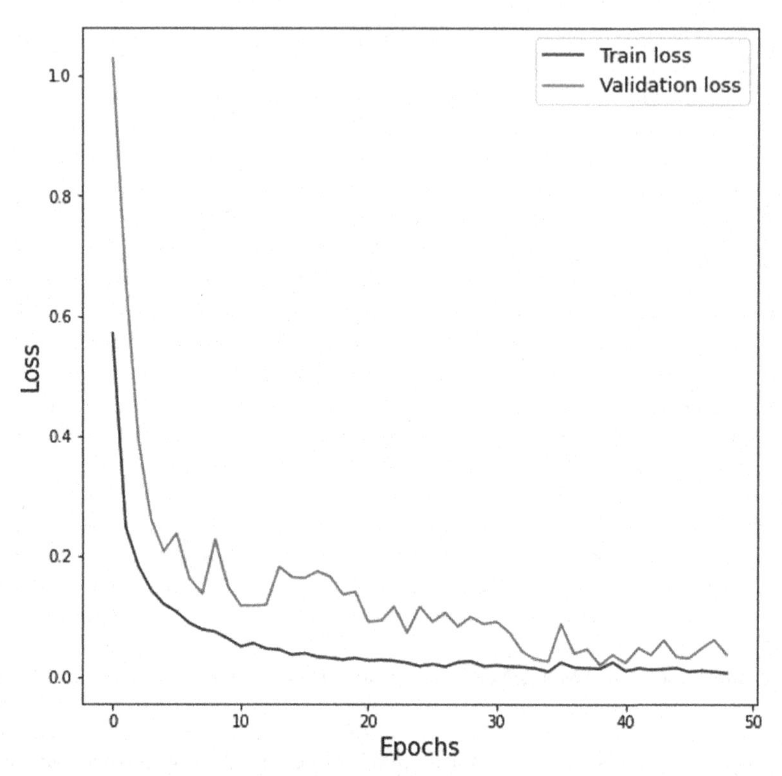

Fig. 6. Loss graph of MobileNetV2 deep learning model with respect to number of epochs.

Table 2. Confusion matrix obtained for VGG-16 model.

	Predicted values				
True Values	Class	Glioma	Meningioma	Non-tumor	Pituitary
	Glioma	187	8	2	1
	Meningioma	29	153	12	12
	Non-tumor	0	1	264	0
	Pituitary	1	5	2	190

Table 3. Confusion matrix obtained for MobileNetV2 Model.

	Predicted values				
True values	Tumor types	Glioma	Meningioma	Non-tumor	Pituitary
	Glioma	183	11	3	1
	Meningioma	0	202	3	1
	Non-tumor	0	0	265	0
	Pituitary	0	3	0	195

Table 4. Accuracy measures for VGG-16 Model.

Evaluation parameters				
Tumor types	Precision	Recall	F1-Score	Specificity
Glioma	85.78%	94.44%	89.90%	95.36%
Meningioma	92.16%	74.27%	82.25%	98.03%
Non-tumor	94.28%	99.62%	96.87%	97.34%
Pituitary	93.59%	95.96%	94.76%	98.05%
Overall Accuracy (OA)	91.46%			

Table 5. Accuracy measures for MobileNetV2 Model.

Evaluation parameters				
Tumor types	Precision	Recall	F1-Score	Specificity
Glioma	100%	92.42%	96.06%	100%
Meningioma	93.51%	98.05%	95.72%	97.88%
Non-tumor	97.78%	100%	98.87%	99.00%
Pituitary	98.98%	98.48%	98.72%	99.70%
Overall Accuracy (OA)	97.46%			

4 Conclusions

The major objective of this study is multi-class brain tumor detection using VGG-16 and MobileNetV2 Deep learning models on MRI images. Results demonstrated that both deep learning models successfully detected brain tumor and classified in the three different types of tumor category i.e. Glioma, Meningioma and Pituitary. It is found that MobileNetV2 obtained highest Overall Accuracy (OA) of 97.46% as compared to VGG-16 Model (91.46%). Results have shown that MobilenetV2 model outperformed and obtained 6.00% higher classification accuracy in comparison to VGG-16 model. On the other hand, in context of implementation of the experiments, it is also observed that training and validation loss reduced rapidly up to 30 epochs and model stabilized at nearly 50 epochs for VGG-16 model. The findings of this study indicates that MobileNetV2 deep learning model is more stable in terms of training and validation accuracy, as well as loss with respect to the number of epochs. It is also observed that MobilenetV2 model training time is significantly lower as compared to VGG-16. Therefore, it can be concluded that performance of deep learning models significantly affected by tuning parameters and selection of deep learning model plays very important role for accurate results.

References

1. Swati, Z.N.K., et al.: Brain tumor classification for MR images using transfer learning and fine-tuning. Comput. Med. Imaging Graph. **75**, 34–46 (2019)
2. Irmak, E.: Multi-classification of brain tumor MRI images using deep convolutional neural network with fully optimized framework. Iranian J. Sci. Technol. Trans. Electr. Eng. **45**(3), 1015–1036 (2021)
3. Mehrotra, R., Ansari, M.A., Agrawal, R., Anand, R.S.: A transfer learning approach for AI-based classification of brain tumors. Machine Learning with Applications **2**, 100003 (2020)
4. Aponte, R.J., Patel, A.R., Patel, T.R.: Brain Tumors. Neurocritical Care for the Advanced Practice Clinician, pp. 251–268. Springer (2018). https://doi.org/10.1007/978-3-319-48669-7
5. Goodenberger, M.L., Jenkins, R.B.: Genetics of adult glioma. Cancer Genet. **205**(12), 613–621 (2012)
6. Weller, M., et al.: Glioma. Nat. Rev. Dis. Primers. **1**(1), 1–18 (2015)
7. Klein, M., et al.: Neurobehavioral status and health-related quality of life in newly diagnosed high-grade glioma patients. J. Clin. Oncol. **19**(20), 4037–4047 (2001)
8. Fathi, A.-R., Roelcke, U.: Meningioma. Curr. Neurol. Neurosci. Rep. **13**(4), 1–8 (2013). https://doi.org/10.1007/s11910-013-0337-4
9. Oya, S., Kim, S.H., Sade, B., Lee, J.H.: The natural history of intracranial meningiomas. J. Neurosurg. **114**(5), 1250–1256 (2011)
10. Chatzellis, E., Alexandraki, K.I., Androulakis, I.I., Kaltsas, G.: Aggressive pituitary tumors. Neuroendocrinology **101**(2), 87–104 (2015)
11. DeAngelis, L.M.: Brain tumors. N. Engl. J. Med. **344**(2), 114–123 (2001)
12. Naser, M.A., Deen, M.J.: Brain tumor segmentation and grading of lower-grade glioma using deep learning in MRI images. Comput. Biol. Med. **121**, 103758 (2020)
13. Ramalho, M., Matos, A.P., Alobaidy, M.: Magnetic resonance imaging of the cirrhotic liver: diagnosis of hepatocellular carcinoma and evaluation of response to treatment—Part 1. Radiol. Bras. **50**(1), 38–47 (2017)
14. Poonam, J.P.: Review of image processing techniques for automatic detection of tumor in human brain. Int. J. Comput. Sci. Mob. Comput. **2**(11), 117–122 (2013)
15. Zhou, L., Zhang, Z., Chen, Y.C., Zhao, Z.Y., Yin, X.D., Jiang, H.B.: A deep learning-based radiomics model for differentiating benign and malignant renal tumors. Translational Oncology **12**(2), 292–300 (2019)
16. Almadhoun, H.R., Abu-Naser, S.S.: Detection of brain tumor using deep learning. Int. J. Acad. Eng. Res. **6**(3) (2022)
17. Sultan, H.H., Salem, N.M., Al-Atabany, W.: Multi-classification of brain tumor images using deep neural network. IEEE Access **7**, 69215–69225 (2019)

18. Amin, J., Sharif, M., Yasmin, M., Fernandes, S.L.: Big data analysis for brain tumor detection: deep convolutional neural networks. Futur. Gener. Comput. Syst. **87**, 290–297 (2018)
19. Deepak, S., Ameer, P.M.: Brain tumor classification using deep CNN features via transfer learning. Comput. Biol. Med. **111**, 103345 (2019)
20. Simonyan, K., Zisserman, A.: Very deep convolutional networks for large-scale image recognition (2014). arXiv preprint arXiv:1409.1556

Five-Year Life Expectancy Prediction of Prostate Cancer Patients Using Machine Learning Algorithms

Md Shohidul Islam Polash(✉), Shazzad Hossen, and Aminul Haque

Department of Computer Science and Engineering, Daffodil International University, Ashulia, Dhaka 1341, Bangladesh
{shohidul15-2523,shazzad15-2420}@diu.edu.bd,
aminul.cse@daffodilvarsity.edu.bd

Abstract. Prostate cancer is the most frequent malignancy and the leading cause of cancer-related mortality globally. A precise survival estimate is required for the effectiveness of treatment to minimize mortality rate. A remedial strategy can be planned under the anticipated survival state. Machine Learning (ML) methods have recently garnered considerable interest, particularly in developing data-driven prediction models. Unfortunately prostate cancer has received less attention to such studies. In this paper, we have built models using machine learning methods to predict whether a patient with prostate cancer would live for five years or not. Compared to prior studies, correlation analysis, a substantial quantity of data, and a unique track with hyperparameter adjustment boost the performance of our model. The SEER(Surveillance, Epidemiology, and End Results) database provided the data for developing these models. The SEER program gathers and disseminates cancer data to mitigate the disease's effect. We analyzed prostate cancer patients' five-year survival state using about seven prediction models. Gradient Boosting, Light Gradient Boosting Machine, and Ada Boost algorithms are identified as top-performed prediction models. Among them, a tuned prediction model using the Gradient Boosting algorithm outperforms others, with an accuracy of 88.45% and found fastest among the other models.

Keywords: Prostate cancer · Five-year survival prediction · Gradient boosting · Parameters tuning · Correlation analysis

1 Introduction

Prostate cancer is the second most common type of malignancy found in men and the fifth most common cause of death worldwide [1]. In terms of prevalence, prostate cancer ranks first, whereas mortality rates put it in third place; this is the most frequent cancer in 105 nations [1,2].

K. K. Patel et al. (Eds.): icSoftComp 2022, CCIS 1788, pp. 314–326, 2023.
https://doi.org/10.1007/978-3-031-27609-5_25

Physicians might design a better treatment plan when treating prostate cancer patients if they know whether or not the patients will live for five years. Sometimes a doctor would try to diagnose a patient based on their physical state by comparing them with the prior patient; however, a doctor can only diagnose a few prostate cancer patients in their lifetime. In this article, we have constructed an artificial prediction model using data from over fifty thousand patients to evaluate the likelihood of patient survival assisting the doctors in preparing the prescription for thousands.

We have collected all of these essential factors from the SEER program, which are supported by the AJCC (American Joint Committee on Cancer) [3]. Through correlation analysis, we found features that had distinct effects and fed the features into the machine for learning. Machine learning (ML) are being used to make medical services more efficient in many sectors. Using machine learning techniques, we sought to forecast whether a patient would live five-year (sixty months) or not. Since the generated target characteristic contains two classes, 0 to 60 months and 61 to more months, it is a binary classification issue. To predict the survival of prostate cancer patients, we employed prominent ML classifiers such as Gradient Boosting Classifier (GBC), Light Gradient Boosting Machine (LGBM), AdaBoost (ABC), Decision Tree (DT), Random Forest (RF), and Extra Trees (ETC). Finally, hyperparameter optimization was used to enhance prediction outcomes. Furthermore, we interpreted our data using accuracy [1,4], precision [1,5,6], sensitivity [4], specificity [4], AUC [5,11], and ROC curves. The new feature added in interpretability is how fast our model predicts. All the boosting methods performed comparably; hence the best model was picked using performance measurements and prediction speed. The optimized GBC performs better than the other classifiers, with an accuracy of 88.45%.

The major contribution of our work is presented below:

1. An optimized five-year life expectancy prediction model using the Gradient boosting technique for prostate cancer patients. Our model performed better compared to Wen et al.'s [1] prediction model.
2. Optimizing parameters for GBC to build survival prediction models.
3. Identified a fast forecast model for five-year life expectancy. This technique can be helpful to conduct other cancer research.

In feature engineering, we employed label encoding to build prediction models with enormous data and characteristics with unique impacts. We improved the performance by optimizing the hyperparameters; also, we examined how quickly the models run on various platforms, which made the prostate cancer life expectancy prediction model superior to others. Additionally, to the best of our knowledge, only a few studies have predicted prostate cancer survival. Wen et al. [1] demonstrated prostate cancer five-year survival; moreover, their accuracy was 85.64%, whereas we achieved 88.45%.

The remains of the paper are categorized into five sections. Previous works concerning planning algorithms are described in Sect. 2, and suitable methods are presented in Sect. 3. The results acquired are shown and discussed in Sect. 4. The key points are presented in Sect. 5, which is the conclusion of the paper. The conflict of interest is presented in section 6 of the article.

2 Literature Review

Many cancer diagnosis studies have made use of machine learning approaches. Wen et al. [1] looked into prostate cancer prognosis. He used ANN, Naive Bayes(NB), Decision Trees(DT), K Nearest Neighbors(KNN), and Support Vector Machines (SVM) as ML methods and applied some preprocessing approaches. His goal survival categories are 60 months or more. With an accuracy of 85.6%, ANN's result is the best.

Cirkovic et al. [4], Bellaachia [6], and Endo et al. [7] used a similar target method to predict breast cancer survival. Cirkovic et al. [4] created an ML model to estimate the odds of survival and recurrence in breast cancer patients. However, they only employed 146 records and twenty attributes to predict 5-year survivorship; the best model was the NB classifier. On the SEER dataset, Bellaachia [6] applied three data mining techniques: NB, back-propagation neural networks, and DT(C4.5) algorithms, with C4.5 performed better. Endo et al. [7] attempted to generate the five-year survival state by contrasting seven models (LR, ANN, NB, Bayes Net, DT, ID3, and J48). The logistic regression model is the most accurate, with an accuracy of 85.8%.

Montazeri et al. [5] and Delen [8] predicted whether or not the patient would live. Montazeri et al. [5] created a rule-based breast cancer categorization technique. He used a dataset of 900 patients, just 24 of whom were men, accounting for 2.7% of the total patient population. He used traditional preprocessing approaches and machine learning algorithms such as NB, DT, Random Forest(RF), KNN, AdaBoost, SVM, RBF Network, and Multilayer Perceptron. He used accuracy, precision, sensitivity, specificity, area under the ROC curve, and 10 fold cross validation to evaluate the model. With a 96% accuracy rate, RF outperformed the earlier methods. Delen [8] used decision trees and logistic regression to develop breast cancer, prediction models. A well-used statistical method with 10-fold cross-validation was demonstrated for comparing exceptional performances. The model with the best performance is the DT model, with 93.6% accuracy.

Regarding survival, the authors of [1] concentrated on estimating whether a patient with prostate cancer will live for 60 months or five years. They built the model using data from 2004 to 2009, where the usage of 15 attributes was observed. No feature impact analysis information was found. Additionally, we also choose five years of survival as our objective quality. However, this trait is heavily utilized to forecast the survival of other cancers [4], [6], and [8]. In order to forecast cancer survival, Montazeri et al. [5], Agrawal et al. [11], and Lundin et al. [12] employed fewer records. For breast, prostate, and lung cancer survival prediction, it appears that Wen et al. [1], Delen [8], and Pradeep [10] utilized a few (fewer than five) different methods. Agrawal et al. [11], Lundin et al. [12], and Endo et al. [7] created models for breast and lung cancer using a limited set of traits. Additionally, none of the models' prediction times was mentioned, and the different quantity of characteristics permitted our attempts to move forward.

3 Methodology

The final model is created by adjusting the hyperparameters and based on a fundamental machine learning life cycle. Data collection, preprocessing, data splitting into train and test sets, model building, cross-validation, and model testing are crucial steps. Finally, the GBC model is the most accurate predictor of a patient's five-year survival.

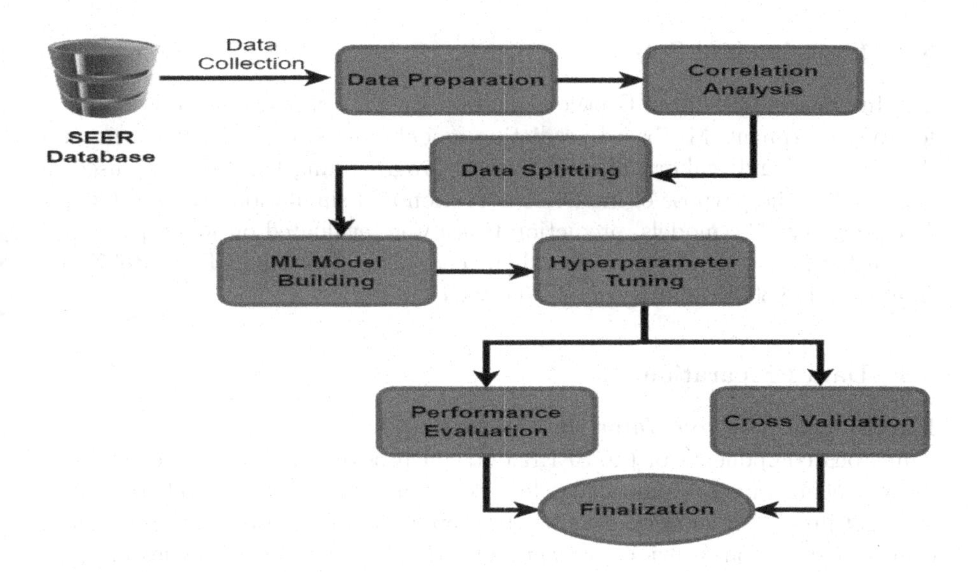

Fig. 1. Procedural Framework

3.1 Data Collection

We mine the SEER database for relevant information for this study. It intends to lessen the burden of sickness on Americans. The collection has 37 features and 187798 entries. The AJCC has found that these elements have a causal connection to cancer and can offer intelligence to machine reasoning inclusion in the SEER database. [3]. As we are predicting five-year survival, we took data of people who died between 0 to 60 months in one class and the rest of the data in another class. The details, such as the patient's ID number and the year of diagnosis, are not crucial to our prediction. The first finding after getting the data was that every patient was a man; hence the sex factor was disregarded. We created our objective attribute, five-year survival or not, based on domain analysis and from the survival months feature. Columns with a single class are excluded. Moreover, 27 facts have included in the data processing. The remaining factors are Age, Laterality, RX Summ Surg Prim Site, CS extension, CS lymph nodes, Derived AJCC N, Histology recode broad groupings, Diagnostic Confirmation, RX Summ Scope Reg LN Sur, Reason no cancer directed surgery, Site recode rare tumours, AYA site recode, Regional nodes examined, RX Summ Surg Oth Reg or Dis, ICD O 3 Hist or behav, First malignant primary indicator, Grade, CS tumour size, CS mets at dx, Derived AJCC Stage Group, Derived AJCC N, Derived AJCC M, Derived AJCC T, Race recode, Total number of benign or borderline tumours for patient, Total number of in situ or malignant tumours for the patient, and Regional nodes positive. Furthermore, we have seen these in other works; hence, they inspired many authors.

3.2 Experimental Setup

The Intel Xeon CPU from Google Colaboratory was utilized to generate models for the experiment. ML-based prediction model created using scikit-learn, pandas, NumPy, and seaborn library. Python programming language was used in general. For the purpose of model construction and simulation, we used Google Colaboratory. The models' prediction times were evaluated on many platforms and utilized a linear process with a single core of the following CPU: Intel Xeon, Intel Core i5-9300H, and Ryzen 7 3700X CPUs.

3.3 Data Preparation

Handling of Missing Value and Data Duplication

Numerous components of the SEER data on prostate cancer are lacking. The development of medical knowledge has led to the emergence of novel traits that were not previously present. As a result, there is no information on these current characteristics from previous patients. Our database has 10349 entries missing, which is about 5.5% of all records. Comparing this 10349 data point to 187798 data points, it is tiny. Patients with prostate cancer seek an accurate diagnosis for better treatment; therefore, adding the missing data could produce inaccurate results. As a result, the data that went missing has been completely removed.

There is a chance that many patients' physiologies are identical. Due to the fact that we found data duplication, which could have led to incorrect findings, we deleted the redundant data. After removing, there were 54731 total records.

Feature Engineering

The only way to build a prediction model is to feed it a numerical data representation. The majority, 19 out of 27, There are eight numeric attributes and ten nominal qualities in the dataset. Thus, 19 attributes need to be converted to a machine-readable format. The Label encoding module from the Python scikit-learn package was utilized to do this. Label encoding is the process of turning labels that are written out in words into numbers.

3.4 Correlation Analysis

Correlation coefficients show how closely two things are related to each other. Strong correlation coefficients of +0.8 to +1 and –0.8 to –1 show the same behaviour [14].

$$r = \frac{\sum(x_i - \bar{x})(y_i - \bar{y})}{\sqrt{\sum(x_i - \bar{x})^2 \sum(y_i - \bar{y})^2}} \tag{1}$$

We found the correlation coefficients(r) using Eq. 1. Xi is the value of one feature, and the x bar is the average of all the values. Yi is the value of another feature, and the y bar is the average of all the values for that feature.

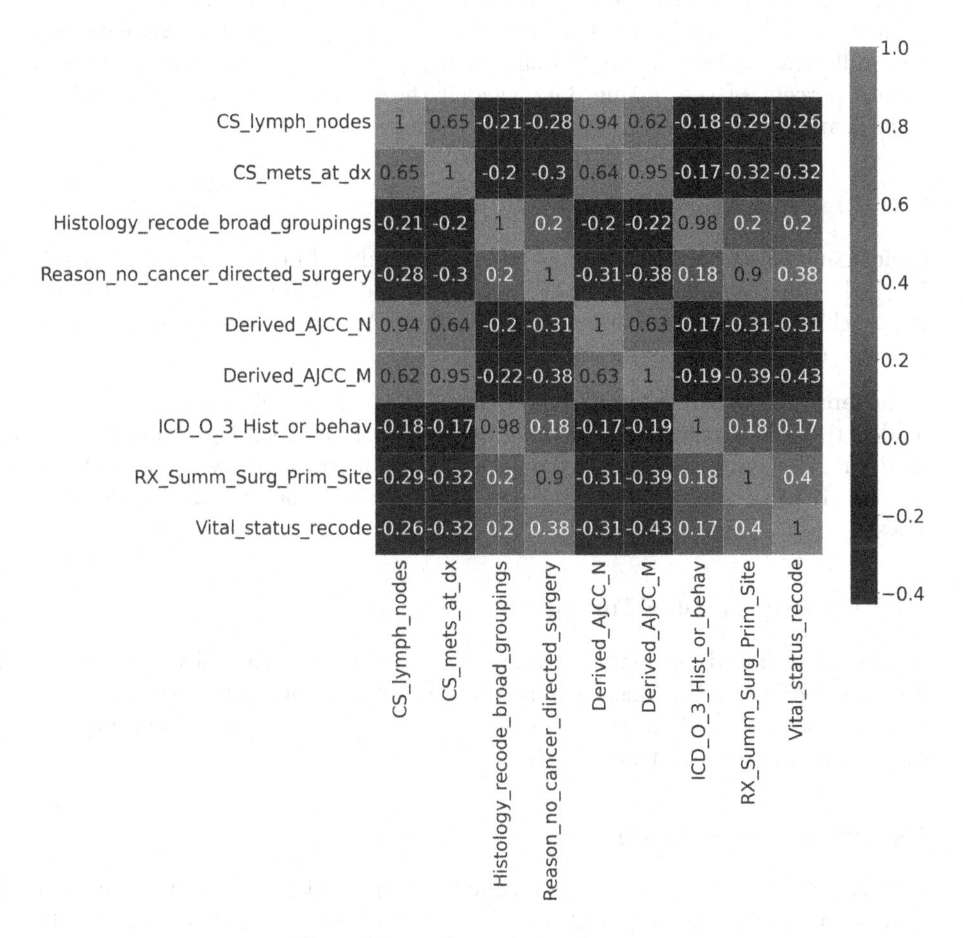

Fig. 2. Zoomed correlation heatmap

Figure 2 shows that four pairs of attributes act in the same way. Since their correlation coefficients are both 0.98, "Histology recode broad groupings" and

"ICD-O-3 Hist or behav" have the same effect. In the same way, "RX Summ Surg Prim Site" and "Reason no cancer-directed surgery" have coefficients of 0.9, "CS Mets at dx" and "Derived AJCC M" have coefficients of 0.95, and "CS lymph node" and "Derived AJCC N" have coefficients of 0.94. In the dataset, the effects of these two are the same. One attribute from each of these four pairs will be used to make the models. "ICD-O-3 Hist or Behavior," "RX Summ Surg Prim Site," "CS Lymph Nodes," and "CS Mets at Diagnosis" were left out. Also, the rest of the attributes are unique in their own ways.

3.5 Data Splitting

To train and test our models, we used a variety of data-to-sample ratios. When comparing 70:30, 75:25, and 80:20 training and testing splits, the average accuracy difference is less than 1%. Using eighty percent of the training data and twenty percent of the testing data yielded the best results. Scikit-"train-test-split" learn's plugin is used for this split testing.

3.6 Machine Learning Model Building

Performance of the six ML models and ANN used to forecast survival. LGBM, RF [11], ETC [7,8], GBC [14],DT [1,4,8,11], and AdaBoost classifier [13] are the machine learning techniques. The Gradient Boosting Classifier was used to generate our best forecasting model.

Gradient Boosting: Gradient boosting classifiers combine weaker learning models to build a powerful predicting model. Gradient boosting techniques are used for identifying challenging datasets. This is used for regression and classification. It is an ensemble of weak prediction models that usually use decision trees.

3.7 Hyperparameter Tuning

We have optimized the parameters of the gradient boosting algorithm using RandomizedSearchCV. Sklearn library's method RandomizedSearchCV examines the results of machine learning models with random parameters and suggests the best model with best parameters.

3.8 Performance Evaluation

Following the development of the machine learning models, specific tests are run to determine their viability. Our model's accuracy, F1 score, Precision, Recall, Cross-Validation [9,14], and time interpretability have all been assessed. A confusion matrix is a required component for measuring these metrics. A confusion matrix, also known as an error matrix, is a table structure used in machine learning to show the effectiveness of a supervised learning system. We get True

Positive (TP), False Positive (FP), True Negative (TN), and False Negative using a confusion matrix (FN).

Equations that need to calculate ML performance measurements:

$$Accuracy = (TP + TN)/(TP + TN + FP + FN) \tag{2}$$

$$Recall = TP/(TP + FN) \tag{3}$$

$$Precision = TP/(TP + FP) \tag{4}$$

$$F1Score = 2 * (Recall * Precision)/(Recall + Precision) \tag{5}$$

AUC: Area Under the Curve (AUC) It has used to measure performance over many criteria. It measures how far apart result is. It also shows how successfully the model classifies data.

ROC Curve: The total classification levels of a categorization model are represented graphically by a receiver operating characteristic (ROC) curve. This curve depicts two variables: True Positive and False Positive Rates.

Sensitivity: Sensitivity is another word for actual positive rate, which is the percentage of positive samples that give a positive result when a specific test is added to a model and does not change the samples.

$$Sensitivity = TP/(TP + FN) \tag{6}$$

Specificity: In the context of an unaffectedly negative model, the true negative rate, sometimes referred to as specificity, is the percentage of samples that test negative when the test is employed.

$$Specificity = TN/(TN + FP) \tag{7}$$

Results of Table 1 calculated using these equations.

3.9 Cross-Validation:

Overfitting is avoided in prediction models by cross-validation, which is especially useful in situations when data are scarce. Cross-validation creates a predefined number of data folds, analyzes each, and averages the error estimate. Stratified k-fold cross-validation has been used to cross-validate our data. We generated ten fold of our data for the cross-validation test.

As part of our study, we have tried several different approaches. We have provided extensive information regarding the methods that allowed us to accomplish our goals.

4 Results and Discussion

Using machine learning techniques, we developed models that could forecast a patient's five-year survival after a diagnosis of prostate cancer. Many prediction models have developed, and the mandatory job is to pick the most effective one.

Table 1. Performance measurements of the ML algorithms

Algorithms	Accuracy (%)	F1 Score	Precision	Recall	AUC	Avg. cross validation
Tuned gradient boosting	88.45	0.844	0.842	0.8470	0.905	0.8811
Gradient boosting	88.386	0.8443	0.8416	0.8471	0.9044	0.8798
LGBM	88.0351	0.8371	0.8396	0.8346	0.8989	0.8794
Ada boost	86.9825	0.8254	0.8229	0.8281	0.8989	0.8758
Random forest	86.5965	0.817	0.8202	0.814	0.8776	0.8596
Extra trees	85.0877	0.7947	0.8005	0.7895	0.8522	0.851
Artificial neural network (ANN)	84.98	0.8033	0.7949	0.8134	0.852	0.824
Decision tree	82.8421	0.767	0.7682	0.7659	0.7914	0.8238

The predictive efficacy of the models is displayed in Table 1. Thus, it is clear that tree-based boosting methods outperformed when applied to our dataset. Each of them is around 86% to 88% accurate. Now we have many potential models, need to choose the best one.

In Table 1, Accuracy, Recall, Precision, and F1 score were calculated using Eqs. 2, 3, 4, and 5, respectively. There we can see that Gradient Boost performs best based on accuracy, and its accuracy is 88.386%. The accuracy difference between the top two models is 0.27%, which implies GBC can accurately predict nearly ten more people than LGBM. The AUC score from Table 1 and the ROC curve in Fig. 3 shows that the top performed model GBC covered almost 90.044% of the data correctly. The rest of the algorithms are also covered near of the GBC model, but the blue curve indicates GBC model covers more data than other models. However, we must consider which algorithm can identify the two target classes more accurately. Thus the sensitivity (green pillars) of Fig. 4 clearly distinguishes the algorithms. The sensitivity of the GBC model has achieved 0.9265, which is more than others. Nevertheless, the specificity (blue pillars) is slightly down from the LGBM model, but the value is in an acceptable range. Using Eqs. 6 and 7, the measurements have calculated. Moreover, we also tried to improve the performance of gradient boosting and found that the model's accuracy improved from 88.38 to 88.45 (Table 1) by tuning the hyperparameters. A 0.07% increase in accuracy resulted in 3 new patients correctly predicting life expectancy. The parameters that made this improvement are: min-samples-split is 10 and n-estimators is 200 rest of the parameters will be the default.

In order to determine whether the GBC model is the most effective, we compared its processing time to that of other models. Figure 5 carries the pre-

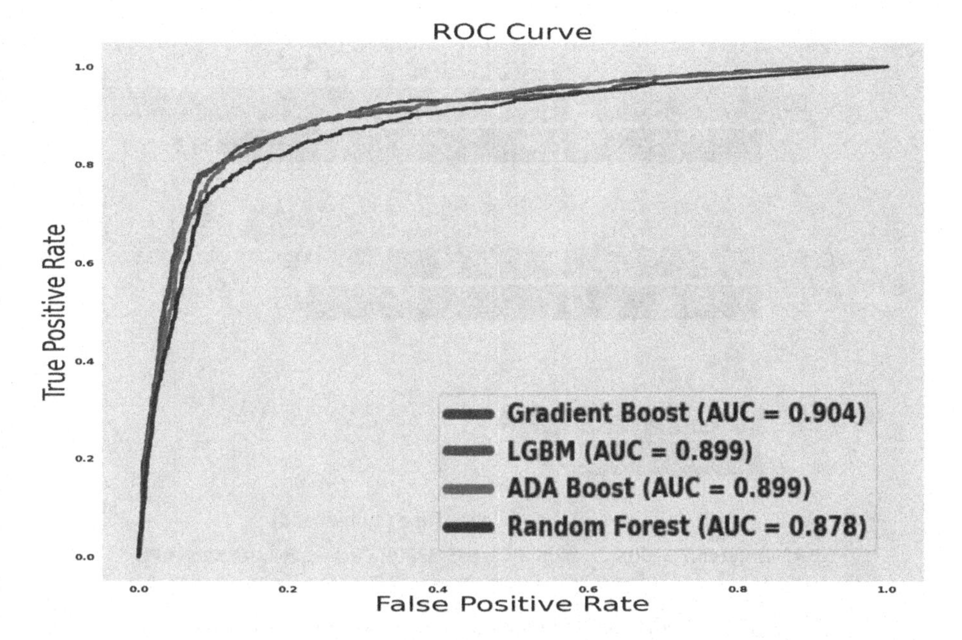

Fig. 3. ROC curve

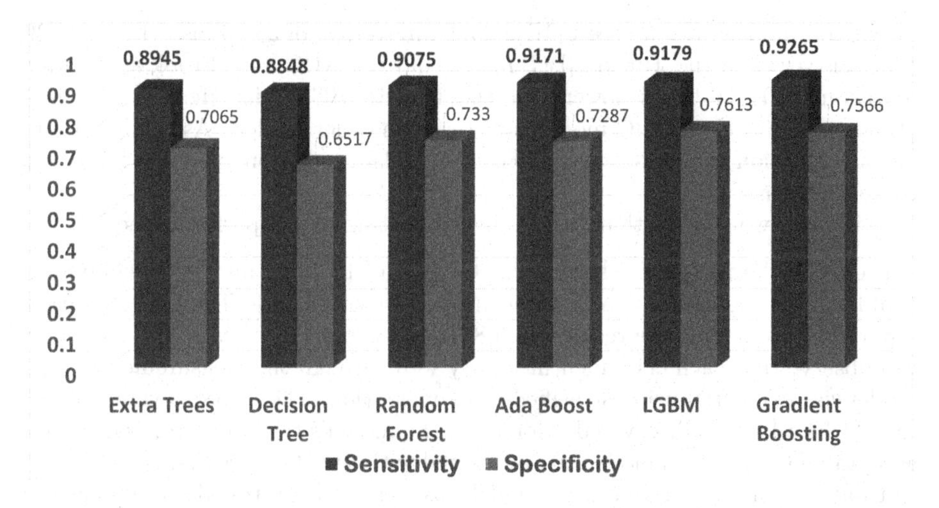

Fig. 4. Sensitivity and specificity of algorithms.

diction time of each algorithm, and it specifies our preferred algorithm. Various ratios of the dataset, including test data, train data, and single-person data, were utilized to examine the prediction time. It is visualized that Tuned GBC has taken 14.7 ms on average on three different platforms. The AMD Ryzen 7 and Intel Core i5 processors took 15.6 ms, and the Intel Xeon processor from colab took 12 ms to predict the test data. Compared with other

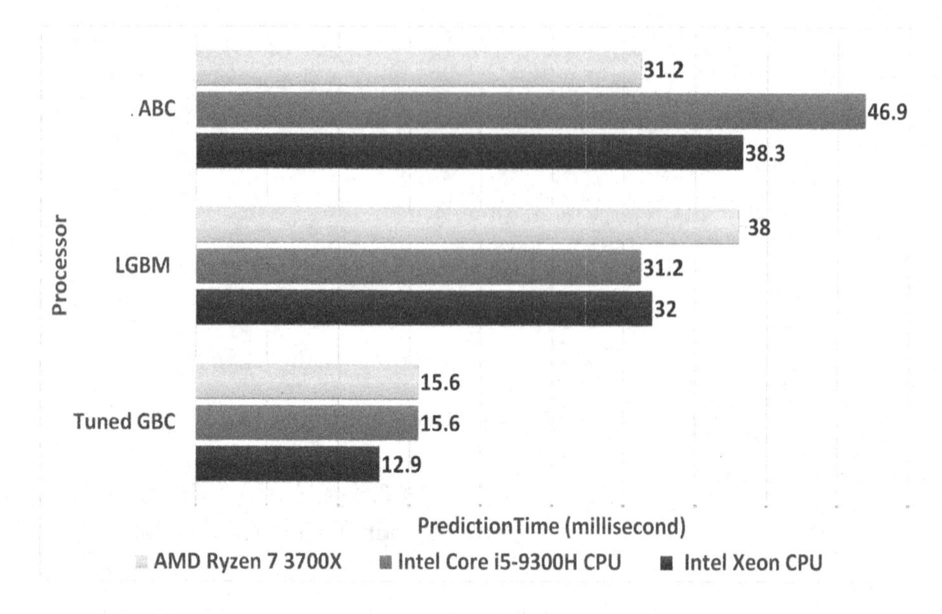

Fig. 5. Prediction time for different models on test data.

prediction models, we can see LGBM took an average of 33.73 ms, which is 2.29 times slower than the Tuned GBC model, and the ABC model took an average of 38.8 ms it is 2.6 times slower than the Tuned GBC model. Here clearly visualized that the tuned GBC model is faster than other models. As other authors did not mention cancer survival forecast models' prediction time, we could not compare them.

Till now, we can see that the gradient boosting model performs the best.

Tuned GBC and GBC Model: The GBC model gave us an F1 score of 0.8443 (Table 1). We were able to get an accuracy rate of 88.38%. The categorization report indicated that the recall was 0.847, and the precision was 0.841. Clearly can observe that each class identified very well through the measurements. This model was evaluated using stratified cross-validation. The 87.98% average accuracy of 10 folds was discovered. Moreover, the model's training error was within acceptable limits. Both models are evaluated with test data; we first set aside the test data, then after training the model, test with unseen test data and get the results. Cross-validation yielded a positive outcome, which is a way to eliminate future data leakage; we may assume that the model will also do well with new data. Consequently, it is evident that this model does not exhibit any overfitting. Therefore, this is a good demonstration. Similarly, tuned GBC increased the accuracy from 88.386% to 88.45%, AUC 0.9044 to 0.905, cross-validation 0.8798 to 0.8811, and the rest of the values are nearly the regular GBC model. Comparing the normalized confusion matrix of Fig. 6, we can see that the value of "61 to more months" has increased from 0.77 to 0.78. As we already know, three more patients' five-year survival becomes accurate for this increase of results, so the tuned gradient boosting model will be our proposed model.

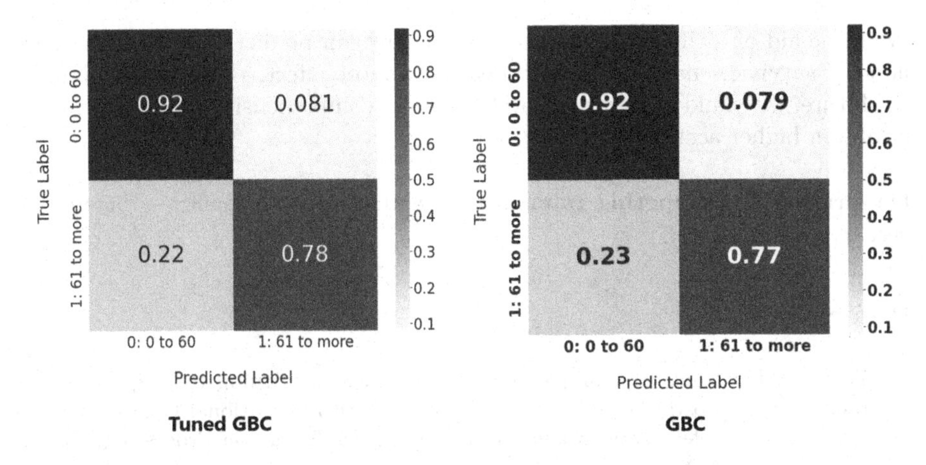

Fig. 6. Normalised confusion matrix of tuned GBC and GBC model.

Comparisons

Limited research has done on prostate cancer survival. Wen et al. [1] has researched prostate cancer's five-year survival by using SEER data from 2004 to 2009, but we used the data from 2004 to 2018. They used 15 attributes, whereas we used 27, including their 15. He got 85.64% accuracy using the neural network, but we got 88.45% with the tuned gradient boosting model. The newest SEER features have a significant impact on prediction results. Moreover, correlation analysis reviewed the unique impact of each feature in our work. The inclusion of the latest data and features, our preprocessing techniques, feature impact analysis, and hyperparameter tuning have differentiated our research from them and yielded improved results. And demonstrated fresh approaches to evaluate models by determining the prediction speed. In contrast, we provided evidence regarding how much time our model needs to make an accurate forecast, but the other authors did not. Our findings were reevaluated using cross-validation, and we applied the unique impacted features found by using correlation.

In summation, for predicting 5-year survival in prostate cancer patients, we propose the tuned prediction model built using the gradient boosting algorithm.

5 Conclusion

In this paper, we seek to assess the five-year life expectancy of prostate cancer patients with computational intelligence. Several strategies are examined based on characteristics with distinct effects. Our customized Gradient Boosting prediction model is proposed to estimate the five-year survival of patients with prostate cancer. This model is the top performer in terms of prediction speed, accuracy, AUC score, and sensitivity. It is also demonstrated that our model is superior to others in terms of performance. Our model would play a revolutionary role in digitising prostate cancer medical diagnosis process. Moreover, it accurately predicted the five-year life expectancy of 88.45% of patients.

With the aid of artificial intelligence, physicians can predict the patient's likelihood of survival, enabling them to develop a more effective treatment plan. In the future, we would attempt to develop a more comprehensive prediction model to obtain higher accuracy.

Declaration of Competing Interest. The authors have no conflicts of interest to declare.

References

1. Wen, H., Li., S, et al.: Comparision of four machine learning techniques for the prediction of prostate cancer survivability. In: 15th International Computer Conference on Wavelet Active Media Technology and Information Processing, vol. I5, pp. 112–116 (2018)
2. Delen, D., Patil, N.: Knowledge extraction from prostate cancer data. In: 39th Annual Hawaii International Conference on System Sciences (HICSS 2006), vol. I5, pp. 92b–92b (2006)
3. Lynch, C.M., et al.: Prediction of lung cancer patient survival via supervised machine learning classification techniques. Int. J. Med. Inf. **108**, 1–8 (2017)
4. Cirkovic, B.R.A., et al.: Prediction models for estimation of survival rate and relapse for breast cancer patients. In: 15th International Conference on Bioinformatics and Bioengineering (BIBE), vol. I5, pp. 1–6 (2015)
5. Montazeri, M., et al.: Machine learning models in breast cancer survival prediction. Technol. Health Care **24**, 31–42 (2016)
6. Bellaachia A, Guven E.: Predicting breast cancer survivability using data mining techniques, pp 10–110 (2006)
7. Endo, A., et al.: Comparison of seven algorithms to predict breast Cancer survival. Int. J. Biomed. Soft Comput. Human Sci. **13**, 11–16 (2008)
8. Delen, D., Walker, G., Kadam, A.: Predicting breast cancer survivability: a comparison of three data mining methods. Artifi. Intell. Med. **34**, 113–127 (2005)
9. Mourad, M., et al.: Machine learning and feature selection applied to SEER data to reliably assess thyroid cancer prognosis. Scient. Reports **10**, 1–11 (2020)
10. Pradeep, K., Naveen, N.: Lung cancer survivability prediction based on performance using classification techniques of support vector machines, C4. 5 and Naive Bayes algorithms for healthcare analytics. Proc. Comput. Sci. **132**, 412–420 (2018)
11. Agrawal, A., et al.: Lung cancer survival prediction using ensemble data mining on SEER data. Sci. Program. **20**, 29–42 (2012)
12. Lundin, M., et al.: Artificial neural networks applied to survival prediction in breast cancer. Oncology **57**, 281–286 (1999)
13. Thongkam, J., et al.: Breast cancer survivability via AdaBoost algorithms. In: Proceedings of the second Australasian Workshop on Health Data and Knowledge Management, vol. 80, pp 55–64, Citeseer (2008)
14. Polash, Md.S.I, Hossen, S., et al.: Functionality testing of machine learning algorithms to anticipate life expectancy of stomach cancer patients. In: 2022 International Conference on Advancement in Electrical and Electronic Engineering (ICA5), pp 1–6 (2022)

An Ensemble MultiLabel Classifier for Intra-Cranial Haemorrhage Detection from Large, Heterogeneous and Imbalanced Database

Bharat Choudhary[1], Akhitha Babu[1], and Upasana Talukdar[2](✉)

[1] Department of Data Science and Analytics, Central University of Rajasthan, Rajasthan, India

[2] Department of Computer Science and Engineering, Indian Institute of Information Technology, Guwahati, Assam, India

upasana@iiitg.ac.in

Abstract. Fast and accurate detection of traumatic brain injuries such as intracranial Haemorrhage (ICH), is substantial as it may lead to death or disability. In this paper, a novel deep learning-based approach is proposed for automated detection of ICH from non-contrast CT scans. The proposed model is a hybrid methodology that amalgamates Convolution Neural Network(CNN) with Support Vector Machine (SVM) and Extreme Gradient Boost (XG-Boost) for detection of ICH. Methodologies for dealing with multi-labeled and imbalanced data are very rarely employed in the literature. This paper put forwards an ensemble multi-label classifier for automatic ICH detection on a huge, heterogeneous, clinical, and imbalanced database of brain CT studies. In the proposed method, CNN is trained and then the extracted feature output received from CNNs is fed to the CNN, XG-Boost, and SVM for classification. Experimental results presents a high performance by the proposed model in the detection of ICH when compared with CNN, CNN with SVM, and CNN with XG-Boost.

Keywords: Intracranial Haemorrhage (ICH). · MultiLabel classification. · Imbalance pre-processing. · Deep Neural Network

1 Introduction

Intracranial Haemorrhage (ICH) or intracranial bleed is any bleeding within the skull vault or calvaria or intracranial vault. ICH is classified into extra-axial that includes bleeding within the skull but outside of the brain tissue. i.e., in the epidural or extradural (EDH), subdural (SDH), and sub arachnoid (SAH) spaces and intra-axial or intracerebral haemorrhage (ICH) which includes bleeding within the brain itself. i.e., intraparenchymal (IPH) and intraventricular (IVH) spaces. Worldwide, the overall occurrence of spontaneous ICH is 24.6 per 100,000 person-years [1]. ICH is a serious medical condition as the growth of blood within the skull

K. K. Patel et al. (Eds.): icSoftComp 2022, CCIS 1788, pp. 327–340, 2023.
https://doi.org/10.1007/978-3-031-27609-5_26

Table 1. Existing approaches on Deep Learning Based ICH Detection

Ref	Year	First author surname	Description
[3]	2018	Arbabshirani et al	CNN was employed. The training and testing set comprised 37,074 and 9499 studies; achieved an AUC of 0.846
[4]	2019	Kuo et. al	CNN was employed and attained AUC of 0.991
[5]	2019	Lee et al	Implemented CNNs (ensemble) and attention maps; attained 95.2% sensitivity, 94.9% specificity, AUC of 0.975,
[6]	2021	Avanija et al	DenseNets were used for the detection of ICH and attained an accuracy of 91%
[7]	2020	Hssayeni et al	U-Net was employed and achieved a Dice coefficient of 0.31
[8]	2021	Watanabe et al	U-Net was employed and attained an accuracy of 89.7%
[9]	2021	Wang et al	2D CNN classifier and two sequence models; attained an AUC of 0.988
[10]	2022	Ganeshkumar et al	CycleGan was used as a data augmentation method; attained F1 score, sensitivity and specificity and sensitivity of 0.91, 0.80, and 0.99 respectively
[11]	2022	Jeong et al	Implemented conditional DCGAN (cDCGAN) as data augmentation approach

can escalate intracranial pressure. This in turn limits the blood supply and can crush brain tissue. Computerized tomography (CT), especially non-contrast CT is often the first diagnostic modality which is the most readily available and widely used technique for the diagnosis and identification of ICH.

Extraction of value from medical imaging calls for high-quality interpretation. But human interpretation is limited and prone to errors. Accounting for over 90% of all medical data, usually medical imaging generates large volumes of data. Emergency room radiologists have to analyze a huge number of medical images, with each medical study involving up to 3K images which are about 250GB of data. Hence, an automatic classification of medical images for ICH detection plays an indispensable role in attaining enhanced clinical outcomes. Machine learning is a widely used technique for enabling computers for automatic learning and detecting patterns. In recent years, there has been a huge interest in the medical field for augmenting diagnostic vision with machine learning for enhanced interpretation [2]. Deep Neural Network (DNN), a class of machine learning algorithms, has the advantage of having the ability for performing a varied automated classification tasks [1]. Literature reported different approaches that applied DNN for ICH detection (See Table 1).

It is seen from the literature that while many research efforts have been publicised till date to detect ICH, most of the existing approaches are designed to handle multi-class instead of multi-label classification. Also, methodologies for dealing with imbalanced data in multi-label classification problems are substantial, but very rarely employed. The problem of the class imbalance is one of the notable and most critical challenges and is described as having an uneven distribution of the data. Because of the dominance of one class, traditional machine learning algorithms may fail to identify ICH cases correctly. Another common drawback of existing brain CT studies for ICH detection is the lack of heterogeneity that is usually encountered in clinical practice as data are collected from common institution instead of varied places. To our knowledge, very few studies have reported the automatic detection of ICH using a large cross-sectional imaging imbalanced database [3,4,6,8]. Nguen et al. [12] proposed a hybrid approach that amalgamates CNN with the LSTM (long short-term memory) model for precise prediction of ICH on the large imbalanced database. However, they have reported individual class accuracy instead of considering multi-class accuracy. Wang et al. [9], Ganeshkumar et al. [10], and Jeong et al. [11] put forwarded different DNN models on large heterogeneous and imbalanced dataset. Ganeshkumar et al. [10], and Jeong et al. [11] proposed data augmentation methodologies while Wang et al. [9] amalgamated two DNNs for ICH detection.

This paper put forward an approach of an ensemble multilabel classifier. The imbalanced pre-processing methodology has been employed to handle class imbalance problem. Experimental results are evaluated in terms of Binary cross-entropy loss, Classification Accuracy (Accuracy), F1-score and Area Under ROC Curve (AUC).

The rest of the paper is organized as follows: The description of the dataset is presented in Sect. 2 while Sect. 3 presents the proposed framework for ICH detection. The experimental results are presented in Sect. 4, while Sect. 5 and Sect. 6 present Discussion and Conclusion respectively.

2 Dataset Description

The brain CT dataset collected by the Radiological Society of North America (RSNA®)[1], members of the American Society of Neuroradiology, and MD.ai is used for the study. The RSNA Brain haemorrhage CT Dataset is the largest public dataset comprising a heterogeneous and huge collection of brain CT studies from varied institutions [13]. The dataset can be also viewed as a *"real-world"* dataset comprising of intricate instances of cerebral haemorrhage in inpatient and emergency settings [13]. It is composed of 25312 examinations, annotated non-contrast cranial CT exams with 21784 for training/validation and 3528 for test. In total, the dataset comprises 752,803 images as training data and 121,232 images as testing data. It consists of a set of image IDs and multi labels indicating the presence and absence of haemorrhage; if present, then its type (subarachnoid, intraventricular, subdural, epidural, and intraparenchymal haemorrhage). The detailed description of the dataset can be found in [13].

3 Proposed Ensemble MultiLabel Classifier for ICH Detection

Figure 1 describes the overall framework of our proposed model. The details are in the following subsections.

[1] https://www.kaggle.com/c/rsna-intracranial-hemorrhage-detection.

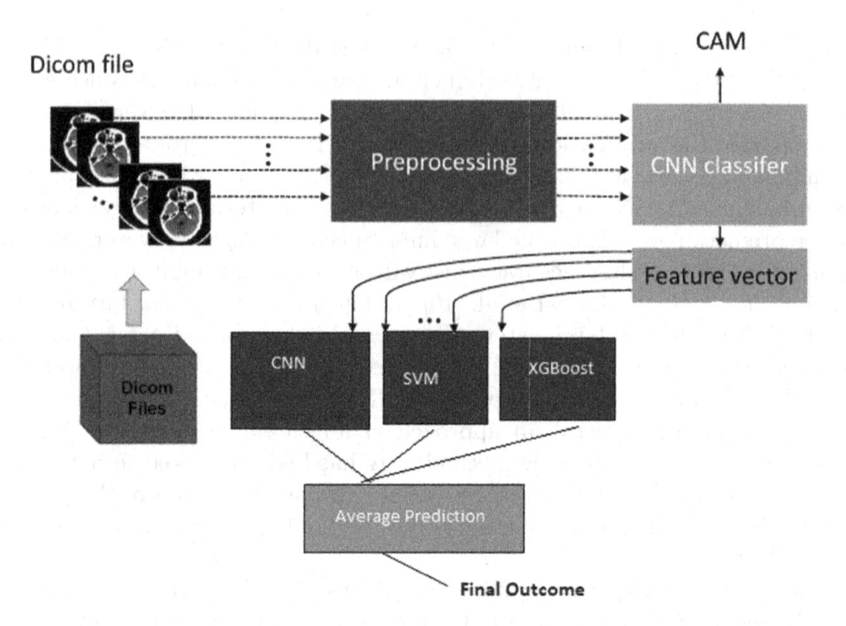

Fig. 1. The overall methodology of ICH Detection

3.1 Data Pre-processing

This module includes procedures of Windowing and Handling Imbalanced Data as discussed in the following subsections.

Windowing: In the first phase, the CT scans have been pre-processed before feeding them into the model. The scans that are stored in the form of DICOM files are reconstructed to form input for the proposed model. At the onset, the threshold of BSD (brain, subdural, bone) windowing of single slice is taken and the CT scan is transformed into three windows: brain window (Level:40, Width: 80), subdural window (Level: 80, Width: 200) and bone window (Level: 600, Width:2000). The three windows are concatenated to form three channel images. This is followed by a resampling technique to get a little more spatial resolution from adjacent slices which kept the pixel spacing consistent, else it may be hard for the model to generalize. Finally, the image is cropped to focus on the informative part as shown in Fig 2. Later, the three adjacent slices of the CT scan with brain window (Level:40, Width: 80) are concatenated to construct RGB (Red, Green, Blue) images. The CT scan metadata enabled the information of the spatially adjacent slices: R = St-1, G = St, B = St+1; where St is the slice of the CT scan. Finally, the image is cropped to focus on the informative part as shown in Fig. 2.

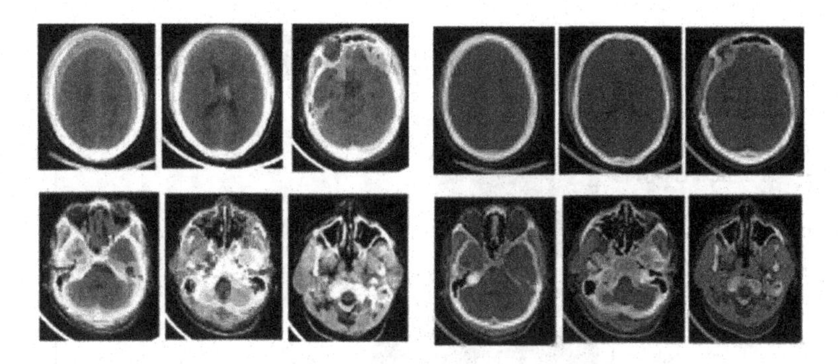

Fig. 2. CT scan images of a patient with intraventricular haemorrhage in which two different preprocessing techniques are used; left-hand side set of images by the first method and right-hand side set of images by the second method of windowing.

Handling Imbalanced Dataset: The dataset is highly skewed and imbalanced (See Fig. 3). Thus, advanced CNN architectures fail to yield good results. Along with the six classes, each class has two subclasses, where 0 represents the negative case and 1 represents the positive case. We have roughly above +6.6 lacs samples from negative class and hardly 86K samples from positive class. It is even extremely difficult for most complicated DNNs to classify instances of an imbalanced dataset correctly. The foremost concern with the imbalanced dataset is because of the biasness of the dominant class (the class with more data) the model will be itching to identify everything as the dominant class.

To tackle the issue of class imbalance problem in multi-label classification, a few effective techniques in the following order are used which resulted in improving performance:

- **Weighted Class Approach** [14]: This approach assigns individual weights for each class instead of a single weight (generally based on the ratio of occurrence of each class) while the data fit the proposed model. Since class-weighting adds up to fine-tune decision thresholds, it has been observed to be useful for estimating class probabilities. This permits to up weight a minority class to achieve better performance on the minority examples.
- **Under Sampling Techniques** [15]: The biasness which knock down the classification performance and increases the number of false negatives are reduced using resampling technique. A simple under-sampling technique is used to include all the samples from the minority class (positive class) selecting randomly twice the number of sampling from the majority classes. It is a simple and effective technique but with some limitations as removing samples will lead to information loss without knowing how important they might be in prediction but still, it gives better results.
- **Data Augmentation:** It helps to substantially enhance the diversity of data of the training models, without acquiring new data. In this process, "new" training samples are generated from the original samples by applying random

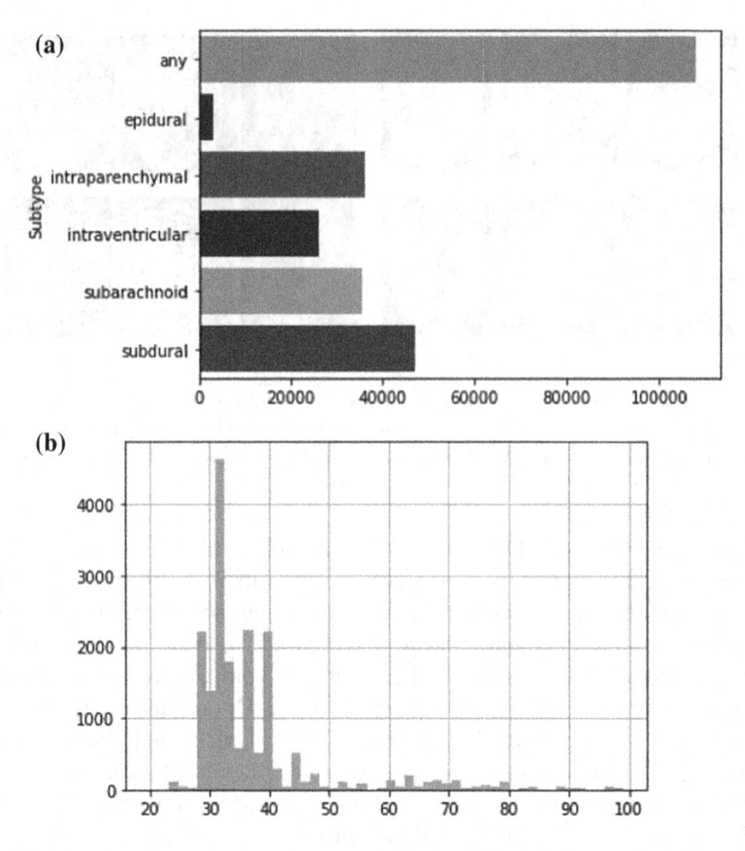

Fig. 3. (a) The distribution of each type of haemorrhage. (b) The distribution of the number of CT-Scan for each Patient ID

jitters and perturbations without altering the data's class labels. The samples which belong to the minority class are increased by up-sampling using various image transformation techniques like sheer, colour saturation, horizontal and vertical flips, translations, rescaling, resizing, rotation, etc. to reinforce the images. The following transformation techniques are implemented:

- Step 1: Set input means to 0 over the dataset.
- Step 2: Applied featurewise and samplewise standard normalization.
- Step 3: Applied ZCA whitening.
- Step 4: Sheared.
- Step 5: Rotated randomly.
- Step 6: Randomly shifted images horizontally and vertically.
- Step 7: Flipped horizontally.
- Step 8: Rescaled.

This increases the model's generalizability by generating additional data. In-place data augmentation or on-the-fly data augmentation is used here which is executed at training time; not generated ahead of time or before training. This leads to a model that performs better on the testing or validation data.

3.2 SX-DNN: The Proposed Model for Automatic ICH Detection

The paper proposes a hybrid model for automatic ICH detection on a large multilabel, imbalanced dataset of head CT studies. The model amalgamates DNN with SVM and XG-Boost (See Fig. 4). SVM, and XG-Boost are ensembled with DNN because they produces the best results among lots of different models which were tried as an experiment. DNN, especially the CNNs enables faster and more accurate detection in various tasks [16]. XG-Boost and SVM achieve extraordinary performance on various machine learning tasks as compared to other traditional machine learning algorithms [17]. Hence, this paper aims to exploit the advantages of Deep CNN approaches along with XG-Boost and SVM. The leverage of CNN architectures like Inception-V3 [18], DenseNet-V3 [19], ResNet-50 [20], VGG19 [21], and MobileNet-V2 [22], are taken in account for haemorrhage detection. The features are extracted from the raw images using CNN. The SVM and XG-Boost take as input the extracted features provided by CNN for the detection of ICH. The output of hybrid CNN, SVM, and XG-Boost is averaged to make final predictions. Due to the statistical property of CT scans, the general methods that work well in image classification did not perform well in these medical images. The stacking of three models gives more prediction confidence and improves accuracy.

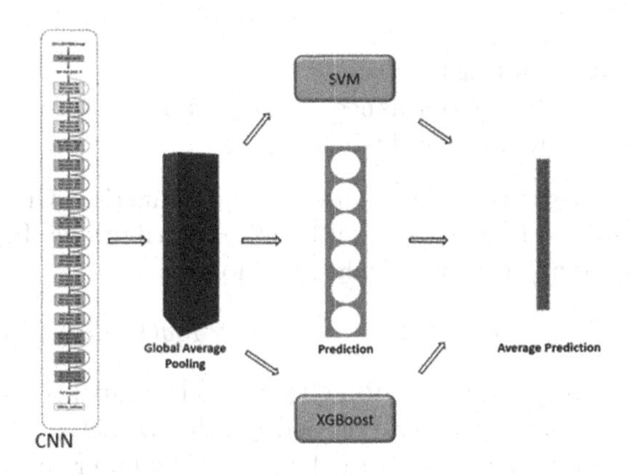

Fig. 4. Framework of the proposed model; SX-DNN

Progressive resizing is also nominated in the model. This technique consists of the progressive, sequential resizing of all images while training the CNN model, from smaller to bigger sizes. It was an effective idea to train by considering one-fourth of the image size followed by taking half of the image size and finally the original image size. It was very effective in the type of neural network (NN) used in this study and in its application. The images used were in square and not rectangular. Different resizing (2 times, 3 times, 4 times) could be used. But the power of 2 gave the best result.

The steps involved in the design of the proposed *SX-DNN* model are shown in Algorithm 1.

Input: RSNA Brain Haemorrhage CT Dataset
Output: RSNA Brain Haemorrhage CT Dataset with the specific class label

Step 1: Backpropagation is used to train CNN to extract features in which binary cross-entropy is taken as the loss function.
Step 2: The extracted features obtained from CNN are fed as an input to the Global Average Pooling (GAP) intermediate layer. A GAP layer reduces the spatial dimensions of a three-dimensional tensor.
Step 3: The GAP intermediate layer output is fed as an input feature vector to the SVM and XG-Boost for classification.
Step 4:: Probabilistic output of SVM and XG-Boost are calculated.
Note: The threshold is set to 0.5 to convert probabilities into binarized class labels. ie, if the predicted value is less than 0.5, it gives out an array with elements from 0 where the condition is True and elements from 1 elsewhere. Hence the probabilities higher than 0.5 is considered ICH positive and the probabilities lower than 0.5 is considered as ICH negative.
Step 5:This output of hybrid SVM and XG-Boost models along with Deep learning are averaged to make final predictions.

Algorithm 1: *SX-DNN*

3.2.1 Evaluation Metrics

For the evaluation of the performance of the proposed approach, four different metrics are used which are described as follows:

- **Loss:** Binary cross-entropy loss is used as a loss function. Cross-entropy measures the difference between two probability distributions. In binary classification, cross-entropy can be calculated as follows:

$$-(x \times log(f) + (1 - x) \times log(1 - f)) \tag{1}$$

  where x is the binary indicator (0 or 1) if the class label is the correct classification for a particular observation and f is the predicted probability that a particular observation is of a particular class. The lower the value of the loss function, the better the performance.
- **Accuracy:** It gives the percentage of correctly classified instances.

$$Accuracy = \frac{No.\_of\_correctly\_classified\_instances}{Total\_number\_of\_instances} \times 100 \tag{2}$$

- **Area Under ROC Curve (AUC):** It measures the two-dimensional area under Receiver Operating Characteristic (ROC) curve. This curve presents the performance of a classification model by plotting the True Positive Rate (TPR) along the Y-axis and the False Positive Rate (FPR) along the X-axis. A higher value of AUC implies the better classification performance.

$$TPR = \frac{True\_Positive}{True\_Positive + False\_Negative} \tag{3}$$

$$FPR = \frac{False\_Positive}{False\_Positive + True\_Negative} \tag{4}$$

- **F1-score:** The F1-score is the harmonic mean of precision (P) and recall (R). The higher value of the F1 score implies better performance of the classification model.

$$F1Score = 2 \times \frac{P \times R}{P + R} \tag{5}$$

$$Precision = \frac{True\_Positive}{True\_Positive + False\_Positive} \tag{6}$$

$$Recall = \frac{True\_Positive}{True\_Positive + False\_Negative} \tag{7}$$

4 Experimental Results

4.1 Training

The training dataset is split into 5 folds by employing 5-fold cross-validation. The training is done using Stochastic Gradient Descent (SGD) for 25 epochs using learning rate annealing with warm restart, known as cycling learning rates [23,24] which helps in rapidly converging to a new and better solution. The network gets to a global minimum value for the loss with each batch of SGD. While it approaches the minimum, it implies that the learning rate becomes smaller so that the algorithm does not overshoot, and settles as close to this point as feasible[2]. To tackle this issue, cosine annealing is employed following the cosine function[3]. A simple warm restart technique for SGD [24] is used to enhance its performance while training CNN. SVM , and XG-Boost classifiers are trained separately with a deep learning model. SVM with RBF kernel is employed. Table 2 summarizes the hyperparameters used in training XG-Boost. Later output of all three models are combined together , and their average turned to be the final prediction.

Table 2. Hyperparameters used in the case of XG-Boost

Hyperparameters	Value
Learning rate	0.1
Max_depth (max depth of the tree)	5
Min split loss	0.1
Subsample	0.8
Min child weight	11
n_estimators	1000
colsample_bytree	0.7

[2] https://mesin-belajar.blogspot.com/2018.
[3] https://mesin-belajar.blogspot.com/2018.

The neural network weights are initialized using He norm initialization [25] to refrain layer activation outputs from vanishing or exploding during the forward pass through DNN. Other hyperparameters are tuned to attain optimum performance on the validation set. 25 no. of epochs, the learning rate of 0.01, and a batch size of 512 are employed.

4.2 Evaluation

The model is customarily designed to mitigate the false negative value. Test-Time Augmentation (TTA) [26] is used to improve the performance of test images. TTA generates multiple amplified copies of each image in test images, making the model to compute a prediction for each, and then return an ensemble of those predictions. Augmentations are generally chosen to give models the opportunity to correctly classify images and a number of copies are selected between 10 and 20.

4.3 Results

The performances of several CNN architectures are evaluated for ICH detection. Five CNN models i.e. VGG19, ResNet-50, DenseNet-V3, MobileNet-V2, Inception-V3 are evaluated in terms of Binary Cross-Entropy Loss, Classification Accuracy, AUC, and F1-Score (See Table 3). It is seen from the table the highest overall performance is achieved by ResNet-50. The implementation of TTA reduced the false-negative rate by 5–10%, and the false-positive rate by 3–5%.

ResNet50 is buckled with SVM and XG-Boost in place of a fully connected layer and trained them over CNN output of ResNet-50. Its performance is further analyzed for prediction as presented in Table 4. The first column of the table presents the model employed, while the second, third, fourth, and the fifth column depicts the performance of the model in terms of Binary Cross-Entropy Loss, Classification Accuracy, AUC, and F1-Score.

It is seen from the table that hybrid models produced an appreciable final prediction output with improved accuracy and AUC score. The robustness of the model has also been enhanced. The hybrid model which comprises CNN and Machine Learning models, performs better than the traditional CNN model, and has fringe benefits. Amalgamating SVM with ResNet-50 (ResNet-50 + SVM) improves the performance as compared to that of ResNet-50, while amalgamating XG-Boost with ResNet-50 (ResNet + XG-Boost) attains better performance as compared to that of ResNet-50 as well as ResNet-50 + SVM. The combination of all three attains the best performance in terms of Loss, Accuracy, AUC, and F1-score. Note that ResNet-50 + SVM, and ResNet-50 + XG-Boost follow the same procedure of feature extraction and classification as that of *SX-DNN*.

Table 3. Performance comparison of different CNN Models using different evaluation metrics on the validation dataset.

Model	Loss	Accuracy(%)	AUC	F1-Score
VGG19	0.2090	94.4	0.84	0.86
Inception-V3	0.1433	96	0.87	0.79
DenseNet-V3	0.703	90.1	0.79	0.81
ResNet-50	0.1103	95.2	0.89	0.84
MobileNet-V2	0.2450	93.6	0.84	0.83

Table 4. Performance comparison of ResNet-50 with hybrid models using different evaluation metrics on the validation dataset.

Model	Loss	Acc.(%)	AUC	F1-Score
ResNet-50	0.1103	95.2	0.89	0.84
ResNet-50 + SVM	0.1298	95.7	0.92	0.87
ResNet-50 + XGB	0.1055	96.2	0.94	0.89
SX-DNN (hybrid)	0.0924	96.77	0.96	0.92

5 Discussion

ICH is a serious health problem that needs expeditious and thorough medical treatment. Nevertheless, in clinical practice, missed diagnosis and misdiagnosis exist because of the strenuousness in delineating the signs of the bleeding regions and the increased workload for radiologists. Hence, an automated ICH detection mechanism holds promise. Machine learning algorithms, especially deep learning models can be employed directly on raw data, thus refraining the tedious approaches of preprocessing and feature engineering and capturing the inherent dependencies and representative high-level features through deep structures. This work put forwarded a deep learning-based formalism for automatic classification and identification of ICH cases and evaluated on the largest multinational and multi-institutional head CT dataset from the 2019-RSNA Brain CT Haemorrhage Challenge. The proposed paradigm attained high accuracy in terms of Binary Cross Entropy Loss, AUC, Accuracy, and F1 score for multi-class learning. Compared with existing algorithms for ICH classification on the same dataset [9], it can be concluded that the proposed method takes better account in terms of computational load. This is because Wang et al. [9] (winner of the 2019-RSNA Brain CT Haemmorhage Challenge) used a hybrid classifier of CNN and LSTM. The limitation is that training a single deep learning model is complex and time-consuming. An ensemble of two deep learning models will increase the computational time. The proposed approach uses an ensemble of CNN and traditional learning methods. Further in terms of results we are getting comparable results as compared to Wang et al. [9] (p value = 1, when tested with the Friedman Statistical Test). Nguen et al. [12] put forward a hybrid method of CNN and long short-term memory (LSTM) on the 2019 RSNA Database of CT scans, however they have reported individual class accuracy instead of considering multi-class accuracy.

Another contribution of this paper is the handling of imbalanced data. Medical databases are usually imbalanced and hence the adoption of proper imbalanced pre-processing formalisms holds promise. The future approaches would include testing the approach with different combinations of undersampling and oversampling techniques.

6 Conclusion

This paper proposes the *SX-DNN* model, a new ensemble multilabel classifier algorithm for automatic identification and classification of ICH on a huge, heterogeneous, and imbalanced database of brain CT studies. The model learns through an interaction between SVM, XG-Boost, and CNN. This model has a comparatively smaller model size, faster diagnosis speed and better robustness as compared to that of Wang et al. [9]. But it is to be noted that we have tested it on a limited test dataset. For better real-world performance, it should be subjected to further experimentations. The performance of the proposed approach also relies on the optimization of the DNN model. Future work would encompass studies on faster optimization techniques for training *SX-DNN*. It also includes testing with different fusion techniques of classifiers in the proposed hybrid model that could give better performance in ICH detection. More diversity in the feature-extracting neural networks could be included. This could be tested on more challenging image classification datasets.

References

1. Caceres, J.A., Goldstein, J.N.: Intracranial hemorrhage. Emerg. Med. Clin. North Am. **30**(3), 771 (2012)
2. Deng, L.: Three classes of deep learning architectures and their applications: a tutorial survey. APSIPA Trans. Signal Inf. Process. **57**, 58 (2012)
3. Arbabshirani, M.R., et al.: Advanced machine learning in action: identification of intracranial hemorrhage on computed tomography scans of the head with clinical workflow integration. NPJ Digital Med. **1**(1), 1–7 (2018)
4. Kuo, W., Häne, C., Mukherjee, P., Malik, J., Yuh, E.L.: Expert-level detection of acute intracranial hemorrhage on head computed tomography using deep learning. Proc. Natl. Acad. Sci. **116**(45), 22737–22745 (2019)
5. Lee, H., et al.: An explainable deep-learning algorithm for the detection of acute intracranial haemorrhage from small datasets. Nature Biomed. Eng. **3**(3), 173–182 (2019)
6. Avanija, J., Sunitha, G., Reddy Madhavi, K., Hitesh Sai Vittal, R.: An automated approach for detection of intracranial haemorrhage using densenets. In: Reddy, K.A., Devi, B.R., George, B., Raju, K.S. (eds.) Data Engineering and Communication Technology. LNDECT, vol. 63, pp. 611–619. Springer, Singapore (2021). https://doi.org/10.1007/978-981-16-0081-4_61

7. Hssayeni, M.D., et al.: Intracranial hemorrhage segmentation using a deep convolutional model. Data **5**(1), 14 (2020)
8. Watanabe, Y., et al.: Improvement of the diagnostic accuracy for intracranial haemorrhage using deep learning-based computer-assisted detection. Neuroradiology **63**(5), 713–720 (2021)
9. Wang, X., et al.: A deep learning algorithm for automatic detection and classification of acute intracranial hemorrhages in head CT scans. NeuroImage Clin. **32**, 102785 (2021)
10. Ganeshkumar, M., Ravi, V., Sowmya, V., Chakraborty, C., et al.: Identification of intracranial haemorrhage (ICH) using resnet with data augmentation using cycleGAN and ICH segmentation using segan. Multimedia Tools Appl. **81**, 1–17 (2022)
11. Jeong, J.J., Patel, B., Banerjee, I.: Gan augmentation for multiclass image classification using hemorrhage detection as a case-study. J. Med. Imaging **9**(3), 035504 (2022)
12. Nguyen, N.T., Tran, D.Q., Nguyen, N.T., Nguyen, H.Q.: A cnn-lstm architecture for detection of intracranial hemorrhage on ct scans. arXiv preprint arXiv:2005.10992 (2020)
13. Flanders, A.E., et al.: Construction of a machine learning dataset through collaboration: the RSNA 2019 brain CT hemorrhage challenge. Radiol. Artif. Intell. **2**(3), e190211 (2020)
14. Xu, Z., Dan, C., Khim, J., Ravikumar, P.: Class-weighted classification: trade-offs and robust approaches. arXiv preprint arXiv:2005.12914 (2020)
15. Durfee, E.H., Lesser, V.R., Corkill, D.D.: Trends in cooperative distributed problem solving. IEEE Trans. Knowl. Data Eng. **1**, 63–83 (1989)
16. Yadav, S.S., Jadhav, S.M.: Deep convolutional neural network based medical image classification for disease diagnosis. J. Big Data **6**(1), 113 (2019)
17. Pardede, J., Sitohang, B., Akbar, S., Khodra, M.L.: Improving the performance of cbir using xgboost classifier with deep cnn-based feature extraction. In: 2019 International Conference on Data and Software Engineering (ICoDSE), pp. 1–6. IEEE (2019)
18. Szegedy, C., Vanhoucke, V., Ioffe, S., Shlens, J., Wojna, Z.: Rethinking the inception architecture for computer vision. In Proceedings of the IEEE Conference on Computer Vision and Pattern Recognition, pp. 2818–2826 (2016)
19. Huang, G., Liu, Z., Van Der Maaten, L., Weinberger, K.Q.: Densely connected convolutional networks. In: Proceedings of the IEEE Conference on Computer Vision and Pattern Recognition, pp. 4700–4708 (2017)
20. He, K., Zhang, X., Ren, S., Sun, J.: Deep residual learning for image recognition. In: Proceedings of the IEEE Conference on Computer Vision and Pattern Recognition, pp. 770–778 (2016)
21. Simonyan, K., Zisserman, A.: Very deep convolutional networks for large-scale image recognition. arXiv preprint arXiv:1409.1556 (2014)
22. Sandler, M., Howard, A., Zhu, M., Zhmoginov, A., Chen, L.C.: Mobilenetv 2: inverted residuals and linear bottlenecks. In: Proceedings of the IEEE Conference on Computer Vision and Pattern Recognition, pp. 4510–4520 (2018)
23. Smith, L.N.: Cyclical learning rates for training neural networks. In: 2017 IEEE Winter Conference on Applications of Computer Vision (WACV), pp. 464–472. IEEE (2017)
24. Loshchilov, I., Hutter, F.: SGDR: stochastic gradient descent with warm restarts. arXiv preprint arXiv:1608.03983 (2016)

25. He, K., Zhang, X., Ren, S., Sun, J.: Delving deep into rectifiers: surpassing human-level performance on imagenet classification. In: Proceedings of the IEEE International Conference on Computer Vision, pp. 1026–1034 (2015)
26. Moshkov, N., Mathe, B., Kertesz-Farkas, A., Hollandi, R., Horvath, P.: Test-time augmentation for deep learning-based cell segmentation on microscopy images. Sci. Rep. **10**(1), 1–7 (2020)

A Method for Workflow Segmentation and Action Prediction from Video Data - AR Content

Abhishek Kumar, Gopichand Agnihotram(✉), Surbhit Kumar, Raja Sekhar Reddy Sudidhala, and Pandurang Naik

Wipro Technology Limited, Wipro CTO Office, Wipro, Bangalore 560100, India
{abhishek.kumar293,gopichand.agnihotram,surbhit.kumar,raja.s54, pradeep.naik}@wipro.com

Abstract. This paper proposes an approach to segment workflows from video content data and outputs textual summary which is nothing but action prediction of these workflows from the speech recognized module. The method of workflow segmentation requires less computation power and achieves good accuracy compared to using other complex deep learning approaches of feature extraction and processing of sequential data. The method provides segmented action frames and video clips based on workflow segmentation on video frames data. The video segmentation module has intelligible video analysis methods which helps the system to provide a summary of important description of workflows in the frames and situation in the frames in order create an easily understandable outline. The workflow content will be helpful to create the AR content for maintenance and repair of different manufacturing systems.

Keywords: Unsupervised methods · Frame matching · Feature descriptors · Workflow segmentation · Action prediction · Speech recognition · Text analytics · Text summarization · LSTM

1 Introduction

Augmented Reality is slowly becoming a very useful platform for developing assisting systems for users in various industries. The AR devices are programmed with a set of instructions and will be guiding the users to perform the task. In many cases, the users may not be having a prior knowledge on the Repair or Maintenance systems and task to be performed but with the guidance provided by the AR devices, users will be able to finish the task. Hence many companies are looking towards AR technologies to develop various systems especially in Repair and Maintenance of devices.

There are systems that are being developed will help the user by providing step wise guidance on how to repair a device or how to do maintenance work on a device. The system will have step wise method on how to perform an action on a device and will be guiding the user to perform that action. To provide the guidance to the user the device

K. K. Patel et al. (Eds.): icSoftComp 2022, CCIS 1788, pp. 341–353, 2023.
https://doi.org/10.1007/978-3-031-27609-5_27

needs to be equipped with the action that needs to be completed for a task, so that it can guide the user and these instruction needs to be programmed to the device.

The existing systems are taking the video of an expert performing a repairing task or a maintenance task and dividing the procedure followed by the expert manually. Once divided, for each step the instruction is also written manually, and these are programmed to the Augmented Reality device. For example, the video of expert repairing a laptop is taken and manually the video is divided into steps and instruction are written manually like remove the battery as step1, removing the back latch as step two, remove the hard disk as step 3 and so on. Now the developers will design the task mentioned above manually. In this way the AR devices will be programmed, and the user will be guided to perform the intended task on the device.

However, the creation of content for AR devices manually will be a very long process as one needs to program it for each and every device by watching the expert's repairing video and divide the video into steps, create the figures virtually and map the instruction to the actions. For example, a system helps to repair mobile phone. To develop such a system, all the phones repair steps need to be created by the developer manually which is very tedious work and from video one needs to extract.

The AR content created from video segmentation data will be ported in the AR devices which will help the users while performing repair and maintenance tasks by giving instruction to the users. Authors experimented on many methods for automatic content creation on video content data, among all methods one is Deep learning approach which will divide the video into dynamic intervals and summarized steps will be extracted from the dynamic intervals. The method generates dynamic steps based on user's actions. The detailed actions will be used to create the AR content and these AR content will be ported on head mounted device for guiding the users or field technicians for repair and maintenance. In this paper, we aim to provide a solution approach for an automated content creation from a video content data (e.g., Laptop repair by a user at field).

In the field of ecommerce, customer experience plays a key role in generating business. AR content created by workflow segmentation and scene description will help customers to get more contextual information about the product which will make purchase more realistic although not physically present in store. Robotics field is also gaining popularity that requires machines to perform different tasks based on different human gestures. The sequential models trained on the content created will help machines to identify different human behavior and based on that aid or guide them with steps to perform a particular task.

Image captioning task is task to generate relevant captions for an image. So, to train the sequential model for such tasks huge amount of image data with their respective captions is required. This workflow segmentation method will help to generate frames for different action in a video and generate text summary for action based on the speech present in the video.

Previous approaches focus on segmenting action form video using supervised method. Here we have unsupervised method to segment actions from a video based on feature matching between frames extracted from frames. For each action segments from video, we can recognize the text and extract summary of that action. It uses state of the art pretrained BERT models for text summarization.

The present paper is organized as follows. Section 2 describes the related work. Section 3 discuss the detailed solution approach on workflow/action segmentation and text summarization using CNN, feature matching and BERT algorithms for extractive summary. The subsection of Sect. 3 discusses about the feature matching, text summarization form speech extracted and real time action detection on video streaming data. Section 4 deals with the application in manufacturing domain. Section 5 describes the conclusions and next steps followed by References.

2 Related Work

This section describes the recent work on action or workflow segmentation form a video using different computer vison and feature matching techniques.

C. Lea et al. [1], used hierarchy of convolutions to determine complex patterns and actions durations from the input data. Here the data with spatiotemporal features for making salad, inspecting supermarket shelf were used to identify actions. In the paper, authors introduced Temporal Convolutional Networks that uses a hierarchy of convolutions to perform fine grained action segmentation or detection. As discussed in the paper, this architecture can segment durations and capture long range dependencies much better, and training is also faster than the LSTM Recurrent architectures. This approach provides a faster way to train model to detect actions from videos.

M. Jogin et al. [2], focused on the method of extracting features using CNN architecture. The idea of extracting image features using CNN was utilised here. These extracted features can be used for various kind of classification task. Here, the network learns the features such as geometry, abstract level, and complex shapes. These learned features are used to implement various object detection, face recognition, surveillance and feature matching task. These features can also be used in transfer learning approach by training head layers to run on different detection and classification tasks.

Obaid, Falah et al. [4], have used sequential neural networks to learn features of video frames in sequence considering time and last hidden layer output. This approach comprises of two models, one is the CNN architecture, and the other is RNN (LSTM or GRU) architecture. Here the frames from video of different hand gestures were used to train these models. CNN model extracts the features from these video frames and these features are fed to a sequential RNN model which learns the gestures from the CNN output considering time stamp and last hidden layer output. This method find application in home appliances and electronic devices.

Hassaballah et al. [7], describes the use of key point and feature descriptors in the field of computer vision. Different descriptors and different type of image distortions are proposed here that helps to extracts the region of interest around key points. Here different methods are used to compare and evaluate these key point descriptors have been discussed. This gives us an idea to select different key point descriptors for feature matching task based on different use case.

Patil, P et al. [6], has proposed text summarization approach which is one the interesting task in Natural language processing domain. They have used transformer-based BERT architecture to perform extractive summarization of the text. The purpose was to get the important information form the textual data without losing important text and at the same time reducing the reading time. This bidirectional training and understanding of the context on CNN daily news dataset reduced reding time and increased accuracy than earlier text summarizers.

Fraga et al. [9], has proposed an approach for automatic summary and video indexing to support queries based on visual content on a video database. Video features extracted helps to create summarized version of videos which were then used for video indexing. This method helps in extracting key frames that can be processed to get statistical features which provides video essence that helps to segmentation of actions and indexing of videos based on these visual features.

3 Solution Approach

The present section discusses about details of solution approach for creating the automatic AR content for repair and maintenance of manufacturing devices. The input to the training model is video content data, and the model will extract all workflows of the video content which will be used for AR training sessions for assisting field technicians in real time.

The proposed algorithm takes the video content data and divide the content into individual frames. The frames (one in every 5 frames) will be used to extract the various features (using pre-trained deep neural networks) and a feature vector is formed for every extracted frame. Next, the distance between every consecutive feature vector is calculated and is compared with a pre-defined threshold. If the distance between consecutive feature vectors is more than the threshold value (this means the frames share dissimilar actions from next frame onwards), then the video is divided at that frame and will be defined as step. This will continue entire video content data. Once all the steps are computed, each of the step's video/frames is taken and with use of predefined models we will be predicting the object detection and action performed on the objects from which storyline of that step can be analyzed and processed to create a summary of that step.

All the frames of the video content can be used to create workflow video clips. In this way, from all the steps the summary of the action in the step are predicted. Once the steps are computed and content creation is completed, the steps will be used to train a LSTM model that will be able to predict the system state and the instruction to the user based on the action of the user. After the LSTM model is trained, the instructions are sent to the AR device. The user gets instruction from the device on how to perform each step and the device will be recording the action performing by the user. Based, on the action performed by the user and a trained LSTM model we will predict the state of the system after an action performed by the user. Based on the state of the system the instruction or steps to be given to the user will be changed. For example, the user performs an action such a way that step 2, step 3 is not needed, the state will be predicted by the LSTM model and based on that the user will directly see the step 4. We will be training the unsupervised models with the video content data and use supervised model to derive the steps and actions associated with each step. In this way, the automatic content creation for AR applications will be achieved for different machines, machine parts repair and maintenance activities.

The high-level features of the approach are given below,

- **Automatic Content creation:** The stepwise procedure for a repair or maintenance work which is extracted directly from a video of expert during the task and the stepwise method is used to instruct the user.
- **Real time Instructions:** Based on the instruction the user will perform a certain action. The action is analyzed using a LSTM method to predict the state of the system and based on the state the instruction or the steps to the user will change.
- **Real time action Prediction:** Once the steps are computed from the video content data, the AR device will instruct the user to perform a task. The user action for each instruction is predicted and tracked in real time using LSTM model.

From the below architecture diagram (Fig. 1) is proposed for the current system for AR content creation. The system will take the video content data as input to the model. At first, frames from these videos are extracted for feature matching. Once we have the frames for the input video, we extract features using different approaches such as CNN architectures or various key point and descriptors available. These features are stored in vector form. Once we have the feature for each of the frame, we can try to find dynamic intervals for each workflow by doing feature matching between the consecutive frames. We will use predefined threshold value for finding similarity between frames using distance measures, Cosine similarity or feature matchers available to segment frames of different actions and create time intervals of different actions.

Time interval helps to get video clips for each action. The speech recognition for each segment is done to get the text from the language. Text summarization techniques is applied on these extracted text and scene summary to derive for each workflow. The AR content that comprises of workflow frames and the corresponding scene descriptions are used to train sequential models such as LSTM so that actions can be identified for the new video data and its summary can be generated easily.

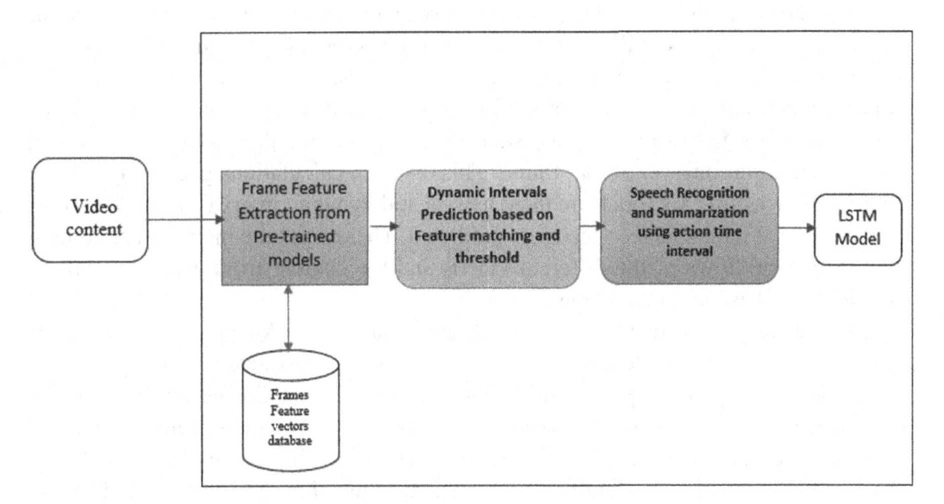

Fig. 1. System architecture diagram for training unit of content creation

3.1 Workflow Creation from Video Content Data

This subsection discus about the algorithms used for video segmentation and text summarization while processing the video content data. The AR content will be created using these processed data to guide the technician in the field to instruct the steps in real time. The video segmentation and text summarization steps are based on pretrained deep learning models for feature matching and NLP techniques are used for summarizing the text.

The following stages are used for workflow creation from video content data and the same stages are given in Fig. 2:

- **Video with speech as input:** The sequence of frames or images along with speech information obtained from the video stream data will be used to create steps/workflows for maintenance or repair of devices. The derived content from this approach can be used for AR content creation.
- **Frame feature extraction from video data:** For extracting features from frames, we will be using pretrained deep learning models and feature matching techniques to segment frames for different actions/workflows from the video.
- **Frames feature vectors Database:** The feature vectors obtained from a frame is stored in a database for further computation. This goes on till feature vectors of all the frames are extracted and all these features will also be stored. These database of feature vectors undergo further computation to create dynamic intervals and time frames which helps to extract the workflow segments. Deep learning CNN based architecture ResNet-101 which is trained on ImageNet can be used for feature extraction [3, 5, 8–10] and different feature key point extraction such as ORB, KAZE etc. [7] and matching techniques such as Cosine similarity can be used for segmentation based on features matched with the subsequent frames.
- **Action interval Segmentation:** The Extracted features are stored; we use these features to perform feature matching and extract time intervals based on for each action. These time intervals are used to create the segmented video clips for different actions identified. The method will be taking the feature vectors of consecutive frames from the database of feature vectors. Using the two vectors, the module will calculate the distance between the two vectors and once computed, it will compare it with the predefined threshold value. If the distance is more than the threshold, the video will be divided as an interval of that frame. This distance calculations goes on for all the consecutive feature vectors from the database and dynamic intervals from the different repair and maintenance videos are predicted. These dynamic intervals are nothing but steps/workflows of the different videos and the content from each step will be extracted and as explained below stages.

  Based on different features extracted, feature matching is done with the new frames from video to identify different actions. These features are extracted from video frames and stored which will help for feature matching. We can use Cosine similarity or any image feature matching algorithm to get the similarity between existing features and new sample frame from a video. Cosine similarity takes the cosine of angle between two vectors and tries to identify actions which is matching to the feature database created after feature extraction.

Key point and feature descriptors such as ORB or KAZE can also be used to get features and matching of features can be done using Brute Force matcher which internally using different distance measures based on threshold to segment frames for different actions. We were able to segment actions from a laptop repair video and Voltmeter operating video which had frames and action description for different segments. Feature matching is done using BF Matcher to segment the action. Action segmentation on laptop repair and Voltmeter operating videos were done with the threshold of feature matching 0.7 and 0.8 respectively for both videos. These threshold values depend on the video features and actions to be segmented from the video.

- **Video Segments creation:** To segment workflows from a video, we need to capture spatial and temporal features of the video content. Time intervals extracted in previous stage will be used to create video segments and speech for each action segment. This will help in segmenting different actions in the video and get the action description based on the recognized speech. The main idea behind this approach is to create AR content that will give assistant in many maintenances activity and enhancing the user experience.
- **Scene Description:** For each action segmented, we will detect the speech from the video segmentation. The speech recognition will extract as text using different types of speech to text libraries. Once we have identified the text part from the video, we will use various text analytics techniques to summarize the action from the text extracted from the speech.

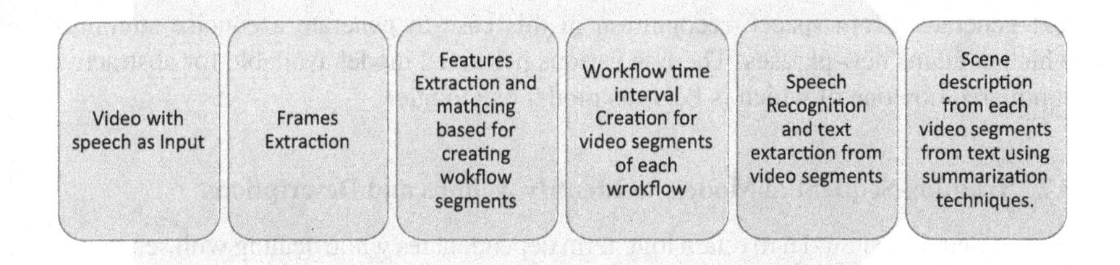

Fig. 2. Action time interval and scene description

- **Summarization (Extractive/Abstractive):** Depending on the use case, Extractive or Abstractive text summarization techniques can be used to obtain the summary of each step. Extractive summarization, Fig. 3 selects top N sentences that best represent the text for each video segment extracted in earlier steps. Abstractive summarization seeks to generate the key points in new words.

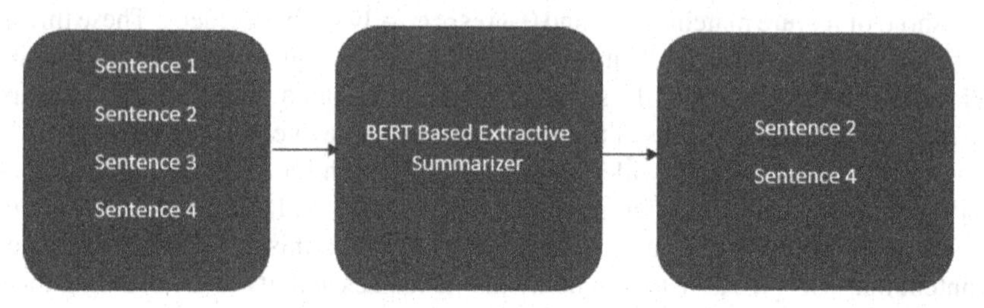

Fig. 3. Extractive text summarization

BERT architecture pertained models can be used to summarize [6] the text recognized from the video, or few layers of the model can be retrained by creating the format of data required to train these NLP based models. The text extracted from the speech can be used as a scene description and its summary can also be extracted. Each workflow frames will have its text content available which can be used while creating AR Content. These architectures are used for text summarization as its pretrained models performs good on text extracted from video data as the model's train on huge corpus of data.
Abstractive summarization is one more technique to get summary that comprises of tokens from the input text but not the same sentences. It contains salient ideas of the text generated from speech recognition in this case to generate a concise summary which contains new phrases. There is various pertained model available for abstractive summarization one of which is Pegasus model by Google.

3.2 Training Sequential Models to Identify Actions and Descriptions

We use Sequential model to retain long term dependencies while dealing with sequential data such as text recognized from video and sequence of frames in an action. These will be acting as an input to our model. Training model on such data can help in identifying actions in a video. These LSTM models (variant of RNN) architecture has cell states and gated mechanism which helps to capture long range dependencies while dealing with sequential data be text or sequence of frames.

Once we have text summary, action frames and videos generated from the earlier steps using CNN based architectures and feature matching techniques such as ORB, KAZE etc., the content created from it can be used to train Encoder-Decoder models using LSTM as its units shown in Fig. 4 to derive actions from video data. In sequence prediction problems this RNN [4, 11] variant is capable to learning feature key point that will be used to identify actions in a video once model is trained on the content created.

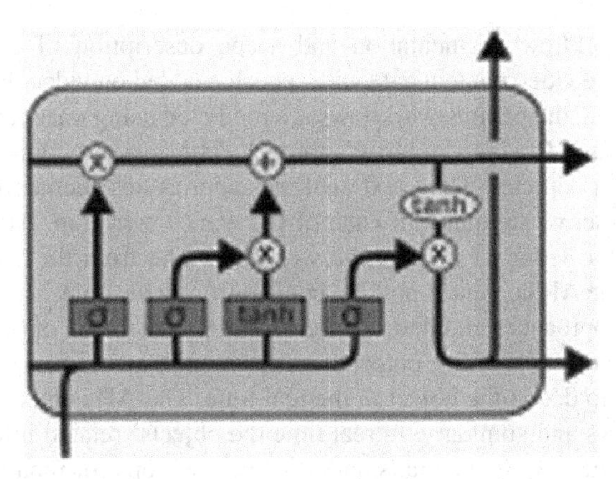

Fig. 4. LSTM architecture

The supervised LSTM models help us to train the content derived unsupervised models to predict the state of the machine using user actions. From a given machine repair video by expert, the system will provide you with the steps and actions associated with the step in automated way. This approach will help to overcome with the lot of manual intervention of machine repair or maintenance. The repair and maintenance steps will be ported into the AR device, the device will interact with the trained LSTM models to derive the state of the machine and actions associated with each state in real time for user repair of multiple machines. The trained models can also handle the skip steps where the action is skipped and predict the next state based on user action and this not always required sequentially.

Training input data for these LSTM models will contain sequence of frames for each action and its summary derived from above stages. Once these models learn the pattern of features for a particular action, it can be used to predict action in a video and suggest next steps based on the requirement such as maintenance, online shopping, video surveillance etc.

This unsupervised method of segmenting action frames and the respective segment textual summary serves purpose of generating image and their textual captions which can be further utilized in various training of sequential model for detection and summarization task. These sequential models whether trained on image data or text tries to learn the context by considering the previous when computing for the current state. Input is sometime bidirectional way in few some deep learning architectures which has significantly improved the performance while dealing with action prediction which comprises of sequence of frames from a video.

3.3 Real Time Action Detection and Suggestions

This section describes the detection of actions on video streaming data in real time and providing the suggestions to the user on field while repair/maintenance or any kind of activity.

The action/workflow segmentation and scene description of each workflow is explained on offline video content data with speech enabled on video data. As explained in the above section, the actions/workflows are predicted using unsupervised method of feature matching and Cosine similarity approach. The corresponding summary of the workflow/action is depicted using text analytics approaches such as BERT algorithm to obtain the extractive summary of each of the workflow/action. AR content will be generated using the video of each of the workflows and from the scene summary of each workflow. The AR content is nothing but identifying the objects in the videos using object detection approaches and describe the scene based on each of the object detected in the sequence frames. An offline object detection models are trained using these frames in the workflows to detect the object in the real time. The AR content is created based on these workflows and summary. In real time the objects' related information is augmented to know the workflow details and sequence of steps for repair/maintenance or any activity.

Along with these approaches, a sequential model such as LSTM is trained on the generated AR content. This approach is used to derive state of actions and give respective text suggestions on the steaming video in real time. The Fig. 5, depicts the step-by-step process of AR content creation using the offline video content data and the state of action/workflows prediction in real time in video streaming data.

The workflows/action predictions, the extractive summary creation using BERT algorithm, the object detection models are trained on workflow video segment data to augment the object in the frames, the AR content creation using workflows along with the summary description, and the state of action prediction models using LSTM models are depending on the offline video content data for a given video procedure. The activities of state prediction with LSTM model, the augmentation of the workflow description and object detection will happen on real time video streaming data as given in Fig. 6. Finally, user will be instructed to set of steps for repair and maintenance activities based on the procedure.

The AR content will be ported in HMD devices and the real time video streaming will happen on the devices to obtain the object detection, to view the augmented content and to instruct the users with different procedures in the field. We have tested this solution multiple devices such as Realwear, HoloLens, Oculus Quest for real time instruction to user for repair and maintenance.

Action segmentation on laptop repair and Voltmeter operating videos were done based on the features extracted from the frames and using BF feature matcher and cosine similarity finding the action segments based on threshold of matching to be decided based on the features of the video.

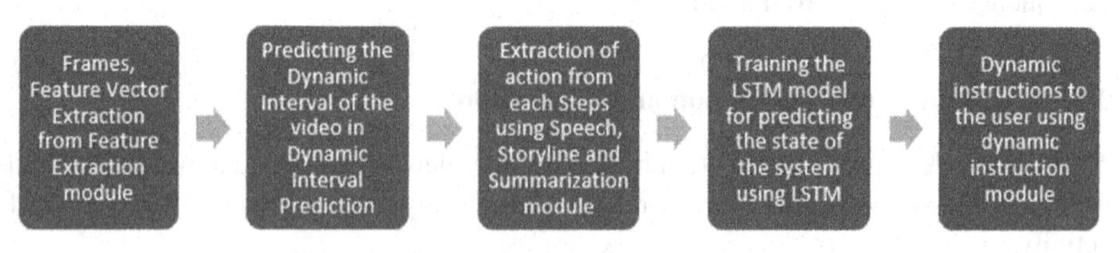

Fig. 5. Video action/workflow segmentation

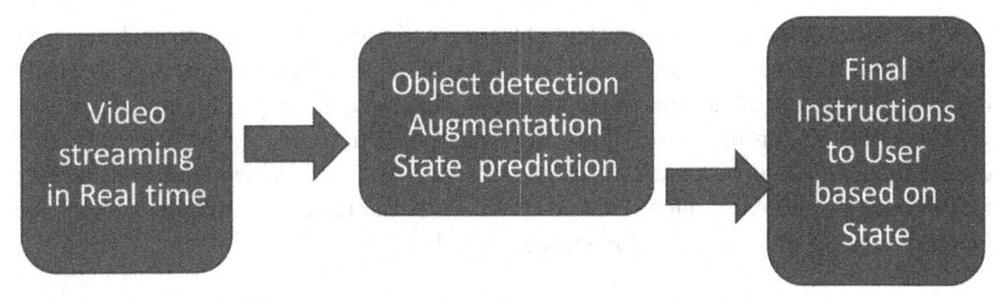

Fig. 6. Real predictions on video streaming data

4 Application in Manufacturing Domain

The solution approach explained in the above sessions, finds applications in many domains such as repairs and maintenance of manufacturing devices, ecommerce, detail, and healthcare. As part of repair and maintenance activity for all kinds of manufacturing devices, anomaly detection and providing various AR instructions in real times for the users to perform the procedures. For example, in case of laptop repair and maintenance there are multiple steps performed by the expert such as keyboard check, latch removal, screw tightening etc. (as shown in Fig. 5) and procedures like changing of battery, checking the battery condition, and ram replacement/adding new ram in the slot etc. The models are trained on these actions of images and corresponding text description can be useful in assisting these maintenance activities.

In Fig. 7, we can see different segmented frames of laptop screw tightening and battery repair videos. These frames are segmented based on feature vector extracted and based on feature matching the frames are segmented. These frames help to decide the time interval of different actions in a video. Each action video has speech that is converted to text and summarization is performed using either extractive or abstractive techniques

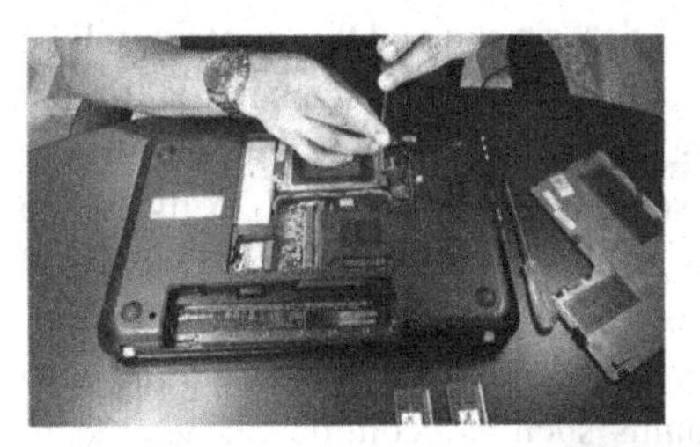

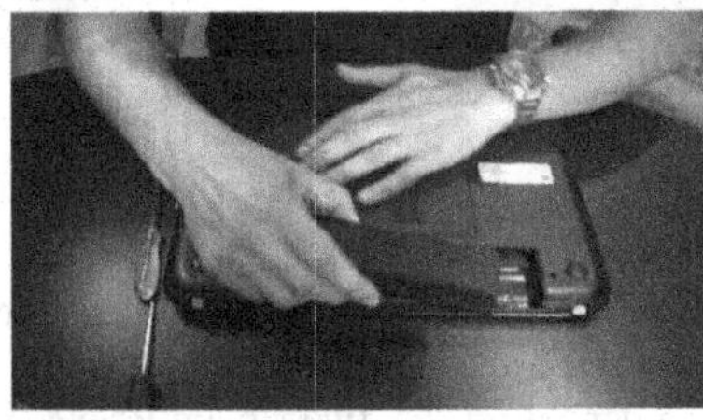

Fig. 7. Laptop repair action/workflow segmentation

based on the requirement. The segmented frames and text associated with each segment will helps in training sequential models which can be ported to android, iOS, or HMD devices to detect different steps of laptop repair and maintenance procedures. A few sets of actions are shown in below Fig. 6. The state prediction model will help to predict different states and give textual description of each state in a video based on the fine tuning done on the AR content created. In this way the repair and maintenance of the laptop will takes place as part of manufacturing domain.

5 Conclusion and Next Steps

In the current state of art there are various methods which talks about video segmentation and each method has some pros and cons when it comes to actual workflows derived from those videos. The video segmentation is still open topic for all the researchers who all are looking for accurate workflow predictions form the video data. In this current paper, we have studied and addressed this with unsupervised methods used to predict different workflow present in the video content data. The workflow prediction is nothing, but segmentation and each segment refers to one workflow here. The unsupervised methods prove to be more accurate, and it takes less time as compared to deep learning approaches in predicting the segments. We have also described the scene description of the workflows segment which will be used to assist the user in the field. The paper also talks about summarization approaches such as extractive summary and abstractive summary where the scene description will be provided to the summarization method or voice to text of the video will be provided to the summarization methods to obtain meaningful each workflow description. The paper also describes the state prediction of the workflows which will be used on real time streaming data to obtain the state of the workflow. In this paper we explained LSTM approach for state prediction on training the images data and textual description offline. We also talked about details of BERT algorithm which will be used on textual data to obtain he extractive summary of the text. We also observed extractive summary is more promising approach comparative to abstractive summary. The AR content will be created using workflows and description and this content will be ported in HMD for instructing the users for repair and maintenance steps in real time.

In continuation to this, we are also exploring on fusion-based techniques where voice and video data can be used to obtain the segments or workflows in the video content data. The voice features and video futures are fused and trained on supervised methods to obtain the multiple workflows from voice enabled video data. The fusion-based techniques will help with more accurate prediction of workflows. We are also extending the solution applications in other domains such as ecommerce, healthcare (patient activity prediction), and retail applications.

References

1. Lea, C., Flynn, M.D., Vidal, R., Reiter, A., Hager, G.D.: Temporal convolutional networks for action segmentation and detection (2016)
2. Zhao, R., Ali, H., van der Smagt, P.: Two-stream RNN/CNN for action recognition in 3D videos (2017)

3. Jogin, M., Madhulika, M.S., Divya, G.D., Meghana, R.K., Apoorva, S.: Feature extraction using convolution neural networks (CNN) and deep learning. In: 2018 3rd IEEE International Conference on Recent Trends in Electronics, Information & Communication Technology (RTEICT), pp. 2319–2323 (2018)
4. Obaid, F., Babadi, A., Yoosofan, A.: Hand gesture recognition in video sequences using deep convolutional and recurrent neural networks. Applied Computer Systems **25**(1), 57–61 (2020)
5. Oprea, S., et al.: A review on deep learning techniques for video prediction (2020)
6. Patil, P., Rao, C., Reddy, G., Ram, R., Meena, S.M.: Extractive text summarization using BERT. In: Gunjan, V.K., Zurada, J.M. (eds.) Proceedings of the 2nd International Conference on Recent Trends in Machine Learning, IoT, Smart Cities and Applications. LNNS, vol. 237, pp. 741–747. Springer, Singapore (2022). https://doi.org/10.1007/978-981-16-6407-6_63
7. Hassaballah, M., Alshazly, H.A., Ali, A.A.: Analysis and evaluation of keypoint descriptors for image matching. In: Hassaballah, M., Hosny, K.M. (eds.) Recent Advances in Computer Vision. SCI, vol. 804, pp. 113–140. Springer, Cham (2019). https://doi.org/10.1007/978-3-030-03000-1_5
8. Krizhevsky, A., Sutskever, I., Hinton, G.E.: ImageNet classification with deep convolutional neural networks. Commun. ACM **60**(6), 84–90 (2017)
9. Pimentel Filho, C.A., Saibel Santos, C.A.: A new approach for video indexing and retrieval based on visual features. J. Inf. Data Manag. **1**(2), 293 (2010)
10. Simonyan, K., Zisserman, A.: Two-stream convolutional networks for action recognition in videos (2014)
11. Ordóñez, F.J., Roggen, D.: Deep convolutional and LSTM recurrent neural networks for multimodal wearable activity recognition. Sensors (Basel, Switzerland) **16**(1), 115 (2016)

Convolutional Neural Network Approach for Iris Segmentation

P. Abhinand[1], S. V. Sheela[2](✉), and K. R. Radhika[2]

[1] Bosch Group, Bangalore, India
[2] BMS College of Engineering, Bangalore, India
sheelasv@hotmail.com

Abstract. Iris segmentation is the initial step for recognition or authentication tasks. In the proposed work, segmentation of iris region is performed using semantic network. A label or category is associated with every pixel in an image. Semantic segmentation is precise since it clearly detects irregular shaped representations. SegNet is the convolutional neural network proposed for semantic image localization. Pixels with similar attributes are grouped together. Labeled images represent categorical identifiers stored as ground truth masks. Encoder decoder blocks followed by pixel-wise classifier constitute the convolutional network. Each block comprises of convolution, batch normalization and rectified linear unit. Indices determine the mapping of encoder and decoder blocks. Encoder depth regulates the number of times image is upsampled or downsampled. Activations of network relate to the features. In the initial layers, color and edges are learnt. Channels in deeper layers learn the complex features. Learnable parameters are used in convolution. The approach determines iris boundaries without the step of preprocessing. Iris and background are the two labels considered. Eye images and ground truth masks are used for training. Testing samples are evaluated using Jaccard index. The experiment has been conducted on UBIRIS and CASIA datasets for segmentation results, obtaining an F-measure value of 0.987 and 0.962, respectively.

Keywords: Semantic segmentation · Image labeling · Ground truth masks · Jaccard index

1 Introduction

Convolutional Neural Network (CNN) has been widely used in image segmentation and classification tasks over the past two decades. Recent developments in terms of network structure and hardware has promoted CNN to be a dominant machine learning approach. LeNet5, AlexNet, VGG, Residual Network, FractalNet, GoogleNet and DenseNet are some of the CNN architectures. CNN has the advantage of good performance based on simple training and fewer parameters. Preprocessing operations required are much lower. Application of appropriate filters capture the spatial and temporal dependencies in an image. Convolution operation extracts high-level features like edges, contours etc., providing wholesome understanding of images. CNNs are trained for rich feature representations, for large collection of images that outperform handcrafted feature

K. K. Patel et al. (Eds.): icSoftComp 2022, CCIS 1788, pp. 354–368, 2023.
https://doi.org/10.1007/978-3-031-27609-5_28

extractors such as local binary patterns or histogram of oriented gradients. Segmentation of scenes, patterns, textures and pixel ranges have implemented deep layered neural network architectures. Handwritten digit recognition, speech analysis, classification of images and detection of objects are few applications. A stack of several convolutional layers and pooling layers followed by fully convolutional layer represent a model for various tasks. The input comprises of images and layers hold pixel information. Convolution layer performs a combination of convolution operation and activation function to extract features. Pooling layer provides downsampling to decrease the number of learning parameters.

Categorization based on pixel-wise labelling is known as semantic segmentation. Each pixel is associated with a label [1]. Network training is performed on pixel basis for the defined class. Precise segmentation is obtained from low resolution features for pixel-wise classification. A symmetry is followed between encoder and decoder blocks. Lower resolution input feature map is non-linearly upsampled by decoder using pooling indices [2].

Shape and boundary information characterize the segmentation mechanism. The computational efficiency in terms of memory and time is considered during inference [3]. Machine translation, object recognition, scene understanding, autonomous driving and medical image analysis are some of the application areas. Accuracy is improved in semantic localization by elimination of background noise [4]. A deep insight on working of visual systems is provided for generalized object detection and recognition tasks. Dense block architecture using complex values extract biological characteristics with intrinsic features based on phase information [5]. The challenging concepts are as follows. In practical scenarios, segmentation algorithms are customized for specific context, in such cases generality is unclear. Large datasets of labeled data are required for best approaches which otherwise would fail with fewer samples. Computational complexity is expensive for training CNN and may require GPUs or supercomputers. The number of parameters increases with depth. Closely labeled objects may lead to incorrect segmentation. Undefined acquisition factors from environment, devices and eyes can obstruct recognition accuracy and discriminative representation [6]. Visible iris images captured using smartphones or surveillance cameras require real-valued feature representation for fast, secure and accurate iris matching in a common embedded latent subspace [7]. Model overfitting and underfitting depends on number of images under consideration [8].

The present research work performs iris segmentation using semantic network. The iris boundaries are accurately determined without preprocessing. Two labels are identified as "iris" and "background". The encoder-decoder architecture of network is used for training. The captured eye images and ground truth masks are considered. The segmentation results are compared using Jaccard coefficient. Following are the novelties in the present research work:

- Semantic segmentation extracts shape and boundary characteristics precisely.
- Pixel-wise categorization avoids region selection and threshold based approaches.
- Preprocessing and similar enhancement techniques are eliminated.
- CNN is trained for encoder depth of 2, 4 and 5 to understand different layered network models.

- CNN training reduces number of parameters with reusability of weights.

The paper is organized as follows. Section 2 provides details on literature survey. Section 3 explains the proposed system for iris segmentation. Experiment and results are discussed in Sect. 4. Section 5 concludes with a note on future work.

2 Related Work

CNNs have attracted more researchers in the field of deep learning and has proved to be effective for visual recognition tasks [9]. In addition to using pretrained models, reformulation of connections between network layers improves the representational and learning properties. The existing models emphasize on network depth, model parameters and feature reuse. Image specific features are encoded into the architecture that are transformed in the layers to produce scores for classification and regression. Context interactive CNN based on spatial-temporal clues is used to develop identification models [10]. CNN training aims to minimize loss function with a set of optimum parameters. Image segmentation involves construction of probability map for subregions and utilizing this map on a global context. Statistical regularities help in designing generalizable model. High performance of CNN is achieved by gradient descent and backpropagation algorithms through self-learning and adaptive ability. The sensitivity of layers is determined by error component in backpropagation approach. Variation in the direction of gradient reduction is observed for each step. Segment-level representations are learnt by CNN to produce fixed-length features. Block-based temporal feature pooling aggregates these features for classification results.

Applications of CNN has been observed in various domains. Few are discussed in this section. A multi-label classification network for varying image sizes has been implemented by Park et al. [11]. Higher resolution feature maps are designed by dilated residual network and aggregation of positional information is obtained by horizontal vertical pooling. Binary relevance is used to decompose multi-label task into multiple independent binary learning tasks thereby neglecting label correlation. Label of each attribute is predicted by an adaptive thresholding strategy developed by Mao et al. [12]. Face detection and attribute classification tasks are performed using multi-label CNN. Task specific features are based on shared features that are extracted using ResNet architecture. The predictions of semantic segmentation are refined by applying appearance-based region growing in the method proposed by Dias et al. [13]. Pixels with low confidence levels are relabeled by aggregating to neighboring areas with high confidence scores. Monte Carlo sampling is used to determine the seed for region growing. Zhang et al. Has made a study on classifying screen content images such as plain text, graphics and natural image regions [14]. CNN model is used for region segmentation and quality related feature extraction. Specific spatial ordering is followed to model multiple parts for synthetic and natural segmentation in composited video applications. Contours of soft tissues are detected on images captured from different devices and techniques by Guo et al. [15]. Fusion network is implemented at feature, classifier and decision-making levels based on input features, pooling layer output and softmax classifier results. An adaptive region growing method based on texture homogeneity is used for breast tumor segmentation in the method by Jiang et al., as an initial process [16]. VGG16 network is later applied to tumor regions for reducing false positive regions in blur and low contrast images.

Segmentation has been used in different forms in the earlier research work. Threshold segmentation used for region localization differentiates pixel values based on specific values. Edge detection segmentation defines boundaries for object separation. Segmentation based on clustering divide the data points into number of similar groups. Semantic segmentation obtains a pixel-wise dense classification by assigning each pixel in the input image to a semantic class. Iris segmentation is an important step for accurate authentication and recognition tasks. In the method by Hofbauer et al., a parameterized approach for segmentation has been deployed [17]. A multi-path refinement network, called RefineNet is used consisting of cascaded architecture with four units, each connecting to output of one residual net and preceding block. Each unit consists of two residual convolution units and pooling blocks. A densely connected fully CNN has been implemented by Arsalan et al. to determine true iris boundary [18]. The algorithm provides information gradient flow considering high frequency areas without pre-processing. The network uses 5 dense blocks for encoding and SegNet for decoding. Each dense block has convolution layer, batch normalization and rectified linear unit. High frequency components are maintained during convolution. Iris segmented using two circular contours and transformed to rectangular image is given as input to off-the-shelf pre-trained CNNs in the method by Nguyen et al. [19]. Alexnet, VGG, Google inception, ResNet and DenseNet are the networks used. Highest recognition accuracy was obtained for DenseNet architecture. Details of images are processed in better way by extracting greater number of global features in the method by Zhang [20]. Dilated convolution is used to obtain larger receptive field information to get more image details. Dense encoder-decoder blocks are used by Chen et al., for iris segmentation [21]. Dense blocks improve the accuracy by solving gradient vanishing issues. Capsule network has been proposed by Liu et al. in which scalar output feature detectors are replaced with vector output capsules to retrieve precise position information [22]. The capsule is computed as weighted sum using prediction vector which is calculated by multiplying input vector with weight matrix. Fuzzified smoothing filters are used to enhance the images. A mixed convolutional and residual network has been proposed by Wang et al. That gain the advantages of both architectures [23]. Learning time is reduced by placing one convolutional layer followed by residual layer. U-Net has proved to be adaptable for segmentation problems in medical imaging [24]. The layers are arranged such that high frequency image features are preserved thereby achieving sharp segmentation. A unified multi-task network is designed based on U-net architecture to obtain iris mask and parameterized inner, outer boundaries [25]. An attention module extracts significant feature signals for prediction targets and suppresses irrelevant noise. A combination of networks are implemented for feature extraction, region proposal generation, iris localization and normalized mask prediction [26]. Fixed size representation is extracted to interface with neural network branches. Spectral-invariant features are generated to extract iris textures using Gabor Trident Network and code device-specific band as residual component [27]. Summary of related work on iris segmentation has been tabulated in Table 1.

Table 1. Summary of Iris segmentation methods

Sl. No.	Author	Findings
1	Hofbauer et al. [17]	Parameterization of iris is based on CNN segmentation using RefineNet; utilized for biometric system or as a noise mask. A parameterized circle for pupil and iris is formulated based on candidate segmentation using circular Hough transform
2	Arsalan et al. [18]	IrisDenseNet determines true iris boundary with inferior-quality images using information gradient flow between the dense blocks. Accurate for segmenting high frequency areas like eyelashes
3	Nguyen et al. [19]	Iris segmentation uses two circular contours that are geometrically normalized using pseudo-polar transformation to obtain a fixed rectangular image. Off-the-shelf CNNs extract features for classification
4	Zhang et al. [20]	Details of images are processed using fully dilated UNet by extracting global features. Network model uses near infrared and visible light illumination images
5	Chen et al. [21]	Adaptive CNN approach uses dense-fully convolutional network for segmentation. Mini-batch Adam algorithm is implemented for optimization to minimize cost function
6	Liu et al. [22]	Fuzzy operations based on smoothing filters enhance images for deep learning model. Faster convergence is obtained with increased recognition accuracy. CNN and Capsule network are used for training
7	Wang et al. [23]	Combination of convolutional and residual network is used. CNN helps in fast learning. Non-saturation features are extracted from residual network
8	Yiu et al. [24]	Fully convolutional neural network is applied on video-oculography images to segment pupil area. Elliptical contour estimation and blink detection are performed
9	Wang et al. [25]	IrisParseNet predicts accurate iris masks using semantic segmentation framework. Iris mask and parameterized boundaries are modeled using multi-task U-Net architecture
10	Feng et al. [26]	Iris R-CNN integrates segmentation and localization to generate normalized images. Double circle regions are extracted in non-cooperative environment
11	Wei et al. [27]	Cross-spectral recognition is based on device-specific bands between near-infrared and visible images. Gabor trident network extracts iris textures under different spectra

3 Proposed System

Iris segmentation is accomplished by semantic network. The process is described as correlating each pixel with a class label. SegNet is a CNN for semantic image segmentation. Pixel classification layer identifies the categorial label for every pixel. The network consists of encoder decoder blocks followed by pixel-wise classification layer. Encoder network has topological identity to convolutional layers in VGG16 [1]. Sampling factor determines the number of times the image is downsampled or upsampled using 2^K, where K is the value of encoder depth. The layers are either configured or pre-trained. Low-resolution encoder feature maps are correlated to input resolution feature maps by the decoder network. A hierarchy of decoders corresponds to each encoder. The pooling indices used in max-pooling step of encoder are used by corresponding decoder to perform non-linear upsampling. Softmax classifier performs pixel-wise classification by considering feature maps obtained from final decoder.

CNN architecture consists of encoder-decoder network. Encoder block (EB) and Decoder block (DB) are structurally similar entities, comprising of Convolution (Conv), Batch Normalization (BN) and rectified linear unit (ReLU). Conv, BN, ReLU and pooling is the order of layers in encoder. The order is unpooling, Conv, BN and ReLU for decoder. The mapping of EB and DB is based on indices. Figure 1 depicts pictorial representation of semantic segmentation for the proposed work.

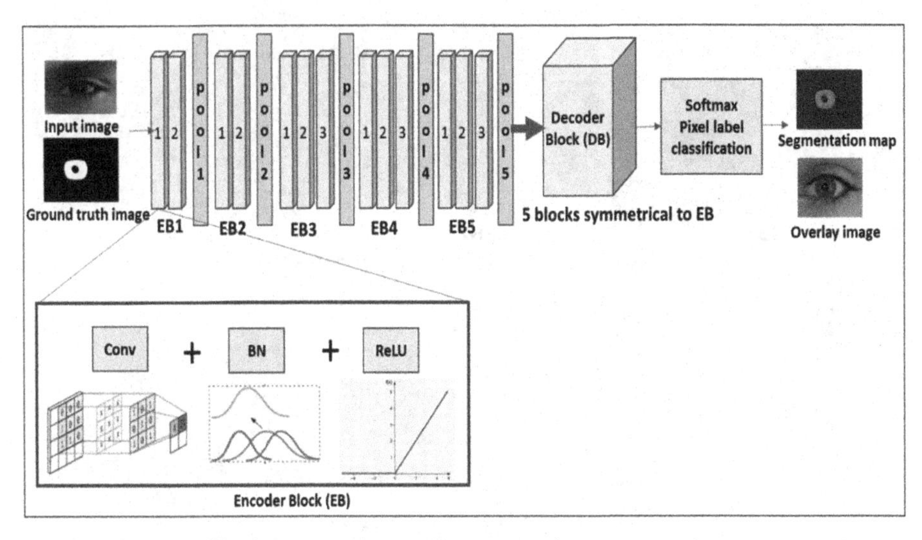

Fig. 1. Proposed semantic segmentation representation

Small convolution filters of size 3x3 are used in VGG architecture which reduces the number of parameters and makes decision function highly discriminatory. Convolution is performed when convolution layer slides filters in horizontal and vertical directions along the input. The features in these regions are learnt by scanning through the image. Suitable weights are applied to regions in the image when filter convolves the input.

Number of filters determine feature maps or channels in the output of convolution layer. The number of weights in a filter is given by $m*n*p$ where m and n are dimensions of the filter, p is the number of channels. The total number of parameters in convolution layer is $(m*n*p + 1)*f$, where f is the number of filters and 1 is the bias. The number of neurons in a feature map is given by product of output height and width, denoted by map size, z. In the convolutional layer, total number of neurons is $z*f$. In dilated convolution, layers are expanded by introducing zeros between each filter element. Dilation factor, df, regulates the step size for sampling. Stride is step size for traversing the input vertically and horizontally. Padding is applied to input borders in horizontal and vertical directions. Padding preserves the spatial resolution after convolution. The output size of convolution layer is given in Eq. 1.

$$os = (iz - ((fs - 1) * df + 1) + 2 * g)/r + 1 \quad (1)$$

where os, iz and fs denote output size, input size and filter size. df is the dilation factor, g and r indicate padding and stride parameters.

Features learnt by the network are represented by activations of CNN. Color and edges are simple features learnt in the initial layers. Eyes are the complex features learnt by channels in deeper layers. Convolutions are performed with learnable parameters. One feature per channel is generally learnt to extract useful features.

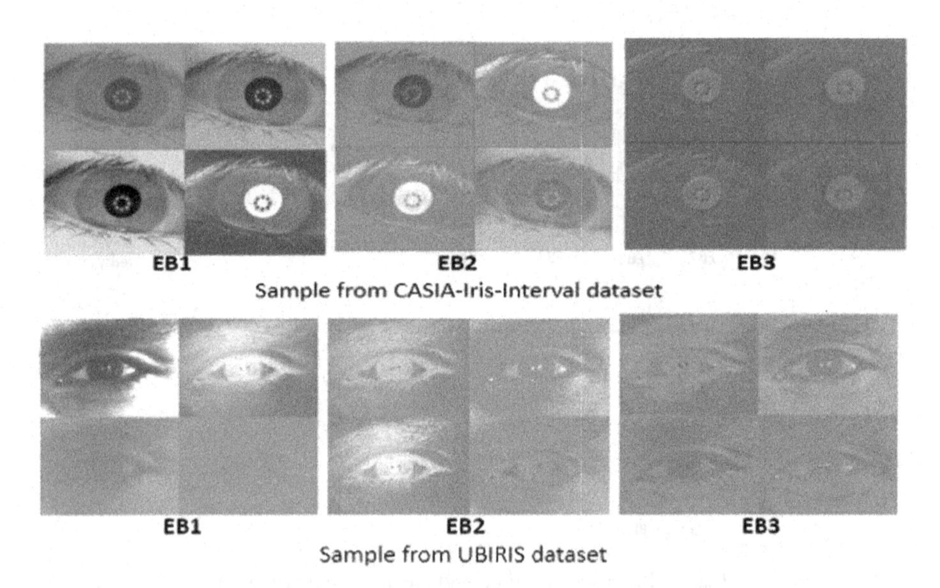

Fig. 2. Visual representation of activations

Figure 2 represents 2x2 grid of activations for each convolution layer. Channel output is depicted as a tile in the grid of activations. Channel in the first layer activates on edges. In the deeper layer, focus is on detailed features, such as iris.

BN layer normalizes activations and gradients propagating through the network. Batch normalization addresses the issues that arise due to changing distribution of hidden neurons by using scale and shift training variables. The mean M_B and variance V_B are calculated over a mini-batch and each input channel. The normalized activations, $\hat{x}_i$ for inputs x_i is computed using Eq. 2.

$$\hat{x}_i = \frac{x_i - M_B}{\sqrt{V_B + \alpha}} \tag{2}$$

where α is a property that improves numerical stability, when mini-batch variance is small. In order to optimize the activations to zero mean and unit variance, the BN layer further performs shifting and scaling according $y_i = T\hat{x}_i + E$, where y_i indicates activation after optimization, T, E are offset and scale properties.

Latest deep learning architectures use ReLU activation function in the hidden layer as it has attained better performance and generalization as compared to other functions such as Sigmoid and tanh [28]. Vanishing gradient issues are eliminated by rectifying the values less than zero. Gradient descent methods with good optimization preserves linear representation. Max-pooling layer performs downsampling. Maximum of each rectangular pooling region is computed. Overfitting is reduced by minimizing the number of hyper-parameters to be trained. Noisy activations are discarded along with dimensionality reduction. Max-pooling is performed over 2x2 pixel window and a stride of 2. Max-unpooling operation performs upsampling and padding with zeros for the output obtained from max-pool layer. Resolution is upsampled by distributing a single value into a higher resolution.

Zero-center data normalization in SegNet architecture improves the model performance. The responsiveness of activation function to weight changes around zero. Image size, number of classes, pretrained model or custom encoder depth are the parameters under consideration.

Increasing the number of feature maps lead to maintaining of expressiveness with deeper networks. An encoder depth of 2 creates a 31-layered network. With encoder depth of 4, 59-layered network is created. Using pretrained model as feature extractor reduces time and effort into training. Models are VGG16 and VGG19. The encoder depth in these models is 5. The proposed work has been carried out with encoder depth set to 2, 4 and 5. Implementation using VGG16 pretrained model creates a network with 91-layers which is depicted in Table 2. Precise segmentation is obtained using this model.

Stochastic gradient descent with momentum is used to train the network. Iterations in gradient descent regulate the number of mini-batches for one epoch. The dataset using mini-batches is passed multiple times depending on number of epochs. The oscillations along the path of steepest descent are reduced by adding a momentum term to the parameter update. The expression for update is given in Eq. 3.

$$P_{k+1} = P_k - \eta \nabla \mathrm{L}(P_k) + \beta(P_k - P_{k-1}) \tag{3}$$

where k specifies the iteration number, $\eta > 0$ is the learning rate. P is the parameter vector and $L(P)$ is the loss function. Gradient of the loss function is denoted as $\nabla L(P)$. The input to current iteration from previous gradient step is determined by β and the difference of parameter vectors.

Table 2. 91-Layered network

Block/Layer	Filters	Feature map	Kernel	Stride	Padding
Image input		300x400x3			
EB1_1	64	300x400x64	3x3	1x1	1x1
EB1_2					
pool1	1	150x200x128	2x2	2x2	0x0
EB2_1	128	150x200x128	3x3	1x1	1x1
EB2_2					
pool2	1	75x100x128	2x2	2x2	0x0
EB3_1	256	75x100x256	3x3	1x1	1x1
EB3_2					
EB3_3					
pool3	1	37x50x256	2x2	2x2	0x0
EB4_1	512	37x50x512	3x3	1x1	1x1
EB4_2					
EB4_3					
pool4	1	18x25x512	2x2	2x2	0x0
EB5_1	512	18x25x512	3x3	1x1	1x1
EB5_2					
EB5_3					
pool5	1	9x12x512	2x2	2x2	0x0
unpool5		18x25x512			
DB5_3	512	18x25x512	3x3	1x1	1x1
DB5_2					
DB5_1					
unpool4		37x50x512			
DB4_3	512	37x50x512	3x3	1x1	1x1
DB4_2					
DB4_1					
unpool3		75x100x256			
DB3_3	256	75x100x256	3x3	1x1	1x1
DB3_2					
DB3_1					
unpool2		150x200x128			

(*continued*)

Table 2. (*continued*)

Block/Layer	Filters	Feature map	Kernel	Stride	Padding
DB2_2	128	150x200x128	3x3	1x1	1x1
DB2_1					
unpool1		300x400x64			
DB1_2	64	300x400x64	3x3	1x1	1x1
DB1_1	64	300x400x2	3x3	1x1	1x1
Softmax		300x400x2			
pixelLabels	Cross-entropy loss with classes 'iris and 'background'				

Step size required to update weights during training is the learning rate. Backpropagation algorithm is used for updation of weights. Optimum learning rate determines the best approximate function. Regularization factor aids in controlled training by reducing overfitting.

Softmax function is normalized exponential that obtains a probabilistic distribution of values summing to one. Dot product of weight and input vector is the net value. Highest probability of each class in a multi-class model determines the target class. The function is predominantly used in output layer of most practical deep learning networks [28].

Network predictions for target classification are evaluated by cross-entropy loss. Difference between two probability distributions, predicted and actual is measured. Each input is assigned to one of N mutually exclusive classes based on softmax function. The 1-of-N coding scheme using cross entropy loss function is given in Eq. 4.

$$C = -\sum_{p=1}^{M}\sum_{q=1}^{N} k_{pq} ln f_{pq} \tag{4}$$

where M is sample count, N is the number of classes. Target function k_{pq} indicates that p^{th} sample belongs to q^{th} class, and f_{pq} is the probability from softmax function that assigns sample p for class q. High value of C indicates that predicted probability deviates from actual value.

Similarity between ground truth mask and categorial image is measured using Jaccard coefficient. Categorial image is the image after semantic segmentation. The elements of categorial array provide the segmentation map that correspond to pixels of the input image. Ratio of intersection of two images to the union of two images provides the score. Equation 5 indicates the computation of Jaccard coefficient.

$$I(G, T) = \frac{|G \cap T|}{|G \cup T|} \tag{5}$$

where G is the ground truth mask and T is the categorial image. I is the Jaccard index. The accuracy is conceptualized for image segmentation. The measurement emphasizes on ratio of overlap area to complete area. F-measure is computed from Jaccard index using Eq. 6.

$$FM = \frac{2 * I(G, T)}{1 + I(G, T)} \tag{6}$$

The workflow of the proposed system is provided in Algorithm 1.

Algorithm 1: Workflow of proposed system

Step 1: Datastore of images and ground truth masks is created. *imds* contains original eye images, *pxds* contains pixel labeled images obtained from ground truth images. Classnames are "iris" and "background" denoted by label identifiers {1 2}.

```
imds = imageDatastore(imageDir);
pxds = pixelLabelDatastore(labelDir,classNames,labelIDs);
```

Step 2: Image and pixel label data stores are partitioned into training and testing samples. 80-20 rule is used for training and testing. Segregation is performed using random number generator.

```
[imdsTrain, imdsTest, pxdsTrain, pxdsTest] = partitionimage(imds,pxds);
```

Step 3: SegNet layers for semantic segmentation using VGG16 as pretrained model is created.

```
datasource = pixelLabelImageSource(imdsTrain,pxdsTrain);
lgraph = segnetLayers(imageSize,numClasses,'vgg16');
```

Step 4: Training options are initialized.

```
options = trainingOptions('sgdm', ''Momentum', 0.9, 'InitialLearnRate', 1e-3,
'L2Regularization', 0.0005, 'MaxEpochs', 50, 'MiniBatchSize', 32, ''Shuffle', 'every-
epoch', 'Plots','training-progress','VerboseFrequency', 10);
```

Step 5: The network is trained.

```
[net, info] = trainNetwork(datasource,lgraph,options);
```

Step 6: Semantic image segmentation for test data to generate categorial image is performed.

```
T = semanticseg(test_img, net);
```

Step 7: Jaccard index, *I*, for ground truth masks, *G*, and categorial image, *T*, is computed.

```
I= jaccard(G,T);
```

Step 8: Semantic segmentation results are evaluated and metrics are analyzed.

```
metrics= evaluateSemanticSegmentation (pxdsResults, pxdstest,' Verbose' ,false);
```

4 Experiment and Results

Initial learning rate is assigned the value 0.001. Regularization term added to the network is set to 0.005. Number of epochs is 50 with mini-batch size of 32. Stride and padding are fixed to 1-pixel. The experiment has been implemented using Matlab 2022a. Training helps to understand sophistication of the image with reduction in number of parameters and reusability of weights.

Semantic segmentation has been carried out on UBIRIS database. The database consists of eye images form 522 subjects. For each subject 30 images are provided in two sessions with 15 images captured per session. The database is intended to study iris patterns captured in visible wavelength under non-ideal imaging conditions. Images were captured in two sessions with an interval of one week. The image acquisition has been performed using Canon EOS 5D with resolution of 300x400 in tiff format.

Similar analysis was performed on CASIA-Iris-Interval database in which the images were captured using self-developed close-up iris camera captured using NIR-LED array with suitable luminous flux for iris imaging. Images thus captured are suitable to study detailed texture features. The database consists of images from 249 subjects with a total of 2639 images and resolution of 320x280. Eye images and ground truth images are concurrently used for training as per 80–20 rule. Segmentation results for sample images are shown in Fig. 3. Jaccard index for ground truth mask and segmentation map is computed.

CNN is trained for different values of encoder depth. Accurate segmentation is obtained using encoder depth of 5 corresponding to VGG16 model. The coarseness of semantic information is reduced using deeper layers in the network. F-measure of 0.987 and 0.962 is obtained for UBIRIS and CASIA-Iris-Interval databases, respectively. State-of-art comparison has been provided in Table 3.

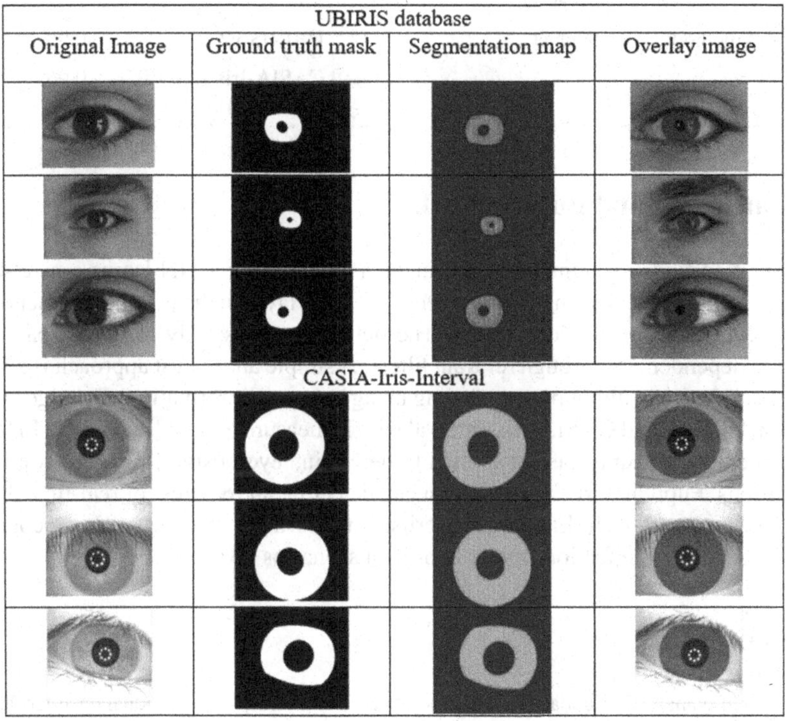

UBIRIS database

Original Image	Ground truth mask	Segmentation map	Overlay image

CASIA-Iris-Interval

Fig. 3. Segmentation results for UBIRIS and CASIA-Iris-Interval databases

Table 3. State-of-art comparison

Author	CNN	Database	F-Measure
Hofbauer et al. [17]	RefineNet	CASIA4i	0.984
Arsalan et al. [18]	IrisDenseNet (DenseNet and Segnet)	CASIAv4	0.975
Zhang et al. [20]	Fully Dilated U-Net	CASIA4i	0.973
Chen et al. [21]	Dense-fully convolutional layer	UBIRISv2	0.960
Yiu et al. [24]	Fully convolutional neural network (U-net and V-net)	VOG images	0.966
Wang et al. [25]	IrisParseNet (Multi-task U-Net with Attention module)	CASIAv4-distance	0.942
		UBIRIS v2	0.917
Proposed method	SegNet	UBIRIS v2	0.987
		CASIA-Iris-Interval (CASIAv4)	0.962

5 Conclusion and Future Work

Segmentation finds its significance in image analysis tasks. CNN learns a pixel-based mapping without considering region selections. Each layer in the network extracts class-salient features from previous layers. The network successfully captures spatial and temporal dependencies through relevant filters. A simple and robust approach is adopted to segment the iris without pre-processing using semantic information. The algorithm is tested on UBIRIS and CASIA-Iris-Interval with F-measure of 0.987 and 0.962. In future, the segmentation results can be applied to determine eye position for gaze contingent applications. Pupil and iris segmentation can detect facial features for real-time 3D eye tracking framework [29]. Embedding periocular information can assist iris recognition tasks for security applications under non-ideal scenarios [30].

References

1. Badrinarayanan, V., Kendall, A., Cipolla, R.: SegNet: a deep convolutional encoder-decoder architecture for image segmentation. IEEE Trans. Pattern Anal. Mach. Intell. **39**(12), 2481–2495 (2017)
2. Minaee, S., Boykov, Y., Porikli, F., Plaza, A., Kehtarnavaz, N., Terzopoulos, D.: Image segmentation using deep learning: a survey. IEEE Trans. Pattern Anal. Mach. Intell. **44**(7) (2022)
3. Khagi, B., Kwon, G.R.: Pixel-label-based segmentation of cross-sectional brain MRI using simplified SegNet architecture-based CNN. Hindawi J. Healthc. Eng. **2018**, 1–8 (2018)
4. Guo, Y., Liu, Y., Georgiou, T., Lew, M.S.: A review of semantic segmentation using deep neural networks. Int. J. Multimedia Inf. Retrieval **7**(2), 87–93 (2018)

5. Nguyen, K., Fookes, C., Sridharan, S., Ross, A.: Complex-valued Iris recognition network (2022). https://doi.org/10.1109/TPAMI.2022.3152857
6. Wei, J., Huang, H., Wang, Y., He, R., Sun, Z.: Towards more discriminative and robust iris recognition by learning uncertain factors. IEEE Trans. Inf. Forensics Secur. **17**, 865–879 (2022)
7. Mostofa, M., Mohamadi, S., Dawson, J., Nasrabadi, N.M.: Deep GAN-based cross-spectral cross-resolution Iris recognition. IEEE Trans. Biometrics, Behav. Identity Sci. **3**(4), 443–463 (2021)
8. Sehar, U., Naseem, M.L.: How deep learning is empowering semantic segmentation. Multimedia Tools Appl. **81**, 30519–30544 (2022). https://doi.org/10.1007/s11042-022-12821-3
9. Wang, W., Yang, Y., Wang, X., Wang, W., Li, J.: Development of convolutional neural network and its application in image classification: a survey. Opt. Eng. **58**(4), 040901 (2019)
10. Song, W., Li, S., Chang, T., Hao, A., Zhao, Q., Qin, H.: Context-interactive CNN for person re-identification. IEEE Trans. Image Process. **29**, 2860–2874 (2020)
11. Park, J.Y., Hwang, Y., Lee, D., Kim, J.H.: MarsNet: multi-label classification network for images of various sizes. IEEE Access **8**, 21832–21846 (2020)
12. Mao, L., Yan, Y., Xue, J.H., Wang, H.: Deep multi-task multi-label CNN for effective facial attribute classification. IEEE Trans. Affect. Comput. **13**, 818–828 (2020)
13. Dias, P.A., Medeiros, H.: Semantic segmentation refinement by monte carlo region growing of high confidence detections. In: Jawahar, C.V., Li, H., Mori, G., Schindler, K. (eds.) ACCV 2018. LNCS, vol. 11362, pp. 131–146. Springer, Cham (2019). https://doi.org/10.1007/978-3-030-20890-5_9
14. Zhang, Y., Chandler, D.M., Mou, X.: Quality assessment of screen content images via convolutional-neural-network-based synthetic/natural segmentation. IEEE Trans. Image Process. **27**(10), 5113–5128 (2018)
15. Guo, Z., Li, X., Huang, H., Guo, N., Li, Q.: Deep learning-based image segmentation on multimodal medical imaging. IEEE Trans. Radiat. Plasma Med. Sci. **3**(2), 162–169 (2019)
16. Jiang, X., Guo, Y., Chen, H., Zhang, Y., Lu, Y.: An adaptive region growing based on neutrosophic set in ultrasound domain for image segmentation. IEEE Access **7**, 60584–60593 (2019)
17. Hofbauer, H., Jalilian, E., Uhl, A.: Exploiting superior CNN-based iris segmentation for better recognition accuracy. Pattern Recogn. Lett. **120**, 17–23 (2019)
18. Arsalan, M., Naqvi, R.A., Kim, D.S., Nguyen, P.H., Owais, M., Park, K.R.: IrisDenseNet: robust iris segmentation 3Vusing densely connected fully convolutional networks in the images by visible light and near-infrared light camera sensors. Sensors **18**, 1501–1530 (2018)
19. Nguyen, K., Fookes, C., Ross, A., Sridharan, S.: Iris recognition with off-the-shelf CNN features: a deep learning perspective. IEEE Access **6**, 18848–18855 (2017)
20. Zhang, W., Lu, X., Gu, Y., Liu, Y., Meng, X., Li, J.: A robust iris segmentation scheme based on improved U-net. IEEE Access **7**, 85082–85089 (2019)
21. Chen, Y., Wang, W., Zeng, Z., Wang, Y.: An adaptive CNNs technology for robust iris segmentation. IEEE Access **7**, 64517–64532 (2019)
22. Liu, M., Zhou, Z., Shang, P., Xu, D.: Fuzzified image enhancement for deep learning in iris recognition. IEEE Trans. Fuzzy Syst. **28**(1), 92–99 (2020)
23. Wang, Z., Li, C., Shao, H., Sun, J.: Eye recognition with mixed convolutional and residual network (MiCoRe-Net). IEEE Access **6**, 17905–17912 (2018)
24. Yiu, Y.H., et al.: DeepVOG: open-source pupil segmentation and gaze estimation in neuroscience using deep learning. J. Neurosci. Methods **324**, 108307–108318 (2019)
25. Wang, C., Muhammad, J., Wang, Y., He, Z., Sun, Z.: Towards complete and accurate iris segmentation using deep multi-task attention network for non-cooperative iris recognition. IEEE Trans. Inf. Forensics Secur. **15**, 2944–2959 (2020)

26. Feng, X., Liu, W., Li, J., Meng, Z., Sun, Y., Feng, C.: Iris R-CNN: accurate iris segmentation and localization in non-cooperative environment with visible illumination. Pattern Recogn. Lett. **155**, 151–158 (2022)
27. Wei, J., Wang, Y., He, R., Sun, Z.: Cross-spectral iris recognition by learning device-specific band. IEEE Trans. Circuits Syst. Video Technol. **32**(6), 3810–3824 (2022)
28. Nwankpa, C.E., Ijomah, W., Gachagan, A., Marshall, S.: Activation functions: comparison of trends in practice and research for deep learning. ArXiv abs/1811.03378 (2018)
29. Wang, Z., Chai, J., Xia, S.: Realtime and accurate 3D eye gaze capture with DCNN-based iris and pupil segmentation. IEEE Trans. Visual Comput. Graphics **27**(1), 190–203 (2021)
30. Wang, K., Kumar, A.: Periocular-assisted multi-feature collaboration for dynamic iris recognition. IEEE Trans. Inf. Forensics Secur. **16**, 866–879 (2021)

Reinforcement Learning Algorithms for Effective Resource Management in Cloud Computing

Prathamesh Vijay Lahande and Parag Ravikant Kaveri(✉)

Symbiosis Institute of Computer Studies and Research, Symbiosis International (Deemed University), Pune, India
{prathamesh.lahande,parag.kaveri}@sicsr.ac.in

Abstract. Cloud has gained enormous significance today due to its computing abilities and service-oriented paradigm. The end-users opt for the cloud to execute their tasks rather than on their local machines or servers. To provide the best results, the cloud uses resource scheduling algorithms to process the tasks on the cloud Virtual Machines (VM). The cloud experiences several inefficiencies and damages due to inappropriate resource scheduling and/or several faults generated by the tasks when they are being processed dynamically, thereby becoming fragile and fault-intolerant. Due to these issues, the cloud produces meager or occasionally no results. The main objective of this research paper is to propose and compare the behavior of the hybrid resource scheduling algorithms – Reinforcement Learning – First Come First Serve (RL – FCFS) and Reinforcement Learning – Shortest Job First (RL– SJF) which are implemented by combining the resource scheduling algorithms FCFS and SJF with Reinforcement Learning (RL) mechanism. To compare the proposed algorithms, heavy tasks are executed on the cloud VMs under various scenarios, and the results obtained are compared with each other with respect to various performance metrics. With the implementation of these algorithms, the cloud will initially go into a learning phase. With proper feedback and a trial-and-error mechanism, the cloud can perform appropriate resource scheduling and also handle the tasks, irrespective of their faults, thereby enhancing the entire cloud performance and making the cloud fault-tolerance. With this, the Quality of Service (QoS) is improved while meeting the SLAs.

Keywords: Cloud computing · Fault tolerance · Performance · Reinforcement learning · Resource scheduling

1 Introduction

Cloud refers to a network that provides several services, such as data storage, applications, computations, etc., over the internet [18]. End-users choose the cloud over their local servers to avail of these services due to its high on-demand availability and high reliability [19]. The number of cloud users has grown tremendously, so much so that users opt for the cloud to perform most of their operations. One of the popular services which

K. K. Patel et al. (Eds.): icSoftComp 2022, CCIS 1788, pp. 369–381, 2023.
https://doi.org/10.1007/978-3-031-27609-5_29

the cloud offers is computing. The end-user submits requests in the form of high-sized tasks to the cloud for computing using the internet. The cloud accepts these tasks and processes them on the cloud Virtual Machines (VM) using resource scheduling algorithms. To provide the best results in terms of both time and cost, resource scheduling becomes a challenge because several times, the larger tasks are scheduled to low-performing VMs and smaller ones to high-performing VMs. This resource scheduling mismatch causes an unnecessary rise in time as well as the cost of task processing. The tasks which are being processed on the cloud VM also generates faults while they are being processed dynamically at run-time making the cloud vulnerable and fault-intolerant [10]. Due to this improper resource scheduling mechanism and fault-intolerance, the cloud is damaged, thereby making the cloud output low performance, especially when no intelligence is provided to it. Hence, providing an intelligence mechanism to the cloud becomes necessary to enhance its performance and address the resource scheduling and fault-intolerance issues of the cloud [19]. The Reinforcement Learning (RL) technique is well known for resolving such problems by enhancing the decision-making of the system, which subsequently facilitates efficient resource scheduling and fault-tolerance mechanisms [21]. The RL mechanism has shown acceptable results when applied to any system since its mechanism is feedback-based, and it can enhance any system with proper feedback over a period of time [20].

The main objective of this research paper is to propose and compare the behavior of two hybrid resource scheduling algorithms: Reinforcement Learning – First Come First Serve (RL – FCFS) and Reinforcement Learning – Shortest Job First (RL – SJF) which are designed by combining Reinforcement Learning (RL) with FCFS and SJF respectively. RL – FCFS and RL – SJF have shown acceptable results since an intelligence mechanism to the cloud to achieve appropriate resource scheduling as well as the fault-tolerance mechanism to make the cloud capable of processing more tasks and outputting better results.

The rest of the paper is organized as follows: Sect. 2 provides the Literature Survey. Section 3 provides the details of the experiment conducted. Section 4 includes the results and implications of the conducted experiment, followed by the conclusion in Sect. 5.

2 Literature Survey

The cloud environment is well-known for its high availability and high-reliability characteristics. Resource scheduling and fault-tolerance mechanisms are vital in making the cloud highly available and reliable. But the cloud faces significant issues due to the faults generated by the tasks during execution, thereby making the cloud fault-intolerant and outputting limited results. Improving the resource scheduling and facilitating fault-tolerance mechanism is considered an NP-hard problem; hence, it is a significant area for researchers to focus upon. Alsadie [6] has proposed a metaheuristic framework named MDVMA for dynamically allocating VMs of task scheduling in the cloud environment. According to simulation results, this algorithm performs better at reducing energy consumption, makespan, and data centre costs. Balla et al. [12] have proposed an enhanced and effective resource management method for achieving cloud reliability. The approach suggested in this research combines queuing theory and the reinforcement learning algorithm to plan user requests. The success of task execution, utilization rate, and reaction time are the performance indicators considered as this strategy is later compared with greedy and random work scheduling policies.

Bitsakos et al. [13] have used a deep reinforcement learning technique to achieve elasticity automatically. When this method is compared in simulation environments, it gains 1.6 times better rewards in its lifetime. In order to choose the optimum algorithm for each VM, Caviglione et al. [9] present a multi-objective method for finding optimal placement strategies that consider many aims. Chen et al. [1] have proposed an effective and adaptive cloud resource allocation scheme to achieve superior QoS in terms of latency and energy efficiency. To maximize energy cost efficiency, Cheng et al. [14] present a novel deep reinforcement learning-based resource provisioning and job scheduling system. This method achieves an improvement of up to 320% in energy cost while maintaining a lower rejection rate of tasks. Guo [7] has put forth a solution for multi-objective task scheduling optimization based on a fuzzy self-defense algorithm that can enhance performance in terms of completion time, rate of deadline violations, and usage of VM resources. Han et al. [11] suggested utilizing the Knapsack algorithm to increase the density of VMs and therefore increase energy efficiency.

Hussin et al. [16] proposed an effective resource management technique using adaptive reinforcement learning to improve successful execution and system reliability. Using the multi-objective task scheduling optimization based on the Artificial Bee Colony algorithm and the Q-Learning technique, Kruekaew & Kimpan [2] introduce the MOABCQ method, an independent task scheduling methodology in the cloud computing environment. This approach yields better outcomes regarding shorter lead times, lower costs, less severe imbalances, higher throughput, and average resource utilization. Praveenchandar & Tamilarasi [3] proposes an improved task scheduling and an optimal power minimization approach for making the dynamic resource allocation process efficient. The simulation results obtained from this proposed method give 8% better results when compared with existing ones.

Shaw et al. [4] explored the use of RL algorithms for the VM consolidation problem to enhance resource management. Compared to the widely used current heuristic methods, this proposed RL methodology increases energy efficiency by 25% while lowering service violations by 63%. Using the reinforcement learning strategy, Swarup et al. [8] suggested a clipped double-deep Q-learning method to resolve the task scheduling problem and reduce computing costs. Xiaojie et al. [15] presented a reinforcement learning-based resource management system to balance QoS revenue and power usage. The experimental results of this approach demonstrate that the suggested algorithm is more robust and consumes 13.3% and 9.6% less energy in non-differentiated and differentiated services, respectively than the existing algorithms. A deep reinforcement learning-based workload scheduling strategy is suggested by Zheng et al. [5] to balance the workload, cut down on service time, and increase task success rates.

Our work proposes and compares the behavior of hybrid resource scheduling algorithms by combining the RL technique with the resource scheduling algorithms FCFS and SJF, respectively, to improve resource scheduling and facilitate the fault-tolerance mechanism to enhance cloud performance.

3 Experiment

This section includes the experiment to compare the performance and behavior of RL – FCFS and RL – SJF under different environmental conditions. This section is further divided into four Subsect. 3.1, 3.2, 3.3, and 3.4. The Subsect. 3.1 includes the experiment configuration and simulation environment. The Subsect. 3.2 provides the dataset used for conducting the experiment. The sub-Sect. 3.3 consists of the VM configuration mentioned scenario-wise. The Subsect. 3.4 provides the rewards and structure of the Q-Table used for the experiment.

3.1 Experiment Configuration and Simulated Environment

The cloud environment is configured using the Java-based WorkflowSim [17] cloud simulation framework. The algorithms RL – FCFS and RL – SJF are incorporated in this WorkflowSim environment for scheduling the tasks on the cloud VMs. A total of 80386 tasks are generated by the Alibaba task event dataset, which are sent to the cloud for processing. The experiment is conducted in two phases: the first phase includes processing all tasks using RL – FCFS algorithm; the second phase includes processing the same tasks using RL – SJF algorithm. To test and compare the behavior of RL – FCFS and RL – SJF thoroughly, each phase consists of a total of ten scenarios. The first scenario consists of processing all tasks on 5 VMs; the second consists of processing all tasks on 10 VMs; the third consists of processing all tasks on 15 VMs, and so on, till the tenth scenario, where all the tasks are processed on 50 VMs. A queue is maintained that contains all the tasks which are in a ready state to be processed but currently are in a waiting state. Once the cloud VM is available, a particular task is selected for processing from this queue using the respective hybrid resource scheduling algorithm. Once a task is allotted to a particular available VM, its starting time is recorded. Similarly, the completion time is also recorded once the task completes its processing. The start time and completion time are later used to evaluate the Turn Around Time (TAT) and Waiting Time (WT) of the task. After all the tasks are processed, the Average Start Time (AST), Average Completion Time (ACT), Average TAT (ATAT), and Average WT(AWT) are computed. In the end, the cost required to process all the tasks is calculated according to the processing cost of the allotted VM. Performance metrics AST, ACT, ATAT, AWT, and average cost are used to compare the behavior of RL – FCFS and RL – SJF. Figure 1. Depicts the entire flowchart of the experiment conducted.

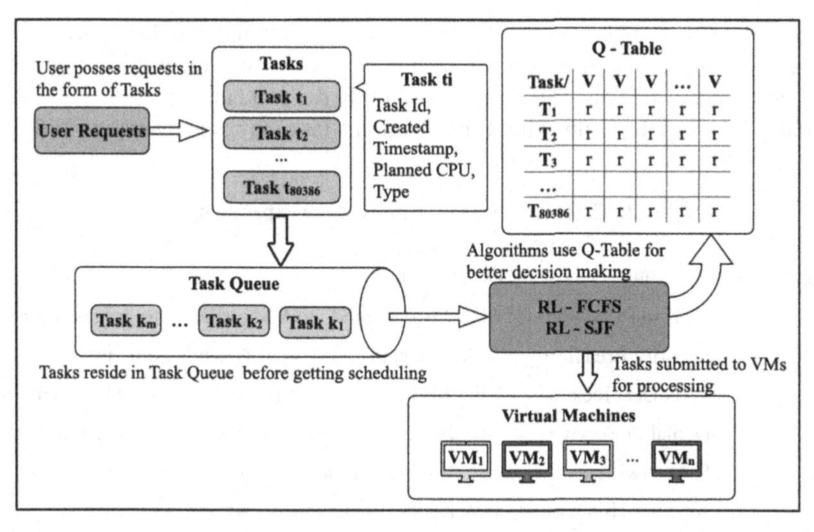

Fig. 1. Flowchart of the experiment

3.2 Experiment Dataset

The cloud is tested when 80386 tasks from the Alibaba task event dataset are submitted for processing, which consists of a total of twelve categories. Every particular task 'ti' is a vector that consists of a 'task id' to uniquely identify a certain task, 'created timestamp' to indicate the task creation time, 'planned CPU' to indicate the total time a task wishes to use cloud VM, and 'type'. The structure of a task is as follows:

$$\text{Task ti} = (\text{Task Id},\ \text{Created Timestamp},\ \text{Planned CPU},\ \text{Type})$$

The planned CPU of the task can be any value from the series: 10, 40, 50, 60, 70, 80, 100, 200, 300, 400, 600, 800. Here, a type of task is categorized as low (L) if its planned CPU is between 10 to 60, moderate (M) if its planned CPU is above 60 to 300, and high (H) if its planned CPU is beyond 300. Table 1 depicts the experimental dataset used.

Table 1. Category-wise task details of the experiment dataset.

Category	Planned CPU	Task type	Task size	Category	Planned CPU	Task type	Task size
1	10	L	9017	7	100	M	17529
2	40	L	372	8	200	M	8
3	50	L	52791	9	300	M	2
4	60	L	199	10	400	H	4
5	70	M	185	11	600	H	5
6	80	M	272	12	800	H	2

Any task that is submitted for processing has the potential to either cause a specific fault at run-time or complete its processing without any issue, regardless of its planned CPU, task type, or its size. This process is entirely dynamic in nature.

Table 2 depicts the faults with its description a task may generate at run-time.

Table 2. Category-wise task details of the experiment dataset.

Fault index	Task fault	Description
1	VMs not available	All VMs occupied, no VM available to process task
2	Security Breach	Task breached cloud security while being processed
3	VMs Deadlocked	All VMs hold and wait for resources held by other VMs
4	Denial of Service	Task denied processing service after waiting long time
5	Data Loss	The cloud undergoes data loss while processing task
6	Account Hijacking	Task hacks cloud accounts while being processed
7	SLA Violations	Task violates SLA of the cloud while being processed
8	Insufficient RAM	RAM is not available for processing tasks

All the tasks will be processed on the cloud VMs in the series: 5, 10, 15, …, 50, i.e., ten scenarios in both phases. Table 3 depicts dynamically generated fault percentages with respect to the fault index for each scenario for both phases. From Table 3, we can observe that all the faults caused by the task have a certain balance, and no particular fault overpowers or dominates any other fault in its generation.

Table 3. Dynamic generated fault percentage for each scenario.

Fault index								
VMs	1	2	3	4	5	6	7	8
5	11.1	11.3	11.2	11.2	11.2	11.1	10.9	11.1
10	11	11.1	11	11.1	11.1	11.1	11.1	11.3
15	11.1	10.9	11.2	11.1	11.2	11.2	11.1	11.1
20	11.1	11	10.9	11.1	11.2	11.1	11.2	11.1
25	11.2	11.1	11.2	11	11.1	10.9	11.2	11.3
30	11	11.1	11.1	11	11.1	11.2	11.1	11.2
35	11.2	11.2	10.9	11	11.1	11	11.1	11.1
40	11.1	11	11.2	11.2	11.2	11.1	11	11.2
45	11.2	11.3	11	10.9	11.2	11.1	11.1	11.2
50	11.1	11.1	11	11.1	11.4	11.1	11.1	10.9

3.3 VM Configuration

Table 4 depicts the scenario-wise cloud VMs sizes, which are categorized into nine cases depending upon their performance configuration. These categories are Low-Low (LL), Low-Medium (LM), Low-High (LH), Medium-Low (ML), Medium-Medium (MM), Medium-High (MH), High-Low (HL), High-Medium (HM), and High-High (HH). The number of VMs are distributed across each scenario in such a way that more VMs of medium configuration are available to process tasks.

Table 4. VM configuration for each scenario

VMs	LL	LM	LH	ML	MM	MH	HL	HM	HH
5	1	–	–	1	1	1	–	–	1
10	1	1	1	1	2	1	1	1	1
15	1	1	1	3	3	3	1	1	1
20	2	2	2	2	4	2	2	2	2
25	2	2	2	4	5	4	2	2	2
30	3	3	3	4	4	4	3	3	3
35	3	3	3	5	7	5	3	3	3
40	4	4	4	5	6	5	4	4	4
45	4	4	4	7	7	7	4	4	4
50	5	5	5	6	8	6	5	5	5

3.4 RL Rewards and Q-Table

A separate data structure called 'Q-Table' is maintained across all ten scenarios in both phases to store the reward 'r' associated with a particular task which is processed on a particular cloud VM. This reward represents the feedback offered by the hybrid resource scheduling algorithms. A smaller reward is offered if the resource scheduling is performed inappropriately or if it causes a certain fault. Similarly, a higher reward is offered with proper resource scheduling, and the task is processed without any faults generated. Rewards play a vital role in making the system adapt and learn from the environment. The below equation is used to update the reward value in the Q-Table:

$$\text{Q-Table}(T_i, VMj) = \text{Q-Table}(T_i, VMj) + \alpha\,[R(T_i, VMj) + \gamma \max \text{Q-Table}'(T_i', VM_j') - \text{Q-Table}(T_i, VMj)]$$

Here, the alpha 'α' and discount rate 'γ' is initialized to 0.9. The structure of the Q-table used is as follows:

$$\text{Q - Table} = \left\{ \begin{array}{ccccc} VM_1 & VM_2 & VM_3 & \ldots & VM_n \\ T_1 & r & r & \ldots & r \\ T_2 & r & r & \ldots & r \\ T_3 & r & r & \ldots & r \\ \ldots & \ldots & \ldots & \ldots & \ldots \\ T_{80386} & r & r & \ldots & r \end{array} \right\}$$

4 Results and Implications

The results and implications of the experiment in included in Subsect. 4.1 and 4.2. The Subsect. 4.1 and 4.2 provides the comparative results and implications of RL – FCFS and RL – SJF for resource scheduling and Subsect. 4.2 for fault tolerance.

4.1 Results and Implications with Respect to Resource Scheduling

From the experiment conducted, the results obtained from using the hybrid resource scheduling algorithms RL – FCFS and RL – SJF are compared scenario-wise with respect to the performance parameters such as AST, ACT, ATAT, AWT, and average cost. Further, empirical analysis is also performed for RL – FCFS and RL – SJF algorithms to extensively compare their behavior under various scenarios and circumstances. Table 5 represents the comparative table of results obtained for the experiment conducted for RL – FCFS and RL – SJF.

Table 6 represents the comparative table of empirical analysis performed for RL – FCFS and RL – SJF with respect to AST, ACT, ATAT, AWT, and Average Cost. Various terminologies used in the below table are: **Para**: Parameter, **LRE**: Linear Regression Equation, **RLS**: Regression Line Slope, **SS**: Slope Sign, **–ve**: Negative, **RYI**: Regression Y-Intercept, **R**: Relationship; + **ve**: Positive.

Following points can be observed from Table 5 and Table 6 across all the scenarios:

- $\uparrow$ VM = $\downarrow$ AST: With increase in the number of VMs, the AST decreases.
- $\uparrow$ VM = $\downarrow$ ACT: With increase in the number of VMs, the ACT decreases.
- $\uparrow$ VM = $\downarrow$ ATAT: With increase in the number of VMs, the ACT decreases.
- $\uparrow$ VM = $\downarrow$ AWT: With increase in the number of VMs, the AWT decreases.
- $\uparrow$ VM = $\uparrow$ Average Cost: With increase in the number of VMs, the average cost increases.
- Performance (RL – SJF) > Performance (RL – FCFS) with respect AST, ACT, ATAT, and AWT
- Performance (RL – FCFS) > Performance (RL – SJF) with respect Average Cost

Table 5. Comparison of RL – FCFS and RL – SJF with respect to AST, ACT, ATAT, AWT and average cost

	AST		ACT		ATAT		AWT		Average Cost	
VMs	RL – FCFS	RL – SJF	RL – FCFS	RL – SJF	RL – FCFS	RL – SJF	RL – FCFS	RL – SJF	RL – FCFS	RL – SJF
5	36513	36505	36570	36551	2506	2308	2449.3	2289.5	16.9	17.1
10	34106	34099	34163	34156	80	73	23.3	15.8	29.2	29.2
15	34130	34129	34186	34185	61	60	4	3	38.4	38.7
20	34113	34113	34170	34170	58	58	1.4	1.2	45.4	45.7
25	34053	34053	34110	34110	57	57	0.7	0.6	50.5	51
30	34177	34177	34234	34234	57	57	0.4	0.3	54.1	54.7
35	34048	34048	34104	34104	57	57	0.2	0.2	56.6	57.3
40	34145	34145	34201	34201	57	57	0.1	0.1	58.6	59.7
45	34166	34166	34222	34222	57	57	0.1	0.1	60.2	61.4
50	34090	34090	34146	34146	57	57	0.1	0.1	61.5	62.7
Result	RL – SJF > RL – FCFS		RL – SJF > RL – FCFS		RL – SJF > RL – FCFS		RL – SJF > RL – FCFS		RL – FCFS > RL – SJF	

Table 6. Empirical analysis of RL – FCFS and RL – SJF with respect to AST, ACT, ATAT, AWT and average cost.

	AST		ACT		ATAT		AWT		Average Cost	
Para	RL – FCFS	RL – SJF	RL – FCFS	RL – SJF	RL – FCFS	RL – SJF	RL – FCFS	RL – SJF	RL – FCFS	RL – SJF
LRE	y = −129x + 35067	y = −128x + 35061	y = −129x + 35124	y = −128x + 35114	y = −134x + 1045	y = −123x + 963	y = −134x + 1045	y = −123x + 963	y = −134x + 1045	y = −123x + 963
RLS	−129	−128	−129	−128	−134	−123	−134	−123	−134	−123
SS	−ve	−ve	−ve	−ve	−ve	−ve	−ve	−ve	−ve	−ve
RYI	35067	35061	35124	35114	1045	963	1045	963	1045	963
R	+ ve	+ ve	+ ve	+ ve	+ ve	+ ve	+ ve	+ ve	+ ve	+ ve
R^2	0.266	0.265	0.267	0.265	0.278	0.276	0.278	0.276	0.278	0.276
Result	RL – SJF > RL – FCFS		RL – SJF > RL – FCFS		RL – SJF > RL – FCFS		RL – SJF > RL – FCFS		RL – FCFS > RL – SJF	

4.2 Results and Implications with Respect to Fault Tolerance

This sub-section includes the results and implications of the conducted experiment with respect to fault tolerance. When the tasks are being processed on the cloud VMs, each task has the potential to generate a fault or get processed without issue. If the cloud encounters faults, then the RL – FCFS and RL – SJF provides a solution and process that particular task. On the other hand, faults that breach the cloud security, make the cloud lose its data, hijacks account, or violates SLAs are tackled and not processed, making the cloud fault tolerant.

Table 7 depicts the execution status of the task with respect to its generated fault.

Table 7. Task execution status with respect to dynamic generated faults.

		Fault Index							
Scenario	VMs	1	2	3	4	5	6	7	8
1	5	✓	✗	✓	✓	✗	✗	✗	✓
2	10	✓	✗	✓	✓	✗	✗	✗	✓
3	15	✓	✗	✓	✓	✗	✗	✗	✓
4	20	✓	✗	✓	✓	✗	✗	✗	✓
5	25	✓	✗	✓	✓	✗	✗	✗	✓
6	30	✓	✗	✓	✓	✗	✗	✗	✓
7	35	✓	✗	✓	✓	✗	✗	✗	✓
8	40	✓	✗	✓	✓	✗	✗	✗	✓
9	45	✓	✗	✓	✓	✗	✗	✗	✓
10	50	✓	✗	✓	✓	✗	✗	✗	✓

From the above table, we can observe that:

- For all the scenarios across all the VMs, the tasks with fault index 1, 3, 4, or 8, i.e., faults such as VMs not available, VMs deadlocked, denial of service, and insufficient amount of RAM, the RL – FCFS and RL – SJF algorithms provide a solution for these faults, and these tasks are executed.
- On the other hand, the tasks with fault index 2, 5, 6, or 7, i.e., faults such as security breaches, data loss, account hijacking, and SLA violations, the RL – FCFS and RL – SJF algorithms do not execute these tasks.

Table 8 shows the task size of successful and failed tasks for all ten scenarios.

Following points can be observed from Table 7 and Table 8 for all the scenarios for RL – FCFS and RL – SJF with respect to fault-tolerance:

- From the 80386 tasks submitted, aggregately, 88.9% of the tasks generated faults.
- Both the algorithms provided fault-tolerance by not processing tasks that generate fault and damage the cloud.
- Initially, both the algorithms worked similarly for the first few VMs, but as the number VMs increased, RL – SJF performed slightly better than RL – FCFS.

Table 8. Successful and failed task percentage for all scenarios.

				RL – FCFS		RL – SJF	
Scenario	VMs	Faults	No faults	Success	Fail	Success	Fail
1	5	71562	8824	44680	35706	44680	35706
2	10	71425	8961	44656	35730	44656	35730
3	15	71420	8966	44741	35645	44745	35641
4	20	71432	8954	44544	35842	44551	35835
5	25	71491	8895	44802	35584	44807	35579
6	30	71501	8885	44542	35844	44546	35840
7	35	71253	9133	44645	35741	44651	35735
8	40	71503	8883	44728	35658	44732	35654
9	45	71589	8797	44475	35911	44482	35904
10	50	71445	8941	44392	35994	44397	35989

Table 9 summarizes the overall comparison of RL – FCFS and RL – SJF algorithms with respect to resource scheduling and fault-tolerance mechanism across all different environmental conditions with respect to performance metrics such as AST, ACT, ATAT, AWT, average cost, and fault-tolerance mechanism.

Table 9. Overall comparison of RL – FCFS and RL – SJF.

Performance metric	Algorithm comparison
AST	RL – SJF > RL – FCFS
ACT	RL – SJF > RL – FCFS
ATAT	RL – SJF > RL – FCFS
AWT	RL – SJF > RL – FCFS
Average cost	RL – FCFS > RL – SJF
Fault-tolerance	RL – SJF > RL – FCFS

5 Conclusions

The main aim of this research paper is to propose and compare the hybrid resource scheduling algorithms RL – FCFS and RL – SJF, which were implemented by combining the RL technique with the resource scheduling algorithms FCFS and SJF, respectively, to improve resource scheduling and facilitate the fault-tolerance mechanism. The main reason for using the RL technique is that no past data is required for the system to learn, and the system adapts from past experiences, just like human beings learn. The RL technique has provided better results when applied to any system, thereby significantly improving the system. An experiment was conducted in the WorkflowSim environment where RL – FCFS and RL – SJF algorithms were implemented in ten different scenarios, and their behavior was compared with each other in terms of resource scheduling and fault-tolerance mechanisms. From Table 9, which depicts the overall results, it can be observed that the performance of the RL – SJF algorithm is better than RL – FCFS algorithm in terms of performance metrics such as AST, ACT, ATAT, AWT, and fault-tolerance mechanism. Whereas the performance of the RL – FCFS algorithm is better than the RL – SJF algorithm with respect to the average cost required. Hence, RL – FCFS algorithm can be opted for in terms of cost, and RL – SJF algorithm can be opted for in terms of time parameters and a better fault-tolerance mechanism. Also, RL – FCFS and RL – SJF algorithms can be opted for instead of the traditional FCFS and SJF to improve resource scheduling and facilitate fault-tolerance mechanisms, thereby improving the overall cloud performance.

References

1. Chen, Z., Hu, J., Min, G., Luo, C., El-Ghazawi, T.: Adaptive and efficient resource allocation in cloud datacenters using actor-critic deep reinforcement learning (2022)
2. Kruekaew, B., Kimpan, W.: Multi-objective task scheduling optimization for load balancing in cloud computing environment using hybrid artificial bee colony algorithm with reinforcement learning (2022)
3. Praveenchandar, J., Tamilarasi, A.: Retraction note to: dynamic resource allocation with optimized task scheduling and improved power management in cloud computing (2022)
4. Shaw, R., Howley, E., Barrett, E.: Applying reinforcement learning towards automating energy efficient virtual machine consolidation in cloud data centers. Inf. Syst. **107**, 101722 (2022)
5. Zheng, T., Wan, J., Zhang, J., Jiang, C.: Deep reinforcement learning-based workload scheduling for edge computing. J. Cloud Comput. **11**(1), 1–13 (2021). https://doi.org/10.1186/s13677-021-00276-0
6. Alsadie, D.: A metaheuristic framework for dynamic virtual machine allocation with optimized task scheduling in cloud data centers (2021)
7. Guo, X.: Multi-objective task scheduling optimization in cloud computing based on fuzzy self-defense algorithm (2021)
8. Swarup, S., Shakshuki, E.M., Yasar, A.: Task scheduling in cloud using deep reinforcement learning (2021)
9. Caviglione, L., Gaggero, M., Paolucci, M., Ronco, R.: Deep reinforcement learning for multi-objective placement of virtual machines in cloud datacenters (2020). https://doi.org/10.1007/s00500-020-05462-x
10. Gonzalez, C., Tang, B.: FT-VMP: fault-tolerant virtual machine placement in cloud data centers (2020)
11. Han, S., Min, S., Lee, H.: Energy efficient VM scheduling for big data processing in cloud computing environments. J. Ambient Intell. Humanized Comput. (2019). https://doi.org/10.1007/s12652-019-01361-8
12. Balla, H.A., Sheng, C.G., Weipeng, J.: Reliability enhancement in cloud computing via optimized job scheduling implementing reinforcement learning algorithm and queuing theory (2018)
13. Bitsakos, C., Konstantinou, I., Koziris, N.: DERP: a deep reinforcement learning cloud system for elastic resource provisioning (2018)

14. Cheng, M., Li, J., Nazarian, S.: DRL-cloud: deep reinforcement learning-based resource provisioning and task scheduling for cloud service providers (2018)
15. Zhou, X., Wang, K., Jia, W., Guo, M.: Reinforcement learning-based adaptive resource management of differentiated services in geo-distributed data centers (2017)
16. Hussin, M., Hamid, N.A.W.A., Kasmiran, K.A.: Improving reliability in resource management through adaptive reinforcement learning for distributed systems (2015)
17. Chen, W., Deelman, E.: WorkflowSim: a toolkit for simulating scientific workflows in distributed environments (2012)
18. Armbrust, M., et al.: A view of cloud computing (2010)
19. Dillon, T., Wu, C., Chang, E.: Cloud computing: issues and challenges (2010)
20. Vengerov, D.: A reinforcement learning approach to dynamic resource allocation (2007)
21. Andrew, A.: Reinforcement Learning (1998)

Corpus Building for Hate Speech Detection of Gujarati Language

Abhilasha Vadesara and Purna Tanna(✉)

GLS University, Ahmedabad, Gujarat, India
abhilashavadesara@gmail.com

Abstract. Social media is a rapidly expanding platform where users share their thoughts and feelings about various issues as well as their opinions. However, this has also led to a number of issues, such as the dissemination and sharing of hate speech messages. Hence, there is a need to automatically identify speech that uses hateful language. Hate speech refers to the aggressive, offensive language that focuses on a specific people or group as far as their ethnic group or race (i.e., racism), gender (i.e., sexism), beliefs, and religion. The aim of this paper is to examine how hate speech contrasts with non-hate speech. A corpus of Gujarati tweets has been collected from Twitter. The dataset was cleaned and pre-processed by removing unnecessary symbols, URLs, characters, and stop words, and the cleaned text was analyzed. Pre-processed data was annotated by twenty-five people and has achieved Fleiss's Kappa coefficient with 0.87 accuracies for agreement between the annotators.

Keywords: Hate speech · Text mining · Kappa's coefficient · Gujarati language · Sentiment analysis

1 Introduction

Expressions that are harassing, abusive, harmful, urge brutality, make hatred or discrimination against groups, target qualities like religion, race, a spot of beginning, race or community, district, individual convictions, or sexual direction are called hate speech. The Ability to spot hate speech has gotten heaps of attention these days. As a result, hate speech has reached new levels in additional advanced and intellectual types.

Social networking sites make it more direct. To provide honest thoughts and feelings to end-users, Twitter provides a site and microblogging service. In this digital age, social media data is increasing daily, where hate speech detection becomes a challenge in provoking conflict among the countries' voters. However, it's impossible to spot hate speech from a sentence without knowing the context.

As we have seen, much research has been accomplished on European, English, and some Indian languages. In any case, little work has been done in Gujarati as it is the primary language most Gujarati people use in speaking and formulating. The purpose of the analysis is to build the corpus of Gujarati language instead of distinctive hate speech.

K. K. Patel et al. (Eds.): icSoftComp 2022, CCIS 1788, pp. 382–395, 2023.
https://doi.org/10.1007/978-3-031-27609-5_30

After collecting tweets, we pre-processed them with the Natural Language Processing technique [21] and implemented annotation by twenty-five different age groups. To check the inter agreement between annotators, we use Fleiss's kappa. In addition, the range of individuals and their backgrounds, cultures, and beliefs will ignite the flames of hate speech [1]. For the Gujarati region, there's a conspicuous magnification within the utilization of social media platforms.

This paper is structured in different sections. In Sect. 2, we describe the short description of related work. The new dataset and the Methodology, which include the data cleaning and Methodology, are defined in Sects. 3 and 4, respectively. In Sect. 5 we discussed the experiments. Section 6 describes the Result and Discussion of the technique. Section 7 finalizes this paper and suggests possible suggestions for future work (Fig. 1).

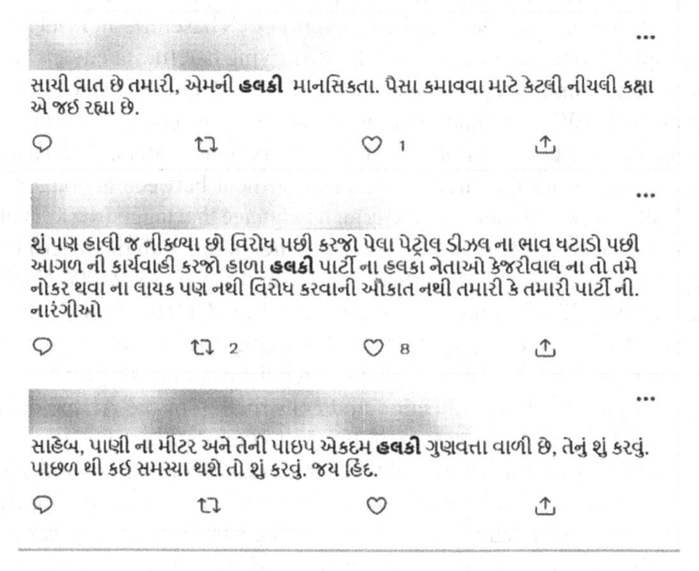

Fig. 1. Post of Twitter tweet

2 Related Forum and Dataset

Collections are an essential quality for any classification method. Several corpora of hate speech were used for analysis. Tremendous work has been done in numerous dialects, particularly for European and Indian. However, standard datasets aren't available for some languages, like Gujarati, and we are trying to make the tagged dataset for such an occasional resource language. Several corpora focus on targets like immigrants, women or racism, religion, politics, celebrities, and community. Others focus on only Hate speech detection or different offensive text types. A recent trend is to classify the data into more fine-grained classification. So, some knowledge challenges need detailed analysis for hate speech like the detection of target, aggressiveness, offensive, stereotype, irony, etc. A recent and attention-grabbing diversity is CONAN. It offers Hate Speech and also the reactions to it [2]. It opens opportunities for detecting Hate Speech by analyzing it collectively with consecutive posts. The researcher summarizes the standard Hate speech dataset attainable at various forums. Karim, Md. Rezaul et al. proposed DeepHateExplainer, which detects different sorts of hate speech with an 88% f1-score on several ML and DNN classifiers. For annotation of the dataset, they used the cohesion kappa technique [13]. Alotaibi, B. Alotaibi et al. have provided an approach for detecting aggression and hate speech through short texts. They have used three models: the bidirectional gated recurrent unit (BiGRU), the second transformer block, and the third convolutional neural network (CNN) based on Multichannel Deep Learning. They used the NLP approach and categorized 55,788 datasets into Offensive and Non-Offensive. They achieved 87.99% accuracy upon evaluating it on trained data 75% and testing data 25% [18]. Karim, Md Rezaul et al. proposed hate speech detection for the under-resourced Bengali language. They Evaluate against DNN baselines and yield F1 scores of 84%. They applied approaches to accurately identifying hateful statements from memes and texts [15]. Gohil and Patel generated G-SWN using Hindi SentiWordNet (H-SWN) and IndoWordNet (IWN) by manipulating synonym relations. The Corpus was annotated for negative and positive polarity classes by two annotators. They used Cohen's kappa Statistical measure for inter-annotator agreement between annotators [16]. The GermEval Task2 2019 is the data set of German language that tagged 4000 Twitter tweets to identify the three levels, hate, type, and implicit/explicit, with the macro F1 0.76 [3]. The racism dataset was used to determine binary and racism on 24000 English tweets with the accuracy 0.72 F1 Score [4]. Arabic social media dataset is on the market to identify Arabic tweets where it focuses on identifying obscene and inappropriate data with the f1 score of 0.60 [5]. Table 1 contains the dataset that is offered. Al-Twairesh, Nora et al. [22] presented the collection and construction of the Arabic dataset. They explained the technique of annotation of 17,573 twitter datasets. For inter agreement, they used Fleiss's Kappa and achieved 0.60 kappa's value, considered moderate. Akhtar, Basile et al. [23] Tested three different Twitter social media datasets in English and Italian language. They annotated the dataset with three annotators and measured Fleiss's kappa value of 0.58. They combined the single classifiers into an inclusive model.

Table 1. Collections of research on hate speech.

Paper reference	Dataset	Task	Example	Font size and style
[6]	Twitter	Binary, Hate	14500	English
[7]	Twitter	Hate, aggression, target	19000	Spain, English
[8]	Twitter	3 levels, Hate, targeted and target type	13200	English
[9]	TRAC COL-ING	3 classes, overtly or covertly Aggressive	15000 each language	English, Hindi
[17]	Facebook	6 classes	5,126	Bengali
[16]	Twitter	2 classes	1120	Gujarati

3 Dataset and Collection

Our main goal was to collect datasets using different techniques. The tweet gathered in the period from January 2020 to January 2021. We gathered the tweets data using the Twitter API with different categories like politics, sports, religion, and celebrity, as shown in Fig. 2. Most of the substance on Twitter isn't offensive, so we attempted different techniques to keep the dissemination of offensive tweets on about 30% of the dataset. Keywords and hashtags used to identify Gujarati hate speech. The Twitter API gives numerous recent tweets with an unprejudiced dataset. Thus, the tweets are acquired with the help of keywords and hashtags containing offensive content. The difficulties during the assessment of hate speech were language registers like irony or indirectness and youth talk, which researchers might not understand. We have collected approximately Twelve thousand tweets on hate and none-hate Gujarati content. The corpus was separated into training and testing categories to perform the classification task (Table 2).

Table 2. Collections of research on hate speech

Categories	Different Target	Example of Gujarati tweet	Translation in English
Religion	Religious people	આ મુ**ઓ ભ**ઓ પોતે દેશ વિરોધી કામ કરે છે.	These mul***s themselves do anti-national work.
Sports	Sports people, Sports	ક્રિકેટ તો માત્ર સટ્ટા વાળ માટે છે.	Cricket is only for betting hair.
Politics	Political party, People	શિ**નાએ કહ્યું, દેશમાં ઘૂસેલા પા***ની, બાં***શી મુસ્લિમોને બહાર ફેંકી દેવા જોઈએ.	The Shi***na said Pa**ni, Ban**shi Muslims who entered the country should be thrown out.
Celebrity	People, Movie, Celebrity	રા* *વંત એવી અટકચાળી અભિનેત્રી જેને કોઇને કોઇ વાતે સોશિયલ મીડિયા પર ચર્ચામાં રહેવાનો શોખ છે.	Ra** **want is a speculation actress who is fond of being in the news on social media for some means.

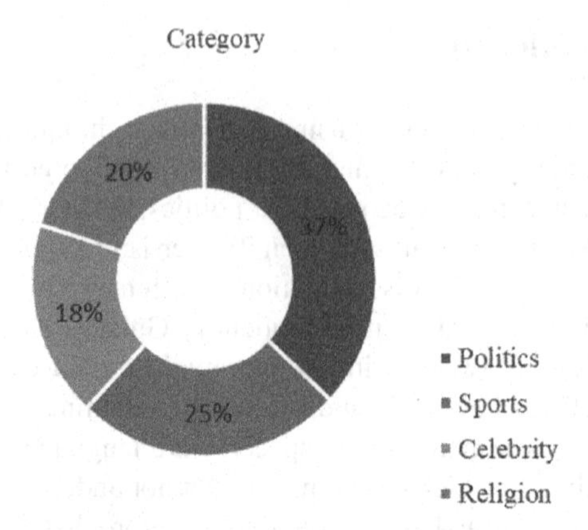

Fig. 2. Distribution diagram of hate speech data in each category

4 Methodology

In this section, we discussed the proposed approach in detail, discussion of preprocessing techniques and its example in the Gujarati language, and the annotation task and its process in detail.

4.1 Data Preprocessing

The dataset from Twitter is very noisy because it is not processed. To extract the model into a better feature, we need to perform the text processing on the actual dataset [21]. Initially, our data was in UTF-8 because of Twitter API responds to it in encoded form. We converted the data into Gujarati with the help of python decode method. Although the data was not clean, it contained many extra characters and Hindi, and English alphabets mixed, so it was necessary to clean it. With the help of python, we implement the preprocessing task which contain removal of URL, hashtags, user mentions, punctuation, Numbers, stop words, tokenizing etc. with the help of python libraries like pandas, re (regular expression), nltk. Below we describe each steps in detail.

Removal of URLs, Hashtags, User Mentions, Other Characters and Noise. Undesirable strings are considered extra information, which creates noise in the data. The tweet contains much extra information, such as URLs (http://www.imrobo.com) which refers to extra information, hashtags symbol (#socialmedia), which denotes the tweet is associated with some particular topic. User mentions (@name) means the post links to a particular user's profile. This information is helpful to human beings, but for a machine, it is considered noise that requires to be handled. Many researchers have presented different techniques to deal with such content [13, 23].

An example is given below:

Before: આગામી છ મહિના સુધી ક્રિકેટ અથવા અન્યકોઈ રમત શક્ય નથી. \nhttps://t.co/DHGGnlLGOi'b' @dhwansdave #cricket
After: આગામી છ મહિના સુધી ક્રિકેટ અથવા અન્ય કોઈ રમત શક્ય નથી.

Removal of Emoticons. Social media users use emojis such as ☹, 😊, 😉, etc., to express their sentiments. Such content is not helpful for tweet classification, so it needs to be removed from the tweet. An example is given below:

Before: અસ્મિતા પર હમાણા તમારો ઇંટરવ્યું જોયો. મજા પડી ગઇ તમને સાંભળવાની😍
After: અસ્મિતા પર હમાણા તમારો ઇંટરવ્યું જોયો .મજા પડી ગઇ તમને સાંભળવાની

Removal of Numbers. The Dataset ordinarily contains undesirable numbers and provides essential information, but they don't provide the information that helps in classification. So many researchers altogether remove it from the corpus. However, Eliminating the number from the Dataset may lead to a loss of information, but it does not impact much on the classification task. So, we eliminate all the numbers from the Dataset. An example is given below:

Before: હિન્દુ ધર્મ ના લોકો એ ફરજીયાત મિસકોલ કરવો88662 88662 પર અને બીજા 10 હિન્દૂ ને આ મેસેજ વિડિઓ સાથે મોકલવો
After: હિન્દુ ધર્મ ના લોકો એ ફરજીયાત મિસકોલ કરવો પર અને બીજા હિન્દૂ ને આ મેસેજ વિડિઓ સાથે મોકલવો

Removal of Stop Words. The tweet contains common words like 'તો,' 'અને,' 'છે,' 'આવે,' 'થતા' etc. are known as stop words in the Gujarati language [19]. It doesn't have complete, meaningful information, which helps in classification. One of the significant advantages of eliminating stop words in NLP text-based handling is decreasing the text by 30–40% from the corpus [20]. In our analysis, we created the list of stop words and eliminated them from the corpus. An example is given below:

Before: આખી દુનિયા ભગવાન ની મરજી થી ચાલે છે
After: આખી દુનિયા ભગવાન મરજી ચાલે

Tokenizing. In this step, tweets are separated using spaces to find the boundaries of words. Splitting a sentence into meaningful parts and recognizing the individual entities in the sentence is called Tokenization. We implement word Tokenization for classification tasks after the annotation of data. An example is given below:

Before: આખી દુનિયા ભગવાન મરજી ચાલે
After: આખી, દુનિયા, ભગવાન, મરજી, ચાલે

4.2 Data Annotation

After collecting the data, the second stage consists of annotating the Gujarati corpus. Before building the corpus, a review of techniques used to detect hate speech [12]. However, we eliminated many tweets from the corpus because of data duplication and off-topic content. At present, the amount of annotated data consists of 10000 tweets. Hate speech is a complex and multi-level concept. The annotation task is tricky and subjective, so we have taken all the initial steps to ensure that all annotators have a general basic knowledge about the task starting with the definition. The annotation process includes a multi-step process. After a fundamental step, it was carried out by 25 annotators manually who the people of different age groups are shown in Fig. 3.

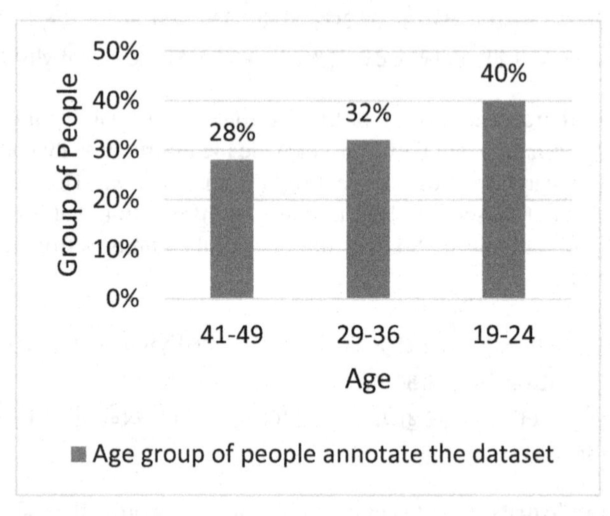

Fig. 3. Annotation different age group people wise

They labeled the corpus based on the relevant definitions, rules, regulations, and examples. The annotators were given the instructions as guidelines to classify the tweet into hate and none-hate category. The following factors are considered for hate tweets.

The first factor considered as a target means that the tweet should address, or refer to, one of the minority groups previously known as hate speech targets or the person considered for membership in that category.

The second is action, or more explicitly pronounced offensive force, in which it is capable of spreading, inciting, promoting, or justifying violence against a target. At whatever point the two factors happen in a similar tweet, we consider it a hate speech case, as in the example below (Table 3):

Table 3. Example of hate and none-hate annotation in tweets

Tweets	Label
કેટલાક મૂર્ખ કહે છે કે મુસ્લિમોને દેશમાં ડરાવવામાં આવે છે.	none-hate
તુ પાકિસ્તાન વયો જા હરામી	hate
આવું જ્ઞાન સરકાર ના નેતા ઓ ને પણ આપો . સાલા બેસર્મો ના લીધે જ આજે આ દિવસો પાછા આવ્યા છે.	hate
સાલા સરકારી કર્મચારી ને મફત નો પગાર જોઈ છે.	hate
હવે શિક્ષકોને શાળાના પુસ્તકો વિદ્યાર્થી પાસે પહોંચાડવાનું પણ ભાડુ જોઇએ છે !	none -hate

Figure 4 illustrate the procedure for the annotation of the corpus. First step is to check the tweets is belong to which category ex. religion, political, ethnicity etc. Then it should be analyzed by few questions Like "Is there any intention to offend someone?" If the answer will be no than it would be considered as none hate. Because that tweet considers normal tweet ex. જન્માષ્ટમી ના દીવસે શ્રદ્ધાળુઓ ભાવુક બન્યા. (Devotees became emotional on the day of Janmashtami.) which doesn't contain any offend towards any religious or person. But if it is yes than the next question would be asked like "Is there any swearing word or expression?" if the answer is yes than it would be consider as hate speech because the swearing word can be used harm the feelings of particular person or religion or group. ex. મુ** સાલા કસાઈ છે(The mullahs are butchers). If it is no means we required to analyzed it in depth like next question will be "Is the post contains any target or any action?" If the answer is yes then the tweets consider as hate ex. તું એક વાર મારા હાથમાં આવ હું તને મારી નાખીશ(you once come into my hands I will kill you) such type of tweets contains some action towards person so it's considered as hate speech. Otherwise, it is non-hate ex. તું એક વાર મારા હાથમાં આવ(you once come into my hands).

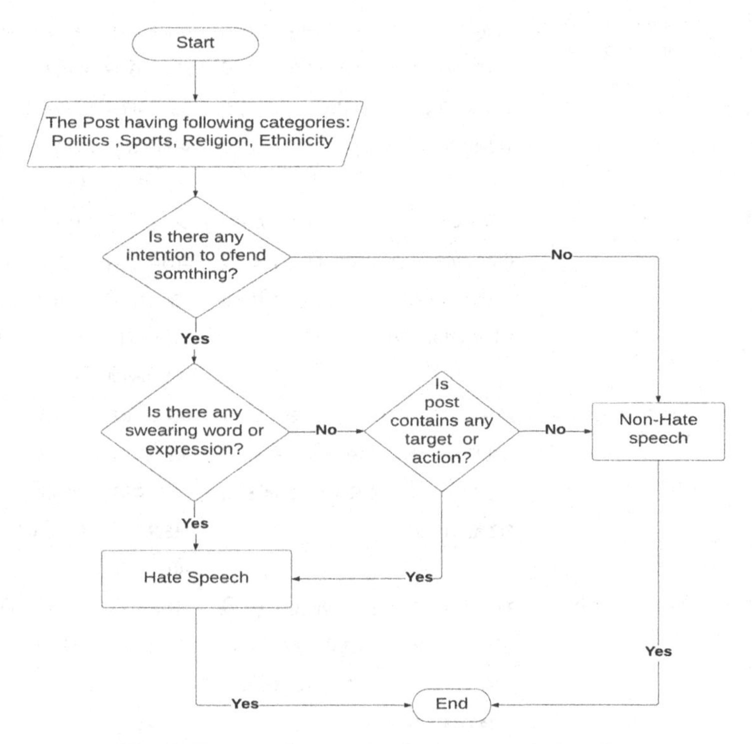

Fig. 4. Process of annotation for hate speech detection

5 Experiments

The 12k Gujarati dataset was raw and mixed with punctuations, URLs, non-Gujarati characters, emoticons, special characters, and stop words and tokenized after the annotation task. We removed the punctuations, stopwords, emoticons, URLs, symbols, non-Gujarati

Table 4. Steps of data cleaning using preprocessing technique

Number	Preprocessing Techniques	Raw data	Clean Data
1.	Removal of URLs, hashtags, user mentions, other characters and noise	@sandeshnews અમદાવાદ મુંબઈ નેશનલ હાઇવે ઉપર અંકલેશ્વર નજીક બે ટ્રક વચ્ચે સર્જાયો ગંભીર અકસ્માત, 2 લોકોના ઘટનાસ્થળે મોત. ☹ #accident #ankleshwar #highway #truckaccident	અમદાવાદ મુંબઈ નેશનલ હાઇવે ઉપર અંકલેશ્વર નજીક બે ટ્રક વચ્ચે સર્જાયો ગંભીર અકસ્માત, 2 લોકોના ઘટનાસ્થળે મોત. ☹
2.	Removal of Emoticons	અમદાવાદ મુંબઈ નેશનલ હાઇવે ઉપર અંકલેશ્વર નજીક બે ટ્રક વચ્ચે સર્જાયો ગંભીર અકસ્માત, 2 લોકોના ઘટનાસ્થળે મોત. ☹	અમદાવાદ મુંબઈ નેશનલ હાઇવે ઉપર અંકલેશ્વર નજીક બે ટ્રક વચ્ચે સર્જાયો ગંભીર અકસ્માત, 2 લોકોના ઘટનાસ્થળે મોત.
3.	Removal of punctuations	અમદાવાદ મુંબઈ નેશનલ હાઇવે ઉપર અંકલેશ્વર નજીક બે ટ્રક વચ્ચે સર્જાયો ગંભીર અકસ્માત,2 લોકોના ઘટનાસ્થળે મોત.	અમદાવાદ મુંબઈ નેશનલ હાઇવે ઉપર અંકલેશ્વર નજીક બે ટ્રક વચ્ચે સર્જાયો ગંભીર અકસ્માત 2 લોકોના ઘટનાસ્થળે મોત
4.	Removal of number	અમદાવાદ મુંબઈ નેશનલ હાઇવે ઉપર અંકલેશ્વર નજીક બે ટ્રક વચ્ચે સર્જાયો ગંભીર અકસ્માત 2 લોકોના ઘટનાસ્થળે મોત	અમદાવાદ મુંબઈ નેશનલ હાઇવે ઉપર અંકલેશ્વર નજીક ટ્રક વચ્ચે સર્જાયો ગંભીર અકસ્માત લોકોના ઘટનાસ્થળે મોત
5.	Removal of Stop-words	અમદાવાદ મુંબઈ નેશનલ હાઇવે ઉપર અંકલેશ્વર નજીક ટ્રક વચ્ચે સર્જાયો ગંભીર અકસ્માત લોકોના ઘટનાસ્થળે મોત	અમદાવાદ મુંબઈ નેશનલ હાઇવે અંકલેશ્વર ટ્રક સર્જાયો ગંભીર અકસ્માત લોકોના ઘટનાસ્થળે મોત
6.	Tokenizing	અમદાવાદ મુંબઈ નેશનલ હાઇવે અંકલેશ્વર ટ્રક સર્જાયો ગંભીર અકસ્માત લોકોના ઘટનાસ્થળે મોત	અમદાવાદ, મુંબઈ, નેશનલ, હાઇવે અંકલેશ્વર, ટ્રક, સર્જાયો, ગંભીર અકસ્માત, લોકોના, ઘટનાસ્થળે, મોત

characters, and tokenization to increase the accuracy of the classification model. Now the dataset is entirely ready to train the model. Table 4 shows the step-by-step process of data cleaning using preprocessing technique.

After the preprocessing task the data we have annotated by 25 different age group people. The training data was hand-coded and manually annotated and admits the potential for hard-to-trace bias within the hate speech categorization [3]. To prove the reliability between annotators we adopt some measures. In addition to the annotation rules, the Kappa call agreement based on the Cohen's letter data points that estimate the data constant between $0 \leq \kappa \leq 1$ is additionally used for the two annotators [11, 21]. For measuring IAA between more than two annotator we used Fleiss's Kappa [26, 27]. Fleiss's Kappa were implemented on ten thousand tweets annotated by twenty-five annotators with classes hate and non-hate. For implementing in python, the algorithm requires the numeric values for that value of non-hate and hate considered as 1 and 0. The kappa's score was measured 0.86. There is no such guideline to assess the value of kappa 0.86 is (i.e., measure the level of agreement between annotators). The Cohen's kappa has been suggested to measure how strong level of agreement annotator have. Table 5 illustrate the lowest value of kappa is between 0 to 20 which is considered as none level of agreement where the value between above 90 considered as almost perfect agreement between annotator. Between 0 to 90 the ranges like 21 to 39, 40 to 59, 60 to 79,80 to 90 are minimal, weak, moderate and strong level of agreement respectively. According to the Table 5 our Fleiss's kappa value is 0.87 which is considered almost perfect agreement between annotator [27].

Table 5. Cohesion Kappa's level of agreement [11]

Value of Kappa	Level of agreement	% of data that are reliable
0–.20	None	0–4%
.21–.39	Minimal	4–15%
.40–.59	Weak	15–35%
.60–.79	Moderate	35–63%
.80–.90	Strong	64–81%
Above.90	Almost Perfect	82–100%

After annotation task, we found 69.3% of all tweets have been considered as hate speech, whereas 30.7% of tweets are none hate across the whole corpus as mentioned in Fig. 5.

As per Table 4, we get the 6930 tweets which belongs to hate speech and 3070 none hate speech among the whole corpus. To implement the classification task, we will be keeping the 80–20 ratio of whole corpus for train and test the model.

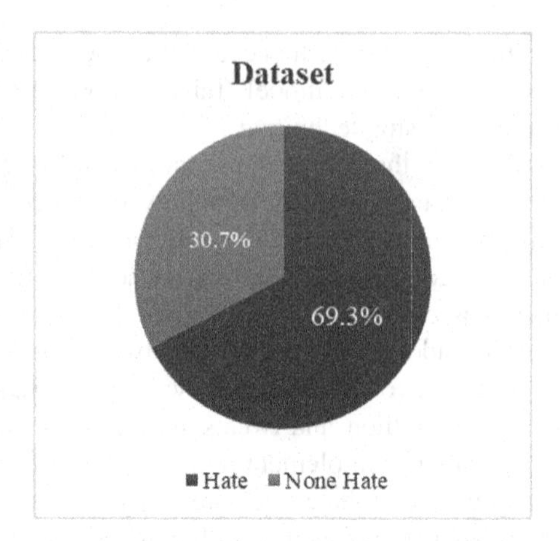

Fig. 5. Distribution of Gujarati hate and none hate tweets in the dataset

Table 6. Total no. of hate and none hate tweets from dataset

Numerical representation	Class	Total instance
0	Hate	6930
1	Non-hate	3070
Total		10000

6 Result and Discussion

Hate speech identification is not just a simple word identification task; it has some complexity, particularly in Gujarati text. As tweets are short texts with a decent number of characters, the capacity to distinguish the hashtags helps enormously in recognizing the subject of the tweet. The collection of tweets was in different categories like politics, sports, religion, and celebrity. We collected the majority of tweets in the politics category, where we get the highest hate data. The initial data was mixed and not clean, so to increase the performance of the dataset, we cleaned the data using preprocessing technique. As discussed in the previous section, extraction and cleaning the tweet is pretty challenging. But We can observe based on Table 4 that the goal was achieved using the preprocessing technique. It shows how data became clean using the different preprocessing techniques. The preprocessing technique executed with the NLTK library is the one that is broadly utilized for preprocessing the other languages texts like Hindi, English, Arabic, etc. We used the RE library for cleaning the Gujarati data, which was quite helpful for the task. Here, we compared the results of preprocessing technique in Waseem et al. and Davidson et al. datasets. We observed on the dataset of Waseem et al. that the performance of the SVM (trigram), CNN, and LR (bigrams) classifier increased using the pre-pressing technique [29]. Because of the username, hashtags, and URLs, the performance of classifiers doesn't increase accuracy in the dataset of Davidson et al. [28]. Therefore, removing URLs, hashtags, and user mentions is required. Removal of punctuation gave the significant performance of the dataset of Davidson et al. The number is not required for the detection of hate speech. In terms of the result of the LSTM classifier, it achieved a good score in Waseem et al. Removal of stop words is the general baseline approach that increases the performance of all the datasets. After the implementation of preprocessing technique, we got the data for annotation. The 25 annotators annotated the whole corpus manually based on the given guideline. We used Fleiss kappa to check their inter agreement and achieved the 0.87 value of k. According to Cohesion Kappa's measure, it is considered a perfect agreement between annotators. After the implementation of the annotation task, we had a clear picture of hate and non-hate data shown in Table 6. The total no of tweets is 10000 after preprocessing, and the 69.3% hate and 30.7% non-hate data after annotation ask. Based on this dataset, we can implement various datasets.

7 Conclusion

Twitter serves as a useful starting point for social media analysis. Through Twitter, people often express their feelings, ideas, and opinions. The major focus of the current contribution is developing and testing a novel schema for hate speech in Gujarati. About 12,000 Gujarati tweets were gathered for the suggested study using the Twitter API. The data was unclean, so we used Python to explore preprocessing methods. After that, twenty-five people of various ages completed the annotating work as class hate and non-hate. We used cohesion kappa's to test the inter-agreement of annotated tweets, and we were able to reach a k value of 0.86, which indicates extremely strong inter-annotator agreement.

In future work, we will extract the features using different NLP technique and implement the machine learning algorithm for the Identifying of Gujarati hate speech. Additionally, we are expanding the annotation process to gather more annotations for one single post and to expand the corpus size.

References

1. Watanabe, H., Bouazizi, M., Ohtsuki, T.: Hate speech on Twitter: a pragmatic approach to collect hateful and offensive expressions and perform hate speech detection. IEEE Access **6**, 13825–13835 (2018)
2. Chung, Y.L., Kuzmenko, E., Tekiroglu, S.S., Guerini, M.: Conan–counter narratives through niche sourcing: a multilingual dataset of responses to fight online hate speech. arXiv preprint arXiv:1910.03270 (2019)
3. Struß, J.M., Siegel, M., Ruppenhofer, J., Wiegand, M., Klenner, M.: Overview of GermEval Task 2, 2019 shared task on the identification of offensive language (2019)
4. Kwok, I., Wang, Y.: Locate the hate: detecting tweets against blacks. In: Twenty-Seventh AAAI Conference on Artificial Intelligence (2013)
5. Mubarak, H., Darwish, K., Magdy, W.: Abusive language detection on Arabic social media. In: Proceedings of the First Workshop on Abusive Language Online, pp. 52–56 (2017)
6. Wang, B., Ding, Y., Liu, S., Zhou, X.: YNU Wb at HASOC 2019: ordered Neurons LSTM with attention for identifying hate speech and offensive language. In: Proceedings of the 11th Annual Meeting of the Forum for Information Retrieval Evaluation, December 2019

7. Basile, V., et al.: Semeval-2019 task 5: multilingual detection of hate speech against immigrants and women in Twitter. In: Proceedings of the 13th International Workshop on Semantic Evaluation, pp. 54–63 (2019)
8. Zampieri, M., Malmasi, S., Nakov, P., Rosenthal, S., Farra, N., Kumar, R.: Predicting the type and target of offensive posts in social media. In: Proceedings of NAACL (2019)
9. Kumar, R., Reganti, A.N., Bhatia, A., Maheshwari, T.: Aggression-annotated corpus of Hindi-English code-mixed data. In: Proceedings of the 11th Language Resources and Evaluation Conference (LREC), Miyazaki, Japan, pp. 1–11 (2018)
10. Viera, A.J.: Understanding inter observer agreement: the Kappa statistic, from the Robert Wood Johnson Clinical Scholars Program, University of North Carolina (2005)
11. Artstein, R., Poesio, M.: Inter-coder agreement for computational linguistics. Comput. Linguist. **34**(4), 555–596 (2008)
12. Abhilasha, V., Tanna, P., Joshi, H.: Hate speech detection: a bird's-eye view. In: Kotecha, K., Piuri, V., Shah, H., Patel, R. (eds.) Data Science and Intelligent Applications, pp. 225–231. Springer, Singapore (2021). https://doi.org/10.1007/978-981-15-4474-3_26
13. Karim, Md.R., et al.: DeepHateExplainer: explainable hate speech detection in under-resourced Bengali language. In: 2021 IEEE 8th International Conference on Data Science and Advanced Analytics (DSAA), pp. 1–10, IEEE (2021). https://doi.org/10.1109/DSAA53316.2021.9564230
14. Chen, B., Zaebst, D., Seel, L.:A macro to calculate kappa statistics for categorizations by multiple raters. In: Proceeding of the 30th Annual SAS Users Group International Conference, pp. 155–230. Citeseer (2005)
15. Karim, Md.R., et al.: Multimodal hate speech detection from Bengali memes and texts. arXiv:2204.10196 [Cs], April 2022
16. Gohil, L., Patel, D.: A sentiment analysis of Gujarati text using Gujarati senti word net. Int. J. Innov. Technol. Explor. Eng. **8**(9), 2290–2292 (2019). https://doi.org/10.35940/ijitee.I8443.078919
17. Ishmam, A.M., Sharmin, S.: Hateful speech detection in public Facebook pages for the Bengali language. In: Proceedings of the 18th IEEE International Conference on Machine Learning and Applications, ICMLA 2019, pp. 555–560 (2019). https://doi.org/10.1109/ICMLA.2019.00104
18. Alotaibi, M., Alotaibi, B., Razaque, A.: A multichannel deep learning framework for cyberbullying detection on social media
19. Rakholia, R.M., Saini, J.R.: A Rule-based approach to identify stop words for Gujarati language. In: Satapathy, S.C., Bhateja, V., Udgata, S.K., Pattnaik, P.K. (eds.) Proceedings of the 5th International Conference on Frontiers in Intelligent Computing: Theory and Applications. AISC, vol. 515, pp. 797–806. Springer, Singapore (2017). https://doi.org/10.1007/978-981-10-3153-3_79
20. Ladani, D.J., Desai, N.P.: Automatic stopword Identification Technique for Gujarati text. In: 2021 International Conference on Artificial Intelligence and Machine Vision (AIMV), 2021, pp. 1–5 (2021) https://doi.org/10.1109/AIMV53313.2021.9670968
21. Effrosynidis, D., Symeonidis, S., Arampatzis, A.: A comparison of pre-processing techniques for Twitter sentiment analysis. In: Kamps, J., Tsakonas, G., Manolopoulos, Y., Iliadis, L., Karydis, I. (eds.) TPDL 2017. LNCS, vol. 10450, pp. 394–406. Springer, Cham (2017). https://doi.org/10.1007/978-3-319-67008-9_31
22. Al-Twairesh, N., et al.: AraSenTi-tweet: a corpus for arabic sentiment analysis of Saudi tweets. Procedia Computer Science **117**, 63–72 (2017). https://doi.org/10.1016/j.procs.2017.10.094
23. Akhtar, B., et al.: Modeling annotator perspective and polarized opinions to improve hate speech detection. Proceedings of the AAAI Conference on Human Computation and Crowdsourcing **8**(1), 151–154 (2020)

24. Landis, J.R., Koch, G.G.: The measurement of observer agreement for categorical data. Biometrics **33**(1), 159 (1977). https://doi.org/10.2307/2529310
25. Ramachandran, D., Parvathi, R.: Analysis of Twitter specific preprocessing technique for tweets. Procedia Computer Science **165**, 245–251 (2019). https://doi.org/10.1016/j.procs.2020.01.083
26. Fleiss, J.L.: Measuring nominal scale agreement among many raters. Psychol. Bull. **76**, 378 (1971)
27. Landis, J.R., Koch, G.G.: The measurement of observer agreement for categorical data. Biometrics **33**, 159–174 (1977)
28. Davidson, T., Warmsley, D., Macy, M.W., Weber, I.: Automated hate speech detection and the problem of offensive language
29. Hovy, D., Waseem, Z.: Hateful symbols or hateful people? Predictive features for hate speech detection on Twitter. In: Proceedings of the NAACL Student Research Workshop (2016)

Intrinsic Use of Genetic Optimizer in CNN Towards Efficient Image Classification

Vaibhav Bhartia[1], Tusar Kanti Mishra[1], and B. K. Tripathy[2](✉)

[1] SCOPE, Vellore Institute of Technology, Vellore 632014, Tamil Nadu, India
[2] SITE, Vellore Institute of Technology, Vellore 632014, Tamil Nadu, India
tripathybk@rediffmail.com

Abstract. The inception of genetic algorithms in the 80's laid a strong foundation in the theory of optimization. Numerous engineering applications are rewarded with wings of optimal and faster solutions through suitable genetic modelling. So far, a handful of evolutionary algorithms and modelling have been introduced by researchers. This has led to vivid applications in numerous domains. In this paper a customized evolutionary framework is proposed that is blended with deep learning mechanism. The widely used convolutional neural networks (CNN) model has been customized whereby optimally informative features are selected through intermittent genetic optimization. The inherent convolution layer outcomes are subjected to the optimizer module that in turn results in optimized set of feature points. The pooling process is abandoned for the purpose; thus, getting rid of uniform feature selection. Now, with this model the feature selection inhibits dynamic process of optimized feature selection. Case study on the usage of the same is shown on classification of facial expression images. Performance of the proposed mechanism is further compared with the simulated outcomes of the generic CNN model. Nevertheless, to say the results show promising rate of efficiency.

Keywords: CNN · Genetic optimization · Soft computing · Deep learning · Fitness function

1 Introduction

Optimization brings in effectiveness and extent of goodness in problem solving [1]. Optimization is a prolonged challenging topic for researchers for centuries. Literature suggests that the foundation for optimization was laid way long before the Christ era. The first ever optimization technique aimed to calculate the closeness among a pair of points [2]. Presently, optimization has its significant existence as a robust tool in almost every field of applications. The omnipresence effect of optimization is well realized in engineering, medical science, agriculture, applied sciences, mathematics, data science, space computation, education management, construction, and many more. The justification behind the vivid utilization of optimization relies on the ease of use and assurance of guaranteed solution with computationally efficient efforts. Researchers are finding it to be an eternal tool whereby upbringings of novel methodologies are still on. Exploring

K. K. Patel et al. (Eds.): icSoftComp 2022, CCIS 1788, pp. 396–405, 2023.
https://doi.org/10.1007/978-3-031-27609-5_31

the scopes towards the usage of optimization has gained further momentum with the introduction of genetic algorithm. Genetic algorithm has been playing major role since 1975. Study on the genetic optimization techniques reveals the advancements in the front of computational efficiency for problem solving especially from the mathematical and computer science approaches. General overview of optimization process is presented in Fig. 1. It starts with identifying the very need for optimization. This enables one to set up the problem specification and goal formulation. Next step is towards choosing parameters/features which is in turn followed by constraint formulation and objective determination.

2 Related Work

In recent years, several optimization methods have been developed that are conceptually different from conventional mathematical programming. These methods are called modern or non-traditional optimization methods. Most of these methods are based on specific properties and behaviors of biological, molecular, insect swarm, and neurological systems such as Particle Swarm Optimization [3], Ant Colony Optimization [4], Honey Bee Optimization [5], Simulated Annealing [6] and Genetic Algorithm [7, 8].

In [9], Jayanthi J et al. used a Particle Swarm Optimization (PSO) based on a CNN model (PSO-CNN) to identify and classify Diabetic retinopathy from other color fundus images. The proposed model consisted of three stages namely pre-processing followed by feature extraction and classification. Firstly, the pre-processing is done for noise removal in images followed by feature extraction using PSO-CNN and then feeding the featured the filtered images as input in decision tree model for classification.

In [10], Abbas et al. used an Ant Colony System (ACS) optimization technique along with CNN model for classification of genders. They carried out a 64-layer architecture named 4-BSMAB obtained from deep AlexNet. The training of the dataset was performed using SoftMax classifier along with optimization of the features using ACS optimization technique.

In [11], Erkan et al. used an artificial bee colony (ABC) optimization algorithm along with deep CNN for identification of plant species. The pre-processing of images are done using various methods such as scaling, segmentation and augmentation. The data is augmented for training and testing. The 20% of the training dataset is used for validation for both the classification and optimization stages.

In [12], Ashok et al. used a deep neural network along with Bidirectional LSTM to undertake hyper-parameter optimization using genetic algorithm following CNN to detect sarcasm. They explored the theoretical part of sarcasm detection by observing not just the semantic features but also its syntactical and semantic properties.

In [13], Chung and Shin proposed a genetic algorithm for optimization of the feature extraction for the CNN model along with comparison between standard neural networks such as ANN and CNN for the effectiveness of the model in the prediction of stock markets.

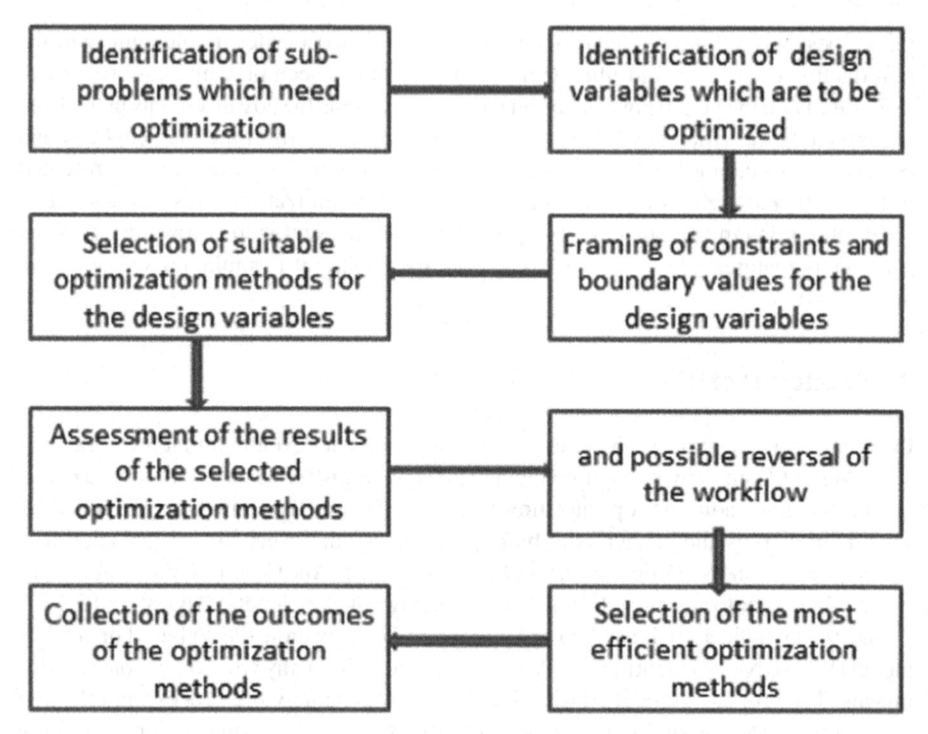

Fig. 1. General overview of optimization process

3 Proposed Work

The proposed work emphasizes on suitable modifications to the popular deep learning model CNN [14] by incorporating the concept of genetic optimization. It is dubbed as Genetic Optimizer for CNN (GO_{CNN}). In general, the CNN processes input through intermittent computations of convolution and pooling and finally carrying out the classification using neural networks. This sequence of process can be viewed as extraction of the key feature points in a fixed manner each time a pooling operation is performed followed by convolution [15]. It exhibits uniformity in selecting feature points. Although it ensures a good level of feature extraction, it does not ensure the most informative feature points extraction. The idea behind the proposed work focuses on this feat. The proposed work applies a mild alteration to the sequence of process of CNN by introducing the genetic optimizer module in place of pooling. The inherent difference is that while pooling draws features at uniform behavior, in contrary the genetic optimizer module compels the model to always extract the optimal feature points at any instance. Thus, the final classification outcomes could be always performing at the rate either equivalent to or greater than that of conventional CNN. A block diagram of this explanation is depicted in Fig. 2. Brief description about each process is presented below in a sequel:

- Input
- Convolution
- Genetic Optimizer for CNN (GO_{CNN})

3.1 Input

The input samples are basically standard images that demand classification. As a case study, facial expression images are considered in this work. However, the proposed mechanism can be suitably utilized for other image samples as well. The facial expression images have been derived from the standard database (FER2013) which is derived from [16].

3.2 Convolution

Most often the inputs to CNN are digital images [15]. Convolution being the first layer performs the convolution operation on the input feeds. A specific size (say d1 × d2) of filtering is applied to the sample. While the filter scrolls over the image, the dot products are computed and are buffered into a matrix. Especially, prominent pixel level changes are obtained while this operation is carried out. This matrix is passed to the next phase where the genetic optimizer works on it.

3.3 Genetic Optimizer for CNN (GO_{CNN})

The use of genetic optimization ensures here not only the finest pixels (as obtained during pooling), but also the nearby value finer pixels which have the potential to contribute towards the precision measure. The idea is to retain that pixel information which is meeting the fitness criteria. Say, talking about the fitness criteria in line with average pooling process, one can idealize the retention of all those pixels whose values are not only exact to the local average but also nearby the local average of the current filtering window. The fitness in this case can be defined as:

$$\mathrm{Fit(Ch)} = \mathrm{Avg_i} - \delta \leq \mathrm{intensity(Ch)} \leq \mathrm{Avg_i} + \delta \quad (1)$$

subject to:

$$0 \leq \mathrm{Avg_i} - \delta \quad (2)$$

and,

$$\mathrm{Avg_i} + \delta \leq \mathrm{Max_{Intensity}} \quad (3)$$

where,

Fit(Ch): Fitness evaluation mapping for chromosome Ch,
Avg_i: Average intensity of the local convolution window resulting i^{th} chromosome,
δ: User specified deviation,
$Max_{Intensity}$: Highest intensity of the entire image.

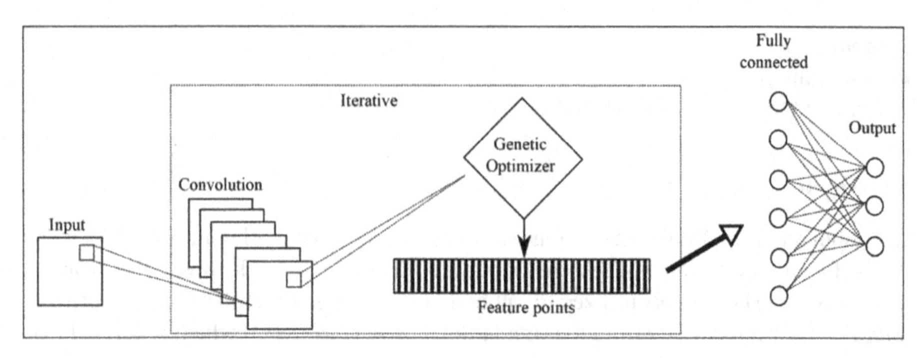

Fig. 2. Block diagram of the proposed model (GO_{CNN})

Instead of applying the pooling operation, here genetic optimization is performed on the input matrix. The entry rows of the input matrix act as population samples for the optimizer. Pseudocode for this is depicted in Algorithm 1.

The intermediate resultant from the convolution layer of the CNN module is the input matrix M_p. This M_p of dimension d1 × d2 acts as the actual input to the GO_{CNN}. The rate of crossover parameter i.e., R_c needs to be fed at the users end. This parameter somehow relies on the length of the feature vector. Hence, providing the choice of value to the user for this is a good option. The algorithm works in accordance with the constraint as specified for a problem. This algorithm is generic in nature to be utilized for vivid variety of scenario. The expected output for this algorithm is to generate a mapped feature matrix M_f that contains the optimal feature points.

The working steps as depicted in this algorithm start with the iteration directive to the constraint satisfaction. Among the available number of rows, certain rows are considered to the pool of population. Suitable genetic processes like the mutation and crossover are applied to pool of population. Fitness evaluation in accordance with the fitness function is implemented and the rows found under-fit are discarded thereby. The complete process is thus implemented for each of the feeds of the convolution process applied on the input matrix. Simultaneously, the resultant fit number of rows forms the optimal feature matrix. The resultant matrix thus obtained is further subjected to minor refinement phase as explained below.

The refinement phase is necessarily implemented to comprehend the penalty of losing certain number of rows that might lead to under-size matrix. To compensate the nearby rank of rows are padded at appropriate positions until the matrix size regains its standard form. While evaluating the fitness, a ranking mechanism is also run in a spontaneous mode that takes care of the rank of each of the rows.

Algorithm 1: GO$_{CNN}$(M$_p$, d_1, d_2, R_c)

Data: M_p : Input matrix,
d$_1$: Number of rows acts as population size,
d$_2$: Number of columns acts as chromosome length,
R$_c$: Rate of crossover.
Result: Mapped feature matrix M_f with optimal features.
while *Constraint NOT satisfied* **do**
  for k_1 *in range 0 to* d_1 **do**
    for k_2 *in range 0 to* d_2 **do**
      Choose n number of chromosomes $\{ch_1, ch_2, \ldots, ch_n\}$
      Mutate $(ch_1, ch_2, \ldots, ch_n)$
      and obtain samples $(ch_1^1, ch_2^1, \ldots, ch_n^1)$
      Apply crossover
      and obtain samples $(ch_1^2, ch_2^2, \ldots, ch_n^2)$
    end
  end
  for p *in range 1 to* n **do**
    if ch_p^2 *is FIT* **then**
      $Append(M_f, ch_p^2)$;
    end
    if ch_p^2 *is !FIT* **then**
      $Discard(ch_p^2)$;
    else
  end
  $Return(M_f)$
end

4 Experimental Analysis

Validation of the same is performed and the overall rate of accuracy is computed following k-fold (k = 10) cross validation. For each of the fold, corresponding performance indicators are mentioned in Table 1. Comparison of the proposed GOCNN with competent scheme (CNN) is carried out. Several performance indicators are considered. The ROC plots obtained from the resultants are presented in Fig. 4. The comparison among the overall accurate rates is presented in Fig. 5. It can be observed that a mild increase in the AOC is noticed for both schemes under consideration.

4.1 Dataset Description

A case study is carried out to validate the working of the proposed scheme. This is accomplished through a pilot experimentation on set of facial expression images. For the purpose, the facial expression dataset (FER2013) is considered ([16]). Benchmark dataset is referred from the FER2013 sample repository. Six distinct set of facial expression images are considered namely Angry, Fear, Happy, Sad, Surprise, and Neutral. Certain samples are presented in Fig. 3 where the labelling is done on integer ranges [0, 1, , 5] for each of the expression classes. The train and test split of the samples are considered to be 1,800 and 600 respectively with uniform number of samples from each category. Modelling the customized GOCNN is suitably carried out using the training samples. The parameters used for the purpose are presented as follows. Mp of dimension 64×64 acts as the actual input to the GOCNN. The rate of crossover parameter i.e. Rc needs to be kept as 6 (in this case 10% of the row size) as in every iterations. The algorithm works in accordance with the constraint and fitness as specified in Eqs. 1 and 2 earlier. As and when needed, the compensation for certain number of rows is accomplished strictly with reference to the ranking mechanism as stated earlier. The expected output feature matrix Mf is generated as per Algorithm 1. Finally, it is subjected to the classifier layer. The classifier layer utilizes conventional back propagation algorithm (BPNN) that classifies out six distinct labels for the six categories of the facial expression considered.

Fig. 3. Sample snapshots of the facial expressions.

5 Discussion

Promising results in favor of the proposed work are obtained that outperforms the competent scheme. In fact, generic CNN, when simulated on the said dataset yields an overall rate of accuracy of 88%, however, the proposed model yields an increased rate of accuracy at 88.5% on the same sample set. With this, suitable justification is built in favor of the proposed work. It strengthens the claim that genetic optimization has been a good choice where performance matters in terms of accuracy. The said work performs satisfactorily where the input sample image resolutions are very high (as nowadays applications on high resolution images are drawing huge demands among researchers). The iterations of convolution to pooling is case of CNN is more when it comes to high resolution images, however, the GOCNN model needs slightly lesser number of iterations. For low resolution image inputs, no difference among the computational time among CNN and GOCNN is observed.

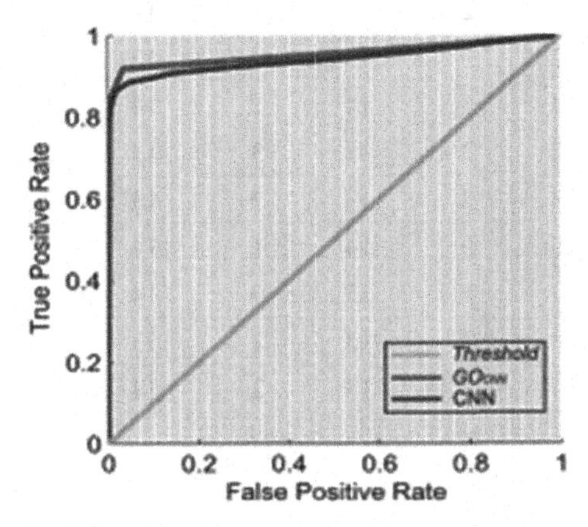

Fig. 4. Comparison among the ROC for CNN and GO_{CNN} respectively.

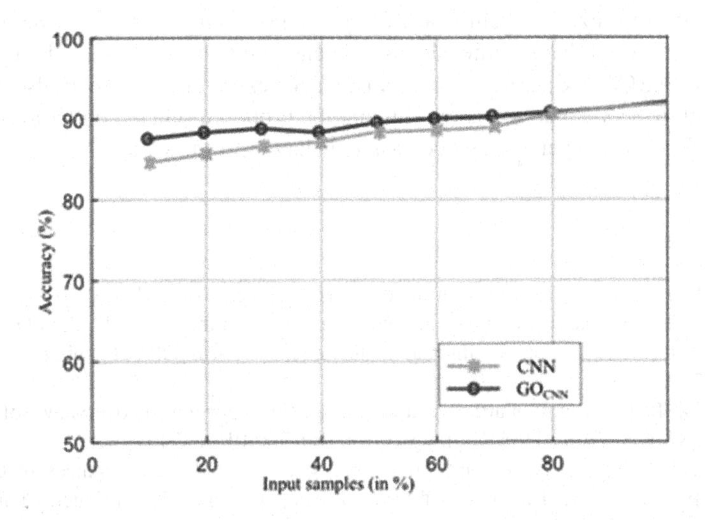

Fig. 5. Comparison among the overall rate of accuracy for CNN and GO_{CNN} respectively

Table 1. Performance indicators for CNN and GOCNN respectively.

Fold#	Size	CNN				GO_{CNN}			
		TP	FP	TN	FN	TP	FP	TN	FN
1	30	12	3	12	3	13	2	12	3
2	30	13	2	13	2	14	1	13	2
3	30	12	3	13	2	12	3	13	2
4	30	14	1	14	1	14	1	13	2
5	30	12	3	12	3	13	2	12	3
6	30	13	2	13	2	13	2	13	2
7	30	15	0	14	1	14	1	14	1
8	30	14	1	15	0	14	1	15	0
9	30	13	2	13	2	13	2	14	1
10	30	14	1	13	2	14	1	13	2

6 Conclusion

In this chapter, milestones on various optimization methods are discussed with an emphasis on genetic optimization techniques. Further, a novel customized scheme dubbed as GOCNN is proposed that signifies the use of genetic optimization in modern soft computing tools like CNN. Case study on a typical facial expression dataset is also presented in brief. Future scope is to focus on utilizing the proposed scheme on various datasets and prove forward the generic and consistent behavior of the same.

References

1. Adby, P.: Introduction to Optimization Methods. Springer, Dordrecht (2013)
2. Sinha, G.R.: Introduction and background to optimization theory. In: Modern Optimization Methods for Science, Engineering and Technology, pp. 1–18. IOP Publishing (2019). ISBN 978-0-7503-2404-5
3. Wang, D., Tan, D., Liu, L.: Particle swarm optimization algorithm: an overview. Soft Comput. **22**(2), 387–408 (2017). https://doi.org/10.1007/s00500-016-2474-6
4. Dorigo, M., Stützle, T.: Ant colony optimization: overview and recent advances. In: Gendreau, M., Potvin, J.Y. (eds.) Handbook of Metaheuristics, pp. 311–351. Springer, Cham (2019). https://doi.org/10.1007/978-3-319-91086-4_10
5. Niazkar, M., Afzali, S.H.: Closure to "Assessment of modified honey bee mating optimization for parameter estimation of nonlinear Muskingum models." J. Hydrol. Eng. **23**(4), 07018003 (2018)
6. Delahaye, D., Chaimatanan, S., Mongeau, M.: Simulated annealing: from basics to applications. In: Gendreau, M., Potvin, J.Y. (eds.) Handbook of Metaheuristics, pp. 1–35. Springer, Cham (2019). https://doi.org/10.1007/978-3-319-91086-4_1
7. Mirjalili, S.: Genetic algorithm. In: Mirjalili, S. (ed.) Evolutionary Algorithms and Neural Networks, pp. 43–55. Springer, Cham (2019). https://doi.org/10.1007/978-3-319-93025-1_4

8. Katoch, S., Chauhan, S.S., Kumar, V.: A review on genetic algorithm: past, present, and future. Multimed. Tools Appl. **80**(5), 8091–8126 (2020). https://doi.org/10.1007/s11042-020-10139-6
9. Jayanthi, J., Jayasankar, T., Krishnaraj, N., Prakash, N.B., Sagai Francis Britto, A., Vinoth Kumar, K.: An intelligent particle swarm optimization with convolutional neural network for diabetic retinopathy classification model. J. Med. Imaging Health Inform. **11**(3), 803–809 (2021)
10. Abbas, F., Yasmin, M., Fayyaz, M., Elaziz, M.A., Songfeng, Lu., Abd El-Latif, A.A.: Gender classification using proposed CNN-based model and ant colony optimization. Mathematics **9**(19), 2499 (2021). https://doi.org/10.3390/math9192499
11. Erkan, U., Toktas, A., Ustun, D.: Hyperparameter optimization of deep CNN classifier for plant species identification using artificial bee colony algorithm. J. Ambient Intell. Hum. Comput. (2022). https://doi.org/10.1007/s12652-021-03631-w
12. Ashok, D.M., Nidhi Ghanshyam, A., Salim, S.S., Burhanuddin Mazahir, D., Thakare, B.S.: Sarcasm detection using genetic optimization on LSTM with CNN. In: 2020 International Conference for Emerging Technology (INCET), pp. 1–4 (2020). https://doi.org/10.1109/INCET49848.2020.9154090
13. Chung, H., Shin, K.-S.: Genetic algorithm-optimized multi-channel convolutional neural network for stock market prediction. Neural Comput. Appl. **32**(12), 7897–7914 (2019). https://doi.org/10.1007/s00521-019-04236-3
14. Albawi, S., Mohammed, T.A., Al-Zawi, S.: Understanding of a convolutional neural network. In: 2017 International Conference on Engineering and Technology (ICET), pp. 1–6. IEEE (2017)
15. Keresztes, P., et al.: An emulated digital CNN implementation. J. VLSI Signal Process. Syst. Signal Image Video Technol. **23**(2), 291–303 (1999). https://doi.org/10.1023/A:1008141017714
16. Zahara, L., Musa, P., Prasetyo Wibowo, E., Karim, I., Bahri Musa, S.: The facial emotion recognition (FER-2013) dataset for prediction system of micro-expressions face using the convolutional neural network (CNN) algorithm based Raspberry Pi. In: 2020 Fifth International Conference on Informatics and Computing (ICIC), pp. 1–9 (2020). https://doi.org/10.1109/ICIC50835.2020.9288560

A Software System for Smart Course Planning

Mohammad Laghari[1(✉)] and Ahmed Hassan[2]

[1] Department of Electrical and Communication Engineering, UAE University, Al Ain, UAE
mslaghari@uaeu.ac.ae

[2] Department of Architectural Engineering, UAE University, Al Ain, UAE

Abstract. The academic course planning at the Department of Electrical and Communication Engineering (ECE), United Arab Emirates University (UAEU), is an important imperative to make certain that students accomplish the bachelor's degree requirements of the university in a structured way, without facing any redundant delays. Although the university has a tailored registration system from Banner as well as a well-defined course planning system called DegreeWorks however, the course 'advising and planning' performed at the faculty and students' level is often short of the required proficiency, aptitude, and skills. Students have encountered problems due to a paucity of a proper advising system as well as an insufficiency of seeking suitable advice. In this paper, a Smart Course Planning System *(SCPS)* is developed to help students prepare needed study plans. The devised system will allow students in selecting the most suitable courses for the next offering semester. The students can also review the selected courses in a 2-D grid for better time management.

Keywords: Academic advising · Course planning · Software system · Study plan

1 Introduction

Academic advising and course planning are expected to trade necessary information to help stakeholders reach their educational and academic goals. It is an understanding and shared responsibility between an academic advisor and the student. Advising is necessary for situations where an academic representative (faculty or staff) advises a university student about academic matters such as requirements needed for degree completion, course and career planning, and typical dialogues on how a course of study fits a particular academic or career interest [1–5].

Course registration intended by a student comprises three main components: knowledge of the necessary and required pre-registration information, instructions, and guidelines for registering for courses, and lastly, guidelines on the next steps after the student has completed the course registration process. The process of registering for courses and what the students are required to do afterward is well established, however, the knowledge of what courses to register for is a noteworthy process and needs attention. A student typically gets confused in selecting suitable courses from a vast pool of available courses.

K. K. Patel et al. (Eds.): icSoftComp 2022, CCIS 1788, pp. 406–418, 2023.
https://doi.org/10.1007/978-3-031-27609-5_32

Typical registration problems that are based on course planning and advising include students missing out on courses specific for alternating semesters, choosing exceedingly more or exceedingly fewer courses than expected as too few courses may increase in graduation duration period and too many may entail a heavy burden with loss of quality and course grades, advised or chosen courses with time conflicts or even greater time apart, for example, classes too early in the morning and too late in the afternoon.

The course advising resulting in study plans is moderately performed before course registration, however, students at the registration time encounter time conflicts which often results in losing other alternatives as classes fill up fast once registration starts, and it is also possible that a student may end up having all courses on two days of the week rather than evenly spread out on all four days of classes.

Students with such problems may suffer from delayed industrial training (one-semester-long training performed in an industry) and graduation duration due to unnecessary course selection or having missed out on important courses, dropping a complete semester because of a minimum number of course requirements (12 credits), or having a heavy semester load and may perform averagely due to a high number of courses, etc.

The SCPS is developed to provide real-time plans for students and to enable them to maximize their opportunities in registering for courses of their interest, as well as advise them to complete their degree requirements optimally. The front-end programming of the software package is done by using HTML, CSS, and JavaScript. Server-side programming is performed by the Node.JS framework and Express.JS library. Study plans are saved by using the NoSQL Database System (MongoDB).

The devised package helps students select suitable courses to prepare a study plan. A typical course selection procedure starts with the student uploading a list of the previously passed and the current semester taken courses. The software package then evolves through several procedures to advise students with an ideal list of the most suitable courses. The course selection is based on a knowledge area built around all courses of the curriculum. The paper describes the complete operation of the course planning package that includes characterized course selection, help menus, restrictions, etc. The software package has been tested to prepare 25 near-perfect study plans. Two examples of such plans are reported in this paper.

The innovation behind this research is to motivate students to use the software package and select the most appropriate courses suitable for a specific semester as well as to be less reliant on advisors who may be time constrained to give error-free advice to a large number of students.

The paper is organized in earlier course planning and scheduling systems in Sect. 2. Courses included with their credit hours, course hierarchy chart, and algorithm with course level characteristics are described in Sect. 3. The Smart Course Planning Software system package includes two test cases that are presented in Sect. 4. The conclusion is Sect. 5 and the last section deals with paper references.

2 Early Systems

In the earlier course registration processes, institutional academic departments were responsible for registering student courses. Those early procedures were marred by delays, a lack of resources, and management & administrative issues. At a later stage, this perception of registration was reviewed from different viewpoints including student course planning, class scheduling, course advising, course registration, etc.

In the late 1990s, when the number of registering students increased, then all the above-mentioned issues also increased to double. With this increment, the registration process also needed restructuring to maximize the allocation of course places for this increased number of registering students. Moreover, many of the institutions started to offer degree programs that were more sophisticated, specialized, and multidisciplinary. Time conflicts while selecting courses for these programs also increased. The course planning and registration systems aim to select the most appropriate courses without encountering time conflicts.

Computerized registration systems came into existence and became popular with time. However, even though these systems had major advantages, the major drawback was enduring machine and network failures. It was anticipated and expected that most of the human errors, such as mistakes in inputting correct data would be detected by the system as they occurred, however, for errors that had not been foreseen, it was expected that some "off-line" data manipulation was necessary. Some of the early systems are discussed in the following paragraphs.

The authors of this reference paper have proposed to design an online course selection system for students who are seeking to apply for a university degree and based on this design, propose improvements to a rule-based decision support system that enables them to choose a reasonably better university when selection is between a public and a private institution [6].

A student information system called MYGJU, developed by the investigators in this reference paper can manage academic and financial information. This online course registration is user-friendly and restricts students from selecting incorrect and unnecessary courses. However, even after the launch of MYGJU, the system did not gain popularity among students which is also discussed in this reference paper [7].

The main authors of the current investigation have proposed, designed, and implemented several course planning and advising systems starting as early as 2005 [8]. Course planning and advising algorithms were investigated with parameters of a varying number of priority fields associated with each course and with different arrangements of these fields. Most of these systems have been phased out due to curriculum changes as well as some unforeseen errors, and due to the percentage of accuracy [9–13].

The authors of this reference paper have surveyed many academic systems to innovate an automated apparatus for academic advising in a typical university system. An impression of the development and implementation of a new model is discussed as a web-based application. A model is proposed to facilitate advisors and advising staff to follow up on student complaints and propositions. The model is a computerized system that helps academic advisors in providing quality, precise and consistent advising services to their students [14].

An online advisor is reported by authors in their reference paper which helps the academic community to improve the in-use university student information system. A feasibility survey conducted on a few faculty and students attributed satisfactory use, effective, efficient, useful, and helpful results [15].

An improvement on "IDiSC" called "IDiSC^{+}" is proposed in this paper. The new system generates not one but a set of near-optimal alternative plans that are structurally different but of similar quality. Alternatives are subsequently analyzed to either approve one plan or to refine the problem settings for generating further solutions. Moreover, "IDiSC^{+}" uses a more sophisticated model compared to the earlier one [16].

Genetic algorithms are used to design a course scheduling information strategy as the course schedule is considered a complex optimization problem and is difficult to handle manually. An optimal solution is sought by looking at potential solutions. The "Codeigniter" framework, "Responsive Bootstrapping" (display), and "MySQL" (database) are the main setup of the genetic algorithm method. Course conflicts are significantly reduced with the help of this algorithm [17].

An Artificial Intelligence Aided course scheduling system is realized in this paper by several authors. Difficulties of the current scheduling system are also discussed in this paper and are compared with the new scheme. A browser/server mode architecture is adopted to execute the prototype of the devised course scheduling system [18].

The authors of this paper have developed a Class Schedule Management System known as the "ClassSchedule" to resolve the course scheduling and planning problem. The system can detect course conflicts, multi-user, and group course planning, and manage many resources together. A lazy loading feature is used for real-time operations [19].

A web-based course scheduling system is developed by Legaspi et al. in which a greedy algorithm is used for course scheduling as well as assigning courses to the faculty [20].

The "Komar University of Science and Technology" developed and instrumented a web-based registration and advisory system to enable their students to select the right courses as well as guide the administration staff for proper logistics. The system help students with many of the prerequisite assignments before the actual registration process starts. The system also relieves academic advisors from performing time-consuming advisory services [21].

The authors of this paper have shown improvements in the academic advising process. Many research articles on electronic academic advising systems have been reviewed. Many different trends and features of electronic advising systems have been surveyed in this investigation. The transformation from a traditional advising system to an AI-based electronic advising system can also be justified by this research [22].

Liberal Arts programs accommodate a large number of course selections and therefore a challenge for students to select appropriate courses based on their academic context and concentration. A "course recommender system" is proposed for the bachelor students at the "University College Maastricht". This smart course selection procedure is recommended to counter traditional academic advising systems and advise students to select the most appropriate courses best suited to their academic interests [23].

A course advising system called "TAROT" is devised in this investigation that proposes a "planning engine" to construct "multi-year course schedules" for challenging setups such as "study-abroad semesters", "course overrides", "transfer credits", "early graduation", and "double majors". This paper also compares the traditional course with TAROT [24].

3 Course Level Characteristics

There are approximately 2,400 students at the College of Engineering (COE). The COE is one of the nine colleges of the United Arab Emirates (UAE) University with an estimated total of 14,000 students. Electrical Engineering is one of the five departments with approximately 300 full-time students. The students complete 147 credit hours to fulfill their bachelor's degree requirements. Based on the quality of their grades and GPA (Grade Point Average), they typically take 4½ years with an average of 17 credits per semester, 5 years with an average of 16 credits, 5½ years with an average of 15 credits, and 6 years with an average of 14 credits per semester to complete their degree requirements. Academically weak students may take even longer as they occasionally fail courses.

Students take 52 courses divided into 6 main sections: General Education Requirements - 7 courses (21 credits), College Requirements - 15 courses (38 credits), Compulsory Specialization Requirements - 23 courses (55 credits) including 7 laboratory courses, Industrial Training - one course (15 credits), Graduation Projects - 2 courses (6 credits), and Elective Specialization Requirements - 4 courses (12 credits).

A course hierarchy chart as shown in Fig. 1 includes all required compulsory and elective courses. The red colored background courses are General Education, the green colored is College Requirements, the blue colors are Compulsory Specialized Requirements, and the uncolored ones are the technical electives. The arrows indicate hierarchies. The asterisk on the course box shows that the specific course is required for industrial training. The numerical value of either '1' (1st semester) or '2' (2nd semester) specifies the course offering semesters.

Industrial Training (IT) is a complete semester course where a student spends the entire time with an industrial unit. The eligibility for IT starts after completing 94 credits or higher and is also dependent on academic standing. However, the students typically aim at completing about 100 credit hours. The students are allowed only two semesters of studies after IT therefore they typically avoid leaving them with more than 16 credit hours per semester for the last two semesters.

To build up the course planning knowledge area, there are two characteristics associated with each course as shown in Table 1. These characteristics help define specific courses in terms of their importance in course selection for preparing study plan purposes. These are prioritized from highest to lowest. The characteristics are described as follows:

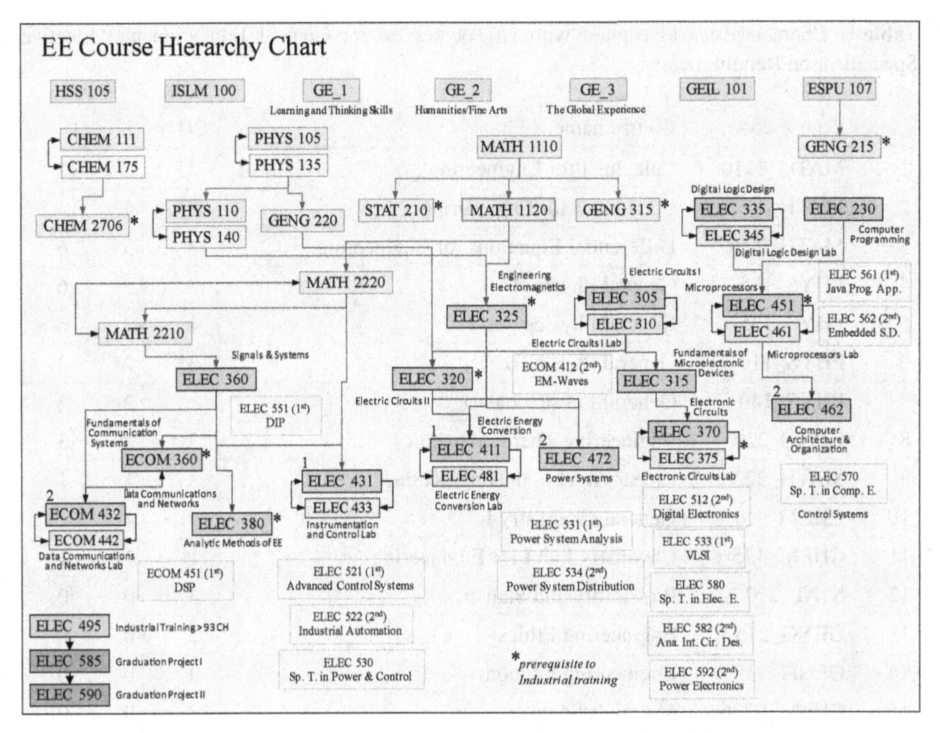

Fig. 1. A course hierarchy chart of the Electrical Engineering program.

3.1 Characteristic 1

This deals with the levels of course hierarchy associated with each course. For example, from Fig. 1, ELEC 305 has two hierarchal levels which are also evident in Table 1. These hierarchies are shown as:

ELEC 305 -> ELEC 315 -> ELEC 370,
ELEC 305 -> ELEC 320 -> ELEC 472, and
ELEC 305 -> ELEC 320 -> ELEC 411.

Students missing courses from long hierarchal levels may delay their industrial training, including graduation.

3.2 Characteristic 2

This is associated with the total number of courses that are linked to a specific course. A numerical value of '1' is considered for each course that is opened by a certain course. For example, ELEC 305 is responsible to open five different courses in all associated hierarchical levels which are ELEC 315, ELEC 320, ELEC 370, ELEC 411, and ELEC 472 as shown in Fig. 1 as well as in Table 1. The associated lab courses are considered as a part of their theory counterpart and therefore not included in the course count.

Table 1. Characteristics associated with all courses except General Education and Elective Specialization Requirements.

	Course code	Course name	CH	I	II
1	MATH_1110	Calculus I for Engineering	3	5	17
2	MATH_1120	Calculus II for Engineering	3	4	14
3	MATH_2210	Differential Equations for Engineering	3	3	6
4	PHYS_105	General Physics I	3	3	6
5	PHYS_135	General Physics Lab I	1	3	6
6	PHYS_110	General Physics II	3	2	3
7	PHYS_140	General Physics Lab II	1	2	3
8	GENG_220	Engineering Thermodynamics	3	2	3
9	MATH_2220	Linear Algebra for Engineering	3	1	1
10	CHEM_111	General Chemistry I	3	1	1
11	CHEM_175	Chemistry Lab I for Engineering	1	1	1
12	STAT_210	Probability and Statistics	3	0	0
13	GENG_215	Engineering Ethics	2	0	0
14	GENG_315	Engineering Economy	3	0	0
15	CHEM_2706	Materials Science	3	0	0
16	ELEC_305	Electric Circuits I	3	2	5
17	ELEC_310	Electric Circuits I Lab	1	2	5
18	ELEC_360	Signals & Systems	3	2	3
19	ELEC_230	Computer Programming	3	2	2
20	ELEC_335	Digital Logic Design	3	2	2
21	ELEC_345	Digital Logic Design Lab	1	2	2
22	ELEC_451	Microprocessors	3	2	2
23	ELEC_325	Engineering Electromagnetics	3	1	2
24	ELEC_320	Electric Circuits II	3	1	2
25	ELEC_315	Fundamentals of Microelectronic Devices	3	1	1
26	ECOM_360	Fundamentals of Communication Systems	3	1	1
27	ELEC_495	Industrial Training	15	2	2
28	ELEC_370	Electronic Circuits	3	0	0
29	ELEC_375	Electronic Circuits Lab	1	0	0
30	ELEC_461	Microprocessors Lab	1	1	1
31	ELEC_380	Analytical Methods in Electrical Engineering	3	0	0
32	ELEC_585	Design and Critical Thinking in Elec. Engg	3	1	1

(*continued*)

Table 1. (*continued*)

	Course code	Course name	CH	I	II
33	ELEC_431	Control Systems	3	0	0
34	ELEC_433	Instrumentation and Control Lab	1	0	0
35	ELEC_411	Electric Energy Conversions	3	0	0
36	ELEC_481	Electric Energy Conversions Lab	1	0	0
37	ECOM_432	Data Communications and Networks	3	0	0
38	ECOM_442	Data Communications and Networks Lab	1	0	0
39	ELEC_462	Computer Architecture and Organization	3	0	0
40	ELEC_472	Power Systems	3	0	0
41	ELEC_590	Capstone Engineering Design Project	3	0	0

4 The SCPS Software

The SCPS software procedure starts with a typical student selecting all completed and current semester courses. The course selection is saved in a database that includes the student's name, university ID, email address, year of study-plan preparation, mobile number (optional), and the name of the college. Three sets of files are initially created in the database. A first file listing all courses of the curriculum included with course id, course name, credit hours, and the courses associated with two characteristics.

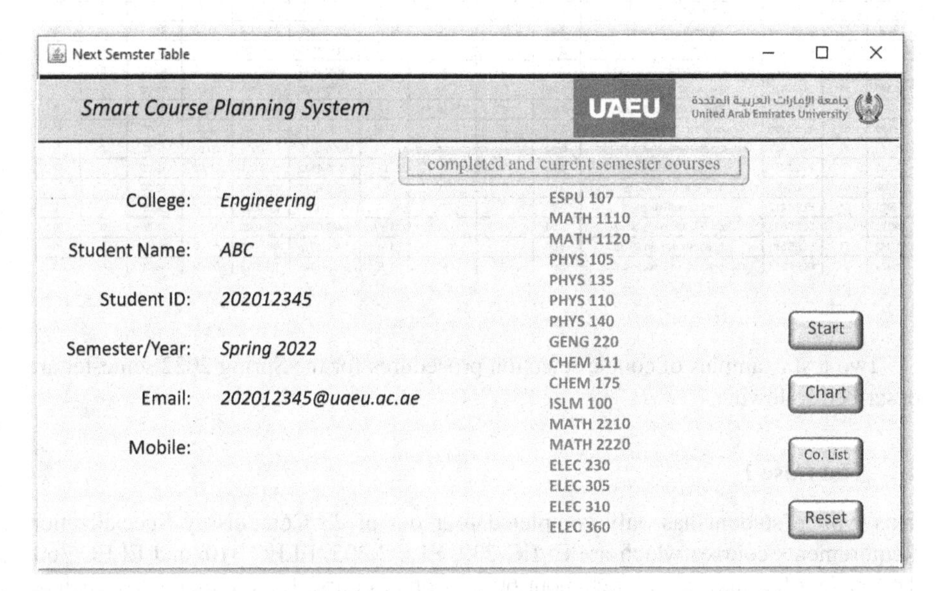

Fig. 2. The SCPS shows the initial information of students and the completed and current semester courses.

A second file lists all courses offered in the Fall semester only, and a third file lists courses offered in the Spring semester. Figure 2 shows the start window of the SCPS package. The course list (completed and current semester) shown here is based on Test Case 1 described later in the paper.

The third file belonging to the Spring 2022 semester is shown in Table 2. This Table shows the complete course offering information for both programs of Electrical Engineering and Communication Engineering, respectively. This information also includes courses offering times and days. The package selects suitable information from Table 2 and converts it to show the required information in a 2-D format as shown in Fig. 3. Courses are shown in black font and laboratories in red font. The two characteristic digits are also appended to each course making it easy for students to select appropriate courses.

Table 2. The class schedule for the Spring 2022 semester.

CRN	Sec.	Course	E-Title	Cr. Hr	Campus	Meet.Sch.	Days	Time	location	Ceiling	Enrolled	Char 1	Char 2
20258	51	ECOM360	Fund of Communication Systems	3.00	G	LL	T R	1100-1215	C6-1001	60.00	59.00	1	1
20329	51	ECOM402	Communication Systems Lab	1.00	G	BI	T	1530-1820	E6-1010	20.00	3.00	0	0
20260	51	ECOM412	Electromagnetic Waves	3.00	G	LL	M W	1230-1345	C6-1012	40.00	6.00	0	0
29590	51	ECOM422	Digital Communication Systems	3.00	G	LL	T R	1230-1345	C6-1009	20.00	6.00	0	0
20327	51	ECOM432	Data Communications & Networks	3.00	G	LL	M W	1530-1645	C6-1001	70.00	53.00	0	0
21652	51	ECOM442	Data Comm. & Networks Lab	1.00	G	BI	F	0800-1050	E6-1008	24.00	22.00	0	0
29584	51	ECOM561	Information Theory & Coding	3.00	G	LL	M W	0930-1045	C6-1012	20.00	7.00	0	0
20255	51	ELEC305	Electric Circuits I	3.00	G	LL	M W	1100-1215	E1-1005	40.00	25.00	2	5
20263	51	ELEC310	Electric Circuits I lab	1.00	G	BI	R	1400-1650	E6-1010	20.00	15.00	2	5
29586	52	ELEC310	Electric Circuits I lab	1.00	G	LB	T	1530-1820	C6-1039	20.00	10.00	2	5
20256	51	ELEC315	Fund. of Microelec. Devices	3.00	G	LL	T R	1230-1345	C6-0002	40.00	25.00	1	1
20265	51	ELEC320	Electric Circuits II	3.00	G	LL	M W	0930-1045	C6-0023	40.00	27.00	1	2
20259	51	ELEC325	Engineering Electromagnetics	3.00	G	LL	T R	0800-0915	C6-0003	40.00	31.00	1	2
20257	51	ELEC335	Digital Logic Design	3.00	G	IL	M W	1400-1515	C6-1015	40.00	29.00	2	2
20262	51	ELEC345	Digital Logic Design Lab	1.00	G	LB	F	0800-1050	C6-1039	20.00	10.00	2	2
22577	52	ELEC345	Digital Logic Design Lab	1.00	G	BI	R	1530-1820	C6-1039	20.00	14.00	2	2
20090	51	ELEC360	Signals & Systems	3.00	G	IL	M W	1230-1345	C6-1037	30.00	21.00	2	3
20344	51	ELEC370	Electronic Circuits	3.00	G	LL	M W	0930-1045	C6-0002	60.00	47.00	0	0
22839	51	ELEC372	Electro-Mechanical Devices	2.00	G	BI	T R	0930-1045	C6-0003	60.00	43.00	0	0
20345	51	ELEC375	Electronic Circuits Lab	1.00	G	BI	R	1530-1820	E6-1013	20.00	21.00	0	0
23479	52	ELEC375	Electronic Circuits Lab	1.00	G	BI	F	0800-1050	E6-1013	20.00	20.00	0	0
27858	51	ELEC380	Analytical Methods for EE	3.00	G	LL	M W	1100-1215	C6-1037	40.00	15.00	0	0
20982	51	ELEC451	Microprocessors	3.00	G	LL	M W	0800-0915	C6-0019	60.00	9.00	2	2
20983	51	ELEC461	Microprocessors Lab	1.00	G	BI	T	1530-1820	E6-1011	20.00	19.00	1	1
29589	51	ELEC462	Comp. Architecture & Organizat	3.00	G	LL	T R	0800-0915	C6-0019	60.00	59.00	0	0
24003	51	ELEC472	Power Systems	3.00	G	LL	M W	1230-1345	C6-0003	60.00	28.00	0	0
29571	51	ELEC521	Advanced Control Systems	3.00	G	LL	M W	0800-0915	C6-0005	60.00	46.00	0	0
20342	51	ELEC562	Embedded System Design	3.00	G	LL	M W	1400-1515	C6-0005	60.00	9.00	0	0
26423	51	ELEC580	Special Top. in Electronic Eng	3.00	G	LL	T R	0930-1045	C6-0019	60.00	46.00	0	0

Two test examples of course selection procedures for the Spring 2022 semester are described following.

4.1 Test Case 1

This typical student has only completed four out of 23 Compulsory Specialization Requirements courses which are ELEC 230, ELEC 305, ELEC 310, and ELEC 360. The student has also completed about 60% of other course Requirements that include courses from General Education and College Requirements, respectively. Figure 4 shows a subset of courses offered in the Spring 2022 semester that are based on the student's eligibility to take. These listed courses are built on the prerequisite courses already completed by the student. The Figure is also missing the advanced level courses because the student is still waiting to take the prerequisites for these courses.

Next Semster Table

Spring 2022	Monday	Tuesday	Wednesday	Thursday	Friday
8:00 – 9:15	ELEC 451 (2,2) ELEC 521 (0,0)	ELEC 325 (1,2) ELEC 462 (0,0)	ELEC 451 (2,2) ELEC 521 (0,0)	ELEC 325 (1,2) ELEC 462 (0,0)	ELEC 345/5 (2,2) ELEC 375/52(0,0) ECOM 442 (0,0)
9:30 – 10:45	ELEC 320 (1,2) ELEC 370 (0,0) ECOM 561 (0,0)	ELEC 372 (0,0) ELEC 580 (0,0)	ELEC 320 (1,2) ELEC 370 (0,0) ECOM 561 (0,0)	ELEC 372 (0,0) ELEC 580 (0,0)	ELEC 345/51(2,2) ELEC 375/52(0,0) ECOM 442(0,0)
11:00 – 12:15	ELEC 305 (2,5) ELEC 380 (0,0)	ECOM 360 (1,1)	ELEC 305 (2,5) ELEC 380 (0,0)	ECOM 360 (1,1)	
12:30 – 13:45	ELEC 360 (2,3) ELEC 472 (0,0) ECOM 412 (0,0)	ELEC 315 (1,1) ECOM 422 (0,0)	ELEC 360 (2,3) ELEC 472 (0,0) ECOM 412 (0,0)	ELEC 315 (1,1) ECOM 422 (0,0)	St. ID:
14:00 – 15:15	ELEC 335 (2,2) ELEC 562 (0,0)		ELEC 335 (2,2) ELEC 562 (0,0)	ELEC 310/51(2,5) ELEC 345/52(2,2) ELEC 375/51(0,0)	St. Name:
15:30 – 16:45 15:30 – 18:20	ECOM 432 (0,0)	ELEC 310/52(2,5) ELEC 416 (1,1) ECOM 402 (0,0)	ECOM 432 (0,0)		Designed by:

Fig. 3. A 2-D chart of all EE and ECOM courses offered for the Spring 2022 semester.

The student's course selection choice is ELEC 335 (2, 2), ELEC 345 (2, 2), ELEC 320 (1, 2), ELEC 325 (1, 2), and ELEC 315 (1, 1). The selection order is prioritized on appended characteristics. Courses with (2, 2) are selected first then the choice of (1, 2) is next and lastly, for the fifth course, it is (1, 1). This selection totals 13 credits therefore, another course is selected from the other Requirements to take 16 credits in the current semester. Most of the students take from 15 to 17 credits per semester to complete their degree requirements in four and a half to five years duration.

The course selection is grey highlighted in the 2-D display of Fig. 4. This display allows fair time management on course selection with two courses distributed on each offering day, respectively. Here, the student has a choice to select courses depending on suitable times and days. It is also evident from the Figure that this typical course selection is optimum concerning the appended characteristics. Although ECOM 360 has similar characteristics compared to ELEC 315 however, a lower course code is selected first.

4.2 Test Case 2

In this test case, the student has completed ten Compulsory Specialization Requirement courses which include ELEC 230, ELEC 305, ELEC 310, ELEC 315, ELEC 320, ELEC 325, ELEC 335, ELEC 345, ELEC 360, and ELEC 451. The student has also completed about 80% of other Requirements that include courses from General Education and College Requirements, respectively. Figure 5 displays all the eligible courses offered in the Spring 2022 semester that the student is entitled to take. These courses are based on the prerequisite courses already completed by the student. Some Specialization courses are missing from this display because these are offered only in the Fall semester schedule.

The student's selected courses are ECOM 360 (1, 1), ELEC 461 (1, 1), ELEC 370 (0, 0), ELEC 375 (0, 0), ELEC 380 (0, 0), and ELEC 562 (0, 0). Again, the selection is subjected to appended characteristic two digits. ECOM 360 and ELEC 461 are a must because of (1, 1) whereas the other course choices are based on offering times and days, respectively. This course selection accumulates 14 credits; therefore, another course is selected from the other Requirements to take a reasonable 17 credits in the current semester.

The course selection is grey color highlighted in the 2-D Figure. There is again a time and day management on course selection as the courses are fairly distributed on each offering day. Here, the student has a choice to select courses depending on suitable times and days. Selecting an elective course of ELEC 562 instead of ELEC 462 or ELEC 472 is based on this fair distribution and time and day management.

Next Semster Table

Spring 2022	Monday	Tuesday	Wednesday	Thursday	Friday
8:00 – 9:15		ELEC 325 (1,2)		ELEC 325 (1,2)	ELEC 345/51(2,2) ECOM 442 (0,0)
9:30 – 10:45	ELEC 320 (1,2)		ELEC 320 (1,2)		ELEC 345/51(2,2) ECOM 442 (0,0)
11:00 – 12:15	ELEC 380 (0,0)	ECOM 360 (1,1)	ELEC 380 (0,0)	ECOM 360 (1,1)	
12:30 – 13:45		ELEC 315 (1,1)		ELEC 315 (1,1)	St. ID:
14:00 – 15:15	ELEC 335 (2,2)		ELEC 335 (2,2)	ELEC 345/52(2,2)	St. Name:
15:30 – 16:45 15:30 – 18:20	ECOM 432 (0,0)	ECOM 402 (0,0)	ECOM 432 (0,0)		Designed by:

Fig. 4. A set of suitable courses only eligible for this specific student.

In this test case, most of the eligible courses with (0, 0) characteristics indicate that there are no hierarchies as well as no courses dependent on these selectable courses. Nevertheless, some courses in this list are important to be taken earlier so that selection of technical electives can be facilitated. For example, ELEC 370 opens four electives ELEC 512, ELEC 533, ELEC 580, and ELEC 592. Similarly, ELEC 472 is also needed as a prerequisite for ELEC 531 and ELEC 534. These hierarchies are evident from Fig. 1.

Next Semster Table

Spring 2022	Monday	Tuesday	Wednesday	Thursday	Friday
8:00 – 9:15		ELEC 462 (0,0)		ELEC 462 (0,0)	ELEC 375/52(0,0) ECOM 442 (0,0)
9:30 – 10:45	ELEC 370 (0,0)		ELEC 370 (0,0)		ELEC 375/52(0,0) ECOM 442 (0,0)
11:00 – 12:15	ELEC 380 (0,0)	ECOM 360 (1,1)	ELEC 380 (0,0)	ECOM 360 (1,1)	
12:30 – 13:45	ELEC 472 (0,0) ECOM 412 (0,0)		ELEC 472 (0,0) ECOM 412 (0,0)		St. ID:
14:00 – 15:15	ELEC 562 (0,0)		ELEC 562 (0,0)	ELEC 375/51(0,0)	St. Name:
15:30 – 16:45 15:30 – 18:20	ECOM 432 (0,0)	ELEC 461 (1,1)	ECOM 432 (0,0)		Designed by:

Fig. 5. A set of suitable courses only eligible for the second test case.

5 Conclusion

Student course planning is a necessary progression to facilitate students to accomplish requirements effortlessly. This paper presented an algorithm that helps students to create a next semester study plan with the most suitable courses offered in a specific semester. A Smart Course Planning System is devised that is constructed around two-course level characteristics to direct students to select from five to seven courses. There are 25 study plans generated by the use of the SCPS software package. Only two of the plans have minor discrepancies resulting in a 92% accuracy of generated results.

References

1. Daly, M., Sidell, N.: Assessing academic advising: a developmental approach. J. Bac. Soc. Work **18**(1), 37–49 (2013)
2. McMahan, B.: An automatic dialog system for student advising. J. Undergrad. Res. Minn. State Univ. Mankato **10**(1), 6 (2010)
3. Muola, J., Maithya, R., Mwinzi, A.: The effect of academic advising on academic performance of university students in Kenyan universities. Int. Multidiscip. J. **5**(5), 332–345 (2011)
4. Egan, D.: Empowerment through advising in individualized major programs. Mentor Acad. Advis. J. **16** (2014)
5. Kumar, P., Girija, P.: A user interface design for the semester course registration system. Int. J. Innov. Res. Comput. Commun. Eng. **4**(7), 14204–14207 (2016)
6. Ghareb, M., Ahmed, A.: An online course selection system: a proposed system for higher education in Kurdistan region government. Int. J. Sci. Technol. Res. **7**(8), 145–150 (2018)
7. Al-hawari, F.: MyGJU student view and its online and preventive registration flow. Int. J. Appl. Eng. Res. **12**(1), 119–133 (2017)

8. Laghari, M., Memon, Q., ur-Rehman, H.: Advising for course registration: a UAE university perspective. In: International Proceedings on Engineering Education, Poland (2005)
9. Laghari, M., Khuwaja, G.: Electrical engineering department advising for course planning. In: IEEE International Proceedings on EDUCON, Morocco, pp. 861–866 (2012)
10. Laghari, M., Khuwaja, G.: Course advising & planning for electrical engineering department. J. Educ. Instruct. Stud. World **2**(2), 172–181 (2012)
11. Laghari, M., Khuwaja, G.: Student advising & planning software. Int. J. New Trends Educ. Their Implic. **3**(3), 158–175 (2012)
12. Laghari, M.: Automated course advising system. Int. J. Mach. Learn. Comput. **4**(1), 47–51 (2014)
13. Laghari, M.: EE course planning software system. J. Softw. **13**(4), 219–231 (2018)
14. Afify, E., Nasr, M.: A proposed model for a web-based academic advising system. Int. J. Adv. Netw. Appl. **9**(2), 3345–3361 (2017)
15. Feghali, T., Zbib, I., Hallal, S.: A web-based decision support tool for academic advising. Educ. Soc. Technol. **14**(1), 82–94 (2011)
16. Mohamed, A.: A decision support model for long-term course planning. Decis. Support Syst. **74**, 33–45 (2015)
17. Wicaksono, F., Putra, B.: Course scheduling information system using genetic algorithms. Inf. Technol. Eng. J. **6**(1), 35–45 (2021)
18. Huang, M., Huang, H., Chen, I., Chen, K., Wang, A.: Artificial intelligence aided course scheduling system. J. Phys. Conf. Ser. **1792**, 012063 (2021)
19. Tuaycharoen, N., Prodptan, V., Srithong, B.: ClassSchedule: a web-based application for school class scheduling with real-time lazy loading. In: 5th International Proceedings on Business and Industrial Research (2018)
20. Legaspi, J., De Angel, R., Lagman, A., Ortega, J.: Web based course scheduling system using greedy algorithm. Int. J. Simul. Syst. Sci. Technol. **20**, 14.1–14.7 (2019)
21. Faraj, B., Muhammed, A.: Online course registration and advisory systems based on student's personal and social constraints. Kurd. J. Appl. Res. **6**(2), 83–93 (2021)
22. Assiri, A., AL-Ghamdi, A., Brdese, H.: From traditional to intelligent academic advising: a systematic literature review of e-Academic advising. Int. J. Adv. Comput. Sci. Appl. **11**(4), 507–517 (2020)
23. Morsomme, R., Alferez, S.: Content-based course recommender system for liberal arts education. In: The 12th International Conference on Educational Data Mining, pp. 748–753 (2019)
24. Ryan Anderson, R., Eckroth, J.: Tarot: a course advising system for the future. J. Comput. Sci. Coll. **34**(3), 108–116 (2019)

Meetei Mayek Natural Scene Character Recognition Using CNN

Chingakham Neeta Devi(✉)

National Institute of Technology Manipur, Imphal 795004, India
neeta.chingakham@gmail.com, neeta.ch@nitmanipur.ac.in

Abstract. Recognition of characters present in natural scene images is a nascent and challenging area of research in computer vision and pattern recognition. This paper proposes a convolutional neural network (CNN) based natural scene character recognition system for Meetei Mayek. Meetei Mayek text present in natural scene images have been detected using maximally stable extremal regions (MSER), geometrical properties, strokewidth and distance. The extracted and manually cropped characters have been used to create a small database. The experiments of the proposed CNN have been conducted on the isolated characters of the Meetei Mayek natural scene character database. The proposed system has been compared with different combinations of feature descriptors, extracted using pretrained CNNs - Alexnet, VGG16, VGG19 and Resnet18 employing three classifiers - support vector machine (SVM), multilayer perceptron (MLP) and k-nearest neighbour (K-NN). The proposed system has achieved better performance with a classification accuracy of 97.57%.

Keywords: Meetei Mayek · Natural scene images · MSER · Classification · CNN

1 Introduction

Researchers have been interested on the topic of character recognition [1,2] and have been carrying out work for recognition of printed and handwritten text in popular languages where they have attained extremely high accuracy. Recognition of Text from Natural Scene Images [3–5] has become a new challenge in the field of character recognition. This is a nascent field that presents many challenges because extracting text from natural scene images and recognising is not an easy process. For tasks like licence plate recognition, image indexing for content-based image retrieval, text-to-speech translation for those with visual impairments and various others, it is important to recognise text present in these images.

Text should be detected and extracted before characters in the natural scene images can be recognised. Text extraction [6,7] entails processes like detection of connected components and removing objects that could be non-text elements using text's inherent properties. An OCR system is used to recognise the text

K. K. Patel et al. (Eds.): icSoftComp 2022, CCIS 1788, pp. 419–431, 2023.
https://doi.org/10.1007/978-3-031-27609-5_33

after it has been extracted from natural scene images. As can be seen from Fig. 1, text detection in natural scene images is a difficult task because such images contains objects other than text. Furthermore, the images contain text in a variety of styles and sizes. After preprocessing, OCR systems perform direct segmentation on images of typical printed or handwritten documents. Therefore, creating an OCR that can recognise text in natural scene images is difficult because doing so requires reliable text detection and extraction.

Fig. 1. Natural scene images consisting of Meetei Mayek

While works for character recognition in widely used languages [8–11] have been recorded in literature, there are very few works for character recognition in Manipuri, an Indian language that employs the Meetei Mayek script. The Meitei people of the northeastern Indian state of Manipur speak Manipuri as their primary language. Approximately 1.7 million people worldwide speak this Tibeto-Burman language. There are two scripts - Bengali and Meetei Mayek for the Manipuri language. The 18th century saw the introduction of Bengali script, however Meetei Mayek has just recently been revived. All signs and writing are currently done in Meetei Mayek, which is the official script in use. Additionally, no work has been reported to this point on text recognition from natural scene images for Meetei Mayek.

The main purpose of this work is development of a system to aid visually impaired people by recognising text and converting the recognised text to speech and translating for the tourists. This paper presents the first stage of the development of a text-to-speech (TTS) conversion system. The proposed system preprocesses the image and detect text using maximally stable extremal regions (MSER), performs geometric filtering, filtering according to strokewidth and distance among detected MSERs which represents connected components. After the text extraction process, using the extracted characters and other characters which have been manually cropped, a small database has been created for Meetei Mayek natural scene characters. A CNN has been proposed for feature extraction and classification and experiments have been carried out on the database developed. The results have been compared with features extracted using pretrained CNN viz. Alexnet, VGG16, VGG19 and Resnet18 using three different classifiers SVM, MLP and KNN.

This paper explains the processes in details for the Meetei Mayek Character Recogniser from Natural Scene Images. In Sect. 1, a brief introduction about

Manipuri language and importance of development of the proposed system has been given. Precedent works have been described briefly in Sect. 2. In Sect. 3, the processes comprising the Text Extraction Algorithm and Databases Creation have been described. Section 4 describes the proposed CNN used for recognition purpose. Section 5 shows the experimental results and comparisons for the proposed Meetei Mayek Recognition System. Section 6 concludes the paper by highlighting the results achieved and future works.

2 Related Work

Research works of Meetei Mayek Optical Character Recognition for both handwritten and printed have not been reported widely in literature. Only a handful of work have been found and till date no work for character recognition for natural scene characters of Meetei Mayek natural scene characters has been found. Therefore, the scope of research is wide which has to be done associated with Manipuri script recognition.

Thokchom et al. [12] proposed the first handwritten OCR for Meetei Mayek script. Their work used Otsu's technique for binarization. Sobel Edge Detection, dilation, and filling methods have been used to segment characters. A total of 79 features, including 31 probabilistic and 48 fuzzy features, have been employed for feature extraction. The training of Backpropagation Neural Network has been done and recognition has been performed using the extracted features. Ghosh et al. [13] have proposed an OCR architecture for printed Meetei Mayek script. Their work accept an image of a printed document's clipped textual section. Preprocessing, segmentation, and classification have all been done. A multistage SVM has been used for classification using the extracted local and global features. Hijam et al. [14] developed an offline CNN based Meitei Mayek character recognition system. The authors compared their proposed CNN using their handwritten dataset with different feature sets and classification techniques.

Text Detection and Recognition is an emerging field in the field of character recognition. Darab et al. [15] proposed a Farsi text localisation system. The text in images of natural scenes have been located by merging edge and colour information. A new pyramid of images addresses orientation text size and variations. The combination of histogram of oriented gradients and wavelet histogram has been used to verify the candidate texts. Their experimental findings have shown that their strategy is efficient and promising. A text detection based recognition system has been presented in [16]. The authors' text detection is based on adaptive local thresholding and MSER detection. To distinguish between characters and non-characters, they have used a variety of simple to compute characteristics. The authors have classified the characters in the images of natural scenes by introducing a new feature called Direction Histogram. After recognition, spelling has been corrected through dynamic programming (DP). Meetei et al. [17] presented a comparative study to detect and recognised Meetei mayek and Mizo scripts. They compared two methods for the said purposed. The first method performed MSERs detection and then strokewidth computation.

The second method used for text detection was using a pretrained text detector - efficient and accurate scene text (EAST) detector. Their work reported that EAST was more effective for detecting text. The extracted text has been then fed to respective OCRs for recognition.

A recognition method based on residual CNN for scene text was developed by Lei et al. [18]. CNN and recurrent neural network (RNN) have been combined to create a convolutional recurrent neural network (CRNN) (RNN). The RNN component handled the encoding and decoding of feature sequences, while the CNN component has been used for extracting features. To train these deep models and obtain the image encoding data, VGG and ResNet were imported. For the purpose of text detection and classification, Khalil et al. [19] proposed developed using fully convolutional networks (FCNs) for classification. Their model increased the accuracy of detection of scene text and has been effective as new branches were integrated in FCN for script recognition. The model additionally included two end-to-end (e2e) ways for combining the training for text detection and identification of script. In contrast to the majority of end-to-end (e2e) methods. Their experiments showed that the system performed well compared to well-known systems, but accuracy outperformed current methods when using the ICDAR MLT 2017 and MLe2e datasets.

3 Meetei Mayek Natural Scene Character Extraction and Database Creation

This section describes the text extraction and database creation processes. The algorithm adopted in this work to detect text is similar to what have been reported in [20]. In this work, the distances between detected connected components have been also calculated and filtering has been done according to distances between maximally stable extremal regions.

3.1 Maximally Stable Extremal Regions (MSER) Detection

Maximally Stable Extremal Regions (MSER) have to be detected after the images have been noise filtered using a bilateral filter [21]. Maximally Stable Extremal Regions [6,7] are connected components with uniform intensity in natural scene images. Due to the fact that text in natural scene images have uniform intensity in comparison to other objects, this technique has been frequently used to identify candidates for basic character candidate. The MSER algorithm accepts a grey level image as input and binarizes the image iteratively using all feasible thresholds. If the values of the pixels below the threshold are higher than the threshold, the iterative binarization procedure sets those pixels as black or white. MSER refers to those connected components whose regions almost remain constant throughout the repetitive binarization process. As a result, the method produces stable MSERs for text regions after the thresholding process. The detection of MSERs in Meetei Mayek natural scene images has been shown in Fig. 2.

Fig. 2. Detected MSERs in Meetei Mayek natural scene images

3.2 Geometric Filtering

For geometric filtering, the areas of each connected component have been calculated following the detection of MSERs in the stage above. The text present in natural scene images are typically neither extremely large nor small. This attribute of text can be taken into account for filtering of non-text components. Because they usually represent non-text things, very small and huge objects are therefore filtered.

Aspect Ratios (AR) are also considered for filtering and are given by the Eq. (1) for each connected component:

$$AR_i = min\left(\frac{Width_i}{Height_i}, \frac{Height_i}{Width_i}\right) \tag{1}$$

The aspect ratios fall within a particular range for Meetei Mayek natural scene characters. The filtering of non-text components using geometric properties have been done easily as shown in Fig. 3.

3.3 Filtering According to Stroke Width

For strokewidth calculation, the algorithm proposed in [20] has been used. The algorithm finds the shortest path from the leftmost foreground pixel to the rightmost backgrounnd ground pixel while scanning the binary image from top to bottom. The MSERs with acceptable strokewidth have been retained while others have been filtered as shown in Fig. 4.

Fig. 3. Text candidates in Meetei Mayek natural scene images after geometric filtering

Fig. 4. Detected text candidates in Meetei Mayek natural scene images after filtering according to stroke width

3.4 Filtering Considering Distance

After geometric and strokewidth filtering, the non-text components can also be filtered considering the distance between MSERs. The distance between the connected components CC_i and CC_j has been determined using Euclidean distance in order to filter additional non-text components:

$$E_D(CC_i, CC_j) = \sqrt{\{CC_{ix} - CC_{jx}\}^2 + \{C_{iy} - CC_{jy}\}^2} \quad (2)$$

The spacing between words' individual characters is typically smaller than the distance between connected components, which represent non-text elements. Connected elements that are farther apart typically represent non-text components. To determine if a connected component may be text or not, an experimental distance threshold is set between the two nearest connected components. The non-text component which are at a greater distance has been filtered as shown in first two images of Fig. 5. The detected Meetei Mayek characters in the natural scene images after the detection process has been bounded by boxes as illustrated in last two images in Fig. 5.

Fig. 5. Detected Meetei Mayek natural scene characters

3.5 Meetei Mayek Natural Scene Character Database Creation

A small database consisting of different 25 characters of Meetei Mayek has been created consisting of 1126 characters in total. Due to time constraint and as Imphal is a small city, more natural scene images could not be captured and as such the database will be extended in future. The database character images consist of extracted images from natural scene images using the algorithm explained in previous section as well as manually cropped images. The images are of size 50×50 and saved in colour .png format.

4 Proposed CNN for Meetei Mayek Natural Scene Character Classification

Convolutional neural network (CNN) [22,23] is one of the most popular technique used for computer vision and pattern recognition problems. Owing to the popularity and robustness, it is highly applied in the field of character recognition.

4.1 CNN Architecture

The input to the proposed Meetei Mayek natural scene classification CNN are grey-level images of size 50×50. The proposed CNN consists four convolutional layers, three maxpool layers and two fully connected layers where the last one is the softmax layer with 25 outputs. The CNN has 589513 trainable parameters and 224 non-trainable parameters totaling to 589737 parameters.

Convolutional Layer: The convolutional layer is the construction lump and main building block of a CNN. Almost all the computationally intensive operation of finding convolutions happens in this layer.

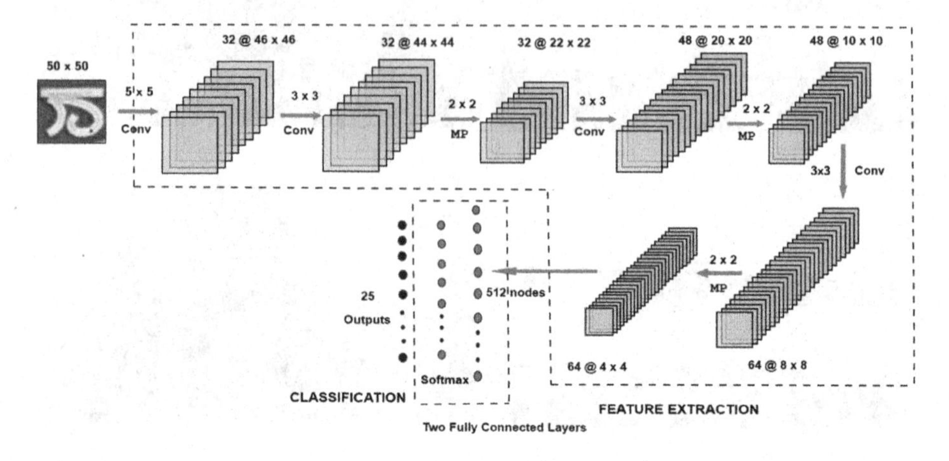

Fig. 6. Proposed CNN for classification of natural scene characters consisting of Meetei Mayek

The formula for discrete convolution for 2-dimension is given as:

$$f * g(x, y) = \sum_{\beta=0}^{N-1} \sum_{\eta=0}^{N-1} f(\beta, \eta) g(x - \beta, y - \eta) \tag{3}$$

The convolutional layers of a CNN are made of parameters normally huge in numbers. The filters of the convolution operation consists of weights and these weights are learned during the process of training.

Max Pool Layer: This layer downsamples the output from previous layer of CNN and thus decreases the number of parameters of CNN to a largescale. The layer finds maximum value by applying a filter to non or nonoverlapping sub-regions of the image. The proposed CNN has three max pool layers after the second, third and fourth convolutional layers. A stride of 2 has been used to obtain the maxpool layer downsampling it to a size of half of its original input.

Rectified Linear Units: Rectified linear units (ReLUs) function is a one such function wgich increasing the non-linearity of its input. ReLUs are used highly in CNNs as they reduce the time needed for training . ReLUs are neurons with non-saturating non-linearity units where all the negative values are reduced to 0.

$$f(x) = max(0, x) \quad (4)$$

Fully Connected Layer: The dense connection of fully connected layer allows the combination of preceding layer features to represent information of the image in a more detailed manner. There are two fully connected layers as shown in Fig. 6. The first one has 512 fully connected nodes and last one is a softmax layer with 25 nodes.

5 Experimental Results

This section give the details of the experiments carried out on the database created using the proposed CNN. The process of choosing the best optimiser for the proposed CNN has been illustrated in details. The comparison of the proposed CNN with existing four pretrained CNN - Alexnet, VGG16, VGG19 and Resnet18 features and classification using three different classifiers SVM, MLP and KNN has also been given.

The database created for the Meetei Mayek natural scene characters is small as more images could not be captured due to time constraint and small geographical area could be only reached. As such, offline data augmentation has been performed and 3652 augmented images have been saved. The augmented image set has been divided into 75% training and 25% testing sets.

In order to obtain the optimal performance of the proposed CNN, the suitable optimiser has been found out by carrying out 20 epoch runs. The learning rate of SGD optimiser has been set to 0.01 with a momentum of 0.9. For ADAM optimiser, the learning rate is 0.001 with β_1 and β_2 set to 0.9 and 0.999 respectively. Keeping the following setting, the accuracy has been computed for minibatch sizes of 8, 16, 32 and 64. Adam optimiser with a learning rate of 0.001 with $\beta_1 = 0.9$ and $\beta_2 = 0.999$ with a minibatch size of 64 has been observed to give better performance. The barchart in Fig. 7 shows the accuracy, precision, recall and f-score values of the two optimisers and ADAM optimiser is slightly better in terms of performance.

The loss and accuracy of the proposed CNN for training and testing data have been shown in Fig. 8. It can be seen from the graphs that high accuracy has been obtained for the proposed model.

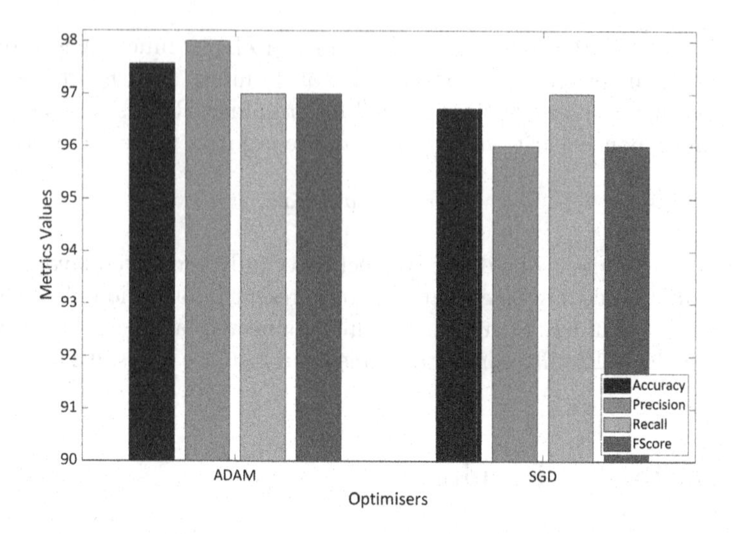

Fig. 7. Comparison of Optimisers for the proposed Meetei Mayek CNN in terms of Accuracy, Precision, Recall and F-Score

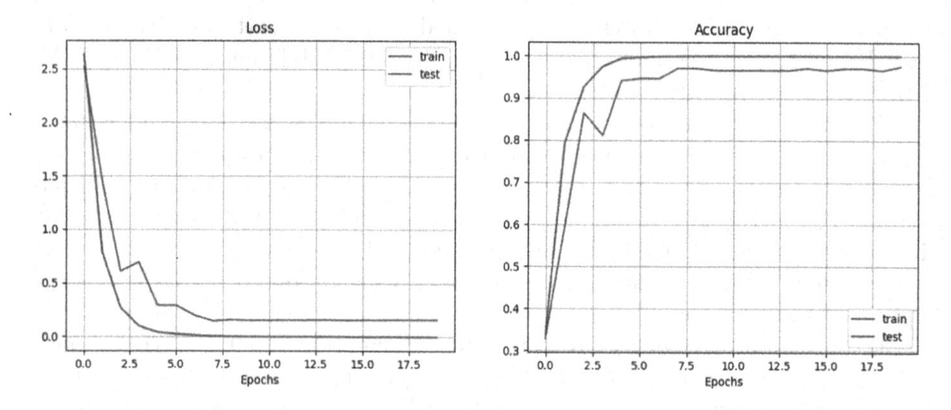

Fig. 8. Train and Test loss and accuracy for proposed Meetei Mayek CNN for natural scene characters

The performance of the proposed CNN for isolated Meetei Mayek natural scene characters has been compared with combination of different feature sets extracted using pretrained CNN - Alexnet, VGG16, VGG19 and Resnet18 employing three different classifiers - SVM, MLP and KNN. The comparison of classification accuracy has been given in Table 1. The results shows that proposed CNN shows better accuracy of 97.57% than the combinations of pretrained CNN feature extraction and classifiers. SVM has given a slightly higher accuracy in one combination with pretrained VGG19 as feature descriptor but the running time of the proposed CNN is faster than the pretrained feature extraction and classification processes.

Table 1. Comparison of Accuracy for different combination of Pretrained CNN features and Classifiers with Proposed CNN

Feature descriptors	Classifiers		
	SVM	MLP	KNN
Alexnet	97.01	95.12	96.10
VGG16	97.17	96.61	95.51
VGG19	97.61	96.63	94.63
Resnet18	96.07	95.61	94.15
Proposed CNN			**97.57**

Table 2. Accuracy for individual Meetei Mayek natural scene (MMNS) characters for the proposed CNN

MMNS Char	Accuracy	MMNS Char	Accuracy	MMNS Char	Accuracy
	100		100	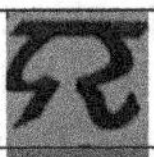	89 .71
	100		100		100
	99.34		88.87		93.33
	90.67		94.31		94.66
	100		92.90	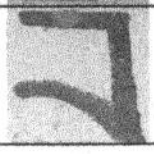	100

In addition, Table 2 shows the individual classification accuracy of some Meetei Mayek natural scene characters for the proposed model. Some Meetei Mayek characters have very similar shapes which have led to low classification accuracy for them while many characters have achieved an accuracy of 100%.

6 Conclusion and Future Works

The paper has presented a convolutional neural network (CNN) for Meetei Mayek natural scene character recognition. Meetei Mayek characters have been extracted from natural scene images using maximally stable extremal regions (MSER) detection, geometric, strokewidth and distance filtering methods. A small database has been created using the extracted and manually cropped character images. The experiments of the proposed CNN have been conducted on the isolated characters of the database. A comparison has been done for the proposed CNN with different combinations of feature sets extracted using pretrained CNNs - Alexnet, VGG16, VGG19 and Resnet18 using three classifiers - SVM,

MLP and KNN. The proposed system has achieved a classification accuracy of 97.57% which has proven better than the feature extraction using pretrained CNNs and classification process.

Deep learning techniques can be explored for Meetei Mayek text extraction from natural scene images. Database size needs to be expanded and more deep learning models can be developed for recognition purpose.

References

1. Govindan, V.K., Shivaprasad, A.P.: Character recognition - a review. Pattern Recogn. **23**(7), 671–683 (1990)
2. Mori, S., Suen, C.Y., Yamamoto, K.: Historical review of OCR research and development. In: Proceedings of the IEEE, vol. 80, pp. 1029–1058, (1992). https://doi.org/10.1109/5.156468
3. Baran, R., Partila, P., Wilk, R.: Automated text detection and character recognition in natural scenes based on local image features and contour processing techniques. In: Karwowski, W., Ahram, T. (eds.) IHSI 2018. AISC, vol. 722, pp. 42–48. Springer, Cham (2018). https://doi.org/10.1007/978-3-319-73888-8_8
4. Chen, X., Jin, L., Zhu, Y., Luo, C., Wang, T.: Text recognition in the wild: a survey. ACM Comput. Surv. **54**(2), 1–35 (2022)
5. Yang, L., Ergu, D., Cai, Y., Liu, F., Ma, B.: A review of natural scene text detection methods. Procedia Comput. Sci. **199**, 1458–1465 (2022)
6. Ephstein, B., Ofek, E., Wexler, E.: Detecting text in natural scene with strokewidth transform. In:18th IEEE International Conference on Computer Vision and Pattern Recognition Proceedings, pp. 2963–2970 (2010). https://doi.org/10.1109/CVPR.2010.5540041
7. Chen, H., Tsai, S.S., Schroth, G., Chen, D.M., Grzeszczuk, R., Girod B.: Robust text detection in natural images with edge- enhanced maximally stable regions. In:18th IEEE International Conference on Image Processing Proceedings, pp. 2609–2612 (2011). https://doi.org/10.1109/ICIP.2011.6116200
8. Zhou, X.-D., Wang, D.-H., Tian, F., Liu, C.-L., Nakagawa, M.: Handwritten Chinese/Japanese text recognition using Semi-Markov conditional random fields. IEEE Trans. Pattern Anal. Mach. Intell. **35**(10), 2413–2426 (2013)
9. Zhou, M.-K., Zhang, X.-Y., Yin, F., Liu, C.-L.: Discriminative quadratic feature learning for handwritten Chinese character recognition. Pattern Recogn. **49**, 7–18 (2016)
10. Supriana, I., Nasution, A.: Arabic character recognition system development. Procedia Technol. **11**, 334–341 (2013)
11. Karimi, H., Esfahanimehr, A., Mosleh, M., Ghadam, F.M.J., Salehpour, S., Medhati, O.: Persian handwritten digit recognition using ensemble classifiers. Procedia Comput. Sci. **73**, 416–425 (2015)
12. Thokchom, T., Bansal, P.K., Vig, R., Bawa, S.: Recognition of handwritten character of manipuri script. J. Comput. **5**(10), 1570–1574 (2010)
13. Ghosh, S., Barman, U., Bora, P.K., Singh, T.H., Chaudhuri, B.B.: An OCR system for the Meetei Mayek script. In: 4th National Conference on Computer Vision, Pattern Recognition and Graphics Proceedings, pp. 1–4 (2013). https://doi.org/10.1109/NCVPRIPG.2013.6776228

14. Hijam, D., Saharia, S.: Convolutional neural network based Meitei Mayek handwritten character recognition. In: Tiwary, U.S. (ed.) IHCI 2018. LNCS, vol. 11278, pp. 207–219. Springer, Cham (2018). https://doi.org/10.1007/978-3-030-04021-5_19
15. Darab, M., Rahmati, M.: A hybrid approach to localize Farsi text in natural scene images. Procedia Comput. Sci. **13**, 171–184 (2012)
16. Gonzalez, A., Bergasa, L.M.: A text reading algorithm for natural images. Image Vision Comput. **31**(3), 255–274 (2013)
17. Meetei, L.S., Singh, T.D., Bandyopadhyay, S.: Extraction and identification of manipuri and mizo texts from scene and document images. In: Deka, B., Maji, P., Mitra, S., Bhattacharyya, D.K., Bora, P.K., Pal, S.K. (eds.) PReMI 2019. LNCS, vol. 11941, pp. 405–414. Springer, Cham (2019). https://doi.org/10.1007/978-3-030-34869-4_44
18. Lei, Z., Zhao, S., Song, H., Shen, J.: Scene text recognition using residual convolutional recurrent neural network. Mach. Vision Appl. **29**(5), 861–871 (2018). https://doi.org/10.1007/s00138-018-0942-y
19. Khalil, A., Jarrah, M., Al-Ayyouba, M., Jararweh, Y.: Text detection and script identification in natural scene images using deep learning. Comput. Electr. Eng. **91** (2021)
20. Devi, C. N., Devi, H. M, Das, D.: Text detection from natural scene images for manipuri meetei mayek script. In: International Conference on Computer Graphics, Vision and Information Proceedings, pp. 248–251. IEEE (2015). https://doi.org/10.1109/CGVIS.2015.7449930
21. Tomasi, C., Manduchi, R.: Bilateral filtering for gray and color images. In: 6th IEEE International Conference on Computer Vision Proceedings, pp. 839–846. IEEE (1998). https://doi.org/10.1109/ICCV.1998.710815
22. Krizhevsky, A., Sutskever, I., Hinton, G. H.: Imagenet classification with deep convolutional netural networks. In: Pereira, F., Burges, C. J., Bottou, L., Weinberger, K.Q. (eds.) Advances in Neural Information Processing Systems, vol. 25 (2012)
23. Simonyan, K., Zisserman, A.: Very deep convolutional networks for large-scale image recognition. arXiv (2015). https://arxiv.org/abs/1409.1556

Utilization of Data Mining Classification Technique to Predict the Food Security Status of Wheat

Mohamed M. Reda Ali[1,2(✉)], Maryam Hazman[2], Mohamed H. Khafagy[1], and Mostafa Thabet[1]

[1] Faculty of Computers and Artificial Intelligence, Fayoum University, Fayoum, Egypt
{mm3655,mhk00,mtm00}@fayoum.edu.eg

[2] Climate Change Information Center, Renewable Energy, and Expert Systems (CCICREES), Agriculture Research Center (ARC), Giza, Egypt
{mreda,m.hazman}@arc.sci.eg

Abstract. Egypt faces wheat insecurity due to the limited cropped area of agricultural lands and the limited horizontal expansion disproportionate to the population increase. The issue of food security, crop consumption rates, and self-sufficiency is considered one of the most important problems facing countries that seek to improve sustainable agriculture and economic development to eliminate poverty or hunger. This research aims to use data mining classification techniques and decision tree algorithms to predict the food security status of strategic agricultural crops (e.g., wheat) as an Agro intelligence technique. Also, the outputs and extracted information from the prediction process will help decision-makers to take an appropriate decision to improve the self-sufficiency rate of wheat, especially in epidemic crises and hard times such as COVID-19, political, and economic disturbances. On the other hand, the research investigates the patterns of wheat production and consumption for the Egyptian population from 2005 to 2020. This research presents a methodology to predict the food security status of strategic agricultural crops through the case study of wheat in Egypt. The proposed model predicts the food security status of wheat with an accuracy of 92.3% to determine the self-sufficiency ratio of wheat in Egypt during the years from 2015 to 2020. Also, it identifies the factors affecting the food security status of wheat in Egypt, their impact on determining and improving the food security state and its rate of self-sufficiency.

Keywords: Data Mining (DM) · Food Balance of Wheat Dataset (FBWD) · Food Security Status of Wheat (FSSW) · Agro Intelligent Decision Support System (AIDSS)

1 Introduction

Wheat is one of the important strategic crops in Egypt. It is milled to extract coarse flour to prepare Baladi Bread (BB), which is a major component of Egyptian food. Egypt is one of the twenty countries that produce wheat, with a production capacity of nine

K. K. Patel et al. (Eds.): icSoftComp 2022, CCIS 1788, pp. 432–445, 2023.
https://doi.org/10.1007/978-3-031-27609-5_34

million tonnes in 2019 [1, 3, 4, 6, 7, 9–11]. Wheat is widely cultivated in four agricultural regions that include 27 Egyptian governorates with a cultivation area was 3.4 Million Feddan (MF) or 1.428 Million hectares (i.e., 1 Hectare = 2.38 Feddan). Whereas, the administrative division of Egyptian agriculture consists of (regions - governorate - center (markaz)/department (qism) - village/residential district). The cultivated area of wheat was approximately 48% of the area of crops grown in winter (7.09MF) and 21% of the total crop area (16.3 MF) in 2020. Egypt suffers from a low rate of self-sufficiency from domestic wheat and seeks to bridge the wheat gap in the local trade market based on the quantities of demand and supply of wheat. Table 1 illustrates the Food Balance Sheet of Wheat (FBSW) in Egypt from 2005 to 2020. Egypt imports wheat from the world trade markets to close the gap of food insecurity of wheat [1–4]. The Average Per Capita of Wheat for Food annually (APCWF) was 67.6 (Kg.) for each person in the world for the years 2018 – 2020. In Egypt, the Average Per Capita of Wheat for Food annually was 188.6 kg [5].

This study uses data mining classification techniques and decision tree algorithms to predict the self-sufficiency status of wheat production in agricultural regions in Egypt as an Agro Intelligent Decision Support System (AIDSS). Data Mining is defined as "a process of discovering various models, summaries, and derived values from a given collection of data". Where, the data mining prediction process aims to produce a model to perform classification, prediction, estimation, or other similar tasks [12]. Descriptive data mining produces useful or nontrivial information from the business problem dataset [12, 13, 28, 31]. Machine learning work to test the process of something to find out how well it performs through using an algorithm without the need for effective formal proof. Figure 1 illustrates the steps of data mining extraction process to discover useful information and knowledge from business datasets, databases, or both [12–14, 28–31].

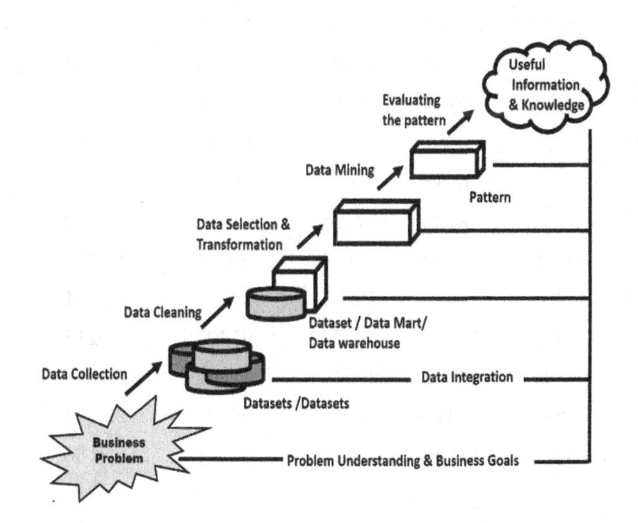

Fig. 1. Information and knowledge discovery process in business data

This paper organizes into eight sections as follows: introductory research in Sect. 1, and Sect. 2 explores the current situation of wheat production and consumption in Egypt.

Section 3 represents previous works of data mining to predict crop yield production. The research objectives are presented in Sect. 4. Section 5 explores the proposed model of research. Section 6 presents a case study to predict the food security status of wheat production in Egypt. Section 7 explores the research results and discussion. Finally, conclusions and future work are in Sect. 8.

2 Current Situation of Wheat Production and Consumption

Table 1 illustrates the food balance sheet for domestic wheat production and consumption from 2005 to 2020. The attributes of the table involve: Year, Domestic Production of Wheat (DPW – 1000 tonnes), Import Wheat (IW -1000 tonnes), Domestic Supply Quantity of Wheat (DSQW-1000 tonnes), Population Volume (Pop. -1000 Citizens), Average Per Capita of Domestic Supply Quantity of Wheat (APC_DSQW – Kg.), Self-Sufficiency Ratio of Wheat (SSRW - % Ratio). Where, SSRW is the ratio of DPW to DSQW to indicate FSSW by (Yes) if SSRW >= 100%, otherwise FSSW indicate by (No) [1–4].

Table 1. Food balance sheet of wheat by CAPMAS & EAS in Egypt.

Year	AW	DPW	IW	DSQW	Pop	APC_DSQW	SSRW
2005	2985	8141	5688	13310	70653	188.3855	61.2%
2006	3064	8274	5817	14667	72009	203.6829	56.4%
2007	2716	7379	5916	13790	74828	184.2893	53.51%
2008	2920	7977	7381	14546	76651	189.7692	54.84%
2009	3179	8523	4061	11450	78522	145.819	74.44%
2010	3066	7169	9805	17685	80443	219.8451	40.54%
2011	3059	8371	9804	17153	82410	208.1422	48.80
2012	3182	8795	6561	15782	84418	186.9507	55.73%
2013	3401	9460	6785	16678	86460	192.8985	56.72%
2014	3414	9280	8105	17825	88530	201.3442	52.1%
2015	3472	9608	9409	19563	90624	215.87	49.11%
2016	3353	9345	10820	19592	92737	211.2641	47.7%
2017	2922	8421	12025	24374	95203	256.0213	34.55%
2018	3157	8349	14892	23549	97147	242.4058	35.45%
2019	3135	8559	12493	21251	98902	212.7185	40.28%
2020	3403	9102	12885	22006	100617	218.7106	41.36%

The value chain flow chart of the average production, imports, and supply quantity of wheat) in Egypt from 2010 to 2013 by FAO composed of the DPW was 8.668 million tonnes, and the imported wheat quantities were 10.15 million tonnes to be (18.818 million tonnes) as an annual supply quantity of wheat in Egypt [11].

3 Related Works

Vogiety, built an intelligent model to predict crop yield by using a random algorithm as a classification decision tree technique to support ranchers with the crop yield before planting the crop to take suitable procedures and decisions. The model anticipated crop yield by climatic boundaries such as precipitation, temperature, overcast spread, fume pressure, and wet day recurrence that influence crop yield. Where his research supported farmers with the predicting yield of many crops in a specific region to select the suitable crop to plant according to climatic boundaries and framer circumstances without any consideration about the local crop production and food security status of it [15]. Rajeswari and Suthendran presented an Advanced Decision Tree (ADT) as a data mining classification model that used classifier algorithm C5 to analyze data on soil nutrients for agricultural land in India. It predicts the fertility level of soil and determined appropriate crops for cultivation through a mobile application that uses the Global Positioning System (GPS) to determine farmer location [16].

Lentz et al. developed a model to determine the food security status for most food-insecure villages in Malawi according to spatial and temporal market data, rain-fall by remote sensing, and geographic and demographic data. Their model predicts food security status by using statistical data and regression relations (regression, lasso regression, linear regression, and Log regression [17].

Dash et al., used support vector machine and decision tree algorithms as data mining classification techniques to predict the type of cultivation crop from three types (Wheat, Rice, and Sugarcane) according to soil macros such as PH and climatic conditions such as rain, humidity, temperature, sunlight [18].

Late and Khan, used data mining classification algorithms with the WEKA tool (Random Tree, J48, …, Bayes Net,) to extract useful information from agriculture data to enhance crop yield prediction according to production, area, crop, and seasons [19]. Akhand et al., used applications of geographical information systems, remote sensing by satellite, and image processing for the agricultural area to predict plant diseases, or crop yields, according to specific parameters such as the Vegetation Condition Index (VCI), Temperature Condition Index (TCI), and Advanced Very High-Resolution Radiometer (AVHRR) sensor. The researcher investigated wheat yield prediction by satellite data, then compared the accuracy of actual yield, and the prediction through an Artificial Neural Network (ANN) as a prediction model [20, 21]. Perez et al. concluded that the impact of climate change will reduce wheat production by 2.3% of total wheat production in Egypt until 2050 [22].

The limitations of the previous studies are summarized in the following point:

- Crop growing and production depends on climate changes, soil fertilization elements, or profit, without concern by national production, and Food Security Status (FSS) of it.
- There is no vision for national crop production and national agricultural rotation.
- There is no support or vision to determine the national FSS of crops.
- There is no prediction process for crop production according to the main agriculture features (i.e., Reg., Area, Yield, …) for cultivation crop area in agriculture regions.

- Not determine crop production patterns to reduce the gap between supply and demand, to reduce the quantities of imported crops.
- Not determined specific or alternative solutions to solve the national crop insecurity.
- Determine the appropriate varieties of crops for growing in specific agricultural areas according to profitable price, without concerning the national Sustainable Development Goals (SDGs).

4 Research Objectives

The proposed technique predicts the food security status of wheat production in agricultural regions. Where, (Yes) indicates the prediction result for wheat sufficient status, and (NO) indicates insufficient wheat status. The patterns and information extracted from the prediction process help agricultural domain experts and decision-makers to take the right decision to enhance the food security status of wheat. Also, it identifies the main factors or features affecting wheat production and consumption patterns to be used to reduce the wheat gap and achieve sustainable agricultural development goals in Egypt. The main objective of the study is "how do adapt the data mining classification technique to predict the food security status for agricultural crops?" the study will try to answer the following sub-questions to be the research objectives:

- Do we have a dataset for crop production in Egyptian regions and governorates?
- Can we collect wheat production data? from where can we collect it? and what are the resources of wheat production data?
- How we can use current wheat production data to create a dataset for wheat production and consumption in Egypt?
- How we can predict the food security status of wheat?
- How we can determine the self-sufficiency ratio of wheat in Egypt to discover useful information and invisible patterns to support decision-makers in agriculture and SDGs 2030?
- What we can do about the food insecurity status of wheat in Egypt?
- What is the architecture of the proposed model or framework to predict the food security status of wheat based on the production and consumption rates in agricultural regions for Egyptian governorates?

5 The Proposed Model and Framework

This section includes two sub-sections, first one presents the proposed model to predict FSSW in the first sub-section. The proposed framework to predict FSSW is presented in the second sub-section. The study utilizes the data mining classification technique to predict FSSW for domestic wheat production and consumption to support decision-makers in the agriculture domain to reduce the wheat gap in Egypt.

5.1 The Proposed Model to Predict FSSW

The proposed Model to Predict the Food Security Status of Wheat (MPFSSW) aims to predict the food security status of regional wheat production in Egypt. It uses decision tree algorithms in the Weka tool as a classification technique to predict the food security status of regional wheat production through decision tree algorithms such as a random tree, J48, and random forest algorithm [23]. The prediction results of MPFSSW support decision-makers to put appropriate strategies and visions to improve SSRW of wheat or crops to fight the hungry and food insecurity in crisis times. The proposed model has the following three phases: (1) the FBW dataset collection phase, (2) the prediction process phase, and (3) the Prediction results, and model accuracy phase. Whereas, Fig. 2 illustrates the phase of the proposed model to predict FSSW in Egypt.

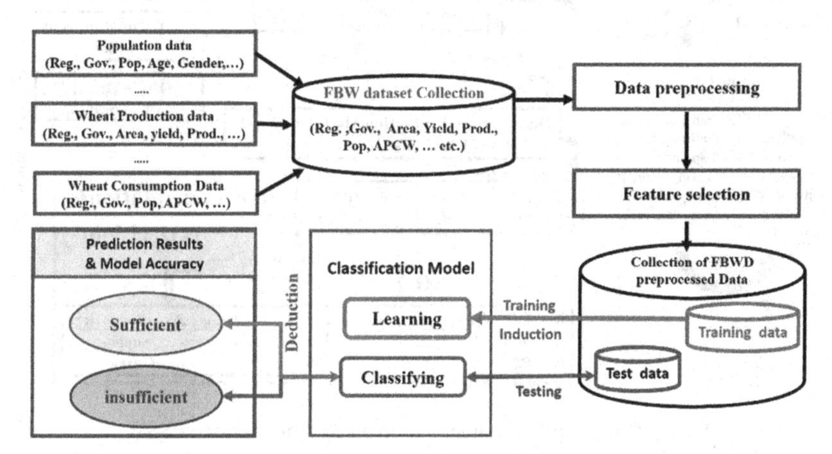

Fig. 2. The proposed model to predict food security status of wheat (MPFSSW).

5.2 The Proposed Framework to Predict and Manage FSSW

The MPFSSW is considered as a core of a proposed framework to predict and manage the food security status of wheat as shown in Fig. 3.

The proposed framework manages and controls FSSW or other crop production and consumption to achieve national SDGs2030, and enhance the self-sufficiency ratio of strategic crops [24–26]. The constraints of MPFSSW on the other models and policy processes in the proposed framework are as follows:

- Use modern agricultural cultivation methods, technologies and machines such as laser land preparation, terraces cultivation, and combine harvest to increase wheat production by 25% [24–27].
- Use modern irrigation techniques such as pivot and sprinkler irrigation to save 25% of uses water quantity in wheat cultivation [24, 25, 27].
- Climate change will reduce wheat production by 2.3% of domestic production in Egypt until 2050 [22].

- Assume the APCW is 145.82 (Kg/citizen) like an APCW in 2009 [1, 3, 4].
- Assume that the current population is the actual consumers of wheat. The APCW that is allocated to infants and the elderly is equal to (10%–13%), which is equivalent to the amount of wheat needed for refugees, residents, and tourists in Egypt [1, 3, 8].
- Update and take the necessary measures to improve FSSW and its self-sufficiency ratio of it, according to the research results and recommendations.

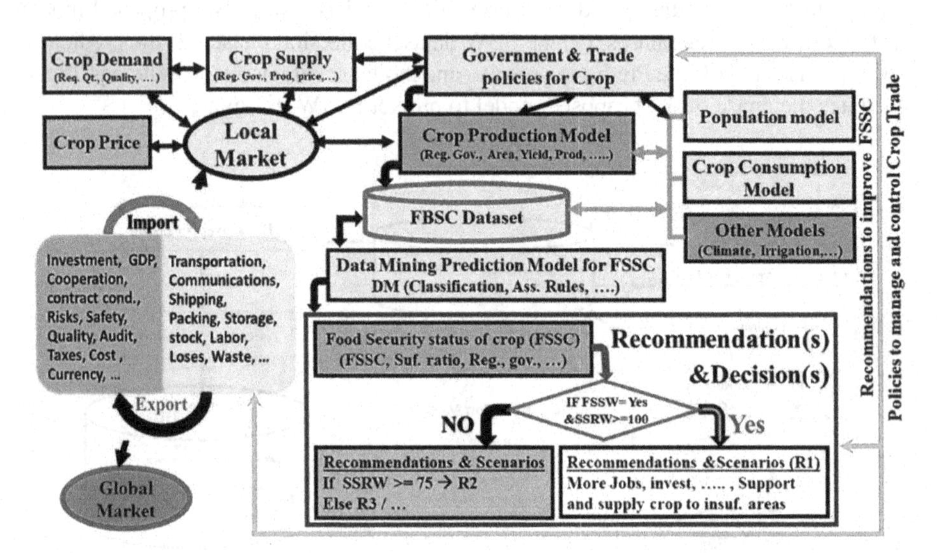

Fig. 3. The proposed framework to predict and manage FSSW.

6 Case Study

The research case study aims to predict wheat self-sufficiency class [Sufficient (Yes) or Insufficient (No)] according to the demographic population of Egypt in agricultural regions and the Food Balance of Wheat Dataset (FBWD). Where the maximum production of wheat was in 2015, and the minimum production of it was in 2018 [1–4]. The following sub-sections illustrate the phases of the proposed model to predict FSSW in Egypt.

6.1 Phase of Food Balance of Wheat Dataset (FBWD)

FBWD has domestic wheat production and consumption in Egypt from 2015 to 2020. FBW data collects from reports and statistics issued by Central Agency for Public Mobilization and Statistics (CAPMAS), Economic Affairs Sector (EAS) in the Ministry of Agricultural and Land Reclamation (MALR), Food and Agriculture Organization of the United Nations (FAO) [1–4, 10]. The study builds FBWD that involves nine columns in comma-separated file format (CSV). The columns are identified as (Year, governorate, AW, Yield, DPW, Pop, and Req. Wheat. Suf. Stat.) to represent 104 instances as shown in a sample of FBWD in Table 2. Where, {Years (is a prediction year for FSSW), Reg. (is a four the agricultural regions in Egypt R1, R2, R3, R4), Gov.(is twenty-seven Egyptian governess in Agricultural regions), AW (are agricultural areas for wheat cultivation by Feddan), Yield (is the productivity of wheat in tonnes from one cultivated Feddan), DPW (is domestic production of wheat in tonnes), Pop. (is a total population of Egypt in citizen), Req. Wheat (are the total requirement quantities of wheat in tonnes for the Egyptian people), Suf. St. (is a sufficiency status of wheat that contained two statuses (Sufficiency status (Yes/No)}.

Table 2. Sample from FBWD for years 2015, 2018, 2019, and 2020.

Year	GOV.	Reg	AW	Yield	DPW	Pop.	Req. wheat	SufSt.
2015	Behera	R1	506712	2.78	1409815	6002000	875092	Yes
2015	Sharkia	R1	416760	3.02	1256532	6680000	973944	Yes
2015	Dakahlia	R1	273311	2.88	787956	6103000	889817	No
2015	Menia	R2	262943	2.9	762009	5356000	780905	No
2015	Fayoum	R2	225753	2.66	600051	3298000	480848	Yes
2015	Kafr-El Sheikh	R1	231814	2.7	623464	3270000	476766	Yes
2015	Suhag	R3	191443	2.8	534317	4775000	696195	No
…	…	…	…	…	…	…	…	…
2018	Gharbia	R1	134963	2.52	340107	5105000	744309	No
…	…	…	…	…	…	…	…	…
2019	Assuit	R3	229084	2.9	662092	4639000	676366	No
2019	New Valley	R4	202176	2.54	513342	251000	36595.8	Yes
…	…	…	…	…	…	…	…	…
2020	Matruh	R4	18824	2.5	47060	493000	71879.4	No
2020	Giza	R2	36744	2.95	108569	9127000	1330717	No
2020	Qalyoubia	R1	51632	2.64	136154	5911000	861824	No

6.2 Prediction Process Phase

The study uses data mining classification techniques through the decision tree algorithms in the Weka tool to predict the food security status of wheat production in Egypt from 2015 to 2020. In the prediction process, explores FBW dataset. CSV in the Weka tool to initial preprocessing and analysis attributes of FBW data to:

- Visualize FBWD features in different two-dimension plots to select relevant features to build a classifier model.

- Select the most relevant attributes (Filter) of the FBW dataset (AW, DPW, Pop, and Req. Wheat. Suf. Stat) to predict the food security status of wheat.
- Then, it performs a prediction for FSSW by using classification decision tree algorithms in the Weka tool such as Random tree, J 48, Random Forest… etc., to predict Sufficient status indicated by (Yes) or In-Sufficient status indicated by (No.). MPFSSW includes a learning (induction) model that learns from the FBW training dataset and a classifying (deduction) model that tests the prediction status for unknown instances through the test, or validation model.
- Visualize a decision tree diagram as shown in Fig. 4 to illustrate the classification of the prediction process for FSSW according to the proposed model.

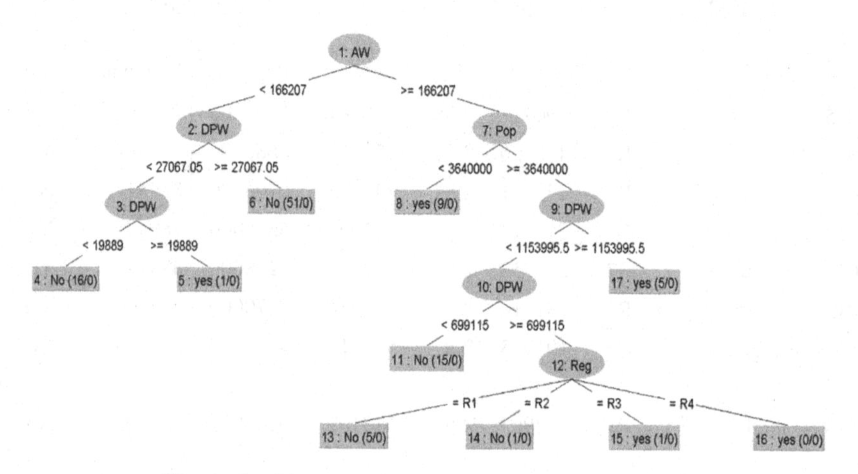

Fig. 4. Decision tree diagram to predict FSSW in Egypt.

6.3 Phase of the Research Results and Model Accuracy

The accuracy and the prediction results of MPFSSW determine by equations from 1 to 5. These equations include True positive (TP = a), True negative (TN = d), False positive (FP = b), and False negative (FN = c) for the prediction process of FSSW to calculate accuracy, precision, recall (sensitivity), specificity, and F measure for the prediction model. Whereas, the accuracy of MPFSSW is equal to all correct predictions divided by the total number of predictions [19].

$$\text{Accuracy} = (\text{TP} + \text{TN})/(\text{TP} + \text{FP} + \text{FN} + \text{TN}) \tag{1}$$

$$\text{Precision} = \text{TP}/(\text{TP} + \text{FP}) \tag{2}$$

$$\text{Recall} = \text{Sensitivity} = \text{TP}/(\text{TP} + \text{FN}) \tag{3}$$

$$\text{Specificity} = \text{TN}/(\text{TN} + \text{FP}) \tag{4}$$

$$\text{F measure} = 2 \times (\text{Precision} \times \text{Recall})/(\text{Precision} + \text{Recall}) \tag{5}$$

7 Results and Discussion

This research presents the Proposed Model to Predict Food Security Status of Wheat (MPFSSW). Where, the proposed model considers a main component in the proposed framework to predict and manage FSSW. The accuracy of the prediction results through the proposed model is 92.3% to predict FSSW in Egypt. Where, SSRW was 41.36% in 2020, compared to the SSRW through MPFSSW which reached 62%. The evaluation of the performance of the data mining prediction process is determined by confusion matrix functions that are illustrated in Table 3 to calculate the accuracy, precision, recall, specificity, and F-measure for prediction model results.

Table 3. Recognition % of data mining prediction process for food security status of wheat

Actual class/predicted class	Suf. St. = Yes	Suf. St. = No	Total	Recognition %
Suf. stat = Yes	a (TP) = 11	b (FN) = 5	16	Recall = a/(a + b) = 11/16 = 68.8%
Suf. stat = no	c (FP) = 3	d (TN) = 85	88	Specificity = d/(c + d) = 96.6%
Total	14	90	104	Acc. = (a + d)/(a + b + c + d) = 92.3% Precision = a/(a + c) = 11/14 = 78.6% F measure = (2 * .7857 * .6875)/(.7857 + 0.6875) = .733

7.1 Comparative Study Between the MPFSSW and the Previous Works

Table 4 presents the comparative study between the proposed framework to predict FSSW and the previous works. The comparison includes the following items: Location, agriculture area of the crop in season (winter, spring,…etc., climate change effect, use modern irrigation methods to save > 25% of water, soil elements, agriculture rotation, crop (types/select/suggest), crop production, food security situation (FSSW), crop self-sufficiency ratio, price and profits, investment, SDGs, Gross Domestic Product (GDP), supply, demand, …etc.

7.2 Recommendations

The study presents the following recommendations to improve wheat production and consumption in Egypt to enhance the food security situation of wheat according to the results of the prediction model and the main features of the case study dataset.

These recommendations support decision-makers to make policies to improve SSRW based on the agricultural development vision and SDGs 2030 As follows:

- Increase wheat cultivation area and wheat productivity in a unit agricultural area.
- Decreases losses and waste of wheat through the different stages of wheat cultivation, harvest, shipment, storage, flour, baking, industry, or both.
- Decrease the consumption quantities of wheat by decreasing the average per capita of wheat (APCW) according to healthy food recommendations
- Determine the population growth rate, the pattern of consumption of wheat (crop) to determine an appropriate APCW, and the required quantities of wheat.
- Determine the case of FSSW and SSRW in agriculture regions as follows:

  Case 1: Completely sufficient of wheat (SSRW >= 100%)
  Case 2: High Partial sufficient (75% <= SSRW < 100%)
  Case 3: Medium Partial sufficient (50% <= SSRW < 75%)
  Case 4: low sufficient Partial sufficient (25% <= SSRW < 50%)
  Case 5: Completely insufficient (0 <= SSRW < 25%)

- Determine and take policies to manage the agriculture processes and practices such as: apply the agriculture rotation, use modern agricultural methods, techniques, technologies…etc., develop early-ripening wheat varieties that tolerated climate change, scarcity, and salinity of the water, … etc.
- Improve the self-sufficiency ratio of DPW according to the features of FBWD that are considered the main agriculture features to predict FSSW.

Table 4. Comparative study between the proposed framework and the previous works.

Items	MPFSSW-framework	Vogiety [15]	Rajeswari and Suthendran [16]	Lentz et al. [17]	Dash et al. [18]	Late and Khan [19]	Akhand et al. [20]
Investment: High (H), Medium (M), Low (L)	H-M-L	M-L	M-L	M-L	M-L	M-L	M-L
Agri. Area in season	> 48% - wheat	Small	Small	Small	Small	Small	Med
Climate change Effect	Constant (5%)	Yes	–	–	Yes	–	–

(continued)

Table 4. (*continued*)

Items	MPFSSW-framework	Vogiety [15]	Rajeswari and Suthendran [16]	Lentz et al. [17]	Dash et al. [18]	Late and Khan [19]	Akhand et al. [20]
Use Modern Irrigation to save > 25% of water	Recommend	–	–	–	–	–	–
Location	Regional	–	GPS	GIS	–	–	Satellite
Soil features	Recommend	–	Yes	–	Yes	–	–
Agriculture rotation	Recommend	–	–	–	–	–	–
Storage & packing	Control rec.	–	–	–	–	–	–
Crop types/select/sug.	Strategic, others	Yes	Yes	Yes	Yes	Yes	Yes
Local crop prod.	Yes (goal)	Yes	Yes	Yes	–	Yes	Yes
Per capita of crop	Yes (goal)	–	–	Yes	–	–	–
Food security situation	Yes (goal)	–	–	Yes	–	–	–
Crop self-sufficiency	Yes (goal)	–	–	Yes	–	–	–
Gross domestic product	Interested	–	–	–	–	–	–
Supply/demand	Control rec.	–	–	–	–	–	–
Price & profits	Fair for all	Yes	Yes	–	–	Yes	–
Food insecurity of crop	Close gap	–	–	Yes	–	–	–
Post-harvest processes	Control rec.	–	–	–	–	–	–
Population/citizens	Yes	–	–	Yes	–	–	–
SDGs	Yes & support	–	–	–	–	–	–

8 Conclusions and Future Work

This study presents a methodology to predict the food security status of wheat production based on the data mining classification technique. In order to apply this methodology, the food security dataset for wheat production and consumption for Egyptian agricultural regions with the APCW = 145.82 (kg) for each Egyptian citizen such as APCW in 2009. The proposed Model Predicts the Food Security Status of Wheat (MPFSSW) was defined based on research assumptions, domestic wheat production, and consumption in years from 2015 to 2020 in Egypt. In addition, a proposed framework to predict and manage FSSW was introduced.

The MPFSSW relates to other models and processes in the proposed framework based on research constraints. The study recommendations for enhancing the food security status of wheat in Egypt are: increasing the agricultural areas that are allocated for wheat cultivation, and the wheat productivity unit (Yield). Also, it recommends decreasing the rate of consumption of wheat, Average Per Capita of Wheat (APCW), loss, and waste of wheat through different stages from cultivation to consumption. The accuracy of the prediction model of food security of wheat is 92.3% through using the random tree algorithm as one of the classification decision tree algorithms. In the future, we will investigate the alternatives and scenarios that support decisions to improve the food security situation of the wheat crop based on the prediction process through data mining techniques to support FSSW and its self-sufficiency ratio according to the pattern of wheat production and consumption.

References

1. CAMPAS: Annual Bulletin of the Movement Production and Foreign Trade and Available for Consumption of Agricultural Commodities in 2020, and Previous Issues. Central Agency for Public Mobilization and Statistics (CAPMAS), Egypt (2022)
2. CAMPAS: Annual Bulletin of Statistical Crop Area and Plant Production in 2019/2020, and Previous Issues. Central Agency for Public Mobilization and Statistics (CAPMAS), Egypt (2022)
3. CAMPAS: Statistical Yearbook 2021, and Previous Issues. Central Agency for Public Mobilization and Statistics (CAPMAS), Egypt (2022). https://www.capmas.gov.eg/Pages/StaticPages.aspx?page_id=5034. Accessed 5 Nov 2022
4. EAS: Bulletin of the agricultural statistical in 2020 & other previous issues, Economic Affairs Sector (EAS) in the Ministry of Agricultural and Land Reclamation (MALR). EAS, Egypt (2022)
5. OECD: OECD-FAO Agricultural Outlook 2021–2030. OECD Publishing, Paris, OECD/FAO (2021). https://www.fao.org/3/cb5332en/cb5332en.pdf. Accessed 5 Nov 2022
6. CAPMAS: Study Self-Sufficiency in Wheat in Egypt. Central Agency for Public Mobilization and Statistics (CAPMAS), Egypt (2015)
7. CAPMAS: Study of Subsidized Baladi in Egypt. Central Agency for Public Mobilization and Statistics (CAPMAS), Egypt (2015)
8. IOM: Refugees in Egypt. Report to International Organization for Migration (IOM). https://egypt.iom.int/news/almnzmt-aldwlyt-llhjrt-fy-msr-tuqdr-aldd-alhaly-llmhajryn-aldhyn-yyshwn-fy-msr-b-9-mlayyn-shkhs-mn-133-dwlt. Accessed 5 Nov 2023
9. FAO: The State of Food Security and Nutrition in the World, and Previous Issues. Food and Agriculture Organization of the United Nations (FAO), Rome (2022). https://www.fao.org/publications/sofi/2022/en/. Accessed 5 Nov 2022
10. FAO: Food Balance sheets (FSB) of Agricultural Commodities and Population from 2010 to 2019. FAO Stat. https://www.fao.org/faostat/en/#data/FBS. Accessed 5 Nov 2022
11. Julian McGill, J., Prikhodko, D., Sterk, B., and Talks, P.: Egypt Wheat Sector Review. Food and Agriculture Organization of the United Nations (FAO), Rome (2015)
12. Kantardzic, M.: Data Mining: Concepts, Models, Methods, and Algorithms, 3rd edn. Wiley-IEEE Press, New York (2020)
13. Campbell, A.: Data Visualization Guide: Clear Introduction to Data Mining, Analysis, and Visualization (2021)
14. Kumar, N., Rohit, R., Sandeep, K., Ramya, L.: Data Mining and Machine Learning Applications. Wiley-Scrivener, New York (2022)
15. Vogiety, A.: Smart agriculture techniques using machine learning. Int. J. Innov. Res. Sci. Eng. Technol. (IJIRSET) **9**(9), 8061–8064 (2020)

16. Rajeswari, S., Suthendran, K.: C5.0: advanced Decision Tree (ADT) classification model for agricultural data analysis on cloud. Comput. Electron. Agric. **56**, 530–539 (2019)
17. Lentz, E., Michelson, H., Baylis, K., Zhou, Y.: A data-driven approach improves food insecurity crisis prediction. World Dev. **122**, 399–409 (2019)
18. Dash, R., Dash, D., Biswal, G.: Classification of crop based on macronutrients and weather data using machine learning techniques. Results Eng. **9**, 100203 (2021)
19. Lata, K., Khan, S.: Experimental analysis of machine learning algorithms based on agricultural dataset for improving crop yield prediction. Int. J. Eng. Adv. Technol. (IJEAT) **9**(1), 3246–3251 (2019)
20. Akhand, K., Nizamuddin, M., Roytman, L.: Wheat yield prediction in Bangladesh using artificial neural network and satellite remote sensing data. Glob. J. Sci. Front. Res. **18**(2), 1–11 (2018)
21. Akhand, K., Nizamuddin, M., Roytman, L., Kogan, F., Goldberg, M.: An artificial neural network-based model for predicting Boro rice yield in Bangladesh using AVHRR-based satellite data. Int. J. Agric. For. **8**(1), 16–25 (2018)
22. Perez, N., Kassim, Y., Ringler, C., Thomas, T., Eldidi, H., Breisinger, C.: Climate Resilience Policies and Investments for Egypt's Agriculture Sector. International Food Policy Research Institute (IFPRI), Washington, DC (2021)
23. Weka software. http://www.cs.waikato.ac.nz/ml/weka. Accessed 5 Nov 2022
24. MALR: The Updated Strategy for Sustainable Agricultural Development in Egypt 2030. Ministry of Agriculture and Land Reclamation (MALR), Egypt (2020)
25. CAPMAS: Egypt Vision 2030. CAPMAS, Egypt (2016). https://www.capmas.gov.eg/Pages/ShowPDF.aspx?page_id=/pdf/Final%20Book%20Mina.pdf. Accessed 5 Nov 2022
26. Bohl, D., Hanna, T., Scott, A., Moyer, J., Hedden, S.: Sustainable Development Goals Report: Egypt 2030, United Nations Development Programme (UNP). UNDP (2018). https://www.undp.org/sites/g/files/zskgke326/files/migration/eg/Sustainable-Development-Goals-Report.-Egypt-2030.pdf. Accessed 5 Nov 2022
27. Al-Minshawi, A.: Advantages of growing wheat on terraces. Akhbarelyom News, 5 November 2021. https://akhbarelyom.com/news/newdetails/3560165/. Accessed 5 Nov 2022
28. Tan, P., Steinbach, M., Karpatne, A., Kumar, V.: Introduction to Data Mining, 2nd edn. Pearson, London (2018)
29. Han, J., Kamber, M., Pei, J.: Data Mining Concepts and Techniques, 3rd edn. Morgan Kaufmann, San Francisco (2011)
30. Sharda, R., Delen, D., Turban, E.: Analytics, Data Science, and Artificial Intelligence: Systems for Decision Support, 11th edn. Pearson, London (2019)
31. Marrè, M.: Intelligent Decision Support Systems. Universitat Politècnica de Catalunya (UPC), Spain (2022). http://www.cs.upc.edu/~idss/idss.html. Accessed 5 Nov 2022

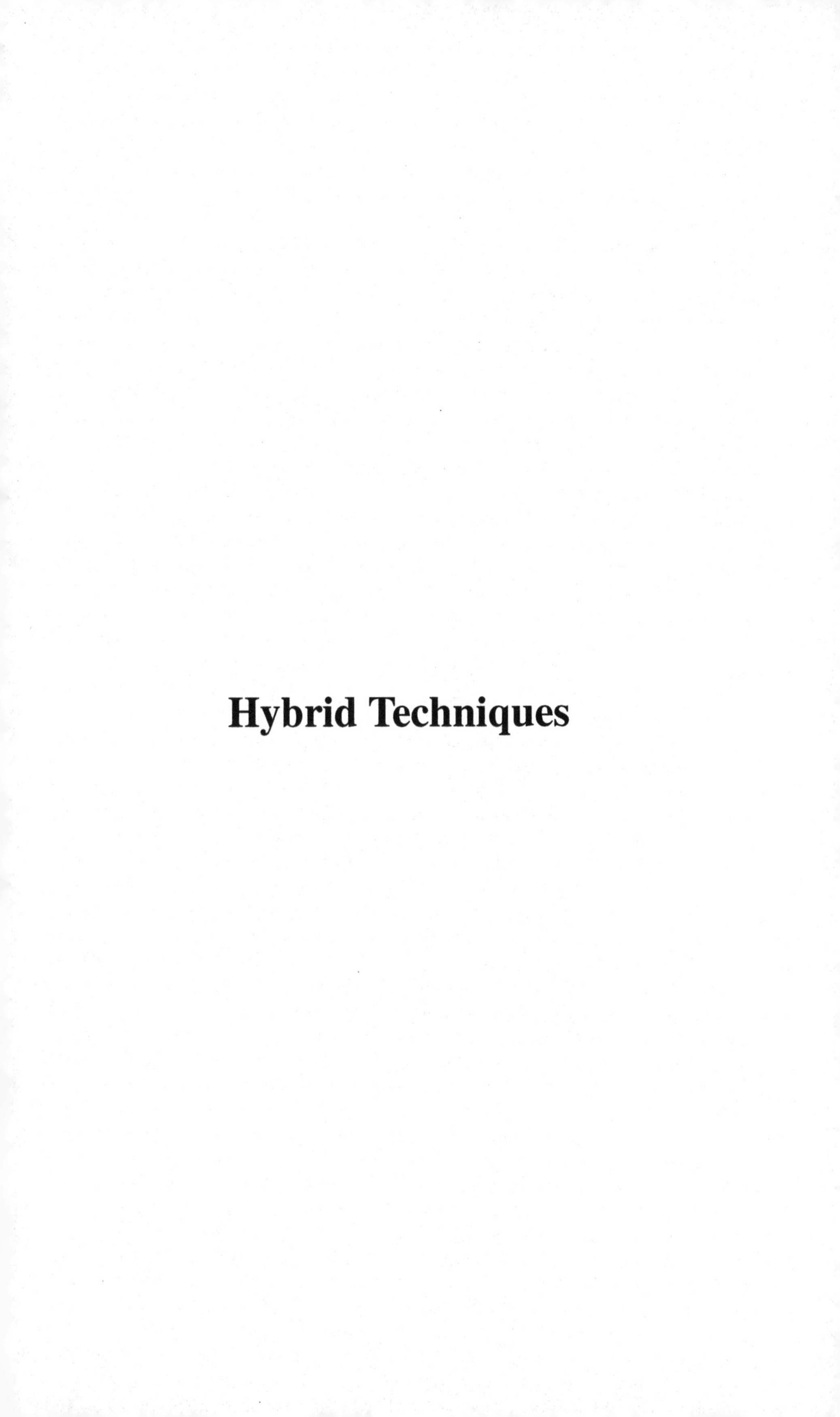

Hybrid Techniques

QoS-Aware Service Placement for Fog Integrated Cloud Using Modified Neuro-Fuzzy Approach

Supriya Singh(✉) and D. P. Vidyarthi

School of Computer and Systems Sciences, Jawaharlal Nehru University, New Delhi 110 067, India
supriy86_scs@jnu.ac.in

Abstract. Fog computing provides an infrastructure for enhancing quality of services (QoS), especially for time-critical applications. The reach of the Internet of things (IoT) has extended to another dimensions, expanding from acquisition of data to device interconnections and to data-processing. This acceleration assimilate fog and cloud compuhting into a single system for improving QoS and resource utilization. Due to the heterogeneity of IoT devices, selecting suitable computation devices and allocating resources are substantial issues that need to be addressed for effective resource utilization. This work proposes a smart decision-making system for service placement based on the various parameters. The proposed work utilizes machine learning based techniques: clustering for the labelling of the services followed by neuro-fuzzy based ANFIS model for offloading the services. A 5-layered neuro-fuzzy inference model is implemented to represent as an intelligent decision-making system. This work provides a solution for the learning phase of ANFIS by employing a meta-heuristic-based algorithm. Three meta-heuristic algorithms, i.e. GA-ANFIS, JAYA-ANFIS and PSO-ANFIS are implemented for the training of the ANFIS model. The effectiveness of the model has been examined for the prediction of computing layer for offloading of the services. The results are compared with each other as well as with the conventional gradient-based ANFIS model. Experiment shows that the evolutionary-based neuro-fuzzy models yield imperative results against gradient-based neuro-fuzzy.

Keywords: Fog computing · IOT · Machine intelligence · Neuro-fuzzy · Resource provisioning

1 Introduction

Internet of Things (IoT) has been gaining popularity to make life easier and innovative. With the expansion of wireless services, many computation-intensive services such as image/speech recognition, video surveillance, virtual reality, and so on, are being processed at the end devices [18]. Millions of IoT devices across a large geographic area produce massive amounts of data. To handle such data,

K. K. Patel et al. (Eds.): icSoftComp 2022, CCIS 1788, pp. 449–462, 2023.
https://doi.org/10.1007/978-3-031-27609-5_35

for analysis and processing, is a non-trivial task often carried out with cloud computing resources [32]. Although numerous benefits are offered by cloud computing, several issues still need proper attention. A centralized cloud is remotely situated and far away from the end users that results in the latency experienced by the network in availing the cloud services. Such delay may become a bottleneck for the services that require ultra-low latency [18,32]. Thus, for time constraint applications, offloading them on the centralized cloud may not be an ideal choice [14]. Location awareness, mobility support, inconsistent latency, and demand for high bandwidth limit the use of cloud [9]. To address this, Cisco in 2011 proposed fog computing, also known "cloud at the edge" as a distributed and decentralized computing technique [14,32]. The fog paradigm extends the cloud computing services and provides service between the end devices and the traditional cloud. It is a cognitive framework that provides services at the edge of the network to assist time-sensitive IoT applications [6]. However, the computational resources in the fog node are not large enough as like the centralized cloud . The significant characteristics of fog computing are its close vicinity to the users, fast response time, and minimum latency. As Fog and Cloud are complementary, integrating these two would suffice to produce the desired result in terms of service and resource provisioning [32]. Figure 1 depicts a distributed multi-layer architecture comprising fog layers, the top layer cloud, and the bottom layer of IoT devices. As per the literature, the Fog layer can be divided into multiple tiers [2,6]. The bottom layer comprises end devices, IoT sensors etc., which produce massive data for computation. The intermediate layer, i.e., fog layer, is the Microdata center, a highly virtualized platform that provides computing service, networking, and storing services between the IoT devices and conventional clouds. The underlying node can be virtualized, like virtual sensor nodes and networks [2]. Since fog devices have poor computational power and limited battery capacity, they cannot accomplish computationally constrained jobs on time. The fundamental challenge with fog computing is that owing to restricted resources, determining which applications should be allotted to fog node and which should be migrated to the cloud necessitates a decision support system [6]. User requests are processed either at the cloud or fog layers, based on the criteria and the policies of the requests [23]. However, the resources required in the dynamic world cannot be predicted. Hence, balancing workload is a critical prerequisite for the efficient resources management [2,27]. Fuzzy logic is a well-known technique for solving such problems and is widely adaptable due to its ability to convey uncertainty [29]. Another technique, neural network, seeks attention because of its self-learning capability. The amalgamation of both has provided a way to process real-time tasks owing to its adaptable intelligent nature [6]. Neuro-fuzzy systems are a recently discovered category that integrates two intelligence groups: fuzzy logic systems and Artificial Neural Networks(ANN), referred as adaptive neuro-fuzzy inference system (ANFIS) [12].

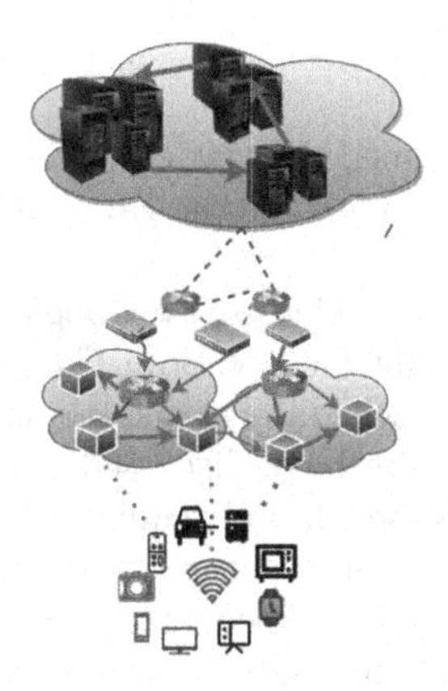

Fig. 1. A multilayer cloud-fog architecture

1.1 Motivation

This work contributes the machine learning based methodology for the clustering and prediction of computing requests in fog-integrated cloud. The work utilize modified ANFIS model for the prediction of suitable layer for the computing request. Conventional ANFIS utilize gradient descent for the training of model. However gradient descent, have the issue of being trapped in a local minimum [4,11]. This work provide a way for the learning phase of ANFIS by utilizing a meta-heuristic-based algorithm like Genetic Algorithm (GA), Particle Swarm Optimization (PSO) and JAYA algorithm. For the training of model, labeled data is required. Labeling is done by fuzzy C-Mean clustering. It is a soft clustering algorithm which calculated a probability score for each data point indicating the likelihood of belonging to every cluster. The effectiveness of the model has been evaluated for the prediction of computing nodes for the offloading of incoming services. The results are compared with each other as well as with the conventional gradient-based ANFIS method. The paper is organized as follows. Section 2 discusses the current research in this domain. Section 3 describes the proposed model for intelligent decision system. Section 4 evaluates all the four models based on real trace data and last section conclude the paper. The primary contributions of this work, are as follows:

1. G-Sutil software retrieves the data from Google Cluster Trace 2019.
2. The raw and unlabeled data is clustered using a fuzzy C Mean clustering technique based on the physical request parameters.
3. Additional QoS features are added to convert unlabeled data into labeled one for the learning phase of the model.
4. To develop and implement a modified neuro-fuzzy workload distribution model, i.e., GA-ANFIS, PSO-ANFIS, and JAYA-ANFIS, to offload the active job for Fog-Integrated Cloud.
5. The performance of the modified ANFIS is evaluated and compared.

2 Related Work

Several researchers have examined the task classification, offloading, and scheduling problems and proposed models for the same. Some recent relevant work in this field is describe as follows:

Google cluster trace is data released by Google publicly. It comprised GBs of workloads processed on eight Google Borg Computer Clusters in May 2019. Task submission, offloading, scheduling, and resource usage statistics for the jobs processed in those clusters have discussed in the google trace [28]. However, trace data are raw, unlabeled, and require a lot of pre-processing to make it worthwhile for our work. Several researchers provide an insight into a trace-2011 dataset [10,15,33]. They have recognized the characteristics and patterns in data and presented a concrete analysis on the trace data set. In Cloud computing, researchers mainly emphasize quality of experience (QoE) management [16] and Service Level Agreements (SLA) [17]. Fog computing studies have highlighted processing and analytics for time-sensitive applications [9], service placement [8,17,20], resource estimation and allocation [6,7,26] for the processing and scheduling of applications on resources [1]. Service placement and scheduling are significant challenges in fog computing [24]. Due to the scarcity of resources at the fog level, it is necessary to utilize them efficiently. Numerous impediments at the fog layer have been investigated [31], concluding workload orchestration is a challenging problem. Workload orchestration refers to distributing service equitably across each tier. Sonmez et al. [25], fuzzy logic introduces as a novel method to deal with service distribution in a cloud-fog architecture. The incoming rate of services are uncertain [5], necessitating an efficient solution to cope with unpredictability. Fuzzy logic is a popular methodology for rapidly changing uncertain systems. However, the fuzzy system does not have the learning power to adapt to the uncertainty in the system. At the same time, the neural network can learn about uncertainty and other learning variables [6]. To cope with real-world situations [29], researchers have integrated the two concepts, fuzzy logic and ANN refered as ANFIS. In [11], authors have implemented ANFIS for dynamic system identification and conclude that in derivative-based algorithms, there is a chance of being stuck in a local minimum. This paper presents another efficient method for the learning phase of ANFIS by utilizing meta-heuristic-based algorithms.

3 Proposed Model

Appropriate offloading of services is essential for the better performance of model. In [3,5], the authors have proposed fuzzy logic for addressing the workload orchestration issue in fog computing systems. Fuzzy logic utilizes a strategy for coping with the uncertainty. However, fuzzy logic has several inherent flaws, such as the inability of fuzzy logic-based systems to learn and the difficulty in locating appropriate membership functions. By integrating the potential of ANN with fuzzy logic, adaptive neuro-fuzzy computing makes decision systems intelligent. This work proposes and trains an ANFIS for efficiently distributing jobs to servers. The systematic flow of work is describe in Fig. 2.

In the proposed model, learning phase of ANFIS utilize the meta-heuristic-based algorithm. Meta-heuristics are not greedy. It generates the solution randomly and the best candidate is selected for the next iteration which minimizes the chances of being trapped in local minima.

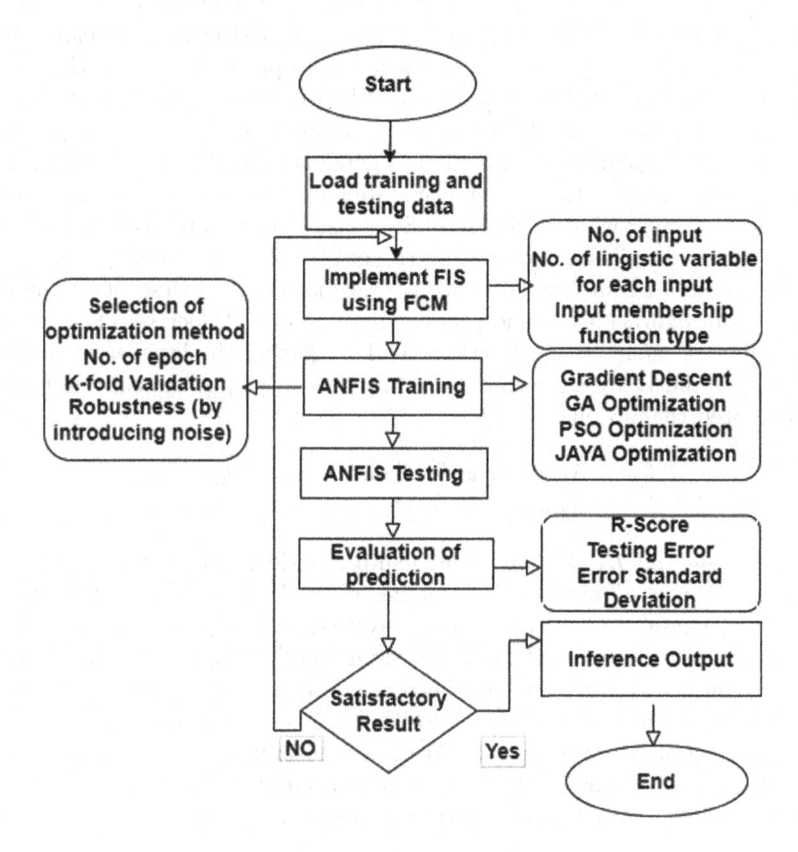

Fig. 2. Flow chart of proposed model

3.1 Architecture of ANFIS

A neuro-fuzzy is simply a 5-layered neural network having input, fuzzification, inference, normalization, and defuzzification layers, as described in Fig. 3. The proposed architecture comprises five input values and one output value. We have used three linguistic values for each input. The chracter used to characterize input is C, the CPU unit required by Service; M, the memory unit required by service; and D, the hard disk required by service. Ds represents the delay sensitivity of the task, and P stands for the priority the jobs. The three linguistic values for input are characterized as high (H), low (L) and medium(M). The output attribute also contains three linguistic classes i.e., cloud layer (C), fog layer (F) and aggregated fog node (AF).

Among five layers, the node of first and forth layer are adaptive while rest are the fixed layer. ANFIS learns by modifying tunable parameters i.e., premise parameter and consequent parameter. Premise parameters are described in first layer and define the shape of the membership functions [30].The adaptable variables have been learned and modified optimally get the minimum error between predicted and actual output. The objective of the ANFIS is to develop a learning ability in a fuzzy system by calculating the optimum value of the premise parameter and consequent on its own. ANFIS utilizes two-pass learning technique. In first phase, ANFIS begins with randomly chosen premise parameter for designing the membership function to generates fuzzy rules. These rules sets have been sent into the fuzzy inference system evaluates the nodes outputs until defuzzification layer where it applied LSS (least square methods) to modify consequent parameters before computing the output. In the second phase, error is back propagated till first layer where ANFIS use gradient descent or any other optimization algorithm to tune premise parameters. Proper selection of shape and number of membership functions lead to getting the least error in predictions. Let A and B are the two input and y is a output as shown in Fig. 3, then the generated rule are like:

$$\text{if x is } A_1 \text{ and y is } B_1 \text{ then y} = p_1\text{x} + q_1 y + r_1$$
$$\text{if x is } A_2 \text{ and y is } B_2 \text{ then y} = p_2\text{x} + q_2 y + r_2$$

where, A_1, A_2 and B_1, B_2 are membership function and p_1, q_1, r_1 and p_2, q_2, r_2 are consequent parameters. The orchestrator decision block anticipates and routes the requests to the appropriate resources. Finding appropriate values of both parameters is essential for the system since it might cause it to overfit. In this paper, this adjustment has been accomplished by nature inspired metaheuristic approach. A traditional algorithm traps in local minima, making the system prediction less appropriate. Minor the discrepancy between ANFIS output and the desired target, the better (more accurate) the ANFIS system. The neuro-fuzzy inference system in detail has presented in the following sections:

Layer 1

It is a fuzzification layer in which each input node i is an adaptive node. The output O_{1i} of this layer as describe in Eq. 1 determines the degree to which given input I satisfy the quantifier I_i according to choosen membership function.

$$O_{1i} = \mu I_i(I) \qquad i = 1, 2, 3 \tag{1}$$

where I represent the input to the node, I_i refer to the linguistic variable associated with Input I, μI_i refers to the membership function. In this paper, we choose $\mu I_i(I)$ to be bell-shaped so that curve is to be smooth. For input attribute memory M and the parameters a_i, b_i, c_i, the MF (membership function) is defined in Eq. 2

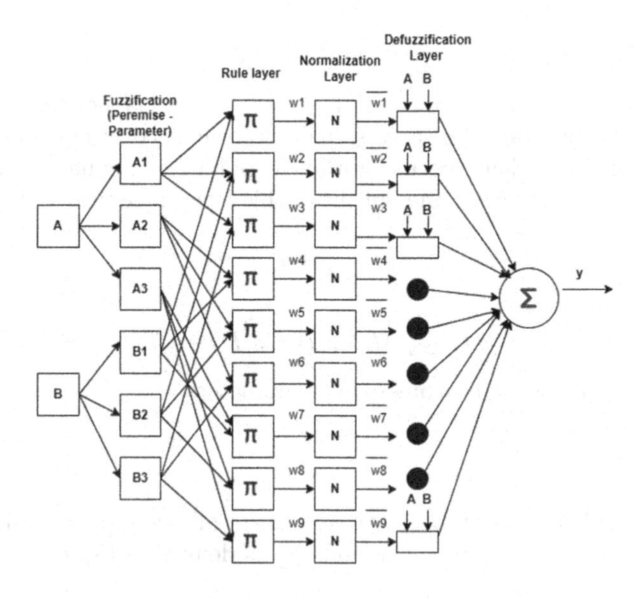

Fig. 3. Architecture of ANFIS for 2 input and 1 output

$$\mu M_i(M) = \frac{1}{1 + \left| \frac{M - c_i}{a_i} \right|^{2b_i}} \tag{2}$$

The parameters a_i, b_i, c_i decides the shapes of MF. Similarly, we can derive the formulas for other input attributes C, D, Ds, and P.

Layer 2

The second layer is called the Rule layer. The output of the previous layer, are used to calculate the firing strength of each rule. The strength of each node has obtained by the product of MF values crossponds to incoming signals at that node as describe in Eq. 3.

$$\begin{aligned} w_i &= \mu M_i(M) * \mu C_i(C) * \mu D_i(D) \\ &* \mu D_{s_i}(D_s) * \mu P_i(P) \end{aligned} \tag{3}$$

Layer 3

The third layer refers to the normalized layer. This layer normalizes the firing strength of each rule with respect to all rules by using Eq. 4. The number of nodes in this layer and the number of fuzzy rules must be same.

$$\overline{w_i} = \frac{w_i}{w_1 + w_2 + \ldots\ldots\ldots\ldots w_n} \tag{4}$$

Layer 4

The forth layer is called defuzzification layer where every node is connected to every input attribute and the corresponding normalized weight. The weighted values of rules have calculated using a linear polynomial equation. If $\overline{w_i}$ represents the normalized weight, the output of this layer is defined in Eq. 5:

$$O_{4i} = \overline{w_i} * \ f_i \tag{5}$$

$$f_i \ = \ p_i C + q_i M + r_i D + s_i Ds + t_i P + u_i \tag{6}$$

where p, q, r, s, t, and u are consequent parameters.

Layer 5

This layer is called the summation layer, where all the input coming from the previous layer is sum up into single node $\sum$ as defined in Eq. 7

$$O_{5i} \ = \ \sum(\ \overline{w_i} * f_i) \tag{7}$$

3.2 Learning Method of Anfis

To deal with the problem of trapping in local minima, three meta-heuristics, i.e., GA, PSO, and JAYA, is employed in the work to find the optimum values of network parameters so that the loss function, defined in Eq. 8, is minimized.

$$RMSE = \sqrt{\frac{\sum_{i=1}^{n} (y_i \ - \ \overline{y_i})^2}{N}} \tag{8}$$

The GA and PSO is well known meta-heuristic-based optimization algorithm for solving multi-parameter problems. It is a population-based search in which each individual could be a candidate for the optimum solution. For more on GA and PSO techniques, structure, and applications to address optimization issues, literature [13,19,21] may be referred. Apart from these two, recently developed JAYA algorithm is a simple meta-heuristic optimization algorithm for constraint and unconstraint problems. All swarm intelligence-based algorithms are probabilistic and need standard regulating parameters such as population size, no. of generations, etc. Apart from the regulating parameters, every algorithm needs its own specific control parameters. The control parameter of GA, PSO are described in Table 1. Inappropriate tuning of algorithm-specific parameters adds unnecessary computation and complexity. Rao et al. [22] introduced JAYA optimization algorithm that does not require any algorithm-specific parameters. Experiments shows that JAYA-ANFIS is computationally faster than GA-ANFIS and PSO-ANFIS. These meta-heuristic algorithms are used to optimize a total of 28 premises and 972 consequent parameters.

Table 1. Control parameter of evolutionary algorithm

GA	Population size	100
	No. of Generation	0
	Crossover rate	0.4
	Mutation rate	0.15
	Selection method	Roulette wheel
PSO	No. of iteration	100
	No. of particle	50
	Initial inertial weight	0.9
	Final inertial weight	0.3
	Coginitive acceleration C1	1
	Social acceleration C2	2
JAYA	Population size	100
	No. of Generation	50

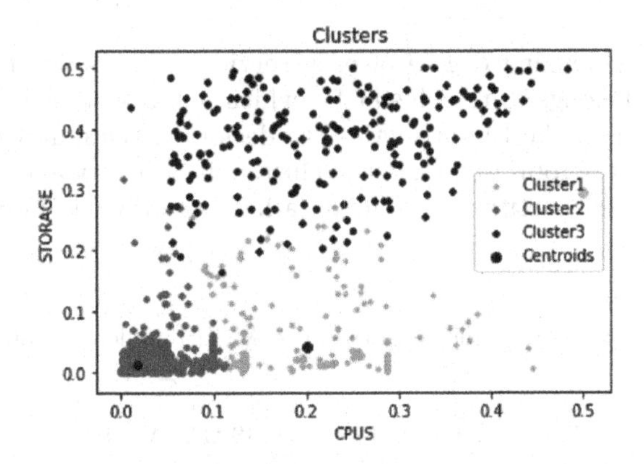

Fig. 4. Service cluster

4 Result and Discussion

Simulations are performed on intel core i5 processor with 8 GB of Ram and 1 TB of storage. The model has been implemented on MATLAB-2019a tool. We have employed Google cluster real traces to portray a high level of realism for the proposed study. This trace has been extracted through Gsuitl package followed by DBMS query. It comprises seven relations but proposed model considered only two relations that is Task Event and Machine Event. The natural join of these two relations generates several attributes, among which few are considered to represent the service set. Services are characterized by the set of five parameters that defined in previous section. Based on these parameters, the model predicts the offloading layer for the services. Initially, we have raw service sets with no label. To attach a label, we have used the fuzzy C-Mean clustering technique

Table 2. Performance evaluation of ANFIS model for batch_1

Measure	Batch 1					
	Epoch	Noise	ANFIS	ANFIS-GA	ANFIS -PSO	ANFIS-JAYA
R-score	250	0%	0.9456	0.9625	**0.96760**	0.96632
		5%	0.9145	0.9245	0.94417	**0.94423**
		10%	0.8923	0.9284	0.93640	**0.93965**
		20%	0.8825	0.9131	0.91728	**0.92632**
Standard variance	250	0%	0.03265	0.02365	0.02354	**0.02311**
		5%	0.1016	0.0144	0.01248	**0.01015**
		10%	0.3693	0.2893	0.23001	**0.13265**
		20%	0.5025	0.4406	0.4169	**0.31020**
Testing error	250	0%	0.1152	0.03516	**0.012360**	0.012510
		5%	0.3012	0.2401	0.2461	**0.2217**
		10%	0.3526	0.2881	0.23191	**0.19562**
		20%	0.4821	0.3428	0.31326	**0.23521**

to find out the belonging degree of each service with respect to each cluster. The service set belongs to the cluster for which it has highest belonging degree score. We have used the Elbow method to decide on the number of clusters in services set. The appropriate number of clusters is arbitrary and depends on the parameters used for partitioning. The data has been divided into three clusters as shown in Fig. 4.

Table 3. Performance evaluation of ANFIS model for batch_2

Measure	Batch 2					
	Epoch	Noise	ANFIS	ANFIS-GA	ANFIS -PSO	ANFIS-JAYA
R-score	250	0%	0.9236	**0.97586**	0.97329	0.97236
		5%	0.8956	0.92320	0.92211	**0.92546**
		10%	0.8625	0.90420	0.904733	**0.909321**
		20%	0.7923	0.87240	0.87446	**0.88253**
Standard variance	250	0%	0.2923	**0.22112**	0.25874	0.25632
		5%	0.4036	0.33189	0.25827	**0.24560**
		10%	0.0.4325	0.0.36923	0.0.39127	**0.0.31523**
		20%	0.4812	0.039429	0.41354	**0.35265**
Testing error	250	0%	0.3825	0.**22510**	0.36071	0.23265
		5%	0.3463	0.30992	**0.27877**	0.29635
		10%	0.4723	0.45850	0.43217	**0.37450**
		20%	0.5936	0.54383	0.54354	**0.52254**

A blue cluster refers to a group of services that require a minimal CPU core, fewer bytes of storage and can process at fog node. In contrast, a green color cluster refers to services that require a large CPU core and significant byte of memory and disk and can process at the cloud node. Further, we have categorized the service based on two nominal attributes: delay sensitivity and priority of the services. Even after considering an additional two nominal attributes, the number of clusters remains the same. Due to the limited resource capacity of fog and fog aggregated nodes, services that are time-sensitive and require fewer resources are placed there.

We have constructed three batches of various sizes to test and train the model. Batch 1 comprises 800 services; batch 2 consist of 1600 services, while batch 3 includes 2560 services. Neuro-fuzzy-based ANFIS is a regression-based model for prediction. It is different from classification and predict the suitable layer for offloading of services. We considered R-Score, testing error, and error variability as a performance measures. R-score measures how close the model makes the prediction. The testing error refers to the mean difference between actual and predicted values.

Table 4. Performance evaluation of ANFIS model for batch_3

Measure	Batch 3					
	Epoch	Noise	ANFIS	ANFIS-GA	ANFIS-PSO	ANFIS-JAYA
R-score	250	0%	0.9336	0.9883	0.9856	**0.99963**
		5%	0.9056	**0.94458**	0.92401	0.94265
		10%	0.8763	0.90439	0.90250	**0.90473**
		20%	0.8198	0.86960	0.84990	**0.87632**
Standard variance	250	0%	0.039023	0.03713	0.03873	**0.03561**
		5%	0.4325	0.32616	0.41977	**0.31632**
		10%	0.5923	0.51395	0.50976	**0.50632**
		20%	0.5636	0.58950	0.53250	**0.52239**
Testing error	250	0%	0.0919	0.003115	**0.00116**	0.002365
		5%	0.2236	**0.14625**	0.17969	0.15523
		10%	0.3296	0.29373	0.30976	**0.26324**
		20%	0.5123	0.49856	0.43854	**0.43452**

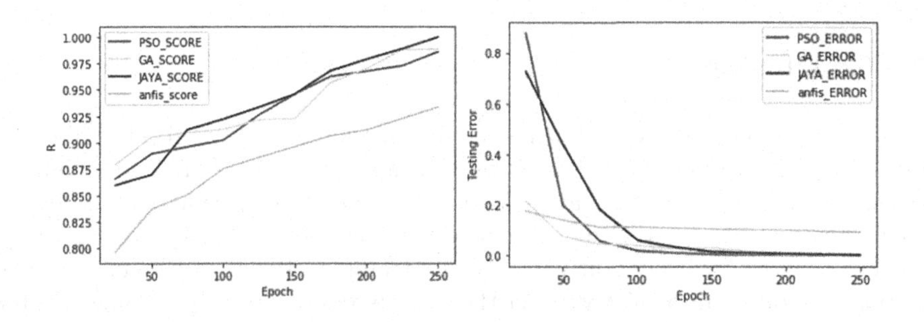

Fig. 5. R-Score and Testing Error for batch 3 at 0% noise (Color figure online)

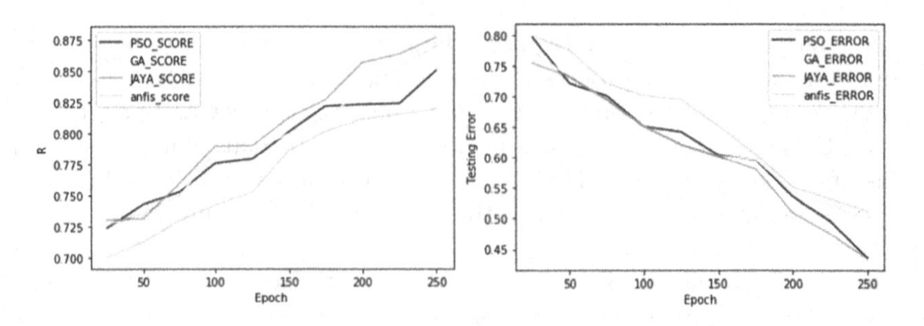

Fig. 6. R-Score and Testing Error for batch 3 at 20% noise (Color figure online)

The performance of all four models has been evaluated on three different batches under two conditions, i.e., noise or without noise. Noise is introduced to check the robustness of the model. For validation, 5-fold cross-validation has been used. It is evident from the tables that the proposed model produces an impressive result against conventional ANFIS with or without noise. [6]. According to the simulations, between three metaheuristics-based ANFIS-model, JAYA-ANFIS yields the best result in less computation time. Table 2, 3 and 4 shows the performance measure of the proposed model for three different batches against conventional ANFIS. To check the robustness and the stability of the model, noise is introduce ranging from 0–20%. Performance of the model increases as the batch size increases. The model grows more accurate as more training samples are feed into it. Without noise, R-score is almost similar for GA-ANFIS, PSO-ANFIS, and JAYA-ANFIS. For batch 3, model makes the prediction more precisely. On comparing all the model, JAYA-ANFIS possesses the highest R-score, 0.01163%, 0.01403% higher than GA-ANFIS and PSO-ANFIS respectively at 0% noise while 0.006% 0.02% higher than GA-ANFIS and PSO-ANFIS respectively at 20% noise. On the other hand JAYA-ANFIS posses lowest testing error i.e. 0.002356% at no noise while 0.43452% at 20% noise, as shown in Fig. 5 and Fig. 6. Additionally, it is noted that as the batch size increases, the model becomes more stable. According to Table 4, for noisy data JAYA-ANFIS is more stable and robust model among all.

5 Conclusion

This work proposes a model which is intelligent enough to recognize the resource requests of the services and offloads them at a suitable tier. A hybrid technique is employed with clustering and classification machine learning methods. To label the services, fuzzy-C means clustering has been utilized. Further, these labeled services have been utilized for the training of the ANFIS model. The suggested model can precisely direct the service requests to suitable one of three types of computing nodes: fog nodes, aggregated fog node, or cloud nodes. Simulations shows that the meta-heuristic-based neuro-fuzzy based ANFIS model yields imperative results against gradient-based neuro-fuzzy. The degree of efficiency has been determined by introducing various levels of noise into different batches. Intelligent decision-making techniques are required to integrate IoT, fog, and cloud and to simplify resource scheduling and allocation. The planned work is a step in that direction. The results are positive, encouraging, and demonstrate the effectiveness of the suggested approach. There is no conflict of interest to disclose.

References

1. Alizadeh, M.R., Khajehvand, V., Rahmani, A.M., Akbari, E.: Task scheduling approaches in fog computing: a systematic review. Int. J. Commun. Syst. (IJCS) **33**(16), e4583 (2020)
2. Asemi, A., Baba, M., Haji Abdullah, R., Idris, N.: Fuzzy multi criteria decision making applications: a review study. In: Proceedings of International Conference, Computer Engineering and Mathematical Sciences (ICCEMS) (2014)
3. Aslinezhad, M., Malekijavan, A., Abbasi, P.: Adaptive neuro-fuzzy modeling of a soft finger-like actuator for cyber-physical industrial systems. J. Supercomput. **77**(3), 2624–2644 (2021)
4. Benmouiza, K., Cheknane, A.: Clustered anfis network using fuzzy c-means, subtractive clustering, and grid partitioning for hourly solar radiation forecasting. Theor. Appl. Climatol. **137**(1), 31–43 (2019)
5. Chauhan, N., Banka, H., Agrawal, R.: Delay-aware application offloading in fog environment using multi-class Brownian model. Wirel. Netw. **27**(7), 4479–4495 (2021)
6. Garg, K., Chauhan, N., Agrawal, R.: Optimized resource allocation for fog network using neuro-fuzzy offloading approach. Arab. J. Sci. Eng. (AJSE) **47**, 1–14 (2022)
7. Gasmi, K., Dilek, S., Tosun, S., Ozdemir, S.: A survey on computation offloading and service placement in fog computing-based IoT. J. Supercomput. **78**(2), 1983–2014 (2022)
8. Goudarzi, M., Wu, H., Palaniswami, M., Buyya, R.: An application placement technique for concurrent IoT applications in edge and fog computing environments. IEEE Tran. Mob. Comput. **20**(4), 1298–1311 (2020)
9. Guevara, J.C., Torres, R.D.S., da Fonseca, N.L.: On the classification of fog computing applications: a machine learning perspective. J. Netw. Comput. Appl. (JNCA) **159**, 102596 (2020)
10. Gupta, S., Dileep, A.D.: Long range dependence in cloud servers: a statistical analysis based on google workload trace. Computing **102**(4), 1031–1049 (2020)
11. Haznedar, B., Kalinli, A.: Training anfis using genetic algorithm for dynamic systems identification. Int. J. Intell. Syst. Appl. Eng. (IJISAE) **4**(Special Issue–1), 44–47 (2016)
12. Jang, J.S.: Anfis: adaptive-network-based fuzzy inference system. IEEE Tran. Syst. Man Cybern. **23**(3), 665–685 (1993)
13. Khandelwal, M., et al.: Implementing an ANN model optimized by genetic algorithm for estimating cohesion of limestone samples. Eng. Comput. **34**(2), 307–317 (2018)
14. Liu, L., Chang, Z., Guo, X., Mao, S., Ristaniemi, T.: Multiobjective optimization for computation offloading in fog computing. IEEE Internet Things J. (IoT-J) **5**(1), 283–294 (2017)

15. Maala, H.H., Yousif, S.A.: Cluster trace analysis for performance enhancement in cloud computing environments. J. Theor. Appl. Inf. Technol. (JTAIT) **97**(7), 2019 (2019)
16. Mahmud, R., Srirama, S.N., Ramamohanarao, K., Buyya, R.: Quality of experience (qoe)-aware placement of applications in fog computing environments. J. Parallel Distrib. Comput. (JPDC) **132**, 190–203 (2019)
17. Mechouche, J., Touihri, R., Sellami, M., Gaaloul, W.: Conformance checking for autonomous multi-cloud SLA management and adaptation. J. Supercomput. **78**, 1–36 (2022)
18. Meng, X., Wang, W., Zhang, Z.: Delay-constrained hybrid computation offloading with cloud and fog computing. IEEE Access **5**, 21355–21367 (2017)
19. Momeni, E., Nazir, R., Armaghani, D.J., Maizir, H.: Prediction of pile bearing capacity using a hybrid genetic algorithm-based ANN. Measurement **57**, 122–131 (2014)
20. Nayeri, Z.M., Ghafarian, T., Javadi, B.: Application placement in fog computing with AI approach: taxonomy and a state of the art survey. J. Netw. Comput. Appl. (JNCA) **185**, 103078 (2021)
21. Qasem, S.N., Ebtehaj, I., Riahi Madavar, H.: Optimizing anfis for sediment transport in open channels using different evolutionary algorithms. J. Appl. Res. Water Wastewater (JARWW) **4**(1), 290–298 (2017)
22. Rao, R.V., Waghmare, G.: A new optimization algorithm for solving complex constrained design optimization problems. Eng. Optim. **49**(1), 60–83 (2017)
23. Shi, W., Cao, J., Zhang, Q., Li, Y., Xu, L.: Edge computing: Vision and challenges. IEEE Internet Things J. (IoT-J) **3**(5), 637–646 (2016)
24. Skarlat, O., Nardelli, M., Schulte, S., Borkowski, M., Leitner, P.: Optimized IoT service placement in the fog. Serv. Oriented Comput. Appl. **11**(4), 427–443 (2017)
25. Sonmez, C., Ozgovde, A., Ersoy, C.: Fuzzy workload orchestration for edge computing. IEEE Tran. Netw. Serv. Manag. **16**(2), 769–782 (2019)
26. Tadakamalla, U., Menasce, D.A.: Autonomic resource management for fog computing. IEEE Trans. Cloud Comput. **10**, 2334–2350 (2021)
27. Tong, L., Li, Y., Gao, W.: A hierarchical edge cloud architecture for mobile computing. In: 35th Annual IEEE International Conference on Computer Communications (INFOCOM), pp. 1–9. IEEE (2016)
28. Verma, A., Pedrosa, L., Korupolu, M., Oppenheimer, D., Tune, E., Wilkes, J.: Large-scale cluster management at google with borg. In: Proceedings of Tenth European Conference on Computer Systems (ECCS), pp. 1–17 (2015)
29. Vlamou, E., Papadopoulos, B.: Fuzzy logic systems and medical applications. AIMS Neurosci. **6**(4), 266 (2019)
30. Walia, N., Singh, H., Sharma, A.: Anfis: adaptive neuro-fuzzy inference system-a survey. Int. J. Comput. Appl. (IJCA) **123**(13), 1–7 (2015)
31. Yi, S., Hao, Z., Qin, Z., Li, Q.: Fog computing: platform and applications. In: Third IEEE workshop on Hot Topics in Web Systems and Technologies (HotWeb), pp. 73–78. IEEE (2015)
32. Yi, S., Li, C., Li, Q.: A survey of fog computing: concepts, applications and issues. In: Proceedings of workshop on Mobile Big Data (MBD), pp. 37–42 (2015)
33. Yousif, S., Al-Dulaimy, A.: Clustering cloud workload traces to improve the performance of cloud data centers. In: Proceedings of The World Congress on Engineering (WCE), vol. 1, pp. 7–10 (2017)

Optimizing Public Grievance Detection Accuracy Through Hyperparameter Tuning of Random Forest and Hybrid Model

Khushboo Shah(✉), Hardik Joshi, and Hiren Joshi

Gujarat University, Ahmedabad-9, India
khushbooshah@gujaratuniversity.ac.in

Abstract. Machine Learning provides an extensive range of supervised learning binary classifiers; Picking up the correct model for your application and dataset is an extremely challenging task as each classifier performs differently for a given dataset. Researchers put their very intense efforts and perform wide range of experiments with various techniques to get the best learner and performer model. In this paper, we are focusing on a binary classification task, to be performed on Indian Railway tweets to identify whether the piece of text is a grievance or not. From our prior work, we found that Random Forest with the Word2Vec word embedding technique perform all around well against 5 other classifiers – Support Vector Machine, Logistic Regression, Decision Tree and K Nearest Neighbor. The core objective of this paper is to study the prediction performance of Random Forest classifier with hyperparameter tuning versus the hybrid model built using the best of all the binary classifiers.

Keywords: Binary classifiers · Hybrid approach · Parameter tuning · Public grievance · Random forest · Signum function

1 Introduction

Machine Learning (ML), a branch of Artificial Intelligence (AI) is the area to study how to make machines learn without explicit programming. It is a powerful field having a wide range of supervised, unsupervised, and reinforcement learning models. ML is deeply rooted in our day-to-day life now which serve us with amazing applications and automatic techniques which work without human intervention [1]. These models get trained first and then they predict the outcomes based on their learning [2]. These models are versatile in nature hence they are used in various applications and are also powerful to handle enormous data [3, 4]. Binary Classification is one of the most performed tasks by the supervised learning models where model predicts the class based on its learning from previously assigned classes, '1' and '0' [5, 6]. Some well-known algorithms for binary classification are: Naive Bayes (NB), Support Vector Machine (SVM), Logistic Regression (LR), Decision Tree (DT), Random Forest (RF) and K-Nearest Neighbors

K. K. Patel et al. (Eds.): icSoftComp 2022, CCIS 1788, pp. 463–476, 2023.
https://doi.org/10.1007/978-3-031-27609-5_36

(KNN) [7]. As they come in variety, it is difficult to identity the best classifier for any dataset [4]. Generally there are two major factors which come into the picture while selecting the model: first is identifying best model for existing dataset and application and second is adjusting corresponding hyperparameters to achieve best prediction result [1]. The main aim of this paper is to evaluate the best binary classification model for our Indian Railway dataset which can predict the piece of text is a grievance or not. These data are tweets posted on social media platform twitter. By identifying the tweet is a grievance or not, these public grievances can be solved rapidly [8]. To accomplish this research, we exercised various experiments.

In our Phase 1 experiment, we integrated six binary classifiers with three word embedding techniques and total 18 combinations tested on 4000 records. These 4000 records are tweets downloaded from Twitter using Twitter API and then manually tagged to assign the class '1' if the tweet is a grievance else '0'. Our experiment showed that Random Forest outperformed among the other binary classifiers [9].

Taking this work further in Phase 2, we increased the dataset of Indian railway tweets with different class ratio of '1' and '0', and observed that RF is continuously performing well against all the other classifiers. Tree predictors are combined in random forests in such a way that each tree is dependent on the values of a random vector that has been sampled independently and has the same distribution for all of the trees in the forest [10]. To observe this beauty of RF, different tests were executed on 10,000 records which brought about 420 results under various conditions. All the results were assessed to identify the performance of each classifier at every stage. And in each condition, RF outperformed against all the other models with the highest accuracy of 0.94 (Fig. 1).

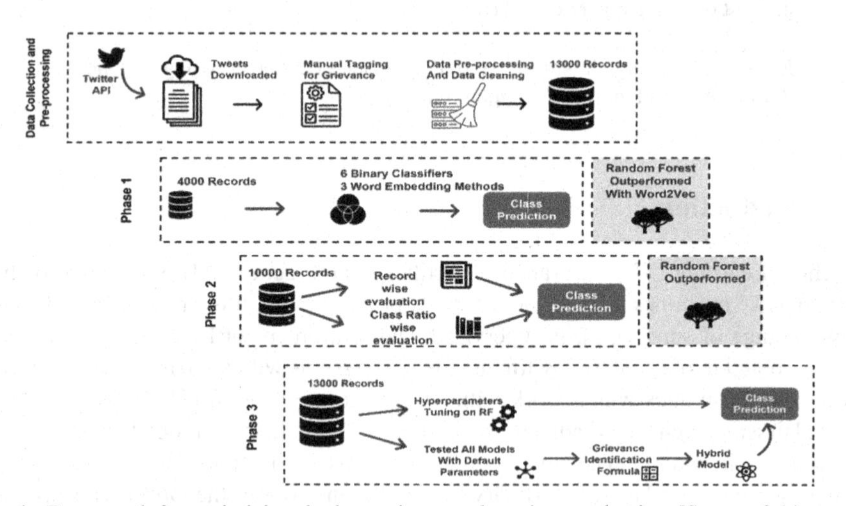

Fig. 1. Framework for optimizing the best grievance detection mechanism [Scope of this paper is exclusive for Phase 3. Other phases of experiments were shared in our previous work.]

For this paper, we carried out our binary classification work on Indian Railway tweets in Phase 3 with 13,000 records. We divided our experiment in 2 parts. In the first part, we applied hyperparameter tuning on RF as it performed well continuously in Phase 1 and 2. In the second part, we trained and tested SVM, LR, DT, RF and KNN with their default parameters. We further built hybrid model with unique mathematical formula of democracy approach to predict the class (1 or 0) of the tweet.

The rest of the paper is organized as follows: Sect. 2 describes the Literature revises for methods used in Phase 3. Section 3 is about the data collection and data preprocessing. Section 4 represents experiments and results of Phase 1 & 2. Section 5 discusses the Phase 3 work for this paper. Section 6 shows the experiment results. Section 7 is the final conclusion from the experiments. Section 8 talks about the future scope of this work.

2 Preliminaries

This section describes the methods those are used in our Phase 3 experiment along with the review of various literature.

2.1 Hyperparameter Tuning on Random Forest

In ML models, hyperparameter tuning is the significant process to get optimal values of specific parameter. The GridSearch function is used to determine a model's best parameters. This makes it possible to quickly iterate over all possible values for any hyper-parameter [11]. Various studies have effectively proposed and carried out the strategies in parameter tuning to get the classification model with best accuracy [12]. In most of the cases, RF works reasonably good with its default parameter values however hyperparameter tuning can enhance the performance of the model [13]. Authors Philipp Probst and Anne-Laure Boulesteix have showed in their research that do we really need to tune parameters of RF or not? It is mentioned in their research that RF model is basically for classification and regression which actually work on the aggregation of huge number T of decision trees. From their empirical studies, they have also observed that in most of the datasets, the highest performance achieved when training the 100 trees which is default value of *estimator* parameter but outcome accuracy may be go higher with the hyperparameter tuning [14]. To achieve hyperparameter specification, it is essential to perform random search on various random combination of hyperparameter values [15].

2.2 Hybrid Approach in Machine Learning

None of the single model is perfect for all the existing problems. Hence, Hybrid Machine Learning (HML) approach in ML came into existence which flawlessly combines various processes and/or algorithms from more than one domains with the goal of supplementing one another [16, 17]. Hybrid models have shown the capability to be the better model for reducing the predicting errors on increasingly complicated training data [18].

2.3 Signum Function

A signum or sign function is considered as piecewise function which determines the sign of real number. Signum function is predominantly used in the fields of engineering, applied physics and artificial intelligence (AI) for prediction. The signum function is usually represented by the symbol *sgn* in mathematics [19]. Also, it is denoted through f(x) where,

$$f(x) = |x|/x \quad (1)$$

f(x) represents Signum Function which means that, If $x < 0$, then $F(x) = -1$, If $x = 0$, then $F(x) = 0$, If $x > 0$, then $F(x) = 1$. In other words, signum function returns +1 for positive input values, −1 for negative input values and 0 where input is zero [20–22].

3 Data Collection and Pre-processing

Total 13000 tweets downloaded from Twitter using twitter API which were posted for Indian Railway handles. We tagged each tweet manually to classify it into two classes. If the tweet is a grievance, then it was tagged as class '1' else '0'. Various NLP techniques like – data cleaning, stop-word removal, information retrieval, stemming etc. applied to clean the data and then finally used Word2Vec word embedding method to make the data machine ready format.

4 Previous Experiments

4.1 Phase 1

In phase 1, we tested six binary classifiers: NB, SVM, LR, DT, RF and KNN with three word embedding techniques: Word2Vec(W2V), TFIDF and BERT on 4000 records (see Fig. 2) [9].

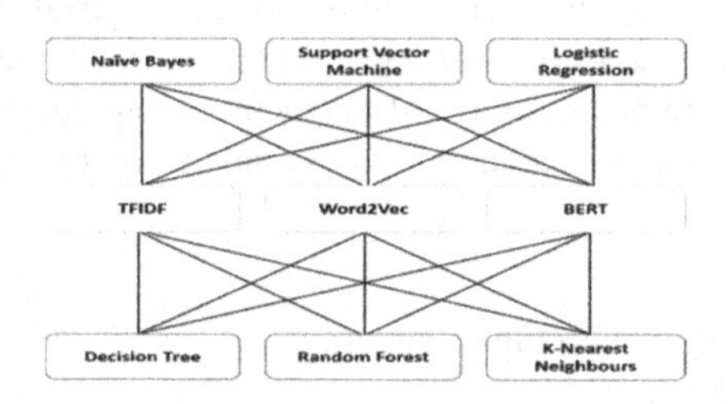

Fig. 2. Three word embedding methods are tested with six binary classifiers

This many to many relationships generated total 18 results and it was observed that RF performed extremely well in collaboration with W2V. Three out of four results are satisfactory with the highest score of 0.93 which was also validated by K-Fold Validation score and AUC score (see Table 1) [9].

Table 1. Test accuracy of six binary classifiers on different no of records.

Records	SVM	LR	DT	RF	KNN
500	0.83	0.84	0.66	0.86	0.81
1000	0.83	0.89	0.78	0.85	0.82
2000	0.8	0.82	0.74	0.83	0.79
4000	0.85	0.84	0.92	0.93	0.82

4.2 Phase 2

After observing good performance of RF in Phase1, we tried to test the RF in controlled environment. Where we created 84 datasets with 7 different class ratios and 12 data record sets. Five binary classifiers: SVM, LR, DT, RF and KNN get trained and tested on these 84 datasets and as a result total 420 scores we got. We did the analysis of these 420 results with various angles. Record wise, Ratio wise and Classifier wise, in all three analysis we found that RF performed best (see Fig. 3).

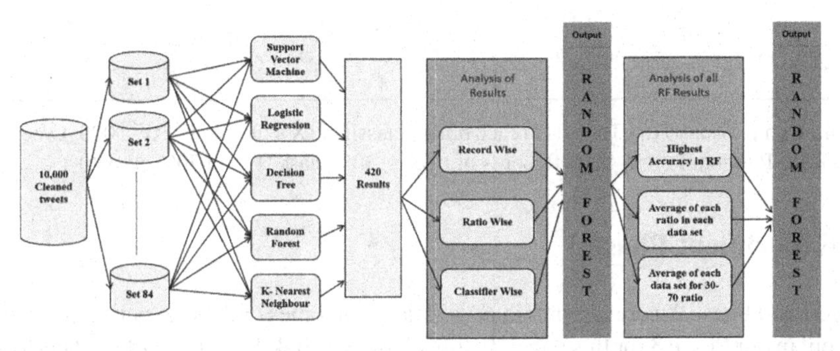

Fig. 3. Total 84 datasets generated with 7 different class ratio and 12 record sets and then all 84 datasets trained and tested on 5 binary classifiers. This way total 420 results analysed and observed that RF performed best out of all

Out of 420 results, Ratio wise analysis depicts that RF scored really well against all the other models (see Table 2). Also, we plotted top 3 results for each classifier and even in that we found that RF performed exceptionally well (see Fig. 4).

Table 2. Binary classifier's highest accuracy scores for each ratio

RATIO (YES-NO)	RECORDS	SVM	LR	DT	RF	KNN
30-70	5000	0.89	0.89	0.9	0.94	0.87
40-60	3500	0.87	0.88	0.89	0.93	0.84
45-55	4000	0.86	0.85	0.91	0.93	0.83
50-50	4000	0.85	0.84	0.9	0.93	0.82
55-45	6000	0.85	0.85	0.93	0.94	0.86
60-40	5000	0.85	0.83	0.9	0.93	0.84
60-40	5500	0.84	0.84	0.92	0.93	0.86
70-30	5500	0.86	0.85	0.92	0.94	0.87

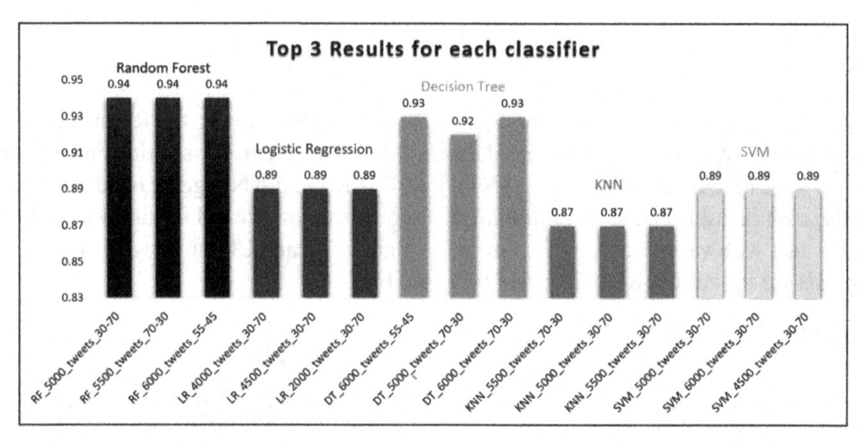

Fig. 4. Comparison of top 3 results in each binary classifier (X axis Label: "RF_5000_tweets_30–70" means Random Forest_5000 Records of tweets_30% class '1' and 70% class '0')

5 Experiment Phase 3

Phase 1 and Phase 2 were our previous work and outcomes of those experiments, we carried out in our Phase 3 for this paper. In this Phase, we took 13000 records and performed two experiments – Random Forest Hyperparameter Tuning and Hybrid Algorithm for Class Determination.

5.1 Random Forest Hyperparameter Tuning

From our previous work we concluded that RF works best for this dataset hence we focused only on RF to enhance the prediction accuracy. We applied hyperparameter tuning on RF. Out of total 18 parameters, tuning done on the parameters which are actually responsible to power the predictions [23]. We have used: *n_estimators* (number of trees in the forest, *min_samples_split* (minimum number of samples required to split an internal node, *min_samples_leaf* (minimum number of samples required to be at a leaf node, *max_depth* (maximum depth of the tree) [23]. Default values are: *n_estimators* = 100, *max_depth = None, min_sample_split = 2, min_samples_leaf = 1* [23].

Table 3. RF hyperparameter tuning results

Parameters / Run	Estimators	Max Depth	Min Sample Split	Min Sample Leaf	Accuracy
Run 1	100	5	5	5	0.81
Run 2	200	5	5	5	0.83
Run 3	200	10	10	10	0.88
Run 4	400	20	20	20	0.87
Run 5	400	15	15	15	0.89
Run 6	800	15	15	15	0.89
Run 7	1600	15	15	15	0.89
Run 8	400	12	12	12	0.89
Run 9	400	11	9	8	0.90
Run 10	400	11	8	7	0.90
Run 11	400	11	7	7	0.90
Run 12	400	11	6	7	0.90
Run 13	400	11	7	6	0.91
Run 14	400	11	6	5	0.91
Run 15	400	11	5	4	0.91
Run 16	400	11	4	3	0.92
Run 17	400	11	3	2	0.92
Run 18	400	11	2	1	0.926
Run 19	400	11	2	2	0.92
Run 20	500	11	2	2	0.92
Run 21	500	11	2	1	0.927
Run 22	100	11	2	1	0.924

To avoid the over-fitting in our model, we optimized tuning parameters to govern the number of features which were used randomly to grow each tree from the root. We applied 3 folds for each candidate of the model to minimize test sample prediction error.

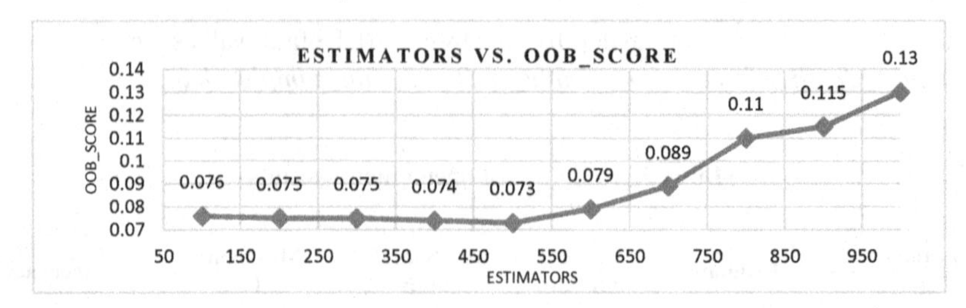

Fig. 5. Estimators vs. oob_score

Figure 5 depicts the Estimators vs oob_score diagram. We witnessed the highest accuracy when Random Forest Hyperparameters set to Max Depth = 11 Min Sample Split = 2 Min Sample Leaf = 1 (see Table 3). We kept those parameters constant and changed the value of the estimators and measured the loss against each value of Estimator. Oob_score i.e. the loss decreases steadily until estimator's value of 500 and steadily increases as estimators value increases.

5.2 Hybrid Algorithm for Class Determination

Accuracy plays potential role in data classification. All the binary classifiers perform differently for different dataset and hence accuracy results also vary from classifier to classifier [5]. In our previous experiments, Random Forest contributed highest accuracy among other binary classifiers on Indian Railway dataset however accuracy score becomes stagnant at certain limit and have very little scope of improvement thereafter. We analyzed the accuracy results of all the Binary Classifiers used during our experiment and observed that a hybrid approach can provide better accuracy if we use best out of each classifier (see Table 3).

Table 4. Accuracy score of each classifier for 13000 records with default parameters

Classifier	SVM	LR	DT	RF	KNN
Accuracy	0.89	0.88	0.87	0.92	0.91

To achieve the higher accuracy, we took experiment further where we applied the model integration method with 'Democracy' approach.

Algorithm Democracy approach for class determination

Input .csv file containing vectors of cleaned tweets
Output predicted class for each tweet in the csv file

1. **Import** libraries – *numpy, pandas, re, spacy, train_test_split*
   Import classifiers *LinearSVC, LogisticRegression, DecisionTreeClassifier, RandomForestClassifier, KNeighborsClassifier* from *sklearn*
   Import *metric*
2. **Open** csv file from location
3. **Read** csv file
4. **Use** *train_test_split* to split the dataset and train the machine with the parameter *test_size*
5 **Variable declaration and initialization**
   f(p) = Predicted class, *sgn* = Signum function, *n* = Total number of classifiers, *Ci* = Classifier's predicted class, *b* = bias value (-0.49), *j* = 1, *x* = tweet, *len* = total_records on csv file
6. **Initialize variables for all the classifiers**
   C_j = classifier1, C_{j+1} = classifier2… C_n = classifierN
7. **Fit the classifier with training dataset**
   C_j*.fit(X_train,y_train)*
8. **Prediction variable**
   $C_jP = C_j$*.predict(X_test)*
9. ***for*** *i*=0 ***to*** *len*
10.   *csv_*C_j*col[i]* = C_j*.predict(x[i])*
        *csv_*C_{j+1}*col[i]* = C_{j+1}*.predict(x[i])* ...
        *csv_*C_n*col[i]* = C_n*.predict(x[i])*
11.   ***for*** *j*=0 ***to*** N
12.   *csv_SumCol[i]* = *csv_SumCol[i]* + *csv_*C_j*col[j]*
13. $fp_i = sgn((csv\_SumCol[i])/n)+b$
14.   ***if*** $fp_{[i]} == 1$
15.   *csv_PreCol[i]* = *'1' // 1 means it is grievance*
16.   ***else***
17.   *csv_PreCol[i]* = *'-1' // -1 means it is not grievance*
18.   ***end if***
19.   ***end for***
20. ***end for***
21. ***update csv***

Grievance Identification Formula

We used label 1 if classifiers identify given tweet as Grievance else 0. Assume that ***Ci*** is the Classifier's predicted class which is either 0 or 1. A bias value ***b*** is used to shift activation function by adding constant value of −0.49. ***sgn*** denotes signum function which presents sign of the real number. Signum function results to 1 if overall value is greater than 0 or −1 if value is less than 0. $\sum$ represents sigma function which sums up the value of each classifiers predicated class for given tweet.

We first add the predicated class value from each classifier and divide the sum with total number of classifiers used. A bias value of −0.49 is added to the total. Bias value is essential when even number of binary classifiers are used as we have one of the possible results where equal number of the classifiers determines 0 or 1. In order to break the tie, bias value switches the activation function to get signum function value either −1 or 1. Signum function determines sign of summate and results to 1 or −1 [19]. If final result is 1 than given text is grievance else it is not a grievance.

For example, in our current experiment we used 5 different binary classifiers and they predicated the class for given tweets (see Table 5). For Tweet ID 1, below calculation suggests given text is grievance.

$$\mathrm{f(p)}_1 = \mathrm{sgn}(1 + 0 + 0 + 1 + 1)/5 + (-0.49)) = \mathrm{sgn}(0.6 - 0.49) = \mathrm{sgn}(0.11) = 1$$

Table 5. Sample data of our experiment

Tweet ID	SVM	LR	DT	RF	KNN	Signum Function value
1	1	0	0	1	1	1
2	1	0	1	1	1	1
3	0	0	1	1	0	−1
4	0	1	0	0	0	−1

$$f(p) = sgn\left(\left[\frac{1}{n}\left(\sum\nolimits_{i=1}^{n} C_i\right)\right] + b\right) \tag{2}$$

where,

f(p) = Predicted class, sgn = Signum function, n = Total number of classifiers, Ci = Classifier's predicted class, b = bias value (−0.49).

6 Results

We executed 21 runs for RF Hyperparameter tuning (see Table 3). In the result it is observed that accuracy is increasing sluggishly. After Run 16, score is quite steady. Considering the highest among all 21 runs are 0.926, 0.927 and 0.924 for the Run 18, 21 and 22 respectively (see Fig. 6).

Looking at the parameters, it is clearly visible that accuracy score does not have significant difference compared to the default values of parameters *max_depth*, *min_sample_split* and *min_sample_leaf*. The highest accuracy in Run 21 with hyperparameter tuning is 0.927 while default parameter in Run 22 provided the accuracy score

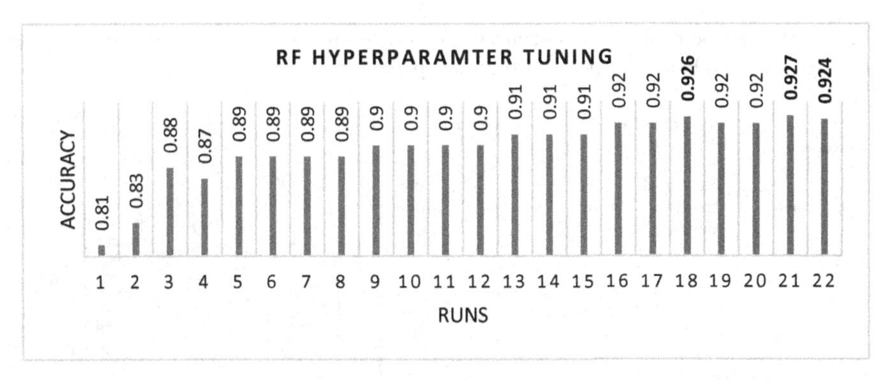

Fig. 6. Hyperparameter tuning done on Random Forest for 13000 records

of 0.924 hence, no potential difference observed between default and tuned parameters results.

Comparing our this experiment with the research work done in the paper [12], there are supporting results that RF works really well compare to other models but it is also noticeable that even after parameter tuning of RF, there is not markable change in the outcome [12].

In second experiment first we observed that individually each classifier performed really well with highest score of RF (see Table 4). To use the power of each model we proposed hybrid method with democracy approach to predict grievance. After applying this model integration technique to the dataset, accuracy reaches to 0.962 which is a remarkable jump (see Fig. 7).

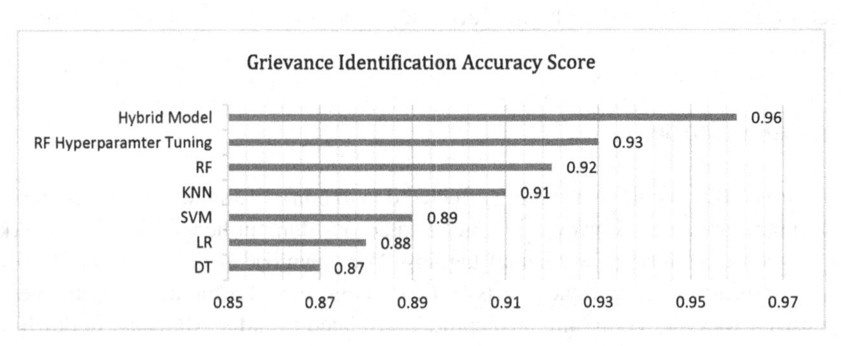

Fig. 7. Grievance Identification Accuracy Score of various models for 13000 records

Table 6 represents summary of the whole experiment which took place in 3 phases. It includes details about number of records, brief description about phase wise experiments, summary of the output and highest final accuracy after each phase. The main intention of this experiment is to improve the accuracy using the best attributes of the different classifiers. Random Forest is the best among all the classifiers however phase 3 results clearly indicates that hybrid model with all 5 binary classifiers.

Table 6. Outputs and final accuracy scores of experiments in all three phases. Phase 1 and 2 are our earlier work and Phase 3 is performed for this paper.

	Phase 1	Phase 2	Phase 3		
Records	4000	6000	13000		
Experiment	6 binary classifiers tested with 3 word embedding methods	Record wise, Ratio wise and Classifier wise analysis	RF		Hybrid Model
Output	RF outperformed with W2V	RF outperformed	Default parame-ter	Hyper parameter Tuning	Model integration method with 'Democracy' approach and Grievance Identification Formula
Final Accuracy	0.93	0.94	0.924	0.927	0.962

7 Conclusion

Identifying the perfect model for any given dataset to improve accuracy is a difficult task. Different binary classifiers present different results for any given dataset. Our focus in this paper was to pick the best single supervised binary classifier for the Indian railway dataset or develop a hybrid model to improve the overall accuracy score. We started our phase 3 experiments with Hyperparameter tuning the Random Forest as RF outperformed all other classifiers in the previous phases, however overall improvement in accuracy was not significant compared to default parameters. In second part, we collected the accuracy score of all 5 binary classifiers with default parameter values that were almost closer to each other. We used the democracy approach and derived a mathematical formula which in turn resulted in the highest accuracy score of 0.962 in an open environment (see Table 6).

8 Future Scope

We observed that accuracy is mainly affected due to non-railway domains tweets which are actually grievances however it is not related our domain hence they were marked as non-grievance during the manual tagging. We download from the Indian Railway handles only, however we found tweets related to non-domain complaints because people have uses it for other complaints which are not related to Indian Railway. In our future experiments, we will filter our non-railway domain tweets during the downloading time only which will also help to enhance the accuracy. We will apply hyperparameter tuning on all the binary classifiers and used best parameters results to build hybrid model to improve accuracy. We will also explore the neural networks and deep learning algorithm in conjunction with our hybrid model to enhance the accuracy score. To extend our work, Hybrid model will also be applied on other domains for binary classifications as well for multi class classification.

References

1. Yao, Q., et al.: Taking human out of learning applications: a survey on automated machine learning. arXiv:1810.13306v4 [cs.AI], pp. 1–20 (2018). http://arxiv.org/abs/1810.13306
2. von Rueden, L., Mayer, S., Sifa, R., Bauckhage, C., Garcke, J.: Combining machine learning and simulation to a hybrid modelling approach: current and future directions. In: Berthold, M.R., Feelders, A., Krempl, G. (eds.) IDA 2020. LNCS, vol. 12080, pp. 548–560. Springer, Cham (2020). https://doi.org/10.1007/978-3-030-44584-3_43
3. Bahel, V., Pillai, S.: A comparative study on various binary classification algorithms and their improved variant for optimal performance. In: IEEE Region 10 Symposium, pp. 5–7, June 2020
4. Patil, T.R., Sherekar, S.S.: Performance analysis of naive Bayes and J48 classification algorithm for data classification. Int. J. Comput. Sci. Appl. **6**(2) (2013). www.researchpublications.org
5. Ranjitha, K.V., Venkatesh.: Classification and optimization scheme for text data using machine learning Naïve Bayes classifier. In: 2018 IEEE World Symposium on Communication Engineering, pp. 33–36 (2018)
6. Kumari, R., Kr, S.: Machine learning: a review on binary classification. Int. J. Comput. Appl. **160**(7), 11–15 (2017). https://doi.org/10.5120/ijca2017913083
7. Isabona, J., Imoize, A.L., Kim, Y.: Machine learning-based boosted regression ensemble combined with hyperparameter tuning for optimal adaptive learning. Sensors **22**(10), 3776 (2022). https://doi.org/10.3390/s22103776
8. Joshi, H., Joshi, H., Shah, K.: Smart approach to recognize public grievance from microblogs. Towar. Excell. UGC HRDC GU **13**(2), 57–69 (2021). https://hrdc.gujaratuniversity.ac.in/Uploads/EJournalDetail/30/1046/6.pdf
9. Shah, H., Joshi, K., Joshi, H.: Evaluating binary classifiers with word embedding techniques for public grievances (2022). https://doi.org/10.1007/978-3-031-05767-0_17
10. Deng, W., Huang, Z., Zhang, J., Xu, J.: A data mining based system for transaction fraud detection. In: 2021 IEEE International Conference on Consumer Electronics and Computer Engineering, ICCECE 2021, pp. 542–545 (2021). https://doi.org/10.1109/ICCECE51280.2021.9342376
11. Hamida, S., Gannour, O.E.L., Cherradi, B., Ouajji, H., Raihani, A.: Optimization of machine learning algorithms hyper-parameters for improving the prediction of patients infected with COVID-19. In: 2020 IEEE 2nd International Conference on Electronics, Control, Optimization and Computer Science, ICECOCS 2020, no. 1 (2020). https://doi.org/10.1109/ICECOCS50124.2020.9314373
12. Ramadhan, M.M., Sitanggang, I.S., Nasution, F.R., Ghifari, A.: Parameter tuning in random forest based on grid search method for gender classification based on voice frequency. DEStech Trans. Comput. Sci. Eng. (CECE) (2017). https://doi.org/10.12783/dtcse/cece2017/14611
13. Probst, P., Wright, M.N., Boulesteix, A.L.: Hyperparameters and tuning strategies for random forest. Wiley Interdiscip. Rev. Data Min. Knowl. Discov. **9**(3) (2019). https://doi.org/10.1002/widm.1301
14. Probst, P., Boulesteix, A.L.: To tune or not to tune the number of trees in random forest. J. Mach. Learn. Res. **18**, 1–8 (2018)
15. Safi, A.A., Beyer, C., Unnikrishnan, V., Spiliopoulou, M.: Multivariate time series as images: imputation using convolutional denoising autoencoder. In: Berthold, M.R., Feelders, A., Krempl, G. (eds.) IDA 2020. LNCS, vol. 12080, pp. 1–13. Springer, Cham (2020). https://doi.org/10.1007/978-3-030-44584-3_1

16. Anifowose, F.: Hybrid machine learning explained in nontechnical terms. J. Pet. Technol. (2020)
17. Bhattacharya, A.: What Is Hybrid Machine Learning And How to Use It? (2022). https://www.analyticsinsight.net/, https://www.analyticsinsight.net/what-is-hybrid-machine-learning-and-how-to-use-it/#:~:text=HML is a progress of, intended to enhance each other. Accessed 28 Jul 2022
18. Dang, C.N., Moreno-García, M.N., De La Prieta, F.: Hybrid deep learning models for sentiment analysis. Hindawi Complex. **2021** (2021). https://doi.org/10.1155/2021/9986920
19. Jerri, A.J.: Signum function. In: Integral and Descrete Transforms with Applications and Error Analysis (1992)
20. Kumar, A.W., Verma, H.K., Singh, S.: Improved relay algorithm for detection and classification of transmission line faults using signum function of instantaneous power. In: Proceedings of 3rd International Conference on Condition Assessment Techniques in Electrical Systems, CATCON 2017, pp. 42–47. IEEE, January 2018. https://doi.org/10.1109/CATCON.2017.8280181
21. Alkatheiri, M.S., Zhuang, Y.: Towards fast and accurate machine learning attacks of feed-forward arbiter PUFs. In: 2017 IEEE Conference on Dependable and Secure Computing, pp. 181–187 (2017). https://doi.org/10.1109/DESEC.2017.8073845
22. Mehlig, B.: Machine learning with neural networks (2021)
23. sklearn.ensemble.RandomForestClassifier. https://scikit-learn.org/, https://scikit-learn.org/stable/modules/generated/sklearn.ensemble.RandomForestClassifier.html. Accessed 29 July 2022

Author Index

K. K. Patel et al. (Eds.): icSoftComp 2022, CCIS 1788, pp. 477–478, 2023.
https://doi.org/10.1007/978-3-031-27609-5

Printed in the United States
by Baker & Taylor Publisher Services